Nonlinear Dynamical Systems in Engineering

Vasile Marinca • Nicolae Herisanu

Nonlinear Dynamical Systems in Engineering

Some Approximate Approaches

Springer

Vasile Marinca
Politehnica University of Timisoara
Department of Mechanics and Vibrations
Bd. Mihai Viteazu 1
300222 Timisoara
Romania

Romanian Academy, Timisoara Branch,
Center for Advanced and Fundamental
Technical Research,
Bd. M.Viteazu, 24
300223 Timişoara
Romania
vmarinca@mec.upt.ro

Nicolae Herisanu
Politehnica University of Timisoara
Department of Mechanics and Vibrations
Bd. Mihai Viteazu 1
300222 Timisoara
Romania

Romanian Academy, Timisoara Branch,
Center for Advanced and Fundamental
Technical Research,
Bd. M.Viteazu, 24
300223 Timişoara
Romania
herisanu@mec.upt.ro

ISBN 978-3-642-22734-9 e-ISBN 978-3-642-22735-6
DOI 10.1007/978-3-642-22735-6
Springer Heidelberg Dordrecht London New York

Library of Congress Control Number: 2011941777

Printed on acid-free paper

Springer is part of Springer Science+Business Media (www.springer.com)

Preface

The practice of science could be described as the process of observation followed by the construction of verbal or mathematical models to explain the observations, or vice-versa. The use of the mathematical models implies that the observations are quantitative, that they involve numbers used to specify the observations. The successful application of any science depends on the use of an appropriate mixture of principles, techniques and approaches. Sometimes principles are easily understood, sometimes so difficult as to be opaque. Often some rejection takes root in the dry mathematical form in which the subject is presented, unrelated to observation, quite irrelevant to life, and lacking any form of interest. Setting problems based on a wide variety of experience should engage interest, challenge and intelligence and even stimulate curiosity. The wealth of detail offered should not lull the reader into thinking that the material can be meaning by leafing through the book. As much time as possible it should be devoted to going through the calculations and solving problems.

Analytical solutions to nonlinear differential equations or linear differential equations with variable coefficients play an important role in the study of nonlinear dynamical systems, but sometimes it is difficult to find these solutions, especially for nonlinear problems with strong nonlinearity. In general, the known analytical methods are restricted to limited cases depending on the parameters which appear in the governing equations and are valid only for nonlinear problems with weak nonlinearity.

Dynamical systems are a vast subject. It is often found that the going gets easier as one goes deeper, learning the mathematical connections tying together the various phenomena. The material of this book can be included in courses covering the theory of nonlinear oscillations, the theory of electrical machines, classical and fluid mechanics, thermodynamics or even biology.

The prerequisites for studying dynamical systems using this book are undergraduate courses in linear algebra, real and complex analysis, calculus, dynamics, and ordinary differential equations, classical physics of oscillations, knowledge of a computer language would be essential. Also, it is assumed that the reader knows basic notions about nonlinear systems of differential equations as well as the plotting of phase portraits, analysis of nonlinear systems, and graphical representation of errors and so on.

This book is informed by the research interest of the authors, which are currently nonlinear differential equations. Some references include recently published research articles. Our work has very hands-on approaches and takes the reader from the basic methods right through to recently published research material.

Most of the material in every chapter is at postgraduate level and has been influenced by the authors' own research interest. These chapters are especially useful as reference material for senior undergraduate project work.

The text is aimed to undergraduate students in accelerated programs, working scientists in various branches of engineering, natural scientists or applied mathematicians.

The whole book consists of concrete examples from various domains of nonlinear dynamical systems. The authors believe that the problem of motion of different dynamical systems can be assimilated only by working with the differential equations applied to concrete examples. Nearly all the sections of this book are followed by comparisons with numerical results or with other known results in the literature.

The aim of this book is to present and extend different known methods in the literature, especially Lindstedt-Poincaré method, the method of harmonic balance, the method of Krylov-Bogolyubov and the method of multiple scales, to solve different types of strong nonlinearities. A better knowledge of these methods lead to a better choice of the so-called "base functions" which are absolutely necessary to obtain the auxiliary functions present in the last Chapters, devoted to some optimal analytical approaches. These auxiliary functions are cornerstone of the optimal methods and also, ensure the conditions of convergence of the solutions obtained by different approaches. Unlike all previous analytic approaches, these few optimal methods provide us with a simple way to control and adjust the convergence region of solutions of nonlinear dynamical systems. These new optimal methods show one step in the attempt to develop a new nonlinear analytical technique working in the absence of small or large parameters. Actually, the capital strength of our optimal procedures is the fast convergence, since after only two or three iterations, or sometimes after only one iteration, it converges to the exact solution, which proves that these optimal methods are very efficient in practice.

The text begins with some known procedures, presented in Chaps. 1–4: the Lindstedt-Poincaré method, the method of harmonic balance, the method of Krylov-Bogolyubov and the method of multiple scales. All these techniques suppose the presence of a small parameter into the governing nonlinear equations. There are presented some alternatives and examples to each of these approaches, such as the use of perturbation method for strong parameter, the rational harmonic balance method, a combination of the method of Krylov-Bogolyubov and iteration method. The last four chapters, from 5 to 8 are devoted to optimal approaches such as: the Optimal Homotopy Asymptotic Method, the Optimal Homotopy Perturbation Method, the Optimal Variational Iteration Method and the Optimal Parametric Iteration Method. The validity of the proposed procedures has been demonstrated on some representative examples and very good agreement was found between the approximate analytic results and numerical simulations. The convergence of the approximate solutions obtained by each of these new methods is greatly influenced by the convergence-control constants which are optimally determined.

The examples presented in this book lead to the conclusion that the accuracy of the obtained results is growing along with increasing the number of constants in the auxiliary functions. These methods are very rapid and effective and show their validity and potential for the solution of nonlinear problems arising in dynamical systems.

Finally, our main aim is to inspire the reader to appreciate the beauty as well as the usefulness of the optimal analytical techniques in the study of nonlinear dynamical systems.

Timişoara

Vasile Marinca
Nicolae Herişanu

Contents

Chapter 1
Introduction

Most nonlinear phenomena are models of our real-life problems. The world around us is inherently nonlinear. A vast body of scientific knowledge has developed over a long period of time, devoted to a description of natural phenomena. Nonlinear evolution of equations are widely used as models to describe complex physical phenomena in various field of sciences, especially in fluid dynamics, solid state physics, plasma physics, mathematical biology and chemical kinetics, vibrations, heat transfer and so on.

Generally speaking, dynamics is a concise term referring to the study of time-evolving processes, and the corresponding system of equations, which describes this evolution, is called a dynamical system. In this book "nonlinear system" refers either to a dynamical process with a physical existence, or to an equation which may be a model of the process.

Initially, the theory of nonlinear systems build up on its own on the foundations of the results of Poincaré (1878–1900), Lyapunov (1893), Birkhoff (1908–1944) and those concerning point mappings, interactions, recurrences obtained at the end of the nineteenth and beginning of twentieth century, by Koenig, Hadamard, Andronov, Krylov-Bogolyubov, etc. Very strong interactions between theoretical researches and practical implications in physical, mathematical or engineering systems were the reason for the success in well-known Soviet schools: Moscow-Gorki and Kiev. The most important component of theory of *dynamical systems* was the theory of *nonlinear oscillations*. In dynamics, these two schools have occupied incontestably the first place, admitted by some of the most famous American mathematicians, as Lasalle and Lefschetz.

From 1960, and especially since 1975, with the explosive growth of researches irregular oscillations in the *chaotic dynamics field*, and with the translation of some Soviet results in Western countries, the study of nonlinear systems has become a subject in vogue out of USSR, with more and more papers on all scientific disciplines. A large part of these papers concerns *abstract nonlinear systems*, and are devoted to nonessential generalizations without any interest for understanding a typical dynamic behaviour, or for practical purposes. Other papers are limited to *concrete nonlinear systems*, a field which is considered to be constituted by two sets

V. Marinca and N. Herisanu, *Nonlinear Dynamical Systems in Engineering*,
DOI 10.1007/978-3-642-22735-6_1, © Springer-Verlag Berlin Heidelberg 2011

of results. The first one is related to the study of problems directly suggested by practice (physics, engineering, ...). The second set concerns the study of equations, directly tied with practice, but having the lowest dimension, and the simplest structure, which permits to isolate in the purest form a "mathematical phenomenon" by eliminating the "parasitic effects" of a more complicated structure (for example the famous Smale's horseshoe). For so wide a scientific field, having given rise to a great lot of publications, an exhaustive presentation of the matter is impossible, and then many results of this time are rediscoveries, or variants of older ones.

The end of the nineteenth century was dominated by Poincaré's works [1] on periodic solutions of ordinary differential equations, which constitute the foundation on the most part of results obtained until now. Briefly these ideas were associated with the notions as: fixed point in the map, limit cycle, bifurcation, node, focus, saddle, in various manifolds, map (point mapping), double asymptotic points, homoclinic, heteroclinic. The Poincaré's contribution of the analytical method of nonlinear dynamics must also be noted with the *Poincaré's method of the small parameter*, and the notion of *generating solution*. It permits to define the nontabulated transcendental functions, solutions of differential equations, by a convergent series expansion. This way was considerably developed and extended (asymptotic methods) later by the dynamics schools of the former Soviet Union.

With Poincaré, *Lyapunov* is the mathematician who has left the most important mark at that time. He gave the basis of the theory of the motion stability [2] and introduced some fundamental notions, for instance: Perturbed motion, characteristic numbers, the functions of the first and of the second kind, critical cases, global stability, and so on.

Researches about the theory of nonlinear systems represent an important part of the Birkhoff's contribution to the mathematical sciences and can be considered in the field of "concrete nonlinear system". The definition and the classification of all possible types of dynamic motions constitute an important contribution by this author. It gives the many aspects of dynamics in which Brikhoff introduced so many new ideas, new theorems and new questions [3].

Under the name of theory of oscillations, the study of nonlinear dynamics had a new phase of growth in the twentieth century, beginning due in particular to the results of Rayleigh [4] and Van der Pol [5]. It was associated with the development of electrical and radio engineering [6]. There are known two relatively independent branches of the nonlinear oscillation theory. So, *the first branch* corresponds to the qualitative methods. Here the complex transcendental functions are defined by the singularities of continuous or discrete dynamic systems such as stationary states which are equilibrium points or periodical solutions (cycles), trajectories passing through saddle singularities, stable and unstable manifold, boundary or separation of the influence domain (domain of attraction or basin) of a stable (attractive) stationary state, homoclinic or heteroclinic singularities, or more singularities of a fractal or nonfractal type, bifurcation, etc.

The second branch corresponds to the *analytical methods*. Here the above mentioned complex transcendental functions are defined by convergent, or at least asymptotically convergent series expansions, or in "the mean". The method of Poincaré's small parameter, the asymptotic method of Krylov-Bogolyulov-Mitropolski are analytical. So are the averaging methods [7, 8] and the method of harmonic linearization in the theory of nonlinear oscillations. These two independent branches of the nonlinear oscillations theory have the same aims: Construction of mathematical tools for the solution of concrete problems, development of a general theory of nonlinear systems.

Since 1960, the important development of computers has meant a large extension to the *numerical approach* of the nonlinear systems problems. Such an approach constitutes a powerful tool when it is associated with the qualitative or analytical methods.

The problem of the construction of mathematical tools fitted to the study of nonlinear oscillations was first formulated by Mandelstham in 1920. This was done in connection with the study of nonlinear systems belonging to radio-engineering. Indeed, with Papaleski, Mandelstham, formulated the fundamental problems solved later by his disciples. Such formulations constituted a decisive step in understanding concrete nonlinear systems.

In the beginning, the most popular approach to nonlinear problems was the *fitting method.* This method is based on the approximation of a nonlinear characteristic, with a piecewise linear characteristic. So the solution of a nonlinear problem is changed into the solution of a set of linear problems corresponding to different linear segments with conditions of continuity at the junction of the segments. The PhD thesis formulated in 1927 by Mandelstham: "The Poincaré's limit cycles and the theory of oscillations" is a first-rank contribution for the evolution of the theory of nonlinear oscillations, because it opens a new way of application for the Poincaré's qualitative theory of differential equations, with a lot of practical consequences. His professor, Andronov amplified his activity with a precise purpose: The elaboration of a *theory of nonlinear oscillations* [9], in order to dispose of mathematical tools, common to different scientific disciplines. For the elaboration of this theory, he used the following fundaments: The Lyapunov stability theory, the Poincaré's qualitative theory of differential equation, the point mappings theory, the Birkhoff's classification of all possible types of dynamic motions. To be *physically significant,* a model of dynamic systems must respect the following conditions: A solution should exist, this solution should be unique, the unique solution should be continuous with aspect to the data contained in the initial conditions, or in the boundary conditions, and the dynamic system should be structurally stable. This last concept was introduced in 1937 by Andronov and Pontrjagin: A dynamic system is structurally stable, if the topological structure of its motion does not change with small changes of the parameters, or the structure of the equation describing these motions.

For Andronov's school, the analytical methods have been always an auxiliary tool for qualitative method studies, or for the understanding of specific problems given by physics or engineering. So, as early as 1932, Mandelstham and Papaleski used the Poincaré's small parameter method to study of nonlinear resonance, subharmonic resonances, synchronization phenomenon, etc.

More recently, in 1956, Malkin published his famous book [10] dealing with the method of small parameter. This book also gives a method of successive approximations in the non-analytic case. Malkin has also given a study of critical cases in the Lyapunov sense via analytical methods.

From the theoretical results of Andronov's school, the practical applications are very large and very important for analysis and synthesis purpose. Now they go beyond the limit of physics or engineering, and concern also natural sciences, dynamics of populations, economy, etc. The applications concern all types of oscillators (electronic type, in particular the theory of multivibrators, mechanical type with the theory of watches, electro-mechanical type with coupling of electrical machines), regulators using relays, steam engines, automatic flight, the dynamics of flight and of gyroscopes, radio-physics, quantum mechanics, dynamics of systems heaving a pure delay, etc.

Since 1952, date of the deterioration of the relation between the two groups (one in Gorki, the other in Moscow) of the school, the group of Moscow has developed a considerable activity in the automatic control field with Chetaev, Lur'e, Aizerman, Petrov, Meerov, Letov, Tsypkin, etc., and in the optimization theory and practice with Pontryagin, Boltjanski, Gamkrelidze, Mischenko, etc. .

The Krylov-Bogolyubov school from Kiev, has developed essentially analytical methods. The foundation of their results is the classical *method of perturbations* which has been generalized to nonconservative systems. In 1932, the Krylov-Bogolyubov method gave a close foundation to the Van der Pol studies about oscillators. Later, the asymptotic method due to Mitropolski constitutes an improvement, with the use of only asymptotically convergent series expansions. It is the same for the *averaging method* and method for accelerating the convergence [7, 8]. With respect to the Poincaré's small parameter method, these methods are such that the "full" determination of the first harmonic and of the following harmonics of a periodic solution does not depend on the determination of the upper harmonics. The contribution of this school is very large and concerns systems with one, or several degrees of freedom, the determination of periodical, or quasi and almost periodical solutions, the determination of transient regimes, the synchronization phenomenon, and the nonlinear systems having pure delays.

The Hayashi's school on nonlinear oscillations developed many studies especially oriented toward electric circuits [11]. Analytical methods as well as qualitative and numerical ones have been intensively used by him and his two disciples Kawakami and Ueda. Complex behaviours of autonomous oscillators and nonautonomous ones with periodical excitation have been the subject of a lot of publication. Studies of *resonance* and *synchronization* phenomena of harmonic, subharmonic, higher harmonic and fractional harmonic types, in phase spaces and in parameter spaces, as well as problems of chaotic behaviours, occupy an important place in their publications.

In the last years efforts were also made for collecting the main results concerning nonlinear vibration in monographs. In this respect we mention the books by Minorski [12], Stoker [13], McLachlan [14], Kauderer [15], Sansone and Conti [16], and Roseau [17].

The theory of random vibration has a comparatively late development. From the publication in 1905 of Einstein's studies of Brownian motion [18], considered by him as a particular type of random vibration, more than two decades elapsed until the appearance of the first work applying the theory of random processes to the study of beam and string solutions. The study of nonlinear vibrations raised great mathematical difficulties. To overcome these difficulties, three different principal tools have been proposed: The use of Fokker-Planck-Kolmogorov equations, the method of statistical linearization and the perturbation method. Crandall [19] proposed and applied a perturbation technique, which is an extension to random vibrations of the perturbation method used for weakly nonlinear deterministic systems. Presently there are a great number of papers concerning the theory of linear and nonlinear random vibrations.

In the study of nonlinear dynamical systems, often are considered equations in the form:

$$\frac{d\mathbf{x}}{dt} = \mathrm{f}(\mathbf{x}, t, \mu), t > t_0 \mathrm{x}_0(t_0) = \mathbf{x}_0 \tag{1.1}$$

where all variables and parameters are real valued, t, t_0 and μ (a parameter) are scalars, $\boldsymbol{x}$ is a vector, and the functions $\boldsymbol{f}$ are at least continuous with respect to t, and analytic with respect to $\boldsymbol{x}$ and μ, at least for $|\ \mathrm{x}| < \bar{x}$ and $|\mu| < \bar{\mu}$.

It is worthy of note that now the well-known Poincaré's analyticity theorem is frequently used in a wrong way to prove the smooth dependence on the parameter μ for determining the periodic solution of Eq. 1.1, this dependence being presumed to follow from the analyticity of $\boldsymbol{f}$ with respect to $\boldsymbol{x}$ and μ. This theorem affirms that the solution of Eq. 1.1 with $\frac{\partial \mathrm{x}_0}{\partial \mu} \equiv 0$, $|\mathrm{x}_0| < \bar{x}$, can be expressed in the form of series :

$$\begin{aligned} &\mathbf{x}(t, \mathbf{x}_0, \mu) = \sum_{i=0}^{\infty} \mathbf{x}_i(t, \mathbf{x}_0)\mu^i, \mathbf{x}_i(t_0, \mathbf{x}_0) = \mathbf{x}_0 \quad for\ i = 0, \\ &\mathbf{x}_i(t_0, \mathbf{x}_0) = 0\ for\ i > 0 \end{aligned} \tag{1.2}$$

which is convergent for $|\boldsymbol{x}_0| \leq \bar{x}_1 < \bar{x}$, $|\mu| < \bar{\mu}_1 < \bar{\mu}$ and $0 < t - t_0 < \bar{t}(t_0, \boldsymbol{x}_0, \mu)$. It is important to stress that $\bar{t}$ depends on t_0, $\boldsymbol{x}_0$ and μ, which is forgotten by many authors, as noted by Gumowski in [20].The limitation of $\bar{t}$ by the properties of $\boldsymbol{f}$, other than smoothness, has a profound influence on any method of determining periodic solutions, because the validity of the Poincaré analyticity theorem in general cannot be extended to the asymptotic limit $\bar{t} \to \infty$, or even to $\bar{t} \geq T$, $T > 0$ being the period of the solution. The fundamental reason is that a periodic solution with a completely defined $\boldsymbol{x}_0$ is parametrically imbedded in a general solution $\boldsymbol{x} = \boldsymbol{F}(t, t_0, \boldsymbol{x}_0, \mu)$ with arbitrary $\boldsymbol{x}_0$ such that $\boldsymbol{F} \approx \bar{\boldsymbol{F}}(\boldsymbol{x}_0, \mu)\boldsymbol{F}_\infty(t, t_0)$ at $t \to \infty$. In general, it happens that $\bar{\boldsymbol{F}}$ is a qualitatively different function of μ for different $\boldsymbol{x}_0$, and for a fixed $\boldsymbol{x}_0$ the precise dependence on μ is not known in advance. Moreover, the qualitative μ-dependence of $\bar{\boldsymbol{F}}$ may exhibit a complex structure. At present, no explicit conditions to be imposed on f are known, which ensure the μ dependence of $\boldsymbol{F}$ to be of specific form for a fixed $\boldsymbol{x}_0$.

It is generally believed that the validity of the asymptotic method constitutes an extension of the Poincaré analyticity theorem, with usual μ-convergence replaced by asymptotic convergence. Many examples show that a mere asymptotic convergence is useless for the determination of a periodic solution unless, for a fixed difference between the exact and the approximate solution, the size of the approximation interval analogous to the convergence interval, $0 < t\text{-}t_0 < \bar{t}\,(t_0, \boldsymbol{x}_0, \mu)$, is not less than the period T. Unfortunately, if the differential equations are reduced to quasi linear standard form, which is the usual practice, and the expansion (1.2) does not apply with $\bar{t} \leq T$, then the asymptotic method either fails completely or yields erroneous results. An analytical method can be improved by using a convenient parametric substitution, which consists in a rewriting of the differential equation parameters. Then each of them presents itself as the product of the former parameter and an auxiliary new one at an entire or fractional power, in general different according to the parameter considered. This leads to series expansions with entire or fraction powers, characterizing the analytical solution of the equation. Such a method leads to guess the "intrinsic parametric dependence" of the solution. This may be facilitated when the equation is the model of a physical process with a well-understood behaviour, or via numerical simulation of the solution.

The future development of analytical methods is hindered by a major difficulty due to the fact that the "intrinsic parameter dependence" does not remain the same for a given equation when its parameters vary. Pure mathematicians accept the characterization of a solution by a series expansion, only if an existence theorem of the solution by a series expansion, has been proved. Such proofs generally present technical difficulties for their establishment, which implies before the understanding of the intrinsic parameter dependence. In the presence of changes of the intrinsic parametric dependence according to the parameter space point, this would imply the demonstration of as many theorems as number of changes, which is non-realistic.

For the future, it is likely that the interest in the analytical methods will not be restricted to limited cases, for which an explicit dependence of a solution with respect to the parameters of the problem is necessary for an application [21].

In this context, it can be written that these recent results will be used to search for better and more efficient solution methods for determining a solution, approximate or exact, analytical or numerical to nonlinear models or linear models with variable coefficients.

Large varieties of engineering, physical, chemical or biological phenomena are governed by nonlinear evolution equations. Obtaining the exact solutions of nonlinear equations, if available it is important because among others, facilitates the verification of numerical solvers and aids in the stability analysis of solutions. Analytical solutions to nonlinear differential equations play an important role in nonlinear science, especially in nonlinear physical science since they can provide much physical information and more inside into the physical aspects of the problem and thus lead to further applications.

In the field of nonlinear dynamical systems, rapid progress in the last two centuries has occurred due in large measure to the ability of investigators to respect physical laws in terms of rather simple equations. In many cases the governing equations were not so simple, therefore certain assumptions, more or less consistent with the physical situation, were employed to reduce the equations to types more easily solvable. Thus, the process of linearization has become an intrinsic part of rational analysis of physical problems. An analysis based on linearized equations, then, may be thought of as an analysis of a corresponding but idealized problem.

In many instances the linear analysis is insufficient to describe the behaviour of physical systems adequately. In fact, one of the most fascinating features of a study of nonlinear problems is the occurrence of new and unsuspected phenomena i.e., new in the sense the phenomena are not predicted, or even hinted at by the linear theory. On the other hand, certain phenomena observed physically are unexplainable, except by some known aspects due to nonlinearities present in the systems. Nonlinearities are commonplace in engineering systems. They result from structural proprieties which give rise to nonlinear elastic forces, geometrical and kinematic configurations which create autoparametric interactions or combined bending and torsional motion, inertial effects including rotational loadings, deformations such as curvatures and buckled states, machine tool chatter vibrations due to tool and work-piece interactions, aerodynamic effects, the elastic foundation, strain displacement of plates, systems with electric deformations, fluid–structure interaction, finite belt stretching, internal combustion, and so on.

As exact solutions for strongly nonlinear system are frequently scarce, at least at the present state of knowledge in nonlinear mathematics, an approach of accurate semi-analytical or approximate analytical solutions is most significant appealing. In this respect, new and innovative approaches capable to solve nonlinear dynamical systems beyond the restriction of the classical methods of perturbation and harmonic balancing are presented here and then applied to acquire accurate approximate higher-order analytical solutions for nonlinear problems. The motivation of this book is to extend the analysis of different known methods in literature, to solve different types of nonlinear problems.

In the last years, some fruitful results have been obtained for solving various nonlinear problems. There exist some well-known analytical approaches applicable for nonlinear problems such as the method of harmonic balance [22], the Lindstedt-Poincaré method [22], a modified Lindstedt-Poincaré method [23], the averaging method [8], a symbolic computation to implement an averaging method with elliptic functions [24], some extensions of the harmonic-balance method, the Krylov-Bogolyubov and elliptic perturbation method to the case of complex strongly nonlinear differential equations [25], the boundary element method [26], the weighted linearization method [27], the artificial parameter method [28], the homotopy analysis method [29], the homotopy perturbation method [30], the variational iteration method [31], a quasi-linearization method [32], and so on. All of the above mentioned methods work very well for weakly nonlinear dynamical systems and some of them work even for strongly nonlinear problems. It is very important in case of strongly nonlinear dynamical systems, to prove and to ensure the condition of convergence of the solutions. Every approach used in the study of nonlinear dynamical systems must be rigorously proved, even if the results obtained by the approach are in good agreement with other known results in literature.

It is also worthy to note that the majority of scientists have not been led to their discoveries by a process of deduction from general postulates, or general principles, but rather by a through examination of properly chosen particular cases [21]. The generalizations have come later, because it is far easier to generalize an established result then to discover a new line of argument. Generalization is the temptation of a lot of researchers working now with nonlinear dynamical systems.

The important development of the theory of nonlinear dynamical systems, during these centuries, has essentially its origins in the studies of the "natural effects" encountered in these systems, and the rejection of non-essential generalizations, i.e. the study of concrete nonlinear systems have been possible due to the foundation of results from the theory of nonlinear dynamical system field.

Chapter 2
Perturbation Method: Lindstedt-Poincaré

We seek an expansion that is valid for small but finite amplitude motions. It is convenient to introduce a small, dimensionless parameter ε which is of the order of the amplitude of the motion and can be used as a crutch, or a bookkeeping device, in obtaining the approximate solution [22, 33–35]. For instance, the equation of motion of an autonomous weakly nonlinear system has the form:

$$\ddot{u} + \omega^2 u = \varepsilon f(u), \quad 0 < \varepsilon \ll 1 \tag{2.1}$$

where in general f is a nonlinear function and $\dot{u} = \frac{du}{dt}$.

The perturbation method aims at getting a *periodic* solution of Eq. 2.1 in the form of a power series with respect to ε. This method, introduced about 1830 by Poisson, was at first applied formally, without any theoretical justification. Nevertheless it has been successfully used to obtain some effective solutions especially in celestial mechanics. By the end of the nineteenth century, the method has been improved as far as calculation is concerned by Gylden, Lindstedt, Bohlin, and others. However, the main contribution to the perturbation method is due to Poincaré [1], who elaborated in 1892 its theoretical grounds and made possible its systematic application to various problems of nonlinear oscillations. We shall discuss in this section a variant of the perturbation method proposed by Lindstedt in 1883.

There is no real and rigorous connection between the way that ε is used in different terms of a nonlinear differential equation and the manner in which it can be applied to the damping, for example. Besides, it forces the assumption onto the problem that there is a genuine strength similarity between damping, excitation, or nonlinear restoring force. In reality this may be a rather vague supposition and almost impossible to prove unequivocally. The use of ε, also introduces the implicit assumption that occasionally certain term is of the second order in ε, whereas other term can be of the first order in ε. Again, such comparison is hard to contemplate accurately and definitively, but if it does turn out to be physically unacceptable, then the particular ordering scheme cannot be used. It is important to proceed with caution and to use one's experience judiciously. The parameter ε can be seen as a

V. Marinca and N. Herisanu, *Nonlinear Dynamical Systems in Engineering*,
DOI 10.1007/978-3-642-22735-6_2, © Springer-Verlag Berlin Heidelberg 2011

convenient universal scaling parameter for different, apparently unrelated quantities within the equation of motion such as damping, excitation amplitude, and coefficients of nonlinear terms. On that basis, one might construct a set of physically reasonable orderings which together convey the requirements for soft excitation, relatively weak damping and subordinate status for geometric nonlinearities when compared with their linear counterparts. The other possibility for introducing ε is based on formal nondimensionalization of the dependent variable (i.e. the chosen coordinate) and the independent variable.

Let us suppose that Eq. 2.1 has a periodic solution $u(t)$ of some period T. We cannot try a solution of the form:

$$u(t) = u_0(t) + \varepsilon u_1(t) + \varepsilon^2 u_2(t) + \dots \tag{2.2}$$

because some functions $u_k(t)$ could be aperiodic. Such a situation occurs, i.e. for the expansion of the periodic function $\sin(1+\varepsilon)t$:

$$\sin(1+\varepsilon)t = \sin t + \varepsilon t \cos t - \frac{1}{2}\varepsilon^2 t^2 \sin t + \dots.$$

whose coefficients are not periodic. Terms like$t \cos t$, $t^2 \sin t$, ... in which the time t appears as "amplitude" are called *secular terms*. It is obvious that the existence of such terms, which grow beyond and bound as $t \to \infty$, destroys the periodicity of the expansion when only a finite number of its terms are considered, which is usually the case.

This difficulty may be avoided by developing the period $T(\varepsilon)$in a power series with respect to ε:

$$T(\varepsilon) = \frac{2\pi}{\omega}\left(1 + \varepsilon h_1 + \varepsilon^2 h_2 + \dots\right) \tag{2.3}$$

and introducing into Eq. 2.1 the new independent variable

$$\tau = \frac{2\pi t}{T} = \frac{\omega t}{1 + \varepsilon h_1 + \varepsilon^2 h_2 + \dots} \tag{2.4}$$

Then, Eq. 2.1 becomes

$$\frac{d^2u}{d\tau^2} + \left(1 + \varepsilon h_1 + \varepsilon^2 h_2 + \dots\right)^2 u = \frac{\varepsilon}{\omega^2}\left(1 + \varepsilon h_1 + \varepsilon^2 h_2 + \dots\right)^2 f(u) \tag{2.5}$$

and has a periodic solution $u(\tau)$ of constant period 2π. Consequently, we may assume for $u(\tau)$ in Eq. 2.5 the power series

$$u(\tau) = u_0(\tau) + \varepsilon u_1(\tau) + \varepsilon^2 u_2(\tau) + \dots \tag{2.6}$$

whose coefficients $u_k(\tau)$ have to be periodic functions of τ of period 2π.

Now, we suppose that the initial conditions are

$$u(0) = a, \quad \dot{u}(0) = 0, \tag{2.7}$$

We satisfy the conditions (2.7) by requiring that

$$u_0(0) = a, \quad u_{k+1}(0) = 0, \quad \frac{du_k(0)}{d\tau}\left(= \frac{du_k(\tau)}{d\tau}|_{\tau=0}\right) = 0 \text{ for } k = 0, 1, 2... \tag{2.8}$$

By substituting Eq. 2.6 into Eq. 2.5 and taking into account that

$$(1 + \varepsilon h_1 + \varepsilon^2 h_2 + \varepsilon^3 h_3 + ...)^2 = 1 + 2\varepsilon h_1 + \varepsilon^2(h_1^2 + 2h_2) + 2\varepsilon^3(h_3 + h_1 h_2) + ... \tag{2.9}$$

and

$$\begin{aligned} f(u) =& f(u_0) + (\varepsilon u_1 + \varepsilon^2 u_2 + \varepsilon^3 u_3 + \ldots)f'(u_0) + \frac{1}{2}(\varepsilon u_1 + \varepsilon^2 u_2 + \\ &+ \varepsilon^3 u_3 + \ldots)^2 f''(u_0) + \ldots. = f(u_0) + \varepsilon u_1 f'(u_0) + \varepsilon^2[u_2 f'(u_0) + \\ &+ \frac{1}{2}u_1^2 f''(u_0)] + \varepsilon^3[u_3 f'(u_0) + u_1 u_2 f''(u_0)] + \ldots \end{aligned} \tag{2.10}$$

by equating coefficients of like powers of ε, we obtain the set of recursive linear differential equations:

$$\begin{aligned} &\frac{d^2u_0}{d\tau^2} + u_0 = 0 \\ &\frac{d^2u_1}{d\tau^2} + u_1 = -2h_1u_0 + \frac{f(u_0)}{\omega^2} := -2h_1u_0 + \varphi_1(\tau) \\ &\frac{d^2u_2}{d\tau^2} + u_2 = -2h_2u_0 - 2h_1u_1 - h_1^2u_0 + \frac{u_1 f'(u_0) + 2h_1 f(u_0)}{\omega^2} := \\ &\qquad - (2h_2 + h_1^2)u_0 + \varphi_2(\tau) \\ &\frac{d^2u_3}{d\tau^2} + u_3 = -2(h_3 + h_1h_2)u_0 - (h_1^2 + 2h_2)u_1 - 2h_1u_2 + \\ &\qquad + \frac{u_2 f'(u'_0) + \frac{1}{2}u_1^2 f''(u_0) + 2h_1u_1 f'(u_0) + (h_1^2 + 2h_2)f(u_0)}{\omega^2} := \\ &\qquad - 2(h_3 + h_1h_2)u_0 + \varphi_3(\tau) \end{aligned} \tag{2.11}$$

and so on, where the symbol := means "note".

The general solution of Eq. 2.11_1 satisfying Eq. 2.8 is

$$u_0(\tau) = a\cos\tau \tag{2.12}$$

In order to integrate Eq. 2.11_2, we develop the function $\varphi_1(\tau)$ in a Fourier series. We obtain

$$\varphi_1(\tau) = C_{10} + C_{11}\cos\tau + C_{12}\cos 2\tau + C_{13}\cos 3\tau + \ldots \tag{2.13}$$

and therefore, Eq. 2.11_2 becomes

$$\frac{d^2u_1}{d\tau^2} + u_1 = -2h_1 a\cos\tau + C_{10} + C_{11}\cos\tau + \sum_{k=2}^{\infty} C_{1k}\cos k\tau \tag{2.14}$$

The solution of this equation does not contain secular terms provided that there are no terms containing $\cos\tau$and $\sin\tau$in the right-hand side. This condition gives $-2h_1a + C_{11} = 0$, whence

$$h_1 = \frac{C_{11}}{2a} \tag{2.15}$$

The general solution of Eq. 2.14 is then

$$u_1(\tau) = C_{10} + \sum_{k=2}^{\infty}\frac{C_{1k}}{1-k^2}\cos k\tau + \alpha_1\cos\tau + \beta_1\sin\tau \tag{2.16}$$

From initial conditions (2.8) we deduce that

$$\begin{aligned} \alpha_1 &= -C_{10} - \sum_{k=2}^{\infty}\frac{C_{1k}}{1-k^2} \\ \beta_1 &= 0 \end{aligned} \tag{2.17}$$

We now have

$$u_1(\tau) = C_{10}(1-\cos\tau) + \sum_{k=2}^{\infty}\frac{C_{1k}}{1-k^2}(\cos k\tau - \cos\tau) \tag{2.18}$$

Introducing Eqs. 2.12 and 2.18 into Eq. 2.11_3 yields $\varphi_2(\tau)$ and hence can be expanded in a cosine series, and so on.

Assume that we have determined in this way the functions $u_k(\tau)$, and the constants h_k, for $k = 1,2,\ldots,m-1$, and that we may calculate $\varphi_m(\tau)$. By expanding $\varphi_m(\tau)$ in a Fourier series, we obtain

$$\varphi_m(\tau) = C_{m0} + C_{m1}\cos\tau + \sum_{k=2}^{\infty} C_{mk}\cos k\tau \tag{2.19}$$

and the m-th equation in Eq. 2.11 becomes

$$\begin{aligned}\frac{d^2u_m}{d\tau^2}+u_m &= -(2h_m+2h_1h_{m-1}+2h_2h_{m-2}+\ldots)a\cos\tau+\\ &+C_{m0}+C_{m1}\cos\tau+\sum_{k=2}^{\infty}C_{mk}\cos k\tau\end{aligned} \tag{2.20}$$

Requiring the periodicity of $u_{\mathrm{m}}(\tau)$ gives

$$h_m=\frac{C_{m1}}{2a}-h_1h_{m-1}-h_2h_{m-2}\ldots \tag{2.21}$$

and we deduce as before, by initial conditions (2.8) that $u_{\mathrm{m}}(\tau)$ may be taken under the form

$$u_m(\tau)=C_{m0}(1-\cos\tau)+\sum_{k=2}^{\infty}\frac{C_{mk}}{1-k^2}(\cos k\tau-\cos\tau) \tag{2.22}$$

We may thus successfully determine the function $u_{\mathrm{k}}(\tau)$ and the constants h_{k}. The solution of Eq. 2.5 is then Eq. 2.6, where τ is given by Eq. 2.4 and the period T is given by Eq. 2.3.

It is apparent from above that there are no special difficulties in applying the perturbation method up to any step. However, the second and following steps do not qualitatively change the approximate solution. They only introduce small quantitative corrections of order ε^2 or higher, which, usually, do not justify the amount of calculation involved.

Poincaré has shown by an example that, in general, the series (2.6) obtained by the method of Lindstedt-Poincaré may not converge. In the variant of the perturbation method devised by him, the solution is obtained once again as a power series with respect to ε, which uniformly converges to $u(t)$ if ε and the initial amplitude $|a|$ are sufficiently small, but which may contain secular terms. Poincaré was interested in astronomical problems, in which the presence of secular terms is harmless, because of the relatively slow motion of the planets.

However, for the study of nonlinear vibration with comparatively high frequencies, a casting-out of the secular terms, as the one of Lindstedt discussed above, seems to be better suited. Moreover, since the expansion (2.6) is practically limited to its first one or two terms, one is mainly interested in the asymptotic behaviour for $\varepsilon \to 0$ of this truncated expansion, and the possible divergence of the whole series is generally immaterial.

2.1 The Oscillator with Cubic Elastic Restoring Force

The oscillator with cubic elastic restoring force occupies an important place in the theory of nonlinear systems, since it is the simplest oscillator displaying specific nonlinear properties. On the other hand, it provides a first approximation for the behaviour of a much large class of oscillators. Indeed, for sufficiently small values of $|u|$, the cubic characteristic may approximate as well as we please an elastic characteristic given by an arbitrary analytic function of u. Finally, another argument in favour of a detailed study of this oscillator is the possibility that it gives to compare an exact solution with the approximate solutions obtained by various analytical methods.

Let us consider the equation of motion of a conservative oscillator with cubic elastic restoring force (well-known Duffing equation)

$$\ddot{u} + \omega^2 u\left(1 + \varepsilon u^2\right) = 0 \tag{2.23}$$

with the initial conditions

$$u(0) = a > 0, \quad \dot{u}(0) = 0 \tag{2.24}$$

where ω, ε and a are constants.

2.1.1 The Exact Solution of Duffing Equation

The potential energy per unit mass for Eq. 2.23 is

$$G(u) = \int_0^u \omega^2 u\left(1 + \varepsilon u^2\right) du = \frac{\omega^2 u^2}{2}\left(1 + \frac{\varepsilon}{2} u^2\right) \tag{2.25}$$

The energy equation reduces in this case to

$$E(t) = \frac{v^2}{2} + G(u) = E_0 \tag{2.26}$$

where $v^2/2$ is the kinetic energy per unit mass ($\dot{u}=v$) and

$$E_0 = \frac{v_0^2}{2} + G(u_0) \tag{2.27}$$

is the initial total energy of the system per unit mass. From initial condition (2.24) we obtain the energy equation

$$\frac{v^2}{2} + \frac{\omega^2 u^2}{2}\left(1 + \frac{\varepsilon}{2}u^2\right) = E_0 \tag{2.28}$$

where

$$E_0 = \frac{\omega^2 a^2}{2}\left(1 + \frac{\varepsilon}{2}a^2\right) \tag{2.29}$$

From Eqs. 2.28 and 2.29 we deduce the velocity as function of the displacement

$$v(u) = \pm\omega\sqrt{(a^2 - u^2)\left[1 + \frac{\varepsilon}{2}(a^2 + u^2)\right]} \tag{2.30}$$

from which the extreme values of the speed

$$v_{1,2} = \pm\omega a\sqrt{1 + \frac{\varepsilon}{2}a^2} \tag{2.31}$$

results for $u = 0$. The extreme values of the displacement may be obtained by putting $v = 0$ in Eq. 2.30. We find, as expected, $u_{1,2} = \pm a$. From Eq. 2.28 we see that the phase trajectories are symmetric with both coordinate axes.

Next, it follows from Eq. 2.30 that the time necessary for the representative point to move in the lower half-plane from $A(a,0)$ to the point of abscissa u is

$$t(u) = -\frac{1}{\omega}\int_a^u \frac{du}{\sqrt{(a^2 - u^2)\left[1 + \frac{\varepsilon}{2}(a^2 + u^2)\right]}} \tag{2.32}$$

By interchanging the limits of integration and putting $u = a\eta$, where η is a new variable, Eq. 2.32 becomes

$$t(u) = \frac{\sqrt{2}}{\varepsilon\omega a}\int_{\frac{u}{a}}^{1} \frac{d\eta}{\sqrt{(1 - \eta^2)\left(1 + \frac{2}{\varepsilon a^2} + \eta^2\right)}} \tag{2.33}$$

This expression may be further transformed by means of the elliptic integral of the first kind $F(\varphi.k)$, [36, 37]. We then obtain

$$t(u) = \frac{F\left(arc\ \cos\frac{u}{a}; k\right)}{\omega\sqrt{1 + \varepsilon a^2}} \tag{2.34}$$

where

$$k = \sqrt{\frac{\varepsilon a^2}{2(1 + \varepsilon a^2)}} \tag{2.35}$$

Finally, by inverting the function (2.34), we find the exact solution of the equation of motion

$$u = a\,cn\left(\omega t\sqrt{1+\varepsilon a^2};k\right) \tag{2.36}$$

where cn is Jacobi's elliptic function (elliptic cosine).

We also notice that the quarter of the vibration period may be calculated by putting $u = 0$ into Eq. 2.36, and hence the period is given by

$$T(a) = \frac{4}{\omega\sqrt{1+\varepsilon a^2}}F\left(\frac{\pi}{2};k\right) = \frac{4}{\omega\sqrt{1+\varepsilon a^2}}K(k) \tag{2.37}$$

where $K(k)$ is the complete elliptic integral of the first kind.

In the special case when ε is sufficiently small, then

$$k^2 = \frac{\varepsilon a^2}{2(1+\varepsilon a^2)} = \frac{\varepsilon a^2}{2}\left[1 - \varepsilon a^2 + \left(\varepsilon a^2\right)^2 - \left(\varepsilon a^2\right)^3 + \ldots\right] \tag{2.38}$$

$$\frac{1}{\sqrt{1+\varepsilon a^2}} = 1 - \frac{\varepsilon a^2}{2} + \frac{3\left(\varepsilon a^2\right)^2}{8} - \frac{5\left(\varepsilon a^2\right)^3}{16} + \frac{35\left(\varepsilon a^2\right)^4}{128}\ldots \tag{2.39}$$

$$K(k) = \frac{\pi}{2}\left(1 + \frac{k^2}{4} + \frac{9k^4}{64} + \frac{25k^6}{256} + \frac{1225k^8}{16384} + \ldots\right)$$

After some manipulation we obtain

$$T(a) = \frac{2\pi}{\omega}\left(1 - \frac{3\varepsilon a^2}{8} + \frac{57\varepsilon^2 a^4}{256} - \frac{315\varepsilon^3 a^6}{2048} + \ldots\right) \tag{2.40}$$

Using the series expansion of the elliptic cosine

$$cn(x,k) = \frac{2\pi}{kK(k)}\left(\frac{\sqrt{q(k)}}{1+q(k)}\cos\frac{\pi x}{2K(k)} + \frac{q(k)\sqrt{q(k)}}{1+q^3(k)}\cos\frac{3\pi x}{2K(k)} + \ldots\right)$$

where

$$q(k) = \exp\left[-\frac{\pi K\left(\sqrt{1-k^2}\right)}{K(k)}\right] \tag{2.41}$$

we obtain the solution (2.36) in the form

$$u(t) = \frac{2\pi a}{kK(k)}\left(\frac{\sqrt{q(k)}}{1+q(k)}\cos\frac{2\pi}{T}t + \frac{q(k)\sqrt{q(k)}}{1+q^3(k)}\cos 3\frac{2\pi}{T}t + \ldots\right) \tag{2.42}$$

We remark that Eqs. 2.40 and 2.42 may be very useful, especially when comparing the exact solution to approximate solutions obtained by other methods.

2.1.2 Use of the Perturbation Method for Duffing Oscillator with Small Parameter

We shall apply the perturbation method to determine an approximate solution of Eq. 2.23 satisfying the initial conditions (2.24), under the assumption that the elastic nonlinearity is weak. We try a solution of the form (2.6) where the variable τ is given by Eq. 2.4 and the period T is given by Eq. 2.3. In this case, Eq. 2.5 becomes

$$\begin{aligned}
&u''_0 + \varepsilon u''_1 + \varepsilon^2 u''_2 + \varepsilon^3 u''_3 + \ldots + \\
&\quad + \left[1 + 2h_1\varepsilon + \left(h_1^2 + 2h_2\right)\varepsilon^2 + 2(h_3 + h_1h_2)\varepsilon^3 + \ldots\right](u_0 + \varepsilon u_1 + \varepsilon^2 u_2 + \\
&\quad + \varepsilon^3 u_3 + \ldots) = -\left[\varepsilon + 2h_1\varepsilon^2 + \left(h_1^2 + 2h_2\right)\varepsilon^2 + \ldots\right]\cdot\left[u_0^3 + 3u_0^2u_1\varepsilon + \right. \\
&\quad \left. + 3\left(u_0u_1^2 + u_0^2u_2\right)\varepsilon^2 + \ldots\right]
\end{aligned} \tag{2.43}$$

where $u' = \frac{du}{d\tau}$.

Equating the coefficients of like powers of ε, we obtain the first three approximations from the following four equations:

$$u''_0 + u_0 = 0 \tag{2.44}$$

$$u''_1 + u_1 = -2h_1u_0 - u_0^3 \tag{2.45}$$

$$u''_2 + u_2 = -\left(h_1^2 + 2h_2\right)u_0 - 2h_1u_1 - \left(3u_0^2u_1 + 2h_1u_0^3\right) \tag{2.46}$$

$$\begin{aligned}
u''_3 + u_3 = &-2(h_3 + h_1h_2)u_0 - 2\left(h_1^2 + 2h_2\right)u_1 - 2h_1u_1 - \\
&- \left[\left(h_1^2 + 2h_2\right)u_0^3 + 6h_1u_0^2u_1 + 3\left(u_0u_1^2 + u_0^2u_2\right)\right]
\end{aligned} \tag{2.47}$$

From Eqs. 2.44 and 2.7 we obtain

$$u_0 = a\cos\tau \tag{2.48}$$

Taking into consideration Eq. 2.48, Eq. 2.45 may be also written in the form

$$u''_1 + u_1 = -\left(2ah_1 + \frac{3a^3}{4}\right)\cos\tau - \frac{a^3}{4}\cos 3\tau \tag{2.49}$$

from which, by Eqs. 2.15, 2.7 and 2.18, it follows that

$$h_1 = -\frac{3a^2}{8} \tag{2.50}$$

and

$$u_1(\tau) = \frac{a^3}{32}(\cos 3\tau - \cos\tau) \tag{2.51}$$

By substituting the last two solutions into Eq. 2.46, we obtain

$$u''_2 + u_2 = \left(\frac{57a^5}{128} - 2ah_2\right)\cos\tau + \frac{3a^5}{16}\cos 3\tau - \frac{3a^5}{128}\cos 5\tau \tag{2.52}$$

and we deduce in the same way as above that

$$h_2 = \frac{57a^4}{256} \tag{2.53}$$

and

$$u_2(\tau) = \frac{3a^5}{128}(\cos\tau - \cos 3\tau) + \frac{a^5}{1024}(\cos 5\tau - \cos\tau) \tag{2.54}$$

Finally, by substituting Eqs. 2.50, 2.51, 2.53 and 2.54 into Eq. 2.47, we have

$$\begin{aligned} u''_3 + u_3 = &\left(-\frac{1107a^7}{4096} - 2ah_3\right)\cos\tau - \frac{705a^7}{4096}\cos 3\tau + \\ &+\frac{13a^7}{512}\cos 5\tau - \frac{a^7}{512}\cos 7\tau \end{aligned} \tag{2.55}$$

from which it follows that

$$h_3 = -\frac{1107a^6}{8192} \tag{2.56}$$

and

$$u_3(\tau) = \frac{705a^7}{32768}(\cos 3\tau - \cos\tau) + \frac{13a^7}{12288}(\cos\tau - \cos 5\tau) + \\ + \frac{a^7}{24576}(\cos 7\tau - \cos\tau) \tag{2.57}$$

Taking into consideration Eqs. 2.51, 2.54 and 2.57, we obtain, from Eq. 2.6, the approximate solution satisfying Eqs. 2.23 and 2.24 to within an error of third-order in ε:

$$u(\tau) = a\cos\tau + \frac{\varepsilon a^3}{32}(\cos 3\tau - \cos\tau) + \frac{\varepsilon^2 a^5}{1024}(\cos 5\tau - 24\cos 3\tau + 23\cos\tau) + \\ + \frac{\varepsilon^3 a^7}{98304}(4\cos 7\tau - 104\cos 5\tau + 2115\cos 3\tau - 2015\cos\tau) \tag{2.58}$$

Next, by introducing Eqs. 2.50, 2.53 and 2.56 into Eq. 2.3, we derive the corresponding approximation for the period

$$T = \frac{2\pi}{\omega}\left(1 - \frac{3\varepsilon a^2}{8} + \frac{57\varepsilon^2 a^4}{256} - \frac{1107\varepsilon^3 a^6}{8192}\right) \tag{2.59}$$

The approximate expression (2.59) of the period, as obtained by the perturbation method, coincides with the exact expansion (2.40), as expected, to within terms of third-order in small parameter ε.

2.1.3 *Use of the Perturbation Method for Duffing Oscillators with Strong Parameter*

Perturbation method provides the most versatile tool available in nonlinear dynamical systems and is constantly developed and applied to ever more complex problems. But, perturbation method has its own limitation: it is based on such assumption that a small parameter must exist in equation. This so called small parameter assumption greatly restricts applications of perturbation techniques, and as it is well-known, an overwhelming majority of nonlinear problems, especially those having strong nonlinearity, have no small parameters at all.

It is even more difficult to determine the so-called small parameter, which seems to be a special art requiring special techniques. An appropriate choice of small parameter may lead to ideal results; however an unsuitable choice of small parameter results in badly effects, sometimes seriously. Even if there exists suitable small parameter, the approximate solutions obtained by the perturbation method are valid, in most cases, only for small values for the parameter.

To overcome this limitation, many novel techniques have been proposed in recent years. For example Cheung et al. [38] presented a modified Lindstedt-Poincaré method. They expand ω^2 instead of ω:

$$\omega^2 = \omega_0^2 + \varepsilon h_1 + \varepsilon^2 h_2 + \ldots \tag{2.60}$$

and a new parameter is defined as

$$\alpha = \frac{\varepsilon h_1}{\omega_0^2 + \varepsilon h_1} \tag{2.61}$$

It is much better to expand u and ω^2 into a power series with respect to α:

$$u = u_0 + \alpha u_1 + \alpha^2 u_2 + \ldots \tag{2.62}$$

$$\omega^2 = \omega_0^2 + \alpha h_1 + \alpha^2 h_2 + \ldots \tag{2.63}$$

The essence of the proposed method is to expand the coefficient 1 from the equation $\ddot{u} + u = f(u)$ rather than the nonlinear frequency into a series of ε:

$$1 = \omega^2 + \varepsilon C_1 + \varepsilon^2 C_2 + \ldots \tag{2.64}$$

where the constants ω^2 and C_i can be identified by means of no secular terms.

Dai [39] introduced another transformation of the independent variable τ:

$$t = \tau + \varepsilon F(u(\tau)) + 0(\varepsilon^2)$$

where $F(u(\tau))$ is an unknown nonlinear functional further determined.

J.H. He [40–43] proposed some perturbation techniques (Taylor expansion, artificial parameter, linear perturbation method, parameterized perturbation method, bookkeeping artificial parameter perturbation method, iteration perturbation method, etc.) which are valid for large parameters. In what follows we present two versions of modified Lindstedt-Poincaré method applied for the Duffing nonlinear oscillator.

(a) In the first version, in 2004 H. Hu [44] presented a classical perturbation technique which is valid for large parameters. Hu considered the Duffing equations in the form

$$\ddot{u} + u + \varepsilon u^3 = 0 \tag{2.65}$$

and the initial condition

$$u(0) = a, \quad \dot{u}(0) = 0 \tag{2.66}$$

where ε is a parameter which does not need be small.

For Eq. 2.65, the solution is assumed to be in the form

$$u(t) = u_0(t) + \varepsilon u_1(t) + \varepsilon^2 u_2(t) + \ldots \tag{2.67}$$

and the fundamental frequency in the form [38], [42], [44]:

$$\Omega^2 = 1 + \varepsilon h_1 + \varepsilon h_2 + \ldots \tag{2.68}$$

where h_i are unknown constants.

By substituting Eqs. 2.67 and 2.68 into Eq. 2.65 we obtain

$$\begin{aligned}\left(\ddot{u}_0 + \varepsilon \ddot{u}_1 + \varepsilon^2 \ddot{u}_2 + \ldots\right) + \left(\Omega^2 - \varepsilon h_1 - \varepsilon^2 h_2 \ldots\right)\left(u_0 + \varepsilon u_1 + \varepsilon^2 u_2 + \ldots\right) + \\ + \varepsilon\left(u_0 + \varepsilon u_1 + \varepsilon^2 u_2 + \ldots\right)^3 = 0\end{aligned} \tag{2.69}$$

By equating coefficients of like powers of ε, and processing as the standard perturbation method, we have

$$\ddot{u}_0 + \Omega^2 u_0 = 0 \tag{2.70}$$

$$\ddot{u}_1 + \Omega^2 u_1 = h_1 u_0 - u_0^3 \tag{2.71}$$

$$\ddot{u}_2 + \Omega^2 u_2 = h_2 u_0 + h_1 u_1 - 3u_0^3 u_1 \tag{2.72}$$

and so on.

The initial conditions are replaced by

$$u_0(0) = a, \quad \dot{u}_0(0) = 0, \quad u_i(0) = \dot{u}_i(0) = 0, \quad i = 1, 2, \ldots \tag{2.73}$$

Solving Eqs. 2.70 and 2.73, we have

$$u_0(t) = a \cos \Omega t \tag{2.74}$$

Substitution of u_0 into Eq. 2.71, results into

$$\ddot{u}_1 + \Omega^2 u_1 = \left(a h_1 - \frac{3}{4} a^3\right) \cos \Omega t - \frac{3}{4} a^3 \cos 3\Omega t \tag{2.75}$$

Eliminating the secular term needs

$$h_1 = \frac{3}{4}a^2 \tag{2.76}$$

and thus the solution of Eq. 2.75 becomes

$$u_1(t) = \frac{a^3}{32\Omega^2}(\cos 3\Omega t - \cos \Omega t) \tag{2.77}$$

Now, by substituting Eqs. 2.76 and 2.77 into Eq. 2.72, we have the following equation

$$\ddot{u}_2 + \Omega^2 u_2 = \left(ah_2 + \frac{3a^5}{128\Omega^2}\right)\cos \omega t - \frac{3a^5}{128\Omega^2}\cos 5\Omega t \tag{2.78}$$

Avoiding the presence of secular terms requires

$$h_2 = -\frac{3a^4}{128\Omega^2} \tag{2.79}$$

The solution of Eq. 2.78 with the initial conditions (2.73) becomes

$$u_2(t) = \frac{a^5}{3072\Omega^4}(\cos 5\Omega t - \cos \Omega t) \tag{2.80}$$

If the second order approximation is sufficient, from relation (2.68) we have

$$\Omega^2 = 1 + \frac{3\varepsilon a^2}{4} - \frac{3\varepsilon^2 a^4}{128\Omega^2} \tag{2.81}$$

Solving for Ω gives

$$\Omega_{app} = \frac{1}{4}\sqrt{8 + 6\varepsilon a^2 + \sqrt{64 + 96\varepsilon a^2 + 30\varepsilon^2 a^4}} \tag{2.82}$$

The second approximate solution of Eq. 2.65 is obtained from Eqs. 2.67, 2.74, 2.77 and 2.80:

$$u(t) = a\cos \Omega t + \frac{\varepsilon a^3}{32\Omega^2}(\cos 3\Omega t - \cos \Omega t) + \frac{\varepsilon^2 a^5}{3072\Omega^4}(\cos 5\Omega t - \cos \Omega t) \tag{2.83}$$

where the frequency Ω is given by Eq. 2.82.

The exact frequency of the periodic motion of the Duffing equation is obtained from Eq. 2.37 with $\omega = 1$, because into Eqs. 2.65 and 2.24 ω is considered equal with 1:

$$\Omega_{ex} = \frac{\pi}{2}\sqrt{1+\varepsilon a^2}\left(\int_0^{\pi/2}\frac{d\theta}{\sqrt{1-m^2\sin^2\theta}}\right)^{-1}, m = \frac{\varepsilon a^2}{2(1+\varepsilon a^2)} \tag{2.84}$$

For comparison, the exact frequency obtained by Eq. 2.84 and the approximate frequency given by Eq. 2.82 are listed in Table 2.1

Table 2.1 indicates that the formula (2.82) can give excellent approximate frequencies for both small and large parameters.

So, for large ε, the present approximate frequencies have the same feature as the exact one, even in case $\varepsilon a^2 \to \infty$. We have

$$\lim_{\varepsilon a^2\to\infty}\frac{\Omega_{app}}{\Omega_{ex}} = \frac{\sqrt{6+\sqrt{30}}}{2\pi}\int_0^{\frac{\pi}{2}}\frac{d\theta}{\sqrt{1-0,5\sin^2\theta}} \approx 0,999691 \tag{2.85}$$

Therefore, for any value of ε, the maximum relative error of the second approximate frequency obtained for the Duffing equation by means of traditional perturbation technique, is less than 0.03%, with respect to the exact solution

It is at least strange that this traditional perturbation method work for large parameters.

If normal expansion is used for the frequency

$$\Omega = 1 + \varepsilon C_1 + \varepsilon^2 C_2 + \dots \tag{2.86}$$

instead of expansion (2.68), the results are different in the cases of small and large parameters. For example, from Eq. 2.58 we derive the corresponding first approximate frequency (for $\omega = 1$) in the case that ε is small:

$$\Omega = 1 + \frac{3\varepsilon a^2}{8} \tag{2.87}$$

Table 2.1 Comparison of approximate frequencies with the corresponding exact frequency for the Duffing equation

εa^2	Ω_{app} (2.82)	Ω_{ex} (2.84)
0.2	1.07200	1.07200
0.6	1.20173	1.20173
1	1.31776	1.31778
5	2.15018	2.15042
10	2.86613	2.86664
100	8.53110	8.53359
1,000	26.8025	26.8107
10,000	84.7013	84.7245

On the other hand, the first approximate frequency, obtained from Eqs. 2.76 and 2.68 for any ε is

$$\Omega = \sqrt{1 + \frac{3}{4}\varepsilon a^2} \tag{2.88}$$

Formula (2.88) can give good approximate frequencies for both small and large parameters, but for large parameters, formula (2.87) is not valid, because

$$\sqrt{1 + \frac{3}{4}\varepsilon a^2} \neq 1 + \frac{3}{8}\varepsilon a^2.$$

We believe that the Duffing equation is one of rather equations in which the expansion (2.68) can be used instead of Eq. 2.86.

In Fig. 2.1 is presented a comparison between the numerical solution and analytic solution (2.83) of Eq. 2.65 in the case $a = \omega = 1$, $\varepsilon = 0.1$ while Fig. 2.2 presents a similar comparison for the analytic solution (2.83) of the Eq. 2.65 in the case $a = \omega = 1$, $\varepsilon = 1$. A very good agreement was found between the numerical and analytical results for Eq. 2.65 in both cases.

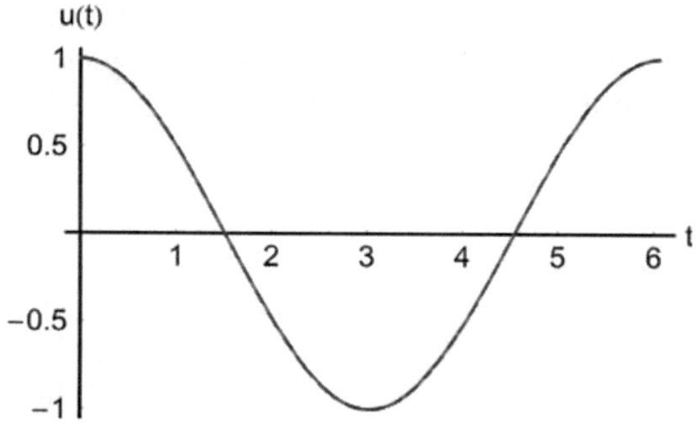

Fig. 2.1 Comparison between the results obtained for Eq. 2.65: ______numerical solution; ---- analytic solution (2.83) for $a = \omega = 1$, $\varepsilon = 0.1$

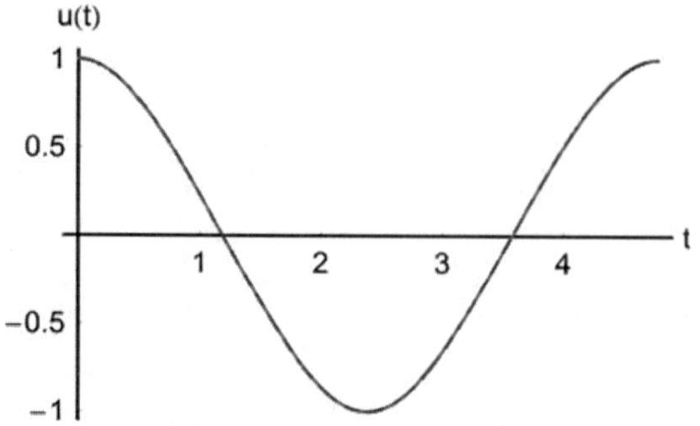

Fig. 2.2 Comparison between the results obtained for Eq. 2.65: ______numerical solution; ---- analytic solution (2.83) for $a = \omega = 1$, $\varepsilon = 1$

It is clear that the Lindstedt-Poincaré procedure produces a periodic expression describing the motion of the system, a frequency-amplitude relationship which is a direct consequence of requiring the expression to be periodic and higher harmonics in the higher-order terms of the expression.

(b) In the second version we expand ω^2 in series (2.60) and define the parameter α by Eq. 2.61 such that

$$\varepsilon = \frac{\alpha}{h_1(1-\alpha)}, \quad (\omega_0 = 1) \tag{2.89}$$

With the transformation $\tau = \Omega t$, Eq. 2.65 becomes

$$\Omega^2 u'' + u + \varepsilon u^3 = 0 \tag{2.90}$$

where primes denotes differentiation with respect to τ. The fundamental frequency Ω given by (2.68) becomes

$$\begin{aligned} \Omega^2 &= \left(1 + \frac{\alpha}{1-\alpha}\right)\left(1 + \lambda_2\alpha^2 + \lambda^3\alpha^3 + \ldots\right) \\ &= \frac{1}{1-\alpha}\left(1 + \lambda_2\alpha^2 + \lambda^3\alpha^3 + \ldots\right) \end{aligned} \tag{2.91}$$

where h_1, λ_2, λ_3... are unknown constants which will be determined in the later by perturbation steps successively.

Substituting Eqs. 2.89 and 2.91 into Eq. 2.90 yields:

$$\begin{aligned} &\left(1 + \lambda_2\alpha^2 + \lambda^3\alpha^3 + \ldots\right)\left(u''_0 + \alpha u''_1 + \alpha^2 u''_2 + \ldots\right) + \\ &+ (1-\alpha)\left(u_0 + \alpha u_1 + \alpha^2 u_2 + \ldots\right) + \frac{\alpha}{h_1}\left(u_0 + \alpha u_1 + \alpha^2 u_2 + \ldots\right)^3 = 0 \end{aligned} \tag{2.92}$$

It can be seen from Eq. 2.61 that the value of α is always kept small regardless to magnitude of εh_1. It is observed that α is a new small parameter which is considered to be better than ε. It will enable a strongly nonlinear system corresponding to ε be transformed into a small parameter system with respect to α.

Equating the coefficients of like terms of α into Eq. 2.92 the following set of linear differential equation can be obtained

$$u''_0 + u_0 = 0, u_0(0) = a, u'_0(0) = 0 \tag{2.93}$$

$$u''_1 + u_1 = u_0 - \frac{u_0^3}{h_1}, u_1(0) = 0, u'_1(0) = 0 \tag{2.94}$$

$$u''_2 + u_2 = u_1 + \lambda_2 u_0 - \frac{3u_0^2 u_1}{h_1}, u_2(0) = 0, u'_2(0) = 0 \tag{2.95}$$

From Eq. 2.93 we obtain

$$u_0 = a\cos\tau \tag{2.96}$$

Equation 2.94 becomes

$$u''_1 + u_1 = a\left(1 - \frac{3a^2}{4h_1}\right)\cos\tau - \frac{a^3}{4h_1}\cos 3\tau, u_1(0) = 0, u'_1(0) = 0 \tag{2.97}$$

In order to ensure that no secular terms appear in the next iteration, resonance must be avoided, so that the coefficient of $\cos\tau$ in Eq. 2.97 requires being zero:

$$h_1 = \frac{3}{4}a^2 \tag{2.98}$$

So, from Eq. 2.97, we have the following solution

$$u_1 = \frac{a}{24}(\cos 3\tau - \cos\tau) \tag{2.99}$$

Substituting Eqs. 2.96 and 2.99 into Eq. 2.95, yields

$$u''_2 + u_2 = a\left(\lambda_2 + \frac{1}{24}\right)\cos\tau - \frac{1}{24}a\cos 5\tau \tag{2.100}$$

Avoiding the secular term into Eq. 2.100, we obtain

$$\lambda_2 = -\frac{1}{24} \tag{2.101}$$

and therefore Eq. 2.91 becomes

$$\Omega^2 = \frac{1}{1-\alpha}\left(1 + \lambda_2\alpha^2 + \ldots\right) = 1 + \frac{3}{4}\varepsilon a^2 - \frac{3\varepsilon^2 a^4}{128 + 96\varepsilon a^2} + \ldots \tag{2.102}$$

To illustrate the remarkable accuracy of the obtained results, we compare the approximate period

$$T_{app} = \frac{2\pi}{\Omega} = \frac{2\pi}{\sqrt{1 + \frac{3}{4}\varepsilon a^2 - \frac{3\varepsilon^2 a^4}{128+96\varepsilon a^2}}} \tag{2.103}$$

with the exact one obtained from Eq. 2.84

$$T_{ex} = \frac{4}{\sqrt{1 + \varepsilon a^2}}\int_0^{\frac{\pi}{2}} \frac{d\theta}{\sqrt{1 - m\sin^2\theta}}, \quad m = \frac{\varepsilon a^2}{2(1 + \varepsilon a^2)} \tag{2.104}$$

What is rather surprising about the remarkable range of validity of Eq. 2.103 is that the actual asymptotic period is also of higher accuracy as $\varepsilon\alpha^2 \to \infty$

$$\lim_{\varepsilon a^2 \to \infty} \frac{T_{ex}}{T_{app}} = \frac{2\sqrt{\frac{17}{24}}}{\pi} \int_0^{\frac{\pi}{2}} \frac{d\theta}{\sqrt{1 - 0,5\sin^2\theta}} = 0,999271 \tag{2.105}$$

Therefore, the approximate analytical solution of the Duffing equation has an error which never exceeds 0.07% with respect to the exact solution.

These extensions of the Lindstedt-Poincaré method, which are simple and easy to use, are effective methods for dealing with strongly non-linear vibration of single degree of freedom systems which cannot be treated by the standard Lindstedt-Poincaré method.

Remark. In general, perturbation methods work very well for weakly nonlinear dynamical systems and there exist cases when this procedure leads to inappropriate results. For instance, we consider the weakly nonlinear system

$$\begin{aligned} \dot{x} &= 0.5x - \varepsilon xy, & x(0) &= 4 \\ \dot{y} &= -0.3y + 2\varepsilon xy & y(0) &= 1 \end{aligned} \tag{2.106}$$

where ε is a small parameter and dot denotes derivative with respect to time t. For Eq. 2.106 we may suppose the power series

$$\begin{aligned} x(t) &= x_0(t) + \varepsilon x_1(t) + \varepsilon^2 x_2(t) + \varepsilon^3 x_3(t) + \ldots. \\ y(t) &= y_0(t) + \varepsilon y_1(t) + \varepsilon^2 y_2(t) + \varepsilon^3 y_0(t) + \ldots.. \end{aligned} \tag{2.107}$$

By substituting Eqs. 2.107 into Eqs. (2.106) and equating the coefficients of like powers of ε, we obtain the following linear differential equations:

$$\begin{aligned} \dot{x}_0 &= 0.5x_0, & x_0(0) &= 4 \\ \dot{y}_0 &= -0.3y_0, & y_0(0) &= 1 \end{aligned} \tag{2.108}$$

$$\begin{aligned} \dot{x}_1 &= 0.5x_1 - x_0 y_0, & x_1(0) &= 0 \\ \dot{y}_1 &= -0.3y_1 + 2x_0 y_0, & y_1(0) &= 0 \end{aligned} \tag{2.109}$$

$$\begin{aligned} \dot{x}_2 &= 0.5x_2 - (x_0 y_1 + x_1 y_0), & x_2(0) &= 0 \\ \dot{y}_2 &= -0.3y_2 + 2(x_0 y_1 + x_1 y_0), & y_2(0) &= 0 \end{aligned} \tag{2.110}$$

$$\begin{aligned} \dot{x}_3 &= 0.5x_3 - (x_0 y_2 + x_1 y_1 + x_2 y_0), & x_3(0) &= 0 \\ \dot{y}_3 &= -0.3y_3 + 2(x_0 y_2 + x_1 y_1 + x_2 y_0), & y_3(0) &= 0 \end{aligned} \tag{2.111}$$

The Eqs. 2.108 yields

$$x_0 = 4e^{0.5t}$$
$$y_0 = e^{-0.3t} \tag{2.112}$$

By substituting Eq. 2.112 into Eq. 2.109 and solving these equations, we obtain

$$x_1 = \frac{40}{3}e^{0.2t} - \frac{40}{3}e^{0.5t}$$
$$y_1 = 16e^{0.2t} - 16e^{-0.3t} \tag{2.113}$$

Now, substituting Eqs. 2.112 and 2.113 into Eq. 2.110 and solving these equations, it results

$$x_2 = \frac{5000}{9}e^{0.5t} - 320e^{0.7t} - \frac{2320}{9}e^{0.2t} + \frac{200}{9}e^{-0.1t}$$
$$y_2 = 48e^{-0.3t} + 128e^{0.7t} - \frac{928}{3}e^{0.2t} + \frac{400}{3}e^{-0.1t} \tag{2.114}$$

From Eqs. 2.111, 2.112, 2.113 and 2.114 we obtain

$$\begin{aligned} x_3 = &-\frac{7502000}{567}e^{0.5t} - \frac{5120}{7}e^{1.2t} + \frac{21760}{3}e^{0.7t} + \frac{12800}{3}e^{0.4t} \\ &+ \frac{86400}{27}e^{0.2t} - \frac{21200}{27}e^{-0.1t} + \frac{2000}{81}e^{-0.4t} \\ y_3 = &\frac{48544}{21}e^{-0.3t} + \frac{2048}{3}e^{1.2t} - \frac{8704}{3}e^{0.7t} + \frac{25600}{21}e^{0.4t} \\ &+ \frac{34592}{9}e^{0.2t} - \frac{42400}{9}e^{-0.1t} - \frac{4000}{9}e^{-0.4t} \end{aligned} \tag{2.115}$$

Finally, from Eqs. 2.107, 2.112, 2.113, 2.114 and 2.115, for $\varepsilon = 0.001$, we obtain an approximate solution of the fourth-order in the form

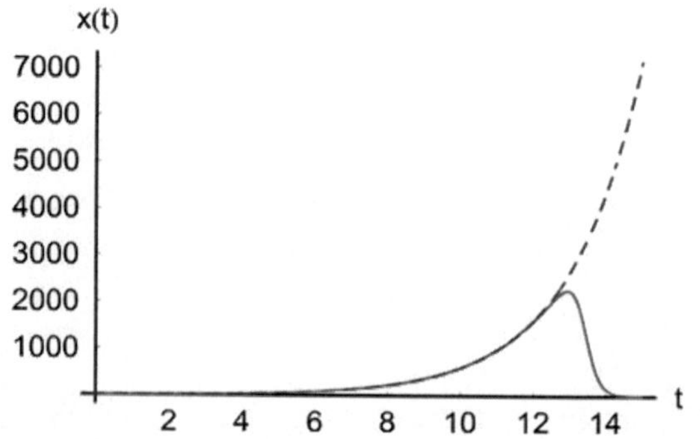

Fig. 2.3 Comparison between the numerical solution x(t) of Eq. 2.106 and approximate solution (2.116): _____numerical solution; _ _ _ _ approximate solution

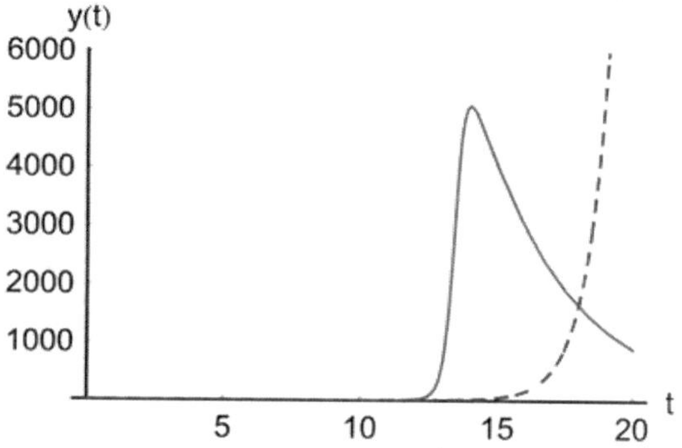

Fig. 2.4 Comparison between the numerical solution y(t) of Eq. 2.106 and approximate solution (2.116): ______numerical solution; _ _ _ _ approximate solution

$$\begin{aligned}
x(t) &= -0.00000073e^{1.2t} - 0.00031274e^{0.71t} + 3.9872e^{0.5t} + \\
&+ 0.00000426e^{0.4t} + 0.01307e^{0.2t} + 0.00002143e^{-0.1t} + 2.46 \cdot 10^{-14}e^{-0.4t} \\
y(t) &= 0.0000006826e^{1.2t} + 0.00012509e^{0.71t} + 0.000001219e^{0.4t} + \\
&+ 0.01569e^{0.2t} + 0.0001286e^{-0.1t} + 0.98405e^{-0.3t} - 0.000000444e^{-0.4t}
\end{aligned} \tag{2.116}$$

In Figs. 2.3 and 2.4 is presented a comparison between the numerical and approximate solutions (2.116) of the Eqs. (2.106) in the case of $\varepsilon = 0.001$, where an important discrepancy between these solutions can be observed.

In conclusion, the perturbation method does not work very well in all cases, even though in the differential equations is involved a small parameter. Therefore, this remark lead to the conclusion that more powerful methods should be developed to overcome these difficulties. Some suitable methods will be presented in Chaps. 6–9.

Chapter 3
The Method of Harmonic Balance

This method may be used to determine the approximate periodic solutions of nonlinear differential equations. If a periodic solution does exist, it may be sought in the form of a Fourier series, whose coefficients are determined by requiring the series to satisfy the equation of motion. However, in order to avoid solving an infinite system of algebraic equations, it is better to approximate the solution by finite sums of trigonometric functions, i.e.

$$u(t) = \sum_{k=0}^{M} A_k \cos(k\omega t + k\varphi) \tag{3.1}$$

or

$$u(t) = \sum_{k=0}^{M} A_k \sin(k\omega t + k\varphi) \tag{3.2}$$

or

$$u(t) = \sum_{k=0}^{M} (A_k \cos k\omega t + B_k \sin k\omega t) \tag{3.3}$$

For the sake of simplicity, we have been considered systems governed by equations having the form

$$\ddot{u} + au + b\dot{u}^2 + cu\dot{u}^2 = 0, a>0 \tag{3.4}$$

where a, b and c are constants.

Substituting for example Eq. 3.1 into the governing Eq. 3.4 and equating the coefficient of each of all harmonics to zero, leads to a set of algebraic equations whose number is generally greater than that of the coefficients A_k and ω. Therefore,

V. Marinca and N. Herisanu, *Nonlinear Dynamical Systems in Engineering*,
DOI 10.1007/978-3-642-22735-6_3, © Springer-Verlag Berlin Heidelberg 2011

to obtain a compatible system, we limit ourselves to imposing a number of conditions equal to the number of coefficients to be determined. This procedure obviously yields only an approximate solution. Usually these equations are solved for ω, A_1, A_2, A_3, ..., A_m in terms of A_1. The accuracy of the resulting periodic solution depends on the value of A_1 and the number of harmonics in the assumed solution (3.1).

First, we consider only one-term expansion into Eq. 3.1:

$$u(t) = A_1 \cos(\omega t + \varphi) := A_1 \cos \phi \tag{3.5}$$

we have

$$\begin{aligned} &\frac{1}{2} b\omega^2 A_1^2 + \left(aA_1 - \omega^2 A_1 + \frac{1}{4} c\omega^2 A_1^3 \right) \cos \phi - \\ &- \frac{1}{2} b\omega^2 A_1^2 \cos 2\phi - \frac{1}{4} c\omega^2 A_1^3 \cos 3\phi = 0 \end{aligned} \tag{3.6}$$

Equating the coefficient of $\cos \phi$ to zero, we obtain

$$\omega^2 = a\left(1 - \frac{1}{4} cA_1^2\right)^{-1} \tag{3.7}$$

which for small A_1 becomes

$$\omega = \sqrt{a}\left(1 + \frac{1}{8} cA_1^2\right) \tag{3.8}$$

From Eq. 3.8 it is clear that only a part of the nonlinear correction to the frequency has been obtained. The reason for the deficiency is that the terms $0(A_1^2)$were neglected in Eq. 3.6, while the terms $0(A_1^3)$were kept. To obtain the rest of the nonlinear correction, we need to include other terms besides the first harmonic in the expression of u.

Therefore, we consider

$$u(t) = A_0 + A_1 \cos \phi \tag{3.9}$$

Substituting Eq. 3.9 into Eq. 3.4, yields

$$\begin{aligned} &aA_0 + \frac{1}{2} b\omega^2 A_1^2 + \frac{1}{2} c\omega^2 A_0 A_1^2 + \left(aA_1 - \omega^2 A_1 + \frac{1}{4} c\omega^2 A_1^3 \right) \cos \phi - \\ &- \frac{1}{2} c\omega^2 A_0 A_1^2 \cos 2\phi - \frac{1}{4} c\omega^2 A_1^3 \cos 3\phi = 0 \end{aligned} \tag{3.10}$$

Equating the constant term and the coefficient of $\cos\phi$ to zero, we have respectively

$$aA_0 + \frac{1}{2}b\omega^2 A_1^2 + \frac{1}{2}c\omega^2 A_0 A_1^2 = 0$$
$$a - \omega^2 + \frac{1}{4}c\omega^2 A_1^2 = 0 \tag{3.11}$$

If we consider A_1 small, the solutions of Eq. 3.11 are

$$\omega^2 = a\left(1 - \frac{1}{4}cA_1^2\right)^{-1}$$
$$A_0 = -4bA_1^2\left(4 + cA_1^2\right)^{-1} \tag{3.12}$$

Hence

$$\omega = \sqrt{a}\left(1 + \frac{1}{8}cA_1^2\right) \tag{3.13}$$

Again, comparing Eq. 3.13 with Eq. 3.8, we establish that the frequencies are the same, but we conclude that the assumption (3.9) has also produced a solution that does not account for all the nonlinear correction to the frequency to $0(A_1{}^2)$. From Eq. 3.10, we find that we still neglected terms $0(A_1{}^2)$ while we kept terms $0(A_1{}^3)$.

Next, let us try to include three terms in the solution, that is:

$$u(t) = A_0 + A_1\cos\phi + A_2\cos 2\phi \tag{3.14}$$

where we consider $|A_0| << |A_1|$ and $|A_2| << |A_1|$. Substituting Eq. 3.14 into Eq. 3.4 yields

$$aA_0 + \frac{1}{2}b\omega^2 A_1^2 + 8b\omega^2 A_2^2 + \frac{1}{2}c\omega^2 A_0 A_1^2 + 8c\omega^2 A_0 A_2^2 + \frac{9}{4}c\omega^2 A_1^2 A_2$$
$$+ \left(aA_1 - \omega^2 A_1 + 4b\omega^2 A_1 A_2 + \frac{3}{4}c\omega^2 A_1^3 + 4c\omega^2 A_0 A_1 A_2 +\right.$$
$$\left. + 12c\omega^2 A_1 A_2^2\right)\cos\phi + \left(aA_2 - 4\omega^2 A_2 + \frac{1}{2}b\omega^2 A_1^2 + \frac{1}{2}c\omega^2 A_0 A_1^2 +\right.$$
$$\left. + \frac{7}{2}c\omega^2 A_1^2 A_2 + 16c\omega^2 A_2^3\right)\cos 2\phi +$$
$$+ \left(4b\omega^2 A_1 A_2 + 10c\omega^2 A_1 A_2^2 + \frac{1}{4}c\omega^2 A_1^3 + 4c\omega^2 A_0 A_1 A_2\right)\cos 3\phi +$$
$$+ \left(8b\omega^2 A_2^2 + \frac{9}{4}c\omega^2 A_1^2 A_2 + 8c\omega^2 A_0 A_2^2\right)\cos 4\phi +$$
$$+ 10c\omega^2 A_1 A_2^2\cos 5\phi + 8c\omega^2 A_2^3\cos 6\phi = 0 \tag{3.15}$$

Equating the constant term and the coefficients of $\cos\phi$ and $\cos 2\phi$ to zero, we obtain:

$$aA_0 + \frac{1}{2}b\omega^2 A_1^2 + 8b\omega^2 A_2^2 + \frac{1}{2}c\omega^2 A_0 A_1^2 + + 8c\omega^2 A_0 A_2^2 + \frac{9}{4}c\omega^2 A_1^2 A_2 = 0 \tag{3.16}$$

$$a - \omega^2 + 4b\omega^2 A_2 + \frac{3}{4}c\omega^2 A_1^2 + 4c\omega^2 A_0 A_2 + 12c\omega^2 A_2^2 = 0 \tag{3.17}$$

$$aA_2 - 4\omega^2 A_2 + \frac{1}{2}b\omega^2 A_1^2 + \frac{1}{2}c\omega^2 A_0 A_1^2 + \frac{7}{2}c\omega^2 A_1^2 A_2 + 16c\omega^2 A_2^3 = 0 \tag{3.18}$$

We suppose that A_1 is small and therefore, through Eqs. 3.16 and 3.18 show that $A_0 = 0(A_1^2)$ and $A_2 = 0(A_1^2)$and hence

$$A_0 = -\frac{1}{2}bA_1^2 + 0(A_1^4) \tag{3.19}$$

$$A_2 = \frac{1}{6}bA_1^2 + 0(A_1^4) \tag{3.20}$$

$$\omega^2 = a - \frac{a(8b^2 + 9c)}{12}A_1^2 + 0(A_1^4) \tag{3.21}$$

Substituting A_0 and A_2 from Eqs. 3.19 and 3.20, respectively into Eq. 3.14 yields

$$u(t) = A_1 \cos\phi + \frac{1}{6}bA_1^2(\cos 2\phi - 3) + 0(A_1^4) \tag{3.22}$$

From Eq. 3.21, it follows that

$$\omega = \sqrt{a}\left(1 - \frac{8b^2 + 9c}{24}A_1^2\right) \tag{3.23}$$

Inspecting the coefficients of the higher harmonics into Eq. 3.15, one finds that they are of the order $0(A_1^3)$and hence the neglected terms are of the order of the error in Eqs. 3.22 and 3.23, which is the reason why it is in agreement with the solutions obtained by other procedures.

It is clear from the development above that, in order to obtain a consistent solution using the method of harmonic balance, one need either to know a great deal about the solution a priori or to carry enough terms in the solution and check the order of the coefficients of all the neglected harmonics. Otherwise one might obtain an inaccurate approximation such as Eqs. 3.13 and 3.23.

On the other hand, in the frame of the harmonic balance method it is very difficult to construct higher-order analytical approximation because it requires analytical solutions of sets of complicated nonlinear algebraic equations. This method can be applied to nonlinear oscillatory problems where the nonlinear terms are not small and no perturbation parameter is required.

Various generalizations of the harmonic balance method have been presented by several other investigators: the unrestricted harmonic balance [45], the rational representation [46], an intrinsic method [47], the use of Jacobi elliptic functions [48], two time scale harmonic balance [49], the incremental harmonic balance [50], etc.

An alternative approach for solving the nonlinear oscillation of a conservative system is obtained by combining the linearization of the governing equation with the harmonic balance method [51]. In this procedure, instead of a set of nonlinear algebraic equations, it results a set of linear algebraic equations, as can be seen in the next Section.

3.1 Free Vibrations of Cantilever Beam

We consider free vibrations of a slender inextensible cantilever beam carrying an intermediate lumped mass with a rotary inertia, described by the nonlinear equation

$$\ddot{u} + u + \alpha u^2 \ddot{u} + \alpha u \dot{u}^2 + \beta u^3 = 0, \quad u(0) = A,\ \dot{u}(0) = 0 \tag{3.24}$$

The third and fourth terms in Eq. 3.24 represent inertia-type cubic nonlinearity arising from the inextensibility assumption. The last term is a static–type nonlinearity associated with the potential energy stored in bending.

With a new independent variable $\tau = \omega t$, where ω is the frequency of system, Eq. 3.24 becomes:

$$\omega^2 \left[\left(1 + \alpha u^2\right) u'' + \alpha u u'^2 \right] + u + \beta u^3 = 0 \tag{3.25}$$

Following the lowest harmonic balance method (see Eq. 3.5), initial approximation with initial conditions (3.24_2) is

$$u_0(\tau) = A \cos \tau \tag{3.26}$$

Substituting Eq. 3.26 into Eq. 3.25, yields

$$\left(A + \frac{3}{4}\beta A^3 - \omega^2 A - \frac{1}{2}\omega^2 \alpha A^3 \right) \cos \tau + \left(\frac{3}{4}\beta A^3 - \frac{1}{2}\omega^2 \alpha A^3 \right) \cos 3\tau = 0 \tag{3.27}$$

Equating the coefficient of $\cos\tau$ to zero, we obtain the first analytical approximate frequency ω_0:

$$\omega_0 = \sqrt{\frac{4+3\beta A^2}{4+2\alpha A^2}} \tag{3.28}$$

At this time, we know the first approximate periodic solution in the form (3.26). Wu et al. [51] introduced a new solution in the form of $u_0(\tau)+v(\tau)$which is composed of the solution (3.26) and the correction $v(\tau)$, which is a periodic function. Making linearization of the governing Eq. 3.25, with respect to the correction $v(\tau)$ at $u=u_0(\tau)$, leads to

$$\begin{aligned}&\omega^2\left[(1+\alpha u_0^2)u''_0+\alpha u_0 u'^2_0\right]+u_0+\beta u_0^3+\omega^2\left[(1+\alpha u_0^2)v''+2\alpha u_0 u'_0 v'+\right.\\&\left.+\left(2\alpha u_0 u''_0+\alpha u'^2_0\right)v\right]+(1+3\beta u_0^2)v=0,\quad v(0)=0,\quad \dot{v}(0)=0\end{aligned} \tag{3.29}$$

Substituting Eq. 3.26 into Eq. 3.29 we obtain an equation in v:

$$\begin{aligned}&\left(A+\frac{3}{4}\beta A^3-\omega^2 A-\frac{1}{2}\omega^2\alpha A^3\right)\cos\tau+\left(\frac{3}{4}\beta A^3-\frac{1}{2}\omega^2\alpha A^3\right)\cos 3\tau+\\&+\omega^2\left[\left(1+\frac{1}{2}\alpha A^2+\frac{1}{2}\alpha A^2\cos 2\tau\right)v''-\alpha A^2\sin 2\tau\cdot v'\right]+\\&+\left[1+\frac{3}{2}\beta A^2+\frac{3}{2}\beta A^2\cos 2\tau-\omega^2\left(\frac{1}{2}\alpha A^2+\frac{3}{2}\alpha A^2\cos 2\tau\right)\right]v=0\end{aligned} \tag{3.30}$$

To obtain the solution v from Eq. 3.30 in the initial conditions (3.29_2), we consider

$$v(\tau)=k(\cos\tau-\cos 3\tau) \tag{3.31}$$

where k is a constant, which can be determined from the condition that Eq. 3.31 be the solution of Eq. 3.30, and therefore yield

$$\begin{aligned}&\left[A+\frac{3}{4}\beta A^3-\omega^2 A-\frac{1}{2}\omega^2\alpha A^3+k\left(-\omega^2+1+\frac{3}{2}\beta A^2\right)\right]\cos\tau+\\&+\left\{\frac{3}{4}\beta A^3-\frac{1}{2}\omega^2\alpha A^3+\left[\left(9+\frac{7}{2}\alpha A^2\right)\omega^2-1-\frac{3}{4}\beta A^2\right]k\right\}\cos 3\tau+\\&+k\left(\frac{9}{2}\omega^2-\frac{3}{4}\beta A^2\right)\cos 5\tau=0\end{aligned} \tag{3.32}$$

Equating the coefficients of $\cos\tau$and $\cos 3\tau$to zero in Eq. 3.32, we obtain

$$A + \frac{3}{4}\beta A^3 - \omega^2 A - \frac{1}{2}\omega^2 \alpha A^3 + k\left(-\omega^2 + 1 + \frac{3}{2}\beta A^2\right) = 0$$
$$\frac{3}{4}\beta A^3 - \frac{1}{2}\omega^2 \alpha A^3 + k\left[\left(9 + \frac{7}{2}\alpha A^2\right)\omega^2 - 1 - \frac{3}{4}\beta A^2\right] = 0 \qquad (3.33)$$

For the nonvanishing solution of the system, the condition

$$M\omega^4 + N\omega^2 + P = 0 \qquad (3.34)$$

must be fulfilled, where

$$\begin{aligned} M &= -144 - 136\alpha A^2 - 28\alpha^2 A^4 \\ N &= 160 + (72\alpha + 124\beta)A^2 + 60\alpha\beta A^4 \\ P &= -16 - 28\beta A^2 - 15\beta^2 A^4 \end{aligned} \qquad (3.35)$$

From Eq. 3.34 we obtain the second approximate frequency ω_1:

$$\omega_1 = \sqrt{\frac{40 + (18\alpha + 31\beta)A^2 + 15\alpha\beta A^4 + \sqrt{\Delta}}{72 + 68\alpha A^2 + 14\alpha^2 A^4}} \qquad (3.36)$$

where

$$\begin{aligned} \Delta = {} & 1024 + (896\alpha + 1472\beta)A^2 + \left(212\alpha^2 + 1364\alpha\beta + 421\beta^2\right)A^4 + \\ & + \left(344\alpha^2\beta + 420\alpha\beta^2\right)A^6 + 120\alpha^2\beta^2 A^8 \end{aligned} \qquad (3.37)$$

From Eq. 3.33_1 we obtain

$$k = \frac{4A + 3\beta A^3 - (4A + 2\alpha A^3)\omega_1^2}{4\omega_1^2 - 4 - 6\beta A^2} \qquad (3.38)$$

and the second approximate periodic solution in the form

$$u_1(\tau) = u_0(\tau) + v(\tau) = (A + k)\cos\tau - k\cos 3\tau \qquad (3.39)$$

where

$$\tau = \omega_1 t \qquad (3.40)$$

The harmonic balance method gives the first approximation as those in Eqs. 3.26 and 3.28, but it is known that to determine higher-order approximations, a set of algebraic equations with third-order nonlinearity has to be solved. The corresponding numerical computation is rather complicated. In contrast, formulas (3.36), (3.38) and (3.39) are simple and easy to be implemented.

In order to illustrate the applicability, accuracy and effectiveness of the proposed approach, we compare the analytical approximate frequency and periodic solution with the exact ones.

The exact frequency is given by [51]

$$\omega_{ex} = \frac{\pi}{2\int_o^{\pi/2}\left[2(1+\alpha A^2\cos^2\theta)/(2+\beta A^2(1+\cos^2\theta))\right]^{1/2}d\theta} \tag{3.41}$$

In Tables 3.1 and 3.2 are compared the exact frequency ω_e obtained by integrating Eq. 3.41 with the first and second approximate frequencies ω_0 and ω_1 computed using Eqs. 3.28 and 3.36, respectively, in two cases: $\alpha = \beta = 0.1$ and $\alpha = \beta = 1$.

Table 3.1 indicates that the formula (3.36) is more accurate than (3.28) and the former provides excellent approximation to exact frequency for small as well as large values of amplitude of oscillation. The comparison of the corresponding approximate frequencies with exact ones for $\alpha = 1$ and $\beta = 1$ is shown in Table 3.2, where again, similar agreement is observed.

For $\alpha = 0.1, \beta = 0.1$, the periodic solution achieved by numerical integration of Eq. 3.24 using a Runge–Kutta scheme and the approximate periodic solution given by Eq. 3.39 are plotted in Figs. 3.1–3.3. The corresponding three solutions are shown in Figs. 3.4–3.6 for the case $\alpha = 1, \beta = 1$. These figures represent, respectively, three different amplitudes $A = 1, 5$ and 10. They show that the approximate periodic solution provides relatively good approximation comparing to the exact periodic solution for small as well as for large amplitude of oscillation. These figures also indicate that discrepancy of solutions widens as the modal constants α and β become larger. The present analytical approximate frequencies and periodic relations apply well to small as well as to large values of amplitude of oscillation. The accuracy obtained is almost independent of the value of the parameter used for solving these oscillations.

Table 3.1 Comparison of frequencies for Eq. 3.24 in the case $\alpha = \beta = 0.1$

A	ω_e from (3.41)	ω_0 from (3.28)	ω_1 from (3.36)
1	1.01202	1.011834731	1.012017273
5	1.16302	1.130388331	1.163583802
10	1.28468	1.190238071	1.276999402

Table 3.2 Comparison of frequencies for Eq. 3.24 in the case $\alpha = \beta = 1$

A	ω_e from (3.41)	ω_0 from (3.28)	Ω_1 from (3.36)
1	1.09035	1.08012345	1.090699384
5	1.34288	1.209530059	1.322166505
10	1.38928	1.220735876	1.350839601

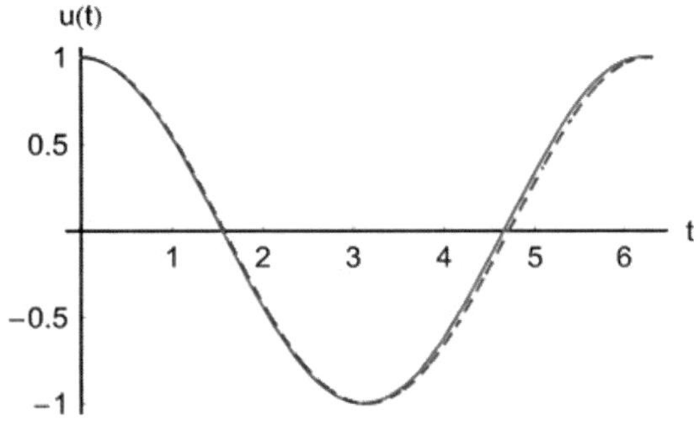

Fig. 3.1 Comparison between analytical approximate solution (3.39) and exact solution of Eq. 3.24 for $\alpha = \beta = 0.1, A = 1$: ______ exact solution; _ _ _ _ _ approximate solution

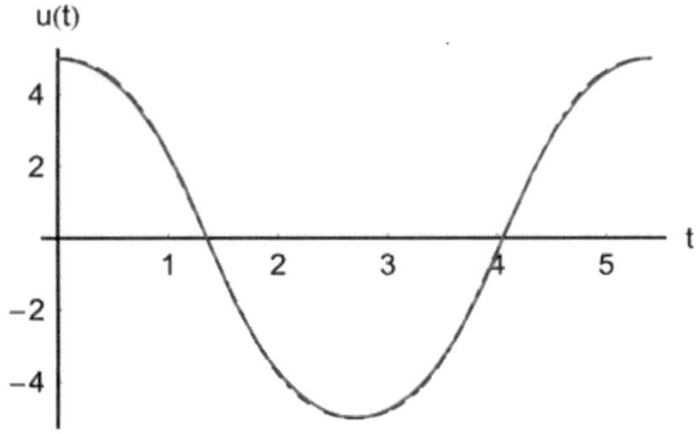

Fig. 3.2 Comparison between analytical approximate solution (3.39) and exact solution of Eq. 3.24 for $\alpha = \beta = 0.1, A = 5$: ______ exact solution; _ _ _ _ _ approximate solution

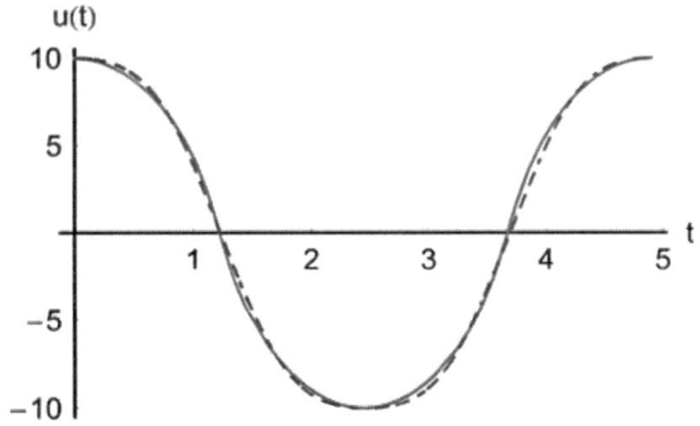

Fig. 3.3 Comparison between analytical approximate solution (3.39) and exact solution of Eq. 3.24 for $\alpha = \beta = 0.1, A = 10$: ______ exact solution; _ _ _ _ _ approximate solution

As a conclusion, unlike the classical harmonic balance method, linearization is performed prior to proceeding with harmonic balancing. As a result, we obtain a set of linear algebraic equations instead of one of nonlinear algebraic equation, which enables us to establish analytical approximate formulas for the frequency and periodic solution. A good agreement of the analytical approximate frequency and periodic solution with the exact solution has been demonstrated.

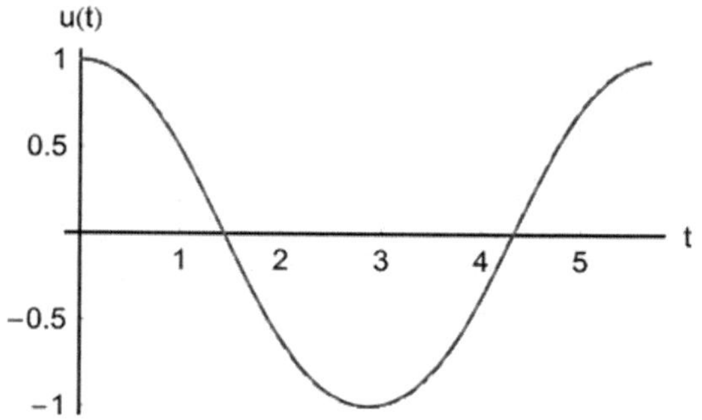

Fig. 3.4 Comparison between analytical approximate solution (3.39) and exact solution of Eq. 3.24 for $\alpha = \beta = 1, A = 1$: ______ exact solution; _ _ _ _ _ approximate solution

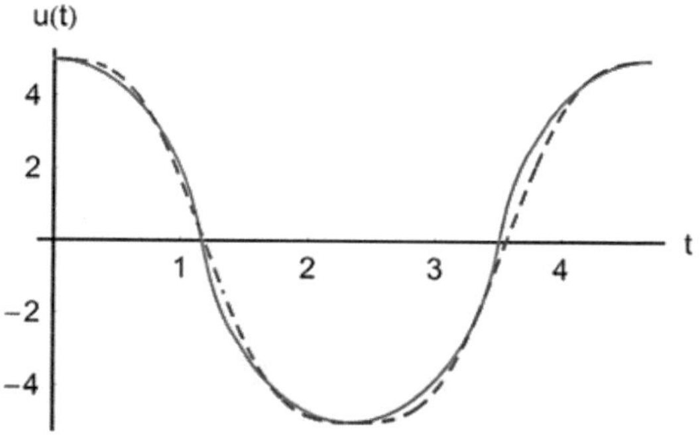

Fig. 3.5 Comparison between analytical approximate solution (3.39) and exact solution of Eq. 3.24 for $\alpha = \beta = 1, A = 5$: ______ exact solution; _ _ _ _ _ approximate solution

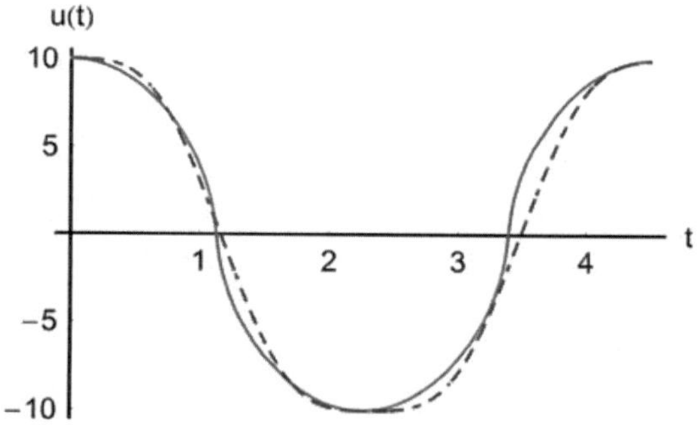

Fig. 3.6 Comparison between analytical approximate solution (3.39) and exact solution of Eq. 3.24 for $\alpha = \beta = 1, A = 10$: ______ exact solution; _ _ _ _ _ approximate solution

3.2 Rational Harmonic Balance Method

In what follows, the rational harmonic balance method is used to determine the appropriate periodic solution of the system in the given initial conditions. We study the stability of the solution of Duffing oscillators by Floquet-Hill method.

We consider a system governed by a nondimensional differential equation [52], [53]

$$\ddot{u} + c\dot{u} + au + bu^3 + d = e \cos \lambda t \tag{3.42}$$

$$u(0) = u_0, \dot{u}(0) = 0 \tag{3.43}$$

where a, b, c, d and e are constant parameters.

After the transformation $\lambda t = \tau$, Eq. 3.42 becomes

$$\lambda^2 u'' + c\lambda u' + au + bu^3 + d = e \cos \tau \tag{3.44}$$

where $u' = \frac{du}{d\tau}$.

With the initial conditions (3.43), we propose, for Eq. 3.44, the solution in the form of rational trigonometric function

$$u(\tau) = \frac{u_0(1+D) - A - C + A\cos\tau + B(\sin 2\tau - 2\sin\tau) + C\cos 2\tau}{1 + D\cos 3\tau} \tag{3.45}$$

where the constants A, B, C and D will be determined in what follows. After the use of trigonometric identities and application of the method of rational harmonic balance to retain only constant terms involving $\cos\tau$, $\cos 2\tau$, $\cos 3\tau$, $\sin\tau$ and $\sin 2\tau$, we obtain equations:

$$\begin{aligned}
&\frac{9}{2}\lambda^2 D(1+2D)[(1+D)u_0 - A - C] + a\left(1 + \frac{D^2}{2}\right)[(1+D)u_0 - A - C] + \\
&+ b\Big\{[(1+D)u_0 - A - C]^3 + \frac{3}{2}[(1+D)u_0 - A - C]\left(A^2 + B^2 + C^2\right) - \\
&- 3AB^2 + \frac{3}{4}A^2C + 3B^2C\Big\} + d + \frac{3}{2}dD^2 = 0
\end{aligned} \tag{3.46}$$

$$\begin{aligned}
&\lambda^2\left(-A + \frac{7}{2}CD + 4AD^2\right) + c\lambda\left(\frac{7}{2}BD - 2B - BD^2\right) + \\
&+ a\left(A + \frac{1}{2}AD^2 + CD\right) + b\Big\{\frac{3}{4}A^3 + \frac{9}{2}AD^2 + \frac{3}{2}AC^2 + \\
&+ 3A[(1+D)u_0 - A - C]^2 - \left(6B^2 + 3AC\right)[(1+D)u_o - A - C]\Big\} - \\
&- e\left(1 + \frac{3}{2}D^2\right) = 0
\end{aligned} \tag{3.47}$$

$$\begin{aligned}
&\lambda^2\left(2B + \frac{1}{2}BD - 14BD^2\right) + c\lambda\left(\frac{7}{2}CD - A - \frac{1}{2}AD^2\right) + \\
&+ a(-2B - BD - BD^2) + b\Big\{-3B[(1+D)u_0 - A - C]^2 - 9B^3 - \frac{3}{2}A^2B - \\
&- 3BC^2 + (3AB + 6BC)[(1+D)u_o - A - C]\Big\} = 0
\end{aligned} \tag{3.48}$$

$$\lambda^2\left(-4C+\frac{1}{2}AD+\frac{23}{2}CD^2\right)+c\lambda(2B-5BD+BD^2)+$$
$$+a\left(C+\frac{1}{2}CD^2+AD\right)+b\Big\{3[(1+D)u_0-A-C]^2C+$$
$$+[(1+D)u_0-A-C]\left(\frac{3}{2}A^2-6B^2\right)+$$
$$+\frac{3}{4}C^3+\frac{3}{2}A^2C+\frac{15}{4}B^2C\Big\}-e\frac{D}{2}\left(3+\frac{D^2}{4}\right)=0 \tag{3.49}$$

$$\lambda^2\left(-4B+BD+\frac{23}{2}BD^2\right)+c\lambda\left(\frac{5}{2}AD-2C-CD^2\right)+$$
$$+a\left(B+2BD+\frac{1}{2}BD^2\right)+b\Big\{\frac{27}{4}B^3+\frac{3}{2}A^2B-\frac{9}{4}BC^2+$$
$$+3B[(1+D)u_0-A-C]-6AB[(1+D)u_0-A-C]\Big\}=0 \tag{3.50}$$

$$9\lambda^2D[(1+D)u_0-A-C]+2aD[(1+D)u_0-A-C]+$$
$$+b\Big\{\frac{1}{4}A^3-\frac{15}{4}AB^2+\frac{3}{4}AC^2-3B^2C+[(1+D)u_0-$$
$$-A-C](6B^2+3AC)\Big\}+3dD\left(1+\frac{1}{4}D^2\right)=0 \tag{3.51}$$

The Eqs. 3.46–3.51 are solved numerically using the Newton-Raphson's iterative procedure and Levenberg–Marquardt algorithm.

In what follows, we consider the Duffing system in the form

$$\ddot{u}+\frac{1}{10}\dot{u}-\frac{1}{2}u+\frac{1}{2}u^3=\frac{2}{5}\cos\lambda t \tag{3.52}$$

with the period $T=2\pi/\lambda$. Assuming that $\lambda=2.1$, and using the value $D=-0.00145312$ obtained from the Eqs. 3.46–3.51, the approximate solution, with the initial conditions $u(0)=-1$, $u'(0)=0$ can be written as

$$u(t)=0.900242-0.108273\cos 2.1t+0.006155\sin 2.1t-0.003077 \times \sin 4.2t \tag{3.53}$$

In [53] is found the approximate solution

$$u(t)=-0.90995-0.11609\cos 2.1t+0.0071020\sin 2.1t \tag{3.54}$$

To ascertain the stability of the periodic solution found above, we examine the time evolution after the application of an infinitesimal arbitrary disturbance $y(t)$ such that

$$\bar{u}(t) = u(t) + y(t) \tag{3.55}$$

where $u(t)$ is the solution (3.53) previously found. The stability of $u(t)$ then depends on whether $y(t)$grows or decays with t. Substituting Eq. 3.55 into Eq. 3.42 and keeping linear terms in $y(t)$, we obtain

$$\ddot{y}(t) + c\dot{y}(t) + \left[a + 3bu^2(t)\right]y(t) = 0 \tag{3.56}$$

which is a linear ordinary differential equation with periodic coefficients. The existence of nontrivial solutions can be shown via Floquet's theorem, which calls for solutions of the form

$$y(t+T) = sy(t), \quad T = \frac{2\pi}{\lambda} \tag{3.57}$$

where s is an eigenvalue (also called a Floquet multipliers [22]) of the monordomy matrix C whose elements are associated with Eq. 3.56 through the relations

$$y_1(t+T) = C_{11}y_1(t) + C_{12}y_2(t) \tag{3.58}$$

$$y_2(t+T) = C_{21}y_1(t) + C_{22}y_2(t) \tag{3.59}$$

where C_{ij} are constants. The functions $y_1(t)$ and $y_2(t)$ are two linearly independent solutions of Eq. 3.56. To generate y_1 and y_2, we use the initial conditions

$$y_1(0) = 1, \quad \dot{y}_1(0) = 0 \tag{3.60}$$

$$y_2(0) = 0, \quad \dot{y}_2(0) = 1 \tag{3.61}$$

The solution $u(t)$ is a stable orbit provided that $y(t)$ does not grow with t. This requires that

$$|s|<1 \tag{3.62}$$

That is why the eigenvalue of monodromy matrix C must remain inside the unit circle in the complex plane. The monodromy matrix C can be obtained by means of rational harmonic balance method subject to the initial conditions (3.60) and (3.61). It follows from Eqs. 3.58, 3.59, 3.60 and 3.61 that

$$C = \begin{bmatrix} y_1(T) & \dot{y}_1(T) \\ y_2(T) & \dot{y}_2(T) \end{bmatrix} \tag{3.63}$$

Therefore,

$$s^2 - [y_1(T) + \dot{y}_2(T)]s + [y_1(T)\dot{y}_2(T) - \dot{y}_1(T)y_2(T)] = 0 \tag{3.64}$$

The values of s determine the stability of the approximate solution $u(t)$ according to Eq. 3.62. The manner in which the eigenvalue s leaves the unit circle characterizes the local qualitative bifurcations occurring to the orbit. For the dissipative one-degree-of-freedom system described by Eq. 3.42 there are two ways in which s can leave the unit circle, each of which creates independent patterns of instability in the periodic orbit. An eigenvalue can leave the unit circle through the real axis at either -1 or $+1$. It follows from Eq. 3.45 that the period of $u(t)$ is $T = 2\pi/\lambda$. Consequently, when s leaves the unit circle through -1, $y(t + 2\,T) = y(t)$ according to Eq. 3.57 and hence solution with the period $2\,T$ is stable, indicating a period doubling of flip bifurcations. On the other hand, when s leaves the unit circle through +1, Eq. 3.57 indicates that $y(t + T) = y(t)$, which implies the coexistence of a stable and an unstable attractor with the period T. The result is a saddle-node or tangent bifurcation, which results in a jump in the response of the system.

With the solution given by Eqs. 3.53, 3.56 becomes

$$\begin{aligned} &\ddot{y} + 0.1\dot{y} + (0.725386 + 0.292415\cos 2.1t - 0.016623\sin 2.1t + \\ &+0.008763\cos 4.2t)y = 0 \end{aligned} \tag{3.65}$$

For Eq. 3.65, we propose the solution by means of the rational harmonic balance method in the form

$$\begin{aligned} &y_i(t) = \frac{A_i + B_i\cos 1.05t + C_i\sin 1.05t + D_i\cos 2.1t + E_i\sin 2.1t}{1 + 2F_i\cos 3.15t} \\ &i = 1, 2 \end{aligned} \tag{3.66}$$

Using of the $F_1 = 0.0002713$ and $F_2 = -0.000156$ determined in the same manner as D from Eqs. 3.46–3.51, the solution y_1 obtained for initial condition (3.60) is found as

$$y_1(t) = 0.243 + 0.758\cos 1.05t + 0.102\sin 1.05t - 0.051\sin 2.1t \tag{3.67}$$

and the solution y_2 obtained for initial conditions (3.61) is found as

$$\begin{aligned} y_2(t) = 0.411 &- 0.533\cos 1.05t + 0.844\sin 1.05t + \\ &+ 0.122\cos 2.1t - 0.094\sin 2.1t \end{aligned} \tag{3.68}$$

The monodromy matrix (3.63) becomes

$$C = \begin{bmatrix} -0.5133 & -0.2142 \\ 1.0661 & -0.9996 \end{bmatrix} \tag{3.69}$$

Equation 3.64 is of the form

$$s^2 + 1.5189s + 0.741586 = 0 \tag{3.70}$$

with the complex conjugates solutions:

$$s_1 = -0.75645 + 0.4115\,i, \quad i = \sqrt{-1} \tag{3.71}$$

$$s_2 = -0.75645 - 0.4115\,i, \tag{3.72}$$

Therefore $|s_1| = |s_2| = 0.8611 < 1$. It follows that the solution (3.53) of the Eq. 3.52 is stable [54].

Remark. The free term in Eq. 3.64 can be obtained in the exact form as shown below. Since $y_1(t)$ and $y_2(t)$ are solutions of Eq. 3.56, it follows that

$$\begin{aligned} \ddot{y}_1 + c\dot{y}_1 + \left(a + 3ba^2\right)y_1 &= 0 \\ \ddot{y}_2 + c\dot{y}_2 + \left(a + 3ba^2\right)y_2 &= 0 \end{aligned} \tag{3.73}$$

Subtracting the first Eq. 3.73 multiplied with y_2 from the second Eq. 3.73 multiplied with y_1 yields

$$y_1\ddot{y}_2 - y_2\ddot{y}_1 = c(\dot{y}_1 y_2 - \dot{y}_2 y_1) \tag{3.74}$$

which can be integrated to yield

$$y_1\dot{y}_2 - \dot{y}_1 y_2 = \exp(-ct) \tag{3.75}$$

It follows from Eq. 3.75 that

$$y_1(T)\dot{y}_2(T) - \dot{y}_1(T)y_2(T) = \exp(-cT) \tag{3.76}$$

In case of Eq. 3.52 we have $c = 0.1$, T $= T = 2\pi/2.1$ and therefore we obtain exact value

$$y_1(T)\dot{y}_2(T) - \dot{y}_1(T)y_2(T) = \exp\left(-\frac{2\pi}{21}\right) = 0.7414 \tag{3.77}$$

This value and the free term in Eq. 3.70 show the efficiency of the rational harmonic balance method and the validity of the approach proposed.

Chapter 4
The Method of Krylov and Bogolyubov

The method of Krylov and Bogolyubov [7, 8, 55, 56] and the analogous method previously developed by Van der Pol in 1934 follows the basic idea of the method of variation of constants of Lagrange.

For $\varepsilon = 0$, the general solution of the weakly nonlinear equation

$$\ddot{u} + \omega^2 u + \varepsilon f(u, \dot{u}) = 0 \tag{4.1}$$

may be written in the form

$$u(t) = a \cos \omega t + b \sin \omega t \tag{4.2}$$

where a and b are constants to be determined by the initial conditions.

In the phase plane, instead of Eq. 4.1 for $\varepsilon \neq 0$, we consider the system

$$\begin{aligned} \dot{u} &= v \\ \dot{v} + \omega^2 u + \varepsilon f(u, v) &= 0 \end{aligned} \tag{4.3}$$

with the solution

$$\begin{aligned} u(t) &= A(t) \cos \omega t + B(t) \sin \omega t \\ v(t) &= -A(t)\omega \sin \omega t + B(t)\omega \cos \omega t \end{aligned} \tag{4.4}$$

Therefore, instead of the constants a and b we consider the functions $A(t)$ and $B(t)$ such that the variable $u(t)$ from Eq. 4.41 verifies Eq. 4.1 or Eq. 4.3. By this condition we obtain

V. Marinca and N. Herisanu, *Nonlinear Dynamical Systems in Engineering*,
DOI 10.1007/978-3-642-22735-6_4,

$$\begin{aligned}
&\dot{A}\cos\omega t - A\omega\sin\omega t + \dot{B}\sin\omega t + B\omega\cos\omega t = -A\omega\sin\omega t + B\omega\cos\omega t,\\
&-\dot{A}\omega\sin\omega t - A\omega^2\cos\omega t + \dot{B}\omega\cos\omega t - B\omega^2\sin\omega t + \omega^2 A\cos\omega t +\\
&+\omega^2 B\sin\omega t + \varepsilon f(A\cos\omega t + B\sin\omega t, -A\omega\sin\omega t + B\omega\cos\omega t) = 0 \qquad (4.5)
\end{aligned}$$

The system (4.5) can be written in the form

$$\begin{aligned}
&\dot{A}\cos\omega t + \dot{B}\sin\omega t = 0\\
&-\dot{A}\omega\sin\omega t + \dot{B}\omega^2\cos\omega t +\\
&\varepsilon f(A\cos\omega t + B\sin\omega t, -A\omega\sin\omega t + B\omega\cos\omega t) = 0 \qquad (4.6)
\end{aligned}$$

Solving this system with respect to $\dot{A}$ and $\dot{B}$ yields

$$\begin{aligned}
\dot{A} &= \frac{\varepsilon}{\omega} f(A\cos\omega t + B\sin\omega t, -A\omega\sin\omega t + B\omega\cos\omega t)\cdot\sin\omega t\\
\dot{B} &= -\frac{\varepsilon}{\omega} f(A\cos\omega t + B\sin\omega t, -A\omega\sin\omega t + B\omega\cos\omega t)\cdot\cos\omega t \qquad (4.7)
\end{aligned}$$

By integrating the system (4.7) we can obtain the exact solution, but this is not easy.

In order to calculate an approximate solution of Eq. 4.1 for $\varepsilon \neq 0$ but small, Van der Pol proposed retaining the same form (4.2) of the solution, but considering the quantities A and B as "slowly varying" functions of time to be determined.

Therefore on a period $T = 2\pi/\omega$, the functions $\dot{A}(t)$ and $\dot{B}(t)$ can be approximated with their mean values given by the method of averaging [22]:

$$\begin{aligned}
\dot{A} &= \frac{\varepsilon}{2\pi}\int_0^{\frac{2\pi}{\omega}} f(A\cos\omega t + B\sin\omega t, -A\omega\sin\omega t + B\omega\cos\omega t)\cdot\sin\omega t dt\\
\dot{B} &= -\frac{\varepsilon}{2\pi}\int_0^{\frac{2\pi}{\omega}} f(A\cos\omega t + B\sin\omega t, -A\omega\sin\omega t + B\omega\cos\omega t)\cdot\cos\omega t dt \qquad (4.8)
\end{aligned}$$

By using this method, Van der Pol obtained a series of important results concerning stationary solutions, the behaviour of the solutions as t increases, and so on. However, this method was initially applied on purely intuitive grounds. Many problems, such as the theoretical justification of the method, its field of applicability, and the way to obtain approximations of higher order, have been solved as late as 1937 by Krylov and Bogolyubov. These authors also modified the form of the first approximation by taking instead of Eq. 4.2 the solution

$$u = a\cos(\omega t + \varphi)$$

with "slightly varying" amplitude and phase. We shall expand in what follows an improved variant of Krylov and Bogolyubov, developed by Bogolyubov and Mitropolsky [55]. For $\varepsilon = 0$, the general solution of Eq. 4.1 can be written in the form

$$u = a\cos\psi,\ \psi = \omega t + \varphi \tag{4.9}$$

and has constant amplitude and constant rate of the total phase, i.e.

$$\dot{a} = 0, \quad \dot{\psi} = \omega \tag{4.10}$$

For $\varepsilon \neq 0$, the quantities $\dot{a}$ and $\dot{\psi}$ generally depend on a and ε, and Eq. 4.10_1 must be completed by terms depending on a, ψ and ε. Consequently, we seek an approximate solution of Eq. 4.1 of the form

$$u = a\cos\psi + \varepsilon u_1(a,\psi) + \varepsilon^2 u_2(a,\psi) + \ldots + \varepsilon^m u_m(a,\psi) \tag{4.11}$$

which is called the approximation of the $(m + 1)$-th order if $u_m(a,\psi) \neq 0$. $u_k(a,\psi)$, k = 1,2,. . .,m are supposed to be periodic functions of ψ with period 2π, and a, ψ are functions of t, which have to satisfy the differential equations with separable variables

$$\dot{a} = \varepsilon A_1(a) + \varepsilon^2 A_2(a) + \ldots + \varepsilon^m A_m(a) \tag{4.12}$$

$$\dot{\psi} = \omega + \varepsilon B_1(a) + \varepsilon^2 B_2(a) + \ldots + \varepsilon^m B_m(a) \tag{4.13}$$

We first try to determine the functions $u_k(a,\psi)$, $A_k(a)$ and $B_k(a)$, $k = 1,2,\ldots,m$ so that Eq. 4.1 be satisfied to within an error of $(m + 1)$-th order in ε. As in the method of Lindstedt-Poincaré , the recursive determination of these functions does not give rise to any qualitative difficulties. However, the calculation is usually performed only for $m = 1$ or $m = 2$, because the results obtained for higher approximations are very intricate.

The practical applicability of the method is determined in the first place not by the convergence of the expansion (4.11) as $m \to \infty$, but by asymptotic proprieties of the approximate solution as $\varepsilon \to 0$, for given m. Therefore, the expansion leading to secular terms may be cast out at each step, even if this procedure could determine, as in the method Lindstedt-Poincaré, the divergence of the expansion (4.11) as $m \to \infty$.

After the determining of the functions u_k, A_k, B_k, and assuming that the initial conditions are given in the form

$$a(0) = a_0, \quad \psi(0) = \varphi \tag{4.14}$$

we may determine the solution by means of two integrations. By integrating Eq. 4.12 we deduce that

$$t(a) = \int_{a_o}^{a} \frac{da}{\varepsilon A_1(a) + \varepsilon^2 A_2(a) + \ldots + \varepsilon^m A_m(a)} \tag{4.15}$$

Solving this equation with respect to a yields

$$a = a(t) \tag{4.16}$$

Substituting Eq. 4.16 into Eq. 4.13 and integrating, it follows that

$$\psi(t) = \omega t + \int_0^t \left[\varepsilon B_1(a) + \varepsilon^2 B_2(a) + \ldots + \varepsilon^m B_m(a)\right] dt \tag{4.17}$$

Finally, by introducing Eqs. 4.16 and 4.17 into Eq. 4.11 we obtain the approximate solution $u(t)$.

In what follows, we consider $m = 2$ in Eqs. (4.11–4.13), i.e.

$$u = a\cos\psi + \varepsilon u_1(a,\psi) + \varepsilon^2 u_2(a,\psi) \tag{4.18}$$

$$\dot{a} = \varepsilon A_1(a) + \varepsilon^2 A_2(a) \tag{4.19}$$

$$\dot{\psi} = \omega + \varepsilon B_1(a) + \varepsilon^2 B_2(a) \tag{4.20}$$

The terms of third or higher order in ε will be systematically omitted in the following formulas.

By differentiating Eq. 4.18 twice with respect to t, we obtain

$$\begin{aligned}\dot{u} = {} & -a\omega\sin\psi + \varepsilon\left(A_1\cos\psi - aB_1\sin\psi + \omega\frac{\partial u_1}{\partial\psi}\right)\\ & + \varepsilon^2\left(A_2\cos\psi - aB_2\sin\psi + A_1\frac{\partial u_1}{\partial a} + B_1\frac{\partial u_1}{\partial\psi} + \omega\frac{\partial u_2}{\partial\psi}\right) + 0(\varepsilon^3)\end{aligned} \tag{4.21}$$

$$\begin{aligned}\ddot{u} = {} & -a\omega^2\cos\psi + \varepsilon\left(-2\omega A_1\sin\psi - 2\omega aB_1\cos\psi + \omega^2\frac{\partial^2 u_1}{\partial\psi^2}\right)\\ & + \varepsilon^2\left[\left(A_1\frac{dA_1}{da} - aB_1^2 - 2\omega aB_2\right)\cos\psi - \left(2\omega A_2 + 2A_1B_1 + A_1a\frac{dB_1}{da}\right)\sin\psi +\right.\\ & \left. + 2\omega A_1\frac{\partial^2 u_1}{\partial a\partial\psi} + 2\omega B_1\frac{\partial^2 u_1}{\partial\psi^2} + \omega^2\frac{\partial^2 u_2}{\partial\psi^2}\right] + 0(\varepsilon^3)\end{aligned} \tag{4.22}$$

On the other hand, by taking into consideration Eqs. 4.21 and 4.18 it follows that

$$\begin{aligned}\varepsilon f(u,\dot{u}) = {} & \varepsilon f(a\cos\psi, -a\omega\sin\psi) + \varepsilon^2\left[u_1\frac{\partial f}{\partial u}(a\cos\psi, -a\omega\sin\psi) +\right.\\ & \left. + \left(A_1\cos\psi - aB_1\sin\psi + \omega\frac{\partial u_1}{\partial\psi}\right)\frac{\partial f}{\partial\dot{u}}(a\cos\psi, -a\omega\sin\psi)\right] + 0(\varepsilon^3)\end{aligned} \tag{4.23}$$

Finally, by substituting Eqs. 4.18, 4.22 and 4.23 into Eq. 4.1 and equating coefficients of ε and ε^2, we obtain the conditions that have to be satisfied if Eq. 4.18 should give the solution of the differential Eq. 4.1 to within an error of third order in ε:

$$\omega^2\left(\frac{\partial^2 u_1}{\partial\psi^2}+u_1\right)=-f(a\cos\psi,-a\omega\sin\psi)+2\omega A_1\sin^2\psi+2\omega a B_1\cos\psi \quad (4.24)$$

$$\begin{aligned}\omega^2\left(\frac{\partial^2 u_2}{\partial\psi^2}+u_2\right)=&-u_1\frac{\partial f}{\partial u}(a\cos\psi,-a\omega\sin\psi)-\\&-\left(A_1\cos\psi-aB_1\sin\psi+\omega\frac{\partial u_1}{\partial\psi}\right)\frac{\partial f}{\partial\dot u}(a\cos\psi,-a\omega\sin\psi)+\\&+\left(aB_1^2-A_1\frac{dA_1}{da}\right)\cos\psi+\left(2A_1B_1+A_1a\frac{dB_1}{da}\right)\sin\psi-\\&-2\omega A_1\frac{\partial^2 u_1}{\partial a\partial\psi}-2B_1\omega\frac{\partial^2 u_1}{\partial\psi^2}+2\omega A_2\sin\psi+2\omega aB_2\cos\psi\end{aligned} \quad (4.25)$$

By developing now the functions $f(a\cos\psi,-a\omega\sin\psi)$ and $u_1(a,\psi)$in a Fourier series with the respect to ψ, we have

$$f(a\cos\psi,-a\omega\sin\psi)=C_0(a)+\sum_{k=1}^{\infty}[C_k(a)\cos k\psi+D_k(a)\sin k\psi] \quad (4.26)$$

$$u_1(a,\psi)=v_0(a)+\sum_{k=1}^{\infty}[v_k(a)\cos k\psi+w_k(a)\sin k\psi] \quad (4.27)$$

where

$$\begin{aligned}C_0(a)&=\frac{1}{2\pi}\int_0^{2\pi}f(a\cos\psi,-a\omega\sin\psi)d\psi\\C_k(a)&=\frac{1}{\pi}\int_0^{2\pi}f(a\cos\psi,-a\omega\sin\psi)\cos k\psi\,d\psi\\D_k(a)&=\frac{1}{2\pi}\int_0^{2\pi}f(a\cos\psi,-a\omega\sin\psi)\sin k\psi\,d\psi\end{aligned} \quad (4.28)$$

The solution $u_1(a,\psi)$ of Eq. 4.24 is a periodic function of ψ only if the right-hand side of the equation does not contain the fundamental harmonic of period 2π. Introducing Eq. 4.26 into Eq. 4.24 and equating to zero the coefficients of $\sin\psi$ and $\cos\psi$, in the right hand it gives

$$A_1(a) = \frac{D_1(a)}{2\omega}, \quad B_1(a) = \frac{C_1(a)}{2\omega a} \tag{4.29}$$

From Eqs. 4.27 and 4.24 we obtain

$$v_0(a) = \frac{-C_0(a)}{\omega^2}, \quad v_k(a) = \frac{C_k(a)}{\omega^2(k^2-1)}, \quad w_k(a) = \frac{D_k(a)}{C_0^2(k^2-1)}, \tag{4.30}$$
$$k = 2, 3, \ldots$$

The solution $u_1(a,\psi)$ given by Eq. 4.27 is

$$u_1(a,\psi) = -\frac{C_0(a)}{\omega^2} + \frac{1}{\omega^2}\sum_{k=2}^{\infty}\frac{C_k(a)\cos k\psi + D_k(a)\sin k\psi}{k^2-1} \tag{4.31}$$

By conveniently dividing the integration interval and using the substitutes $\sin a = x$ and $\cos a = x$, respectively, it may be shown that the relations

$$\int_0^{2\pi} F(\cos\alpha)\sin\alpha\, d\alpha = 0, \quad \int_0^{2\pi} F(\sin\alpha)\cos\alpha\, d\alpha = 0 \tag{4.32}$$

hold for every function which are integrable on the interval $[-1,1]$. These relations may be used in order to simplify the expression of $A_1(a)$ and $B_1(a)$ from Eqs. 4.24, 4.26, 4.28 and 4.29.

$$\begin{aligned} A_1(a) &= \frac{1}{2\pi\omega}\int_0^{2\pi} f(a\cos\psi, -a\omega\sin\psi)\sin\psi\, d\psi \\ B_1(a) &= \frac{1}{2\pi\omega}\int_0^{2\pi} f(a\cos\psi, -a\omega\sin\psi)\cos\psi\, d\psi \end{aligned} \tag{4.33}$$

The functions $A_2(a)$, $B_2(a)$ and$u_2(a,\psi)$may be similarly obtained from Eq. 4.25. However, $u_2(a,\psi)$does not contribute to the second approximation. Therefore, we content ourselves with calculating $A_2(a)$ and $B_2(a)$ requiring that the terms containing $\sin\psi$ and $\cos\psi$in Eq. 4.25 vanish. It follows that

$$\begin{aligned} A_2(a) = &-\frac{1}{2\omega}\left(2A_1B_1 + aA_1\frac{dB_1}{da}\right) + \frac{1}{2\pi\omega}\int_0^{2\pi}\left[u_1(a,\psi)\frac{\partial f}{\partial u}(a\cos\psi, -a\omega\sin\psi) + \right. \\ &\left. + \left(A_1\cos\psi - aB_1\sin\psi + \omega\frac{\partial u_1}{\partial\psi}\right)\frac{\partial f}{\partial \dot{u}}(a\cos\psi, -a\omega\sin\psi)\right]\sin\psi\, d\psi \end{aligned} \tag{4.34}$$

$$B_2(a) = -\frac{1}{2\omega}\left(B_1^2 - \frac{A_1}{a}\frac{dA_1}{da}\right) + \frac{1}{2\pi a\omega}\int_0^{2\pi}\left[u_1(a,\psi)\frac{\partial f}{\partial u}(a\cos\psi, -a\omega\sin\psi)\right.$$
$$\left. + \left(A_1\cos\psi - aB_1\sin\psi + \omega\frac{\partial u_1}{\partial\psi}\right)\frac{\partial f}{\partial\dot{u}}(a\cos\psi, -a\omega\sin\psi)\right]\cos\psi\, d\psi \tag{4.35}$$

We summarize the formulas which are necessary for the determination of the first two approximations by the method of Krylov and Bogolyubov.

(a) First approximation

$$u = a\cos\psi$$
$$t(a) = \frac{1}{\varepsilon}\int_{a_0}^{a}\frac{da}{A_1(a)}$$
$$\psi(t) = \omega t + \varepsilon\int_0^t B_1(a)dt \tag{4.36}$$

where a_0, A_1 and B_1 are given by Eqs. 4.141 and 4.33, respectively.

(b) Second approximation

$$u = a\cos\psi + \varepsilon u_1(a,\psi)$$
$$t(a) = \frac{1}{\varepsilon}\int_{a_0}^{a}\frac{da}{A_1(a) + \varepsilon A_2(a)}$$
$$\psi(t) = \omega t + \varepsilon\int_0^t [B_1(a) + \varepsilon B_2(a)]dt \tag{4.37}$$

where u_1, A_2 and B_2 are given by Eqs 4.31, 4.34 and 4.35 respectively.

4.1 Oscillator with Linear and Cubic Elastic Restoring Force and Weak Asymmetric Quadratic Damping

The equation of motion of nonlinear nonconservative system with linear and cubic elastic restoring force and weak asymmetric quadric damping is of the form

$$\ddot{u} + \omega^2 u + d_0(\dot{u}) + \varepsilon u^3 = 0 \tag{4.38}$$

where

$$d_0(\dot{u}) = \begin{cases} c_1\dot{u}^2 & \textit{if } \dot{u} \geq 0 \\ -c_2\dot{u}^2 & \textit{if } \dot{u} < 0 \, , \end{cases}$$

with c_1 and c_2 constants. By introducing the dimensionless small parameter

$$\varepsilon = a_0 c_1 \tag{4.39}$$

where $a_0 = a(0)$ and a is the amplitude of the fundamental harmonic, with $a > 0$, and comparing Eqs. 4.1 and 4.38, it follows that

$$f(u, \dot{u}) = d(\dot{u}) + u^3 \tag{4.40}$$

where

$$d(\dot{u}) = \begin{cases} \dfrac{1}{a_0}\dot{u}^2 & \textit{if } \dot{u} \geq 0 \\ -\dfrac{c_2}{a_0 c_1}\dot{u}^2 & \textit{if } \dot{u} < 0 \end{cases} \tag{4.41}$$

Therefore, from (4.40) yields

$$f(a\cos\psi, -a\omega\sin\psi) = \frac{a^3}{4}(\cos 3\psi + 3\cos\psi) + d(-a\omega\sin\psi) \tag{4.42}$$

By expanding in a cosine Fourier series the even function $d(a\cos\psi)$ we have

$$d(a\cos\psi) = \sum_{k=0}^{\infty} d_k(a)\cos k\psi \tag{4.43}$$

Now, by replacing ψ by $\psi + \pi/2$ and a by ωa in Eq. 4.43, we obtain

$$d(-a\omega\sin\psi) = \sum_{k=0}^{\infty} d_k(a\omega)\cos k\left(\psi + \frac{\pi}{2}\right) \tag{4.44}$$

It is interesting that we will determine the coefficients from Eq. 4.43:

$$d_k(a\omega) = \frac{\omega^2 a^2}{\pi a_0}\left[\int_0^{\frac{\pi}{2}} \cos^2\psi \cos k\psi \, d\psi - \frac{c_2}{c_1}\int_{\frac{\pi}{2}}^{\frac{3\pi}{2}} \cos^2\psi \cos k\psi \, d\psi + \right.$$
$$\left. + \int_{\frac{3\pi}{2}}^{2\pi} \cos^2\psi \cos k\psi \, d\psi \right] = \frac{\omega^2 a^2}{\pi a_0}\left\{F_k\left(\frac{\pi}{2}\right) - F_k(0) - \right.$$
$$\left. - \frac{c_2}{c_1}\left[F_k\left(\frac{3\pi}{2}\right) - F_k\left(\frac{\pi}{2}\right)\right] + F_k(2\pi) - F_k\left(\frac{3\pi}{2}\right)\right\} \tag{4.45}$$

where

$$F_k(\psi) = \int \cos^2\psi \cos k\psi \, d\psi = \frac{\sin k\psi}{2k} + \frac{\sin(k+2)\psi}{4(k+2)} +$$
$$+ \frac{\sin(k-2)\psi}{4(k-2)} \quad \text{for k} \neq 0, \text{ k} \neq 2 \tag{4.46}$$

and

$$F_0(\psi) = \frac{\psi}{2} + \frac{\sin 2\psi}{4}, \quad F_2(\psi) = \frac{\sin 2\psi}{4} + \frac{\psi}{4} + \frac{\sin 2\psi}{16} \tag{4.47}$$

Substituting Eq. 4.46 into Eq. 4.45 for $k \neq 2$, yields

$$d_k(a\omega) = \frac{a^2\omega^2}{\pi a_0}\left(1 + \frac{c_2}{c_1}\right)\left[\frac{\sin\frac{k\pi}{2} - \sin\frac{3k\pi}{2}}{2k} + \frac{\sin\frac{(k+2)\pi}{2} - \sin\frac{3(k+2)\pi}{2}}{4(k+2)} + \right.$$
$$\left. + \frac{\sin\frac{(k-2)\pi}{2} - \sin\frac{3(k-2)\pi}{2}}{4(k-2)}\right] = \frac{a^2\omega^2}{\pi a_0}\left(1 + \frac{c_2}{c_1}\right)\left[-\frac{1}{k} + \frac{1}{2(k+2)} + \right.$$
$$\left. + \frac{1}{2(k-2)}\right] \sin\frac{k\pi}{2} \cos k\pi \tag{4.48}$$

and

$$d_0(a\omega) = \frac{a^2\omega^2}{2a_0}\left(1 - \frac{c_2}{c_1}\right), \quad d_2(a\omega) = \frac{a^2\omega^2}{4a_0}\left(1 - \frac{c_2}{c_1}\right) \tag{4.49}$$

From Eqs. 4.48 and 4.49 we deduce that

$$d_1(a\omega) = \frac{4a^2\omega^2}{3\pi a_0}\left(1 + \frac{c_2}{c_1}\right) \tag{4.50}$$

$$d_0(a\omega) = \frac{a^2\omega^2}{2a_0}\left(1 - \frac{c_2}{c_1}\right),\ d_2(a\omega) = \frac{a^2\omega^2}{4a_0}\left(1 - \frac{c_2}{c_1}\right)$$
$$d_{2p}(a\omega) = 0,\ p = 2, 3... \tag{4.51}$$

$$d_{2p+1}(a\omega) = \frac{4a^2\omega^2}{\pi a_0}\left(1 + \frac{c_2}{c_1}\right)\frac{(-1)^{p+1}}{(2p-1)(2p+1)(2p+3)},\quad p = 1, 2, ... \tag{4.52}$$

On the other hand we can write that

$$\cos(2p+1)\left(\psi + \frac{\pi}{2}\right) = -\sin[p\pi + (2p+1)\psi] = (-1)^{p+1}\sin(2p+1)\psi$$
$$\cos 2\left(\psi + \frac{\pi}{2}\right) = -\cos 2\psi \tag{4.53}$$

such that, substituting Eqs. 4.50–4.53 into Eq. 4.44 yields

$$d(-a\omega\sin\psi) = \frac{a^2\omega^2}{2a_0}\left(1 - \frac{c_2}{c_1}\right) + \frac{a^2\omega^2}{4a_0}\left(\frac{c_2}{c_1} - 1\right)\cos\psi +$$
$$+\frac{4a^2\omega^2}{\pi a_0}\left(1 + \frac{c_2}{c_1}\right)\sum_{p=0}^{\infty}\frac{\sin(2p+1)\psi}{(2p-1)(2p+1)(2p+3)} \tag{4.54}$$

and therefore, from Eqs. 4.42 and 4.54 we obtain

$$f(a\cos\psi, -a\omega\sin\psi) = \frac{a^2\omega^2}{2a_0}\left(1 - \frac{c_2}{c_1}\right) + \frac{3a^3}{4}\cos\psi +$$
$$+\frac{a^2\omega^2}{4a_0}\left(\frac{c_2}{c_1} - 1\right)\cos\psi + \frac{a^3}{4}\cos 3\psi +$$
$$+\frac{4a^2\omega^2}{\pi a_0}\left(1 + \frac{c_2}{c_1}\right)\sum_{p=0}^{\infty}\frac{\sin(2p+1)\psi}{(2p-1)(2p+1)(2p+3)} \tag{4.55}$$

From Eqs. 4.29 and 4.55 it follows that

$$A_1(a) = -\frac{2a^2\omega}{3\pi a_0}\left(1 + \frac{c_2}{c_1}\right) \tag{4.56}$$

$$B_1(a) = \frac{3a^2}{8\omega} \tag{4.57}$$

and from Eqs. 4.55 and 4.31 yields

$$u_1(a,\psi) = \frac{a^2}{2a_0}\left(1-\frac{c_2}{c_1}\right) + \frac{a^2\omega^2}{12a_0}\left(\frac{c_2}{c_1}-1\right)\cos 2\psi - \frac{a^3}{24}\cos 3\psi - \\ -\frac{4a^2\omega^2}{\pi a_0}\left(1+\frac{c_2}{c_1}\right)\sum_{p=1}^{\infty}\frac{(2p+3)!\sin(2p+1)\psi}{(2p-2)!} \tag{4.58}$$

where $p! = 1,2,3,\ldots,p$, $0! = 1$

The first approximate solution of the oscillator with linear and cubic elastic restoring force and weak asymptotic quadratic damping can be obtained from Eqs. 4.36. For this, we have

$$t(a) = -\frac{1}{a_0c_1}\int_{a_0}^{a}\frac{3\pi a_0 da}{2a^2\omega(1+\frac{c_1}{c_2})} = \frac{3\pi}{2\omega(c_1+c_2)}\left(\frac{1}{a}-\frac{1}{a_0}\right)$$

and by inversion

$$a(t) = \frac{3\pi a_0}{3\pi + 2\omega a_0(c_1+c_2)t} \tag{4.59}$$

From Eq. 4.36_3 we obtain

$$\psi(t) = \omega t + a_0c_1\int_0^t \frac{3a^2}{8\omega}dt = \omega t + \frac{9\pi a_0^4 c_1(c_1+c_2)\,t}{12\pi + 8\omega a_0(c_1+c_2)\,t} \tag{4.60}$$

such that the first approximate solution becomes

$$u(t) = \frac{3\pi a_0}{3\pi + 2\omega a_0(c_1+c_2)\,t}\cos\left[\omega t + \frac{9\pi a_0^4 c_1(c_1+c_2)\,t}{12\pi + 8\omega a_0(c_1+c_2)\,t}\right] \tag{4.61}$$

From Eqs. 4.37, 4.39 and 4.58 we can determine the second approximate solution in the form

$$u = a\cos\psi + \frac{a^2}{2}(c_1-c_2) + \frac{a^2\omega^2}{12}(c_1-c_2)\cos 2\psi - \frac{a_0a^3c_1}{24}\cos 3\psi - \\ -\frac{4a^2\omega^2(c_1+c_2)}{\pi}\sum_{p=1}^{\infty}\frac{(2p+3)!\sin(2p+1)\psi}{(2p-2)!} \tag{4.62}$$

In order to determine the values of the functions $A_2(a)$ and $B_2(a)$, from Eqs. 4.34 and 4.35 respectively, we first calculate the following integrals

$$\int_0^{2\pi} u_1(a,\psi)\frac{\partial f}{\partial u}(a\cos\psi, -a\omega\sin\psi)\sin\psi\, d\psi =$$

$$= \frac{3a^2}{4}\int_0^{2\pi} u_1(a,\psi)(\sin\psi + \sin 3\psi)\, d\psi = -\frac{360a^2\omega^2}{a_0}\left(1+\frac{c_2}{c_1}\right) \tag{4.63}$$

$$\int_0^{2\pi} u_1(a,\psi)\frac{\partial f}{\partial u}(a\cos\psi, -a\omega\sin\psi)\cos\psi\, d\psi =$$

$$= \frac{3a^2}{4}\int_0^{2\pi} u_1(a,\psi)(3\cos\psi + \cos 3\psi)\, d\psi = -\frac{\pi a^5}{32} \tag{4.64}$$

$$\int_0^{2\pi} \frac{\partial f}{\partial \dot{u}}(a\cos\psi, -a\omega\sin\psi)\sin\psi\ \cos\psi\, d\psi =$$

$$= -\frac{2a\omega c_2}{a_0 c_1}\int_0^{\pi} \sin^2\psi\cos\psi d\psi + \frac{2a\omega}{a_0}\int_{\pi}^{2\pi} \sin^2\psi\cos\psi d\psi = 0 \tag{4.65}$$

$$\int_0^{2\pi} \frac{\partial f}{\partial \dot{u}}(a\cos\psi, -a\omega\sin\psi)\sin^2\psi\, d\psi = -\frac{2a\omega c_2}{a_0 c_1}\int_0^{\pi}\sin^3\psi\, d\psi +$$

$$+\frac{2a\omega}{a_0}\int_{\pi}^{2\pi}\sin^3\psi\, d\psi = -\frac{8a\omega}{3a_0}\left(1+\frac{c_2}{c_1}\right) \tag{4.66}$$

$$\int_0^{2\pi} \frac{\partial f}{\partial \dot{u}}(a\cos\psi, -a\omega\sin\psi)\cos^2\psi\, d\psi = -\frac{2a\omega c_2}{a_0 c_1}\int_0^{\pi}\cos^2\psi\sin\psi\, d\psi +$$

$$+\frac{2a\omega}{a_0}\int_{\pi}^{2\pi}\cos^2\psi\sin\psi\, d\psi = \frac{4a\omega}{3a_0}\left(1+\frac{c_2}{c_1}\right) \tag{4.67}$$

$$\int_0^{2\pi} \frac{\partial u_1(a,\psi)}{\partial \psi}\frac{\partial f}{\partial \dot{u}}(a\cos\psi, -a\omega\sin\psi)\sin\psi\, d\psi = -\frac{a^4\omega^2 c_2}{4a_0 c_1}\int_0^{\pi}\sin 3\psi\sin^2\psi\, d\psi$$

$$+\frac{a^4\omega}{4a_0}\int_{\pi}^{2\pi}\sin 3\psi\sin^2\psi\, d\psi = \frac{2a^4\omega}{45a_0}\left(1+\frac{c_2}{c_1}\right) \tag{4.68}$$

$$\int_0^{2\pi} \frac{\partial u_1(a,\psi)}{\partial \psi}\frac{\partial f}{\partial \dot{u}}(a\cos\psi, -a\omega\sin\psi)\cos\psi\, d\psi =$$

$$-\frac{a^2\omega^2}{6a_0}\left(\frac{c_2}{c_1}-1\right)\left[-\frac{2a\omega c_2}{a_0 c_1}\int_0^{\pi}\sin 2\psi\sin\psi\cos\psi\, d\psi - \right.$$

$$\left.+\frac{2a\omega}{a_0}\int_{\pi}^{2\pi}\sin 2\psi\sin\psi\cos\psi\, d\psi\right] - \frac{4a^2\omega^2}{\pi a_0}$$

$$\left(1+\frac{c_2}{c_1}\right)\sum_{p=1}^{\infty}\frac{(2p+3)!(2p+1)}{(2p-2)!}\left[\frac{2\omega a c_2}{a c_1}\int_0^{\pi}\cos(2p+1)\psi\sin\psi\cos\psi\, d\psi - \right.$$

$$\left.-\frac{2a\omega}{a_0}\int_{\pi}^{2\pi}\cos(2p+1)\psi\sin\psi\cos\psi\, d\psi\right] = \frac{a^3\omega^3}{3a_0^2}\left(\frac{c_2}{c_1}-1\right)^2 +$$

$$+\frac{16a^3\omega^3}{\pi a_0^2}\left(1+\frac{c_2}{c_1}\right)^2\sum_{p=1}^{\infty}2p(2p+1)^2(2p+2) \tag{4.69}$$

From Eqs. 4.34, 4.35, and 4.60–4.66 we deduce $A_2(a)$ and $B_2(a)$ respectively

$$A_2(a) = \frac{45a^4 - 45a^3 - 16198a^2\omega}{90\pi a_0} \tag{4.70}$$

$$B_2(a) = -\frac{9a^4}{128\omega^3} - \frac{a^4}{64\omega} + \frac{a^3}{2\pi\omega a_0}\left(1+\frac{c_2}{c_1}\right) + \frac{a^2\omega^3}{6\pi a_0^3}\left(\frac{c_2}{c_1}-1\right)^2 +$$

$$+\frac{8a^2\omega^3}{\pi^2 a_0^2}\left(1+\frac{c_2}{c_1}\right)^2\sum_{p=1}^{\infty}2p(2p+1)^2(2p+2) \tag{4.71}$$

The amplitude a and the total phase ψ have to be calculated from the system

$$t(a) = 90\pi a_0\int_{a_0}^{a}\frac{da}{a^2\left[45\varepsilon^2 a^2 - 45\varepsilon^2 a - 16198\varepsilon^2\omega - 60\varepsilon\left(1+\frac{c_2}{c_1}\right)\right]} \tag{4.72}$$

$$\psi(t) = \omega t + \int_0^t \left[\frac{3\varepsilon a^2}{8\omega} - \frac{9\varepsilon^2 a^4}{128\omega^3} - \frac{\varepsilon^2 a^4}{64\omega} + \frac{\varepsilon^2 a^3}{2\pi\omega a_0}\left(1 + \frac{c_2}{c_1}\right) + \right.$$
$$\left. + \frac{\varepsilon^2 a^2 \omega^3}{6\pi a_0^3}\left(\frac{c_2}{c_1} - 1\right)^2 + \frac{8\varepsilon^2 a^2 \omega^3}{\pi^2 a_0^2}\left(1 + \frac{c_2}{c_1}\right)^2 \sum_{p=1}^{\infty} 2p(2p+1)^2(2p+2) \right] dt \tag{4.73}$$

Therefore, the second approximate solution is given by Eq. 4.62 where $a(t)$ and $\psi(t)$ have to be replaced by Eqs. 4.72 and 4.73 respectively.

In Fig. 4.1 is given a comparison between the analytic solution (4.61) in the first approximation and the numerical solution for $C_1 = 0.02$, $C_2 = 0.01$, $\varepsilon = 0.003$, $\omega = 1$ and $a_0 = 0.1$. An excellent agreement was found between the analytical and numerical solutions.

One can conclude that the method of Krylov-Bogolyubov provides very accurate results in the first approximation in the case of cubic elastic restoring force and asymmetric quadratic damping.

4.2 Use of the Method of Krylov-Bogolyubov and Iteration Method to Weakly Nonlinear Oscillators

Consider the following equation [56, 57].

$$\ddot{u} + \omega^2 u = \varepsilon f(\Omega t, u, \dot{u}) \tag{4.74}$$

where ω and Ω are positive constants, in general f is assumed to be a nonlinear function of both u and $\dot{u}$, periodic of Ωt with period 2π which may be expanded in a Fourier series and the parameter ε is assumed to be small.

According to Ref. [57], we can construct the following iteration formula (named correction functional, see Chap. 8)

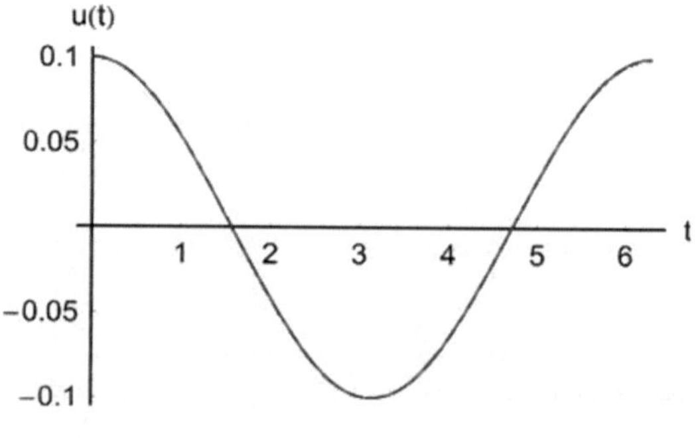

Fig. 4.1 Comparison between the analytical approximate solution (4.61) and the exact solution of Eq. 4.38 for $c_1 = 0.02$, $c_2 = 0.01$, $\varepsilon = 0.003$, $\omega = 1$, $a_0 = 0.1$: ________ numerical results; _ _ _ _ _ analytical results

$$u_{n+1}(t) = u_n(t) + \frac{1}{\omega}\int_0^t \sin\omega(\tau - t)[u_n''(\tau) + \omega^2 u_n(\tau) - \varepsilon f(\Omega\tau, u_n(\tau), u'(\tau))]\, d\tau \tag{4.75}$$

In this method, the problems are initially approximated with possible unknowns and so far do not depend on small parameters, such that it can find wide application in nonlinear problems without linearization or small perturbation. The method reveals that the approximate solutions obtained by the proposed method rapidly converge to the exact solution.

We distinguish between two cases, which will be separately treated: the "resonance" case, when $\omega = \frac{p}{q}\Omega$ with p and q integers and relatively prime, and the "nonresonance" case, for other values of Ω sufficiently far from "resonance". In what follows we study general and periodic solution of Eq. 4.74.

4.2.1 "Nonresonance" Case ($\omega \neq \frac{p}{q}\Omega$)

We have seen earlier, that for $\varepsilon = 0$, Eq. 4.74 has the solution $u(t) = a\cos\psi$, where $\dot{a}=0$, $\dot{\psi} = \omega$. For $\varepsilon \neq 0$, we try a solution of the form

$$u_0(t) = a\cos\psi + \varepsilon X_1(t) + \varepsilon^2 X_2(t) + 0(\varepsilon^3) \tag{4.76}$$

which is called the approximation of the third order if $X_2 \neq 0$ ($u_0 = 0(\varepsilon^3)$) and of the second order if $X_2 = 0$ ($u_0 = 0(\varepsilon^2)$), $X_1(t)$ and $X_2(t)$ are supposed to be periodic functions of period 2π in both Ωt and ψ. However, the amplitude and the total phase of the first harmonic are supposed to satisfy the differential equations with separable variables ("slowly varying")

$$\dot{a} = \varepsilon A_1(a) + \varepsilon^2 A_2(a) \tag{4.77}$$

$$\dot{\psi} = \omega + \varepsilon B_1(a) + \varepsilon^2 B_2(a) \tag{4.78}$$

We first try to determine the functions $X_1(t)$, $X_2(t)$, $A_1(a)$, $A_2(a)$, $B_1(a)$, $B_2(a)$ so that Eq. 4.74 be satisfied to within an error of third order in ε.

The terms of third or higher order in ε will be omitted in the following formulas. We emphasize that X_1 and X_2 are functions depending of t instead of $u_1(a,\psi)$ and $u_2(a,\psi)$.

Like at the precedent section, by differentiating Eq. 4.76 twice with respect to t, we obtain

$$\dot{u}_0 = \dot{a}\cos\psi + a\sin\psi\dot{\psi} + \varepsilon\dot{X}_1 + \varepsilon^2\dot{X}_2 \tag{4.79}$$

$$\ddot{u}_0 = \ddot{a}\cos\psi - 2\dot{a}\dot{\psi}\sin\psi - a\cos\psi\dot{\psi}^2 - a\sin\psi\cdot\ddot{\psi} + \varepsilon\ddot{X}_1 + \varepsilon^2\ddot{X}_2 \tag{4.80}$$

From Eqs. 4.77 and 4.78 we deduce that

$$\begin{aligned}
\ddot{a} &= \left(\varepsilon\frac{dA_1}{da} + \varepsilon^2\frac{dA_2}{da}\right)\left(\varepsilon A_1 + \varepsilon^2 A_2\right) = \varepsilon^2 A_1\frac{dA_1}{da}\\
\ddot{\psi} &= \left(\varepsilon\frac{dB_1}{da} + \varepsilon^2\frac{dB_2}{da}\right)\left(\varepsilon A_1 + \varepsilon^2 A_2\right) = \varepsilon^2 A_1\frac{dB_1}{da}\\
\dot{a}\dot{\psi} &= (\varepsilon A_1 + \varepsilon^2 A_2)(\omega + \varepsilon B_1 + \varepsilon^2 B_2) = \varepsilon\omega A_1 + \varepsilon^2(A_1 B_1 + \omega A_2)\\
\dot{\psi}^2 &= (\omega + \varepsilon B_1 + \varepsilon^2 B_2)^2 = \omega^2 + 2\varepsilon\omega B_1 + \varepsilon^2(B_1^2 + 2\omega B_2)
\end{aligned}$$

Introducing these expressions, as well as Eqs.4.77 and 4.78 into Eqs. 4.79 and 4.80 yields

$$\begin{aligned}
\ddot{u}_0 = -a\omega^2\cos\psi + \varepsilon\left(-2\omega A_1\sin\psi - 2a\omega B_1\cos\psi + \ddot{X}_1\right) + \varepsilon^2\Bigg[\Bigg(A_1\frac{dA_1}{da} - \\
- aB_1^2 - 2\omega a B_2\Bigg)\cos\psi - \left(2\omega A_2 + aA_1\frac{dB_1}{da} + 2A_1B_1\right)\sin\psi + \ddot{X}_2\Bigg]
\end{aligned} \tag{4.81}$$

and

$$\begin{aligned}
f(\Omega\tau, u_0, \dot{u}_0) &= f(\Omega\tau, a\cos\psi, -a\omega\sin\psi) +\\
&+ \varepsilon\left[\frac{\partial f}{\partial u}(\Omega\tau, a\cos\psi, -a\omega\sin\psi)X_1 + \right.\\
&\left. + \frac{\partial f}{\partial \dot{u}}(\Omega\tau, a\cos\psi, -a\omega\sin\psi)\right]\left(A_1\cos\psi - aB_1\sin\psi + \dot{X}_1\right)
\end{aligned} \tag{4.82}$$

where

$$\frac{\partial f}{\partial u}(\Omega\tau, a\cos\psi, -a\omega\sin\psi) = \left.\frac{\partial f(\Omega\tau, u, u')}{\partial u}\right|_{(u=a\cos\psi, u'=-a\omega\sin\psi)}$$

Finally, by substituting Eqs. 4.76, 4.81 and 4.82 into Eq. 4.75 for $n = 0$, we obtain

$$
\begin{aligned}
u_1(t) &= u_0(t) + \frac{\varepsilon}{\omega}\int_0^t \sin(\omega\tau - \omega t)\big[X''_1(\tau) + \omega^2 X_1(\tau) - 2\omega A_1 \sin(\omega\tau + \theta) - \\
&- 2\omega a B_1 \cos(\omega\tau + \theta) - f(\Omega\tau, a\cos(\omega\tau + \theta), -a\omega\sin(\omega\tau + \theta))\big]d\tau + \\
&+ \frac{\varepsilon^2}{\omega}\int_0^t \sin(\omega\tau - \omega t)\Big\{X''_2(\tau) + \omega^2 X_2(\tau) + \left[\left(A_1\frac{dA_1}{da} - aB_1^2 - \right.\right. \\
&\left. - 2a\omega B_2\right)\cos(\omega\tau + \theta) - \left(2\omega A_2 + aA_1\frac{dB_1}{da} + 2A_1B_1\right)\sin(\omega\tau + \theta) - \\
&- \frac{\partial f}{\partial u}(\Omega\tau, a\cos(\omega\tau + \theta), -a\omega\sin(\omega\tau + \theta))X_1(\tau) - \\
&- \frac{\partial f}{\partial \dot{u}}(\Omega\tau, a\cos(\omega\tau + \theta), -a\omega\sin(\omega\tau + \theta))\Big] \left[A_1\cos(\omega\tau + \theta) - \right. \\
&\left. - aB_1 sin(\omega\tau + \theta) + X'_1(\tau)\right]\Big\}\, d\tau
\end{aligned}
\tag{4.83}
$$

By developing the function f in a Fourier series with respect $\Omega\tau$ and $\psi = \omega\tau + \theta$, we have

$$
f(\Omega\tau, a\cos(\omega\tau + \theta), -a\omega\sin(\omega\tau + \theta)) = \sum_{k=0}^{\infty}\sum_{j=0}^{\infty} f_{kj}\cos(k\Omega\tau + j\omega\tau + j\theta)
$$

$$
+ \sum_{k=0}^{\infty}\sum_{j=0}^{\infty} g_{kj}\sin(k\Omega\tau + j\omega\tau + j\theta) \tag{4.84}
$$

where

$$
\begin{aligned}
f_{00}(a) &= \frac{1}{4\pi^2}\int_0^{2\pi}\int_0^{2\pi} f(\alpha, a\cos\beta, -a\omega\sin\beta)d\alpha\, d\beta \\
f_{kj}(a) &= \frac{1}{2\pi^2}\int_0^{2\pi}\int_0^{2\pi} f(\alpha, a\cos\beta, -a\omega\sin\beta)\cos(k\alpha + j\beta)d\alpha\, d\beta, k + j \geq 1 \\
g_{kj}(a) &= \frac{1}{2\pi^2}\int_0^{2\pi}\int_0^{2\pi} f(\alpha, a\cos\beta, -a\omega\sin\beta)\sin(k\alpha + j\beta)d\alpha\, d\beta, k, j \geq 0
\end{aligned}
\tag{4.85}
$$

In what follows, we should propose that $u_1(t) = u_0(t)$ in Eq. 4.83. In this case it is clear that the sequence $u_n(t)$ is a constant and therefore the solution of Eq. 4.74 can be obtained by only one iteration step, due to the fact the Lagrange multipliers [57], $\lambda(\tau, t) = \frac{1}{\omega}\sin\omega(\tau, t)$ can be exactly identified.

The coefficient of $\frac{\varepsilon}{\omega}$ in Eq. 4.83 would have to vanish and can be written as follows

$$\int_0^t \sin(\omega\tau-\omega t)\left[X''_1(\tau)+\omega^2 X_1(\tau)-2\omega A_1\sin(\omega\tau+\theta)-\right.$$
$$\left.-2a\omega B_1\cos(\omega\tau+\theta)-f(\Omega\tau,a\cos(\omega\tau+\theta),-a\omega\sin(\omega\tau+\theta))\right]d\tau=$$
$$=X'_1(0)\sin\omega t+\omega X_1(0)\cos\omega t-\omega X_1(t)-(2\omega A_1+g_{01})\left[\frac{1}{2}t\cos(\omega t+\theta)-\right.$$
$$\left.-\frac{\sin(\omega t+\theta)+\sin(\omega t-\theta)}{4\omega}\right]-(2a\omega B_1+f_{01})\left[\frac{\cos(\omega t-\theta)-\cos(\omega t+\theta)}{4\omega}-\right.$$
$$\left.-\frac{1}{2}t\sin(\omega t+\theta)\right]-\sum_{k=0}^{\infty}\sum_{j=0}^{\infty}f_{kj}\left[\frac{\omega\cos(k\Omega t+j\omega t+j\theta)}{(k\Omega+j\omega)^2-\omega^2}+\frac{\cos(\omega t-j\theta)}{2k\Omega+2(j+1)\omega}-\right.$$
$$\left.-\frac{\cos(\omega t+j\theta)}{2k\Omega+2(j-1)\omega}\right]-\sum_{k=0}^{\infty}\sum_{j=0}^{\infty}g_{kj}\left[\frac{\omega\sin(k\Omega t+j\omega t+j\theta)}{(k\Omega+j\omega)^2-\omega^2}-\frac{\sin(\omega t-j\theta)}{2k\Omega+2(j+1)\omega}-\right.$$
$$\left.-\frac{\sin(\omega t+j\theta)}{2k\Omega+2(j-1)\omega}\right]=0 \quad (k,j)\neq(0,1) \tag{4.86}$$

Avoiding the presence of secular term, needs

$$A_1(a)=-\frac{1}{2\omega}g_{01},\quad B_1(a)=-\frac{1}{2a\omega}f_{01} \tag{4.87}$$

Observe that for

$$X_1(t)=-\sum_{k=0}^{\infty}\sum_{j=0}^{\infty}\frac{f_{kj}\cos(k\Omega t+j\theta)}{(k\Omega+j\omega)^2-\omega^2}-\sum_{k=0}^{\infty}\sum_{j=0}^{\infty}\frac{g_{kj}\sin(k\Omega t+j\theta)}{(k\Omega+j\omega)^2-\omega^2},\ (k,j)\neq(0,1) \tag{4.88}$$

the Eq. 4.86 is valid. The conditions (4.87) are equivalent with the condition that the solution $X_1(t)$ of Eq. 4.86 may not contain terms in $\sin(\omega t+\theta)$ and $\cos(\omega t+\theta)$. We may, however to assure the vanishing of the coefficient of ε^2/ω in Eq. 4.83. The functions A_2 and B_2 may be similarly obtained by requiring that the terms containing $\sin(\omega t+\theta)$ and $\cos(\omega t+\theta)$ vanish. Taking into account Eq. 4.87 it follows that:

$$A_2(a)=\frac{g_{01}}{8a\omega^2}\left(f_{01}-a\frac{df_{01}}{da}\right)-$$
$$-\frac{f_{01}g_{01}}{4a\omega^3}-\frac{1}{4\pi^2\omega}\int_0^{2\pi}\int_0^{2\pi}\left[\frac{\partial f}{\partial u}(\alpha,a\cos\beta,-a\omega\sin\beta)X_1+\right.$$
$$\left.+\frac{\partial f}{\partial u}(\alpha,a\cos\beta,-a\omega\sin\beta)X_1(A_1\cos\beta-aB_1\sin\beta+X'_1)\right]\sin\beta\,d\alpha\,d\beta \tag{4.89}$$

$$B_2(a) = \frac{g_{01}}{8a\omega^2}\frac{dg_{01}}{da} - \frac{f_{01}^2}{8a\omega^3} +$$
$$+ \frac{1}{4\pi^2 a\omega}\int_0^{2\pi}\int_0^{2\pi}\left[\frac{\partial f}{\partial u}(\alpha, a\cos\beta, -a\omega\sin\beta)X_1 +\right.$$
$$\left. + \frac{\partial f}{\partial u'}(\alpha, a\cos\beta, -a\omega\sin\beta)(A_1\cos\beta - aB_1\sin\beta + X_1')\right]\cos\beta\, d\alpha\, d\beta \tag{4.90}$$

The solution X_2(t) does not contain terms in $\sin(\omega t + \theta)$ and $\cos(\omega t + \theta)$, and therefore its expansion is readily found to be

$$X_2(t) = -\frac{1}{2\pi^2}\left\{\sum_{k=0}^{\infty}\sum_{j=0}^{\infty}\frac{\cos(k\Omega t + j\theta)}{(k\Omega + j\omega)^2 - \omega^2}\int_0^{2\pi}\frac{\partial f}{\partial u}(\alpha, a\cos\beta, -a\omega\sin\beta)X_1 +\right.$$
$$\left. + \frac{\partial f}{\partial u'}(\alpha, a\cos\beta, -a\omega\sin\beta)\left(A_1\cos\beta - aB_1\sin\beta + X_1'\right)\right]\cos(k\alpha + j\beta)\, d\alpha\, d\beta -$$
$$- \sum_{k=0}^{\infty}\sum_{j=0}^{\infty}\frac{\sin(k\Omega t + j\theta)}{(k\Omega + j\omega)^2 - \omega^2}\int_0^{2\pi}\int_0^{2\pi}\left[\frac{\partial f}{\partial u}(\alpha, a\cos\beta, -a\omega\sin\beta)X_1 +\right.$$
$$+ \frac{\partial f}{\partial \ddot{u}}(\alpha, a\cos\beta, -a\omega\sin\beta)(A_1\cos\beta - aB_1\sin\beta + X'_1)]\sin(k\alpha + j\beta)d\alpha d\beta,$$
$$(k,j) \neq (0,1) \tag{4.91}$$

The amplitude a and the phase ψ has to be determined from the Eqs. 4.77 and 4.78 respectively. When studying periodic solutions only, we have to set

$$\dot{a} = 0, \quad \dot{\psi} = 0 \tag{4.92}$$

and then, by (4.77) it follows that the constant values of the amplitude, say $a = a_0$ may be found from the algebraic equation

$$\varepsilon A_1(a) + \varepsilon^2 A_2(a) = 0 \tag{4.93}$$

Next, introducing a_0 into (4.78), and integrating with respect to t, gives the total phase

$$\psi = \psi_0 + \left[\omega + \varepsilon B_1(a_0) + \varepsilon^2 B_2(a_0)\right] t \tag{4.94}$$

4.2.2 "Resonance" Case $\omega \approx \frac{p}{q}\Omega$

Let us first assume that $\omega \approx \frac{p}{q}\Omega$, with p and q integers, relatively prime, and let us denote

$$\omega^2 = \left(\frac{p}{q}\Omega\right)^2 + \varepsilon\sigma \tag{4.95}$$

where σ is so-called detuning parameter. The equation of motion (4.74), may then be written

$$\ddot{u} + \left(\frac{p}{q}\Omega\right)^2 u = \varepsilon[f(\Omega t, u, u') - \sigma u] \tag{4.96}$$

For Eq. 4.96, we try a solution of the form

$$u_0(t) = a\cos\psi + \varepsilon X_1(t) + \varepsilon^2 X_2(t) \tag{4.97}$$

where $X_1(t)$ and $X_2(t)$ are periodic functions with period 2π in both Ωt andψ, with $\dot{\psi} = \frac{p}{q}\Omega$. In the "resonance" regions, the phase difference θ between the solution and the external force exerts a strong influence on the amplitude and frequency of the forced vibration. Therefore, we write

$$\psi = \frac{p}{q}\Omega t + \theta \tag{4.98}$$

and assume that the functions A_k and B_k, $k = 1{,}2$ in Eqs. 4.77 and 4.78 depend on both a and θ.

$$\dot{a} = \varepsilon A_1(a, \theta) + \varepsilon^2 A_2(a, \theta) \tag{4.99}$$

$$\dot{\psi} = \frac{p}{q}\Omega + \varepsilon B_1(a, \theta) + \varepsilon^2 B_2(a, \theta) \tag{4.100}$$

Substituting Eq. 4.98 into Eq. 4.100 gives

$$\dot{\theta} = \varepsilon B_1(a, \theta) + \varepsilon^2 B_2(a, \theta) \tag{4.101}$$

Equations 4.99 and 4.101 will be used for calculating the functions $a(t)$ and $\theta(t)$ after the functions $A_k(a,\theta)$ and $B_k(a,\theta)$, $k = 1{,}2$ have been found.

The Lagrange multiplier for Eq. 4.96 is given by the relation

$$\lambda(\tau, t) = \frac{q}{p\Omega} \sin\frac{p}{q}\Omega(\tau - t) \tag{4.102}$$

We obtain the following iteration formula, corresponding to Eq. 4.96:

$$u_{n+1}(t) = u_n(t) + \frac{q}{p\Omega} \int_0^t \sin\frac{p}{q}\Omega(\tau - t) \left[u''_n(\tau) + \left(\frac{p\Omega}{q}\right)^2 u_n(\tau) \right.$$
$$\left. - \varepsilon f(\Omega\tau, u_n(\tau), u'_n(\tau) + \varepsilon\sigma u_n(\tau)) \right] d\tau \tag{4.103}$$

By introducing now Eq. 4.97 into Eq. 4.103, considering Eqs. 4.99 and 4.100, we obtain the second order approximation

$$u_1(t) = u_0(t) + \frac{\varepsilon q}{p\Omega} \int_0^t \sin\frac{p}{q}\Omega(\tau - t) \left[X''_1(\tau) + \left(\frac{p}{q}\Omega\right)^2 X_1(\tau) - \right.$$
$$- 2\frac{p}{q}\Omega A_1 \sin\left(\frac{p}{q}\Omega\tau + \theta\right) - 2\frac{p}{q}\Omega a B_1 \cos\left(\frac{p}{q}\Omega\tau + \theta\right) -$$
$$\left. - f\left(\Omega\tau, a\cos\left(\frac{p}{q}\Omega\tau + \theta\right), -\frac{p}{q}\Omega a \sin\left(\frac{p}{q}\Omega\tau + \theta\right)\right) \right] d\tau +$$
$$+ \frac{\varepsilon^2 q}{p\Omega} \int_0^t \sin\frac{p}{q}\Omega(\tau - t) \{ X''_2(\tau) + \left(\frac{p}{q}\Omega\right)^2 X_2(\tau) -$$
$$- \left(2\frac{p}{q}\Omega A_2 + 2A_1 B_1 + a A_1 \frac{\partial B_1}{\partial a} + a B_1 \frac{\partial B_1}{\partial \theta} \right) \sin\left(\frac{p}{q}\Omega\tau + \theta\right) +$$
$$+ \left(A_1 \frac{\partial A_1}{\partial a} - 2\frac{p}{q}\Omega a B_2 + B_1 \frac{\partial A_1}{\partial \theta} - a B_1^2 \right) \cos\left(\frac{p}{q}\Omega\tau + \theta\right) + \sigma X_1 -$$
$$- \frac{\partial f}{\partial u}\left(\Omega\tau, a\cos\left(\frac{p}{q}\Omega\tau + \theta\right), -a\frac{p}{q}\Omega \sin\left(\frac{p}{q}\Omega\tau + \theta\right) \right) X_1 -$$
$$- \frac{\partial f}{\partial u'}\left(\Omega\tau, a\cos\left(\frac{p}{q}\Omega\tau + \theta\right), -\frac{p}{q}\Omega a \sin\left(\frac{p}{q}\Omega\tau + \theta\right)\right) \times$$
$$\times \left[A_1 \cos\left(\frac{p}{q}\Omega\tau + \theta\right) - a B_1 \sin\left(\frac{p}{q}\Omega\tau + \theta\right) + X'_1(\tau) \right] \Big\} d\tau \tag{4.104}$$

By developing the function f in a Fourier series, we obtain

$$f\left(\Omega t, a\cos\left(\frac{p}{q}\Omega t+\theta\right), -\frac{p}{q}\Omega a\sin\left(\frac{p}{q}\Omega t+\theta\right)\right) = \\ = \sum_{k=0}^{\infty}\sum_{j=0}^{\infty} f_{kj}(a)\cos\left(k\Omega t + j\frac{p}{q}\Omega t + j\theta\right) + \\ + \sum_{k=0}^{\infty}\sum_{j=0}^{\infty} g_{kj}(a)\sin\left(k\Omega t + j\frac{p}{q}\Omega t + j\theta\right) \tag{4.105}$$

where

$$f_{00}(a) = \frac{1}{4\pi^2}\int_0^{2\pi}\int_0^{2\pi} f\left(\alpha, a\cos\beta, -\frac{p}{q}\Omega a\sin\beta\right) d\alpha\, d\beta \\ f_{kj}(a) = \frac{1}{2\pi^2}\int_0^{2\pi}\int_0^{2\pi} f\left(\alpha, a\cos\beta, -\frac{p}{q}\Omega a\sin\beta\right)\cos(k\alpha + j\beta)\, d\alpha\, d\beta, \\ k+j \geq 1 \\ g_{kj}(a) = \frac{1}{2\pi^2}\int_0^{2\pi}\int_0^{2\pi} f\left(\alpha, a\cos\beta, -\frac{p}{q}\Omega a\sin\beta\right)\sin(k\alpha + j\beta)\, d\alpha\, d\beta, \\ k,j \geq 0 \tag{4.106}$$

Taking into account Eq. 4.105 and vanishing the coefficients of $\frac{\varepsilon q}{p\Omega}$ and $\frac{\varepsilon^2 q}{p\Omega}$, we obtain first, as in the preceding section, the coefficient of $\frac{\varepsilon q}{p\Omega}$ for $q \neq 1$, in the form

$$\int_0^t \sin\frac{p}{q}\Omega(\tau - t)\left[X''_1(\tau) + \left(\frac{p}{q}\Omega\right)X_1(\tau) - 2\frac{p}{q}\Omega A_1\sin\left(\frac{p}{q}\Omega\tau+\theta\right) - \right. \\ - 2\frac{p}{q}\Omega a B_1\cos\left(\frac{p}{q}\Omega\tau+\theta\right) + a\sigma\cos\left(\frac{p}{q}\Omega\tau+\theta\right) - \\ \left. - f\left(\Omega\tau, a\cos\left(\frac{p}{q}\Omega\tau+\theta\right) - \frac{p}{q}a\Omega\sin\left(\frac{p}{q}\Omega\tau+\theta\right)\right)\right] d\tau = \\ = X'_1(0)\sin\frac{p}{q}\Omega t + \frac{p}{q}\Omega X_1(0)\cos\frac{p}{q}\Omega t - \frac{p}{q}\Omega X_1(t) - \\ - \left(2\frac{p}{q}\Omega A_1 + g_{01}\right)\left[\frac{1}{2}t\cos\left(\frac{p}{q}\Omega t+\theta\right) - q\frac{\sin\left(\frac{p}{q}\Omega t+\theta\right)+\sin\left(\frac{p}{q}\Omega t-\theta\right)}{4p\Omega}\right] +$$

$$+\left(2\frac{p}{q}\Omega a B_1-\sigma a+f_{01}\right)\left[q\frac{\cos\left(\frac{p}{q}\Omega t-\theta\right)-\cos\left(\frac{p}{q}\Omega t+\theta\right)}{4p\Omega}-\right.$$

$$\left.-\frac{1}{2}t\sin\left(\frac{p}{q}\Omega t+\theta\right)\right]-\sum_{k=0}^{\infty}\sum_{j=0}^{\infty}f_{kj}\left[\frac{\frac{p}{q}\Omega\cos\left(k\Omega t+j\frac{p}{q}\Omega t+j\theta\right)}{\left(k\Omega+j\frac{p}{q}\Omega\right)^2-\left(\frac{p}{q}\Omega\right)^2}+\right.$$

$$\left.+\frac{\cos\left(\frac{p}{q}\Omega t-j\theta\right)}{2k\Omega+2(j+1)\frac{p}{q}\Omega}-\frac{\cos\left(\frac{p}{q}\Omega t+j\theta\right)}{2k\Omega+2(j-1)\frac{p}{q}\Omega}\right]-$$

$$-\sum_{k=0}^{\infty}\sum_{j=0}^{\infty}f_{kj}\left[\frac{\frac{p}{q}\Omega\sin\left(k\Omega t+j\frac{p}{q}\Omega t+j\theta\right)}{\left(k\Omega+j\frac{p}{q}\Omega\right)^2-\left(\frac{p}{q}\Omega\right)^2}-\right.$$

$$\left.-\frac{\sin\left(\frac{p}{q}\Omega t-j\theta\right)}{2k\Omega+2(j+1)\frac{p}{q}\Omega}-\frac{\sin\left(\frac{p}{q}\Omega t+j\theta\right)}{2k\Omega+2(j-1)\frac{p}{q}\Omega}\right]=0,\quad (k,j)\neq(0,1)\tag{4.107}$$

In order to ensure that no secular terms appear in Eq. 4.107, resonance must be avoided, and thus, for $q \neq 1$ we obtain

$$\begin{aligned} A_1(a) &= -\frac{q g_{01}}{2p\Omega} \\ B_1(a) &= \frac{q(\sigma a - f_{01})}{2pa\Omega} \end{aligned}\tag{4.108}$$

So, from Eq. 4.107, we have the following expression for X_1:

$$X_1(t)=-\sum_{k=0}^{\infty}\sum_{j=0}^{\infty}\frac{f_{kj}\cos(k\Omega t+j\theta)}{\left(k\Omega+j\frac{p}{q}\Omega\right)^2-\left(\frac{p}{q}\Omega\right)^2}-$$

$$-\sum_{k=0}^{\infty}\sum_{j=0}^{\infty}\frac{g_{kj}\sin(k\Omega t+j\theta)}{\left(k\Omega+j\frac{p}{q}\Omega\right)^2-\left(\frac{p}{q}\Omega\right)^2}\qquad (k,j)\neq(0,1)\tag{4.109}$$

The coefficient of ε for $q = 1$, by the same manipulation as before is

$$\int_0^t \sin p\Omega(\tau - t)[X''_1(\tau) + (p\Omega)^2 X_1(\tau) - 2p\Omega A_1 \sin(p\Omega\tau + \theta) -$$
$$- 2pa\Omega B_1 \cos(p\Omega\sigma + \theta) + \sigma a \cos(p\Omega\tau + \theta) -$$
$$- f(\Omega\tau, a\cos(p\Omega\tau + \theta), -pa\Omega \sin(p\Omega\tau + \theta))]\, d\tau =$$
$$= X'(0) \sin p\Omega t + p\Omega_1 X_1(0) \cos p\Omega t - p\Omega X_1(t) -$$
$$- \frac{1}{2} t \cos p\Omega t[(2p\Omega A_1 + g_{01}) \cos\theta + (\sigma a - f_{01} - 2p\Omega a B_1)\sin\theta + g_{10}] +$$
$$+ \frac{1}{2} t \sin p\Omega t[(2p\Omega A_1 + g_{01}) \sin\theta + (f_{01} - \sigma a + 2pa\Omega B_1)\cos\theta + f_{10}] -$$
$$- \sum_{k=0}^{\infty}\sum_{j=0}^{\infty} f_{kj} \left[\frac{p\Omega \cos(k\Omega t + jp\Omega t + j\theta)}{(k\Omega + jp\Omega)^2 - (p\Omega)^2} + \frac{\cos(p\Omega t - j\theta)}{2k\Omega + 2(j+1)p\Omega} - \right.$$
$$\left. - \frac{\cos(p\Omega t + j\theta)}{2k\Omega + 2(j-1)p\Omega}\right] - \sum_{k=0}^{\infty}\sum_{j=0}^{\infty} g_{kj} \left[\frac{p\Omega \sin(k\Omega t + jp\Omega t + j\theta)}{(k\Omega + jp\Omega)^2 - (p\Omega)^2} - \right.$$
$$\left. - \frac{\sin(p\Omega t - j\theta)}{2k\Omega + 2(j+1)p\Omega} - \frac{\sin(p\Omega t + j\theta)}{2k\Omega + 2(j-1)p\Omega}\right] = 0, \ (k,j) \neq (0,1), (k,j) \neq (1,0) \tag{4.110}$$

No secular terms in Eq. 4.110, require that

$$A_1(a,\theta) = -\frac{f_{10}\sin\theta + g_{10}\cos\theta + g_{01}}{2p\Omega},$$
$$B_1(a,\theta) = \frac{\sigma a - f_{01} - f_{10}\cos\theta + g_{10}\sin\theta}{2pa\Omega} \tag{4.111}$$

The expansion of X_1 in this case is

$$X_1(t) = -\sum_{k=0}^{\infty}\sum_{j=0}^{\infty} \frac{f_{kj}\cos(k\Omega t + j\theta)}{(k\Omega + jp\Omega)^2 - (p\Omega)^2} - \sum_{k=0}^{\infty}\sum_{j=0}^{\infty} \frac{g_{kj}\sin(k\Omega t + j\theta)}{(k\Omega + jp\Omega)^2 - (p\Omega)^2},$$
$$(k,j) \neq (0,1)$$
$$(k,j) \neq (1,0) \tag{4.112}$$

Next, let us examine the coefficient of ε^2 in Eq. 4.104. The functions A_2 and B_2 will be determined by requiring the vanishing of coefficients of $t\sin(\frac{p}{q}\Omega t + \theta)$ and $t\cos(\frac{p}{q}\Omega t + \theta)$ for$q \neq 1$, by taking into account that $X_1(t)$ and $\dot{X}_1(t)$ may contribute to such terms. However, by Eqs. 4.109 and 4.112, it follows that X_1 and $\dot{X}_1$ do not involve the terms $t\, sin(\frac{p}{q}\Omega t + \theta)$ and $t\, cos(\frac{p}{q}\Omega t + \theta)$ and for $q = 1$ neither the terms $t\sin p\Omega t$ and $t\sin p\Omega t$. Hence we may write for $q \neq 1$:

$$A_2(a,\theta) = -\frac{q}{2p\Omega}\left(2A_1B_1 + aA_1\frac{\partial B_1}{\partial a} + aB_1\frac{\partial B_1}{\partial \theta} + G_{01}\right)$$
$$B_2(a,\theta) = \frac{q}{2p\Omega}\left(A_1\frac{\partial A_1}{\partial a} + B_1\frac{\partial A_1}{\partial \theta} - aB_1^2 - F_{01}\right) \tag{4.113}$$

where

$$F_{kj} = \frac{1}{2\pi^2}\int_0^{2\pi}\int_0^{2\pi}\left\{\frac{\partial f}{\partial u}\left(\alpha, a\cos\beta, -\frac{p}{q}a\Omega\sin\beta\right)X_1 + \right.$$
$$+\frac{\partial f}{\partial \dot{u}}\left(\alpha, a\cos\beta, -\frac{p}{q}a\Omega\sin\beta\right)[A_1\cos\beta - aB_1\sin\beta + X'_1] -$$
$$\left. - \sigma X_1\right\}\cos(k\alpha + j\beta)\,d\alpha\,d\beta \qquad k+j \geq 1$$

$$F_{00} = \frac{1}{4\pi^2}\int_0^{2\pi}\int_0^{2\pi}\left\{\frac{\partial f}{\partial u}\left(\alpha, a\cos\beta, -\frac{p}{q}a\Omega\sin\beta\right)X_1 + \right.$$
$$+\frac{\partial f}{\partial \dot{u}}\left(\alpha, a\cos\beta, -\frac{p}{q}a\Omega\sin\beta\right)[A_1\cos\beta - aB_1\sin\beta + X'_1] -$$
$$\left. - \sigma X_1\right\}d\alpha\,d\beta$$

$$G_{kj} = \frac{1}{2\pi^2}\int_0^{2\pi}\int_0^{2\pi}\left\{\frac{\partial f}{\partial u}\left(\alpha, a\cos\beta, -\frac{p}{q}a\Omega\sin\beta\right)X_1 + \right.$$
$$+\frac{\partial f}{\partial \dot{u}}\left(\alpha, a\cos\beta, -\frac{p}{q}a\Omega\sin\beta\right)[A_1\cos\beta - aB_1\sin\beta + X'_1] -$$
$$\left. - \sigma X_1\right\}\sin(k\alpha + j\beta)\,d\alpha\,d\beta \qquad k,j \geq 0 \tag{4.114}$$

For $q = 1$, A_2 and B_2 become

$$A_2(a,\theta) = -\frac{1}{2p\Omega}\left(2A_1B_1 + aA_1\frac{\partial B_1}{\partial a} + aB_1\frac{\partial B_1}{\partial \theta} + G_{01} + \right.$$
$$\left. + G_{10}\cos\theta - F_{10}\sin\theta\right)$$
$$B_2(a,\theta) = \frac{1}{2pa\Omega}\left(A_1\frac{\partial A_1}{\partial a} + B_1\frac{\partial A_1}{\partial \theta} - aB_1^2 - F_{01} + G_{10}\sin\theta + F_{10}\cos\theta\right) \tag{4.115}$$

For $q \neq 1$, the expression of X_2 is

$$X_2(t) = -\frac{q}{p\Omega}\sum_{k=0}^{\infty}\sum_{j=0}^{\infty}\frac{F_{kj}\cos(k\Omega t + j\theta)}{\left(k\Omega + j\frac{p}{q}\Omega\right)^2 - \left(\frac{p}{q}\Omega\right)^2} - $$
$$-\frac{q}{p\Omega}\sum_{k=0}^{\infty}\sum_{j=0}^{\infty}\frac{G_{kj}\sin(k\Omega t + j\theta)}{\left(k\Omega + j\frac{p}{q}\Omega\right)^2 - \left(\frac{p}{q}\Omega\right)^2} \tag{4.116}$$

and for $q = 1$, similar with Eq. 4.116 but moreover $(kj) \neq (1,0)$

Summarizing the above results, we see that the third approximation for the solution of Eq. 4.74 is given by Eq. 4.76 with the functions X_1 and X_2 are given by Eqs. 4.109 or 4.112and respectively by Eq. 4.116. The amplitude a and the phase ψ have to be obtained from Eqs. 4.77 and 4.78 or by the averaging method [34]:

$$\dot{a} = \frac{\varepsilon}{2\pi}\int_0^{2\pi} A_1(a,\theta)\,d\theta + \frac{\varepsilon^2}{2\pi}\int_0^{2\pi} B_1(a,\theta)\,d\theta \tag{4.117}$$

$$\dot{\theta} = \frac{\varepsilon}{2\pi}\int_0^{2\pi} A_2(a,\theta)\,d\theta + \frac{\varepsilon^2}{2\pi}\int_0^{2\pi} B_2(a,\theta)\,d\theta \tag{4.118}$$

where A_1, A_2, B_1, and B_2 are given by Eqs. 4.108, 4.111, 4.113, 4.115 respectively, according to $q \neq 1$ or $q = 1$.

To study periodic vibrations we require that$\dot{a} = \dot{\theta} = 0$.

For real systems, the expansion of the function $f(\Omega t, \alpha\cos\psi, -\alpha\omega\sin\psi)$ usually contains only a small number of harmonics, and hence the number of the values m, n for which "resonance" can actually occur is also small; it generally suffices to take as values of m and n a few natural numbers close to 1.

In what follows, the responses of single-degree-of-freedom, linear and nonlinear systems are investigated.

4.2.3 Numerical Examples

Example 1. To better understanding how this method really works, let us apply it to a linear oscillator with weak viscous damping, where equation of motion is

$$\ddot{u} + u = -2\varepsilon\mu\dot{u} + \varepsilon k\cos\Omega t, \quad 0<\varepsilon<<1, \quad \mu, k, \Omega>0 \tag{4.119}$$

As it is well-known, the general exact solution of this equation for $\varepsilon\mu < 1$ is

$$u(t) = a_0 \exp(-\varepsilon\mu t)\cos\left(\theta_0 + t\sqrt{1-\varepsilon^2\mu^2}\right) + \\ + \frac{\varepsilon k(1-\Omega^2)}{\left(1-\Omega^2\right)^2 + 4\varepsilon^2\mu^2\Omega^2}\cos\Omega t + \frac{2\varepsilon^2\mu k\Omega}{\left(1-\Omega^2\right)^2 + 4\varepsilon^2\mu^2\Omega^2}\sin\Omega t \tag{4.120}$$

where a_0 and θ_0 are constants to be determined from the initial conditions.

The only periodic solution of the equation is

$$u_{per}(t) = \frac{\varepsilon k(1-\Omega^2)}{\left(1-\Omega^2\right)^2 + 4\varepsilon^2\mu^2\Omega^2}\cos\Omega t + \frac{2\varepsilon^2\mu k\Omega}{\left(1-\Omega^2\right)^2 + 4\varepsilon^2\mu^2\Omega^2}\sin\Omega t \tag{4.121}$$

Case 1a (nonresonance case). By assuming $\Omega \neq 1$, let us apply the above method for Eq. 4.119 and we propose an initial approximate solution in the form

$$u_0(t) = a\cos(t+\theta) + \varepsilon X_1(t) + \varepsilon^2 X_2(t) \tag{4.122}$$

with

$$\begin{aligned} \dot{a} &= \varepsilon A_1(a) + \varepsilon^2 A_2(a) \\ \dot{\theta} &= \varepsilon B_1(a) + \varepsilon^2 B_2(a) \end{aligned} \tag{4.123}$$

Eq. 4.83 becomes

$$X_1(t) = u_0(t) + \varepsilon\int_0^t \sin(\tau - t)[X''_1 + X_1 + 2(A_1 + \mu a)\sin(\tau+\theta) - \\ - 2aB_1\cos(\tau+\theta) - k\cos\Omega\tau]\,d\tau + \varepsilon^2\int_0^t \sin(\tau-t)[X''_2 + X_2 - \\ - \left(2A_2\omega + 2A_1B_1 + aA_1\frac{dB_1}{da} + 2\mu aB_1\right)\sin(\tau+\theta) + \\ + \left(A_1\frac{dA_1}{da} - 2aB_2 - aB_1^2 + 2\mu A_1\right)\cos(\tau+\theta) + 2\mu X']\,d\tau \tag{4.124}$$

and by integrating, the coefficient of ε becomes

$$X'_1(0)\sin t + X_1(0)\cos t - X_1(t) - (A_1+\mu a)\left[t\cos(t+\theta) - \\ - \frac{\sin(t+\theta)+\sin(t-\theta)}{2}\right] + aB_1\left[t\sin(t+\theta). + \frac{\cos(t+\theta)-\cos(t-\theta)}{2}\right] - \\ - \frac{k}{\Omega^2 - 1}(\cos\Omega t - \cos t) = 0 \tag{4.125}$$

Avoiding the presence of a secular term in Eq. 4.125 needs:

$$A_1 = -\mu a, \quad B_1 = 0 \tag{4.126}$$

With this requirement, the solution of Eq. 4.125 is

$$X_1(t) = X_1(\Omega t) = \frac{k}{1-\Omega^2}\cos\Omega t \tag{4.127}$$

By integrating and by replacing Eq. 4.127 into Eq. 4.124, the coefficient of ε^2 in Eq. 4.124 must be vanished:

$$\begin{aligned}
&X'_2(0)\sin t + X_2(0)\cos t - X_2(t) - \left(A_2 + A_1B_1 + \frac{1}{2}aA_1\frac{dB_1}{da} + \right.\\
&\left. + \mu aB_1\right)\left[t\cos(t+\theta) - \frac{\sin(t+\theta)+\sin(t-\theta)}{2}\right] - \\
&- \left(\frac{1}{2}A_1\frac{dA_1}{da} - aB_2 - \frac{1}{2}aB_1^2 + \mu A_1\right)\left[t\sin(t+\theta) + \right.\\
&\left. + \frac{\cos(t+\theta)-\cos(t-\theta)}{2}\right] + \frac{2\mu k\Omega}{(\Omega^2-1)^2}(\sin\Omega t - \Omega\sin t) = 0
\end{aligned} \tag{4.128}$$

Again, elimination of secular terms requires

$$A_2 = 0, \quad B_2 = -\frac{1}{2}\mu^2 \tag{4.129}$$

We, hence, obtain a solution of Eq. 4.128, which reads

$$X_2(t) = \frac{\varepsilon\mu k\Omega}{(\Omega^2-1)^2}\sin\Omega t \tag{4.130}$$

From Eqs. 4.19, 4.20, 4.126 and 4.129 it follows that

$$\begin{aligned}
\dot{a} &= -\varepsilon\mu a\\
\dot{\theta} &= -\frac{1}{2}\varepsilon^2\mu^2
\end{aligned} \tag{4.131}$$

The amplitude a, respectively the phase θ are given in the form

$$\begin{aligned}
a &= a_0\exp(-\varepsilon\mu t)\\
\theta &= \theta_0 - \frac{1}{2}\varepsilon^2\mu^2 t
\end{aligned} \tag{4.132}$$

where a_0 and θ_0 are constants. Hence, we may write that

$$u(t) = a_0 \exp(-\varepsilon\mu t) \cos\left[\theta_0 + t\left(1 - \frac{1}{2}\varepsilon^2\mu^2\right)\right] - \\ - \frac{\varepsilon k}{\Omega^2 - 1} \cos \Omega t + \frac{2\varepsilon^2 \mu k \Omega}{\left(\Omega^2 - 1\right)^2} \sin \Omega t \tag{4.133}$$

This result is very similar to Eq. 4.120. We obtain the frequency of the damped part of exact solution (4.120) up to terms of second order in ε. This can be easily verified by developing the exact expression of the frequency namely $\sqrt{1 - \varepsilon^2\mu^2}$ in a power series of ε and retaining only the first term of the expression.

On the other hand, the periodic solution of Eq. 4.119 may be found from the algebraic equation (4.93). We obtain $a = 0$ and therefore, for $\Omega \neq 1$, the periodic solution of Eq. 4.119 is given by

$$u_{per}(t) = -\frac{\varepsilon k}{\Omega^2 - 1} \cos \Omega t + \frac{2\varepsilon^2 \mu k \Omega}{\left(\Omega^2 - 1\right)^2} \sin \Omega t \tag{4.134}$$

Case 1b (resonance case) Suppose now that Ω is close to 1 and denote $1 = \Omega^2 + \varepsilon\delta$, p = q = 1. We obtain the equation

$$u(t) = a\cos(\Omega t + \theta) + \varepsilon X_1(t) + \varepsilon^2 X_2(t) \tag{4.135}$$

Equation 4.110 becomes

$$X'_1(0)\sin\Omega t + \Omega X_1(0)\cos t - \Omega X_1(t) - t\cos\Omega t\left[(\Omega A_1 + \mu a\Omega)\cos\theta + \right. \\ \left. + \left(\frac{1}{2}\sigma a - \Omega a B_1\right)\sin\theta\right] + t\sin\Omega t\left[(\Omega A_1 + \mu a\Omega)\sin\theta + \right. \\ \left. + \left(a\Omega B_1 - \frac{1}{2}a\sigma\right)\cos\theta + \frac{1}{2}k\right] + \frac{1}{2}(A_1 + \mu a)[\sin(\Omega t + \theta) + \sin(\Omega t - \theta)] + \\ + \frac{a(\sigma - 2\Omega B_1)}{4\Omega}[\cos(\Omega t - \theta) - \cos(\Omega t + \theta)] = 0 \tag{4.136}$$

The coefficients of $t\cos\Omega t$ and $t\sin\Omega t$ in Eq. 4.136 must be zero, i.e.

$$A_1 = -\mu a - \frac{k}{2\Omega}\sin\theta \\ B_1 = \frac{\sigma}{2\Omega} - \frac{k}{2a\Omega}\cos\theta \tag{4.137}$$

So, from Eq. 4.136 we have $X_1(t) = 0$. The coefficient of ε^2 in Eq. 4.104 must be 0 i.e.

$$\begin{aligned} A_2 &= -\frac{\mu k}{2\Omega}\cos\theta - \frac{k^2}{2a\Omega^2}\sin 2\theta \\ B_2 &= -\frac{\mu^2}{2\Omega} - \frac{\sigma^2}{8\Omega^3} + \frac{\sigma k}{8a\Omega^3}\cos\theta - \frac{\mu k}{4a\Omega^2}\sin\theta, \\ X_2(t) &= 0 \end{aligned} \tag{4.138}$$

The amplitude a and the phase θ of the first harmonic of the Eq. 4.122 have to be calculated from the system of differential equations, by considering only A_1 and B_1:

$$\begin{aligned} \dot{a} &= -\varepsilon\mu a - \frac{\varepsilon k}{2\Omega}\sin\theta \\ \dot{\theta} &= \frac{\varepsilon\sigma}{2\Omega} - \frac{\varepsilon k}{2a\Omega\cos\theta} \end{aligned} \tag{4.139}$$

The integration of system (4.139) is rather difficult. That is way we assume $X_1 = X_2 = 0$ and we suppose the solution in the condensed form:

$$x(t) = a\cos(\Omega t + \theta) \tag{4.140}$$

which implies a tedious calculation. In exchange, the periodic vibration can be determined by a straightforward algebraic calculation by putting $\dot{a} = \dot{\theta} = 0$ into Eq. 4.139 and solving the equations obtained for a and θ. It then follows that:

$$\begin{aligned} a &= \frac{k}{\sqrt{4\mu^2\Omega^2 + \sigma^2}}, \quad \sin\theta = -\frac{2\mu a\Omega}{k}, \quad \cos\theta = \frac{a\sigma}{k} \\ u_{per}(t) &= \frac{k\sigma}{4\mu^2\Omega^2 + \sigma^2}\cos\Omega t + \frac{2\mu k\Omega}{4\mu^2\Omega^2 + \sigma^2}\sin\Omega t \end{aligned} \tag{4.141}$$

Taking into account that $\delta = \dfrac{1 - \Omega^2}{\varepsilon}$, the periodic solution becomes

$$u_{per}(t) = \frac{\varepsilon k\left(1 - \Omega^2\right)}{\left(1 - \Omega^2\right)^2 + 4\varepsilon^2\mu^2\Omega^2}\cos\Omega t + \frac{2\varepsilon^2\mu k\Omega}{\left(1 - \Omega^2\right)^2 + 4\varepsilon^2\mu^2\Omega^2}\sin\Omega t \tag{4.142}$$

i.e. one recovers the exact expression (4.121) of the steady-state vibration and therefore, the convergence of the method is more rapid in the "resonance" case than in the "nonresonance" case.

From the above solution process, we can see clearly that the approximate solution converge to the exact solution relatively rapid.

Example 2

We consider the forced vibration of an oscillator with linear viscous damping and cubic elastic restoring force, governed by the differential equation:

$$\ddot{u} + \omega^2 u + 2\varepsilon\mu\dot{u} + \varepsilon\alpha u^3 = P\cos\Omega t \tag{4.143}$$

where

$$0 < \varepsilon << 1, \quad \mu, k, \Omega, P > 0, \quad \omega \neq \Omega \tag{4.144}$$

which is also called generalized Duffing's equation with damping.

Case 2a ("nonresonance" case)

By assuming that $\omega \neq \frac{1}{3}\Omega$ and $\omega \neq 3\Omega$ and applying the transformation

$$u = x + \frac{P}{\omega^2 - \Omega^2}\cos\Omega t \tag{4.145}$$

Eq. 4.143 becomes

$$\ddot{x} + \omega^2 x + 2\varepsilon\mu\left(\dot{x} - \frac{\Omega P}{\omega^2 - \Omega^2}\sin\Omega t\right) + \varepsilon\alpha\left(x + \frac{P}{\omega^2 - \Omega^2}\cos\Omega t\right)^3 = 0 \tag{4.146}$$

For Eq. 4.146 we propose an initial approximate solution in the form

$$x_0(t) = a\cos(\omega t + \theta) + \varepsilon X(t) \tag{4.147}$$

with

$$\begin{aligned} \dot{a} &= \varepsilon A_1(a) \\ \dot{\theta} &= \varepsilon B_1(a) \end{aligned} \tag{4.148}$$

Equation 4.83 becomes

$$\begin{aligned} x_1(t) = x_0(t) &+ \frac{\varepsilon}{\omega}\int_0^t \sin\omega(\sigma - t)\{X''(\tau) + \omega^2 X(\tau) - 2\omega(A_1 + \\ &+ \mu a)\sin(\omega t + \theta) + \left[\frac{3}{4}\alpha a^3 + \frac{3\alpha a P^2}{2(\omega^2 - \Omega^2)^2} - 2\omega a B_1\right]\cos(\omega\tau + \theta) - \\ &- \frac{2\mu\Omega P}{\omega^2 - \Omega^2}\sin\Omega t + \frac{3\alpha a^2 P}{2(\omega^2 - \Omega^2)}\cos\Omega t + \frac{3\alpha a^2 P}{2(\omega^2 - \Omega^2)}\cos(2\omega\tau + \\ &+ 2\theta)\cos\Omega\tau + \frac{3\alpha a P^2}{2(\omega^2 - \Omega^2)^2}\cos(\omega\tau + 2\theta)\cos 2\Omega\tau + \\ &+ \frac{\alpha P^3}{4(\omega^2 - \Omega^2)^3}(\cos 3\Omega\tau + 3\cos\Omega\tau)\}\, d\tau \end{aligned} \tag{4.149}$$

Integrating the last equation and avoiding the presence of secular term, we obtain

$$A_1 = -\mu a$$
$$B_1 = \frac{3\alpha a^2}{8\omega} + \frac{3\alpha P^2}{4\omega\left(\omega^2 - \Omega^2\right)^2} \tag{4.150}$$

The parameters $a(t)$ and $\theta(t)$ are to be calculated by solving the equations

$$\dot{a} = -\varepsilon\mu a, \quad \dot{\theta} = \varepsilon\left[\frac{3\alpha a^2}{8\omega} + \frac{3\alpha P^2}{4\omega\left(\omega^2 - \Omega^2\right)^2}\right] \tag{4.151}$$

for which, by integration, it follows that

$$a(t) = a_0 \exp(-\varepsilon\mu t)$$
$$\theta = \theta_0 - \frac{3\varepsilon\alpha P^2}{4\omega\left(\omega^2 - \Omega^2\right)^2} t - \frac{3\alpha a_o^2}{16\mu\omega} \exp(-2\varepsilon\mu t) \tag{4.152}$$

where a_0 and θ_0 are constants.

Taking into account Eq. 4.150, and imposing that $x_1(\mathrm{t}) = x_0(\mathrm{t})$ into Eq. 4.149, we obtain

$$\begin{aligned} x(t) = {} & \frac{2\mu\Omega P}{\left(\omega^2 - \Omega^2\right)^2} \sin\Omega t - \left[\frac{3\alpha a^2 P}{2\left(\omega^2 - \Omega^2\right)^2} + \frac{3\alpha P^3}{4\left(\omega^2 - \Omega^2\right)^4}\right] \cos\Omega t - \\ & - \frac{\alpha P^3}{4\left(\omega^2 - \Omega^2\right)^3\left(\omega^2 - 9\Omega^2\right)} \cos\Omega t - \frac{3\alpha a P^2}{8\left(\omega^2 - \Omega^2\right)^3} \cos(\omega t + \theta) \cos 2\Omega t - \\ & - \frac{3\alpha a\omega P^2}{8\Omega\left(\omega^2 - \Omega^2\right)^3} \sin(\omega t + \theta) \sin 2\Omega t + \\ & + \frac{3\alpha a^2 P\left(3\omega^2 + \Omega^2\right)}{\left(\omega^2 - \Omega^2\right)^2\left(9\omega^2 - \Omega^2\right)} \cos 2(\omega t + \theta) \cos\Omega t + \\ & + \frac{12\alpha\omega\Omega a^2 P}{\left(\omega^2 - \Omega^2\right)^2\left(9\omega^2 - \Omega^2\right)} \sin 2(\omega t + \theta) \sin\Omega t \end{aligned} \tag{4.153}$$

where a and θ are given by Eq. 4.152. Finally by substituting the relations (4.153) and (4.147) into (4.145) we obtain the solution

$$u(t) = a\cos(\omega t + \theta) + \frac{P}{\omega^2 - \Omega^2}\cos\Omega t + \varepsilon x(t) + 0(\varepsilon^2) \tag{4.154}$$

In the case of periodic vibration ($\dot{a} = 0$)we obtain $a = 0$ and follows that

$$\begin{aligned} u_{per}(t) = & \frac{P}{\omega^2 - \Omega^2}\cos\Omega t + \\ & + \varepsilon\left[\frac{2\mu\Omega P}{\left(\omega^2 - \Omega^2\right)^2}\sin\Omega t - \frac{3\alpha P^3}{4\left(\omega^2 - \Omega^2\right)^3}\cos\Omega t\right] \end{aligned} \tag{4.155}$$

In order to compare our results with the numerical solution, we select $\omega = \alpha = \mu = 1$, $\Omega = 2$, $P = 0.2$, $\varepsilon = 0.01$. The comparison of the corresponding analytical approximate solution (4.154) and approximate periodic solution (4.155) is shown in Figs. 4.2 and 4.3 respectively. They show that this procedure provide excellent approximations comparing to numerical solutions obtained by a fourth-order Runge-Kutta method for small oscillations in different initial conditions.

There are several studies that treated forced oscillations of systems with cubic nonlinearities, under the influence of slight viscous damping [7, 8, 34, 35, 55, 56] and so on. The results obtained here can be compared with those given by the method of successive approximations, the method of harmonic balance, etc. The solution (4.145) in the second approximation, coincides with the first approximation ($x = 0$ into Eq. 4.154) given by the method of multiple scales [34] and Chap. 5.

Case 2b ("resonance" case)

By assuming $\omega \approx 3\Omega$, we denote $\omega^2 = 9\Omega^2 + \varepsilon\delta$. The governing equation becomes

$$\ddot{u} + 9\Omega^2 u + 2\varepsilon\mu\dot{u} + \varepsilon\alpha u^3 + \varepsilon\sigma u = P\cos\Omega t \tag{4.156}$$

By transformation

$$u = x + \frac{P}{8\Omega^2}\cos\Omega t \tag{4.157}$$

we obtain from Eq. 4.156

$$\begin{aligned} & \ddot{x} + 9\Omega^2 x + 2\varepsilon\mu\left(\dot{x} - \frac{P}{8\Omega}\sin\Omega t\right) + \varepsilon\alpha\left(x + \frac{P}{8\Omega^2}\cos\Omega t\right)^3 + \\ & + \varepsilon\sigma\left(x + \frac{P}{8\Omega^2}\cos\Omega t\right) = 0 \end{aligned} \tag{4.158}$$

Initial approximate solution of Eq. 4.158 is in form:

$$x_0(t) = a\cos(3\Omega t + \theta) + \varepsilon X(t) \tag{4.159}$$

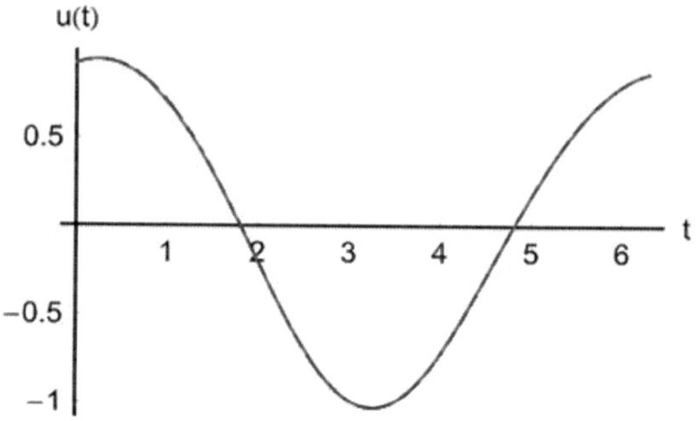

Fig. 4.2 Comparison between approximate solution (4.154) and numerical solution of Eq. 4.143 for $\omega = \alpha = \mu = a_0 = 1$, $\Omega = 2$, $P = 0.2$, $\varepsilon = 0.01$, $u(0) = 0.916$, $\dot{u}(0) = 0.177$: _____ numerical results; ____ approximate results

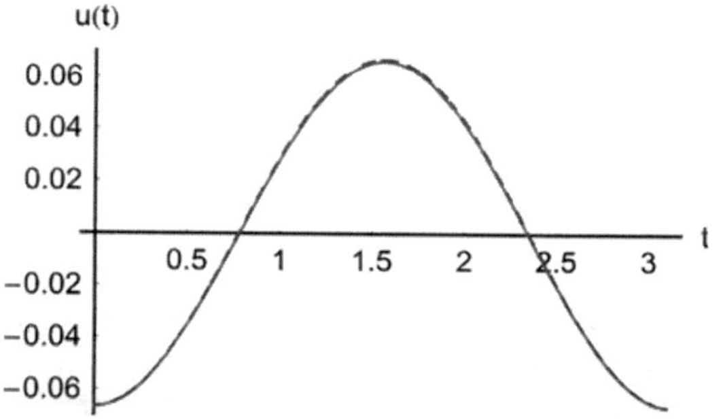

Fig. 4.3 Comparison between approximate solution (4.155) and numerical solution of Eq. 4.143 for $\omega = \alpha = \mu = a_0 = 1$, $\Omega = 2$, $P = 0.2$, $\varepsilon = 0.01$, $u(0) = -0.066$, $\dot{u}(0) = 0.0008$: _____ numerical results; ____ approximate results

By the same manipulation as before, we have the following solution, neglecting terms of second order in ε:

$$\begin{aligned} u(t) = {} & a\cos(3\Omega t + \theta) + \frac{P}{8\Omega^2}\cos\Omega t + \varepsilon\left[\frac{\mu P}{64\Omega^3}\sin\Omega t + \right. \\ & + \left(\frac{3\alpha a^2}{128\Omega^4} + \frac{3\alpha P^2}{1638\Omega^8} - \frac{\sigma P}{64\Omega^4}\right)\cos\Omega t - \frac{3\alpha P^2}{2048\Omega^4}\cos(\Omega t + \theta) + \\ & + \frac{3\alpha P^2}{4096\Omega^4}\cos(5\Omega t + \theta) - \frac{3\alpha a^2}{32\Omega^2}\cos(5\Omega t + 2\theta) - \\ & \left. - \frac{3\alpha a^2}{32\Omega^2}\cos(7\Omega t + 2\theta) + \frac{\alpha a^3}{4}\cos(9\Omega t + 3\theta)\right] \end{aligned} \tag{4.160}$$

where $a(t)$ and $\theta(t)$are given by the equations

$$\begin{aligned} \dot{a} &= -\varepsilon\mu a + \frac{\varepsilon\alpha P^3}{12288\Omega^7}\sin\theta \\ \dot{\theta} &= \varepsilon\left(\frac{\alpha a^2}{8\Omega} + \frac{\alpha P^2}{256 a\Omega^3} + \frac{\sigma}{6\Omega} - \frac{\alpha P^3}{12288 a\Omega^7}\cos\theta\right) \end{aligned} \tag{4.161}$$

with the approximate solution given by averaging method

$$a(t) = a_0 \exp(-\varepsilon\mu t)$$
$$\theta(t) = \theta_0 - \frac{\alpha a_0^2}{16\mu\Omega}\exp(-2\varepsilon\mu t) + \frac{\alpha P^2}{256\mu a_0 \Omega^3}\exp(\varepsilon\mu t) + \frac{\varepsilon\sigma}{6\Omega}t \tag{4.162}$$

The periodic solution of Eq. 4.156 is obtain for $\dot{a} = \dot{\theta} = 0$ and therefore we deduce that a and θ are given by equations

$$256a^2\Omega^8\left[589824\mu^2\Omega^6 + \left(96\alpha a^2\Omega^2 + \frac{3\alpha p^2}{a} + 128\sigma\Omega^2\right)^2\right] - \alpha^2P^6 = 0 \tag{4.163}$$

$$\sin\theta = \frac{12288\mu a\Omega^7}{\alpha P^3}$$
$$\cos\theta = \frac{1536a^3\Omega^6}{P^3} + \frac{48\Omega^4}{P} + \frac{2048a\sigma\Omega^6}{\alpha P^3} \tag{4.164}$$

The curve described by Eq. 4.163 is named, "resonance curve". The system (4.164) has solutions in the conditions

$$\frac{12888\mu a\Omega^7}{\alpha P^3} \le 1$$
$$\frac{1536a^3\Omega^6}{P^3} + \frac{48\Omega^4}{P} + \frac{2048a\sigma\Omega^6}{\alpha P^3} \le 1 \tag{4.165}$$

The first order approximate solution in the case of "resonance" for Eq. 4.156 is obtained from Eq. 4.159 with $a(t)$ and $\theta(t)$ given by Eqs. 4.162.

Fig.4.4 illustrates the resonance curves obtained for $\Omega = 1/3$, $a_0 = 1$, $\theta_0 = 0.18328125$, $\varepsilon = 0.01$, $\mu = a = 1$, $P = 0.2$, $\sigma = 1.0025$, $u(0) = 1.23185466$, $\dot{u}(0) = 0.00028125$. These curves are plotted using numerical integration results and results from Eq. 4.160. One can observe that the results are nearly identical.

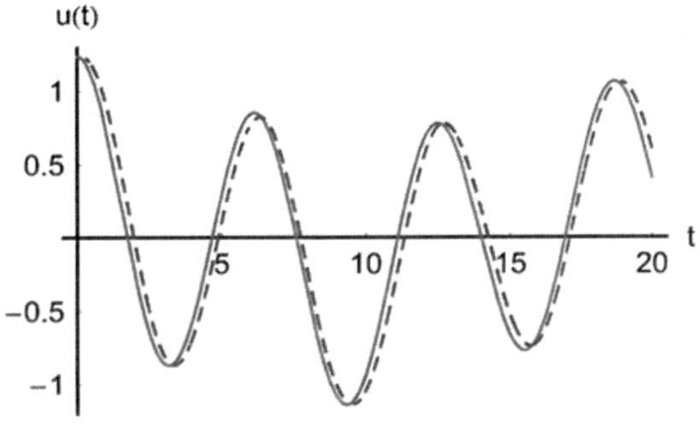

Fig. 4.4 Comparison between the approximate solution (4.160) and numerical solution of Eq. 4.156 for $\Omega = 1/3$, $a_0 = 1$, $\theta_0 = 0.18328125$, $\varepsilon = 0.01$, $\mu = a = 1$, $P = 0.2$, $\sigma = 1.0025$, $u(0) = 1.23185466$, $\dot{u}(0) = 0.00028125$: _____ numerical results; _ _ _ _ analytical results

This combination of iteration method and method of Krylov-Bogolyubov is a kind of powerful tool for treating weakly nonlinear problems. Due to the very high accuracy of the first order (initial) approximate solution, we always stop at the second iteration step. We can obtain higher order approximation if in the initial approximation we seek terms in ε^m ($m \geq 1$). However, the calculation is usually performed only for $m = 1$ or $m = 2$, because the contribution of higher order terms is generally unimportant and because the formulas obtained for higher approximations are very intricate.

It is interesting to note that by a certain transformation, the original equation of motion can be rewritten in the form (4.74). Nonlinear systems of type (4.74) have been known to arise in many problems related to dynamical systems. It is hoped therefore that the obtained results demonstrate the applicability of this method but also underline the importance of the periodic solutions in gaining a better understanding of these physically relevant models.

Chapter 5
The Method of Multiple Scales

The origins of the method of multiple scales go back to Krylov and Bogolyubov in 1932. The general principle behind the method is that the dependent variable is uniformly expanded in terms of two or more independent variables, nominally referred to as scales.

This obviously requires that the time derivatives of the dependent variable are similarly expressed, with the general consequence that uniformity is relatively well preserved even for nonconservative problems incorporating excitations and dissipations. The multiple independent variables T_k are generated with respect to real time t such that $T_k = \varepsilon^k t$, $k = 0, 1, 2, \ldots$. Clearly, then, the time derivatives will be expanded, in their own right, in terms of partial derivatives, each with respect to the T_k as follows:

$$\frac{d}{dt} = D_0 + \varepsilon D_1 + \varepsilon^2 D_2 + \ldots.. \tag{5.1}$$

$$\frac{d^2}{dt^2} = D_0^2 + 2\varepsilon D_0 D_1 + \varepsilon^2\left(D_1^2 + 2D_0 D_2\right) + \ldots.. \tag{5.2}$$

where $D_k = \frac{\partial}{\partial T_k}$, and the assumption is made that the parameter ε is small ($\varepsilon << 1$).

If we consider the equation

$$\ddot{u} + \alpha_1 u + \alpha_2 u^2 + \alpha_3 u^3 = 0 \tag{5.3}$$

then, one assumes that the solution of Eq. 5.3 can be represented by an expansion having the form

$$\begin{aligned} u(t,\varepsilon) = \varepsilon u_1(T_0, T_1, \ldots, T_m) + \varepsilon^2 u_2(T_0, T_1, \ldots, T_m) \\ + \ldots + \varepsilon^{m-1} u_{m-1}(T_0, T_1, \ldots, T_m) \end{aligned} \tag{5.4}$$

V. Marinca and N. Herisanu, *Nonlinear Dynamical Systems in Engineering*,
DOI 10.1007/978-3-642-22735-6_5, © Springer-Verlag Berlin Heidelberg 2011

Substitution of Eq. 5.4 and the time derivative expansion into the governing Eq. 5.3 results in a set of perturbation equations, hierarchically ordered to ε, which can then be solved successively. The solution can be determined only avoiding secular terms [22, 34, 58]. Uniformity is assured if $\varepsilon^{k} u_{k} = 0(\varepsilon^{k-1} u_{k-1})$ for all k and $T_0, T_1, T_2..., T_m$. Secular terms are routinely identified, by means of recognizable resonance conditions, removed and then equated to zero. This process is invariably used to determine the amplitude of zero-th order perturbation solution (sometimes called the generating solution) corresponding to $k = 0$ and expressed in complex form (sometimes real form). We obtain for $m = 2$:

$$D_0^2 u_1 + \alpha_1 u_1 = 0 \tag{5.5}$$

$$D_0^2 u_2 + \alpha_1 u_2 = -2D_0 D_1 u_1 - \alpha_2 u_1^2 \tag{5.6}$$

$$D_0^2 u_3 + \alpha_1 u_3 = -2D_0 D_1 u_2 - D_1^2 u_1 - 2D_0 D_2 u_1 - 2\alpha_2 u_1 u_2 - \alpha_3 u_1^3 \tag{5.7}$$

With this approach it turns out to be convenient to write the solution of Eq. 5.5 in the forms

$$u_1 = A(T_1, T_2)\exp(i\sqrt{\alpha_1}T_0) + \bar{A}(T_1, T_2)\exp(-i\sqrt{\alpha_1}T_0) \tag{5.8}$$

or

$$u_1 = B(T_1, T_2)\cos[\sqrt{\alpha_1}T_0 + \varphi(T_1, T_2)] \tag{5.9}$$

where $A(T_1,T_2)$ is an unknown complex function, $\bar{A}(T_1,T_2)$ is the complex conjugate of $A(T_1,T_2)$, $B(T_1,T_2)$ and $\varphi(T_1,T_2)$are unknown real functions. The governing equations for A (respectively B) are obtained by requiring u_2 and u_3 be periodic in T_0.

Substituting for example Eq. 5.8 into Eq. 5.6 leads to

$$\begin{aligned} D_0^2 u_2 + \alpha_1 u_2 = & -2i\sqrt{\alpha_1}D_1 A\exp(i\omega_0 T_0) - \\ & - \alpha_2\left[A^2\exp(2i\sqrt{\alpha_1}T_0) + A\bar{A}\right] + c.c. \end{aligned} \tag{5.10}$$

where $c.c.$ denotes the complex conjugate of the preceding terms. One can ignore the initial conditions and the homogeneous solutions in all u_k for $k \geq 2$, until the last step [34].

Any particular solution of Eq. 5.10 has a secular term containing the factor $T_0\exp(i\sqrt{\alpha}, T_0)$ unless

$$D_1 A(T_1, T_2) = 0 \tag{5.11}$$

Therefore A must be independent of T_1: $A = A(T_2)$. In this case, the solution of Eq. 5.10 is

$$u_2 = \frac{\alpha_2 A(T_2)^2}{3\alpha_1} \exp(2i\sqrt{\alpha_1}T_0) - \frac{\alpha_2}{\alpha_1} A\bar{A} + c.c. \tag{5.12}$$

where the solution of the homogeneous equation is not needed.

Now, substituting Eq. 5.9 into Eq. 5.6, it follows that

$$\begin{aligned} D_0^2 u_2 + \alpha_1 u_2 = & -\sqrt{\alpha_1} D_1 B(T_1, T_2) \sin[\sqrt{\alpha_1}T_0 + \varphi(T_1, T_2)] + \\ & + B D_1 \varphi(T_1, T_2) \cos[\sqrt{\alpha_1}T_0 + \varphi(T_1, T_2)] - \\ & - \frac{1}{2}\alpha_2 B^2(T_1, T_2)(1 + \cos(2\sqrt{\alpha_1}T_0 + 2\varphi(T_1, T_2))) \end{aligned} \tag{5.13}$$

To eliminate secular terms from Eq. 5.13, the following conditions must hold

$$D_1 B(T_1, T_2) = 0 \tag{5.14}$$

$$D_1 \varphi(T_1, T_2) = 0 \tag{5.15}$$

which imply that $B = B(T_2)$ and $\varphi = \varphi(T_2)$.

Having satisfied the conditions (5.14) and (5.15) (called the *solvability conditions*), from Eq. 5.13, we obtain

$$u_2 = -\frac{1}{2}\frac{\alpha_2}{\alpha_1} B^2(T_2) + \frac{\alpha_2}{6\alpha_1} B^2(T_2) \cos[2\sqrt{\alpha_1}T_0 + 2\varphi(T_2)] \tag{5.16}$$

Substituting u_1 and u_2 from Eqs. 5.8 and 5.12 into Eq. 5.7 and recalling that $D_1 A = 0$, we obtain

$$\begin{aligned} D_0^2 u_3 + \alpha_1 u_3 = & - \left[2i\sqrt{\alpha_1} D_2 A + \frac{9\alpha_1\alpha_3 - 10\alpha_2^2}{3\alpha_1} A\bar{A} \right] \exp(i\sqrt{\alpha_1}T_0) - \\ & - \frac{3\alpha_1\alpha_3 + 2\alpha_2^2}{3\alpha_1} A^3 \exp(3i\sqrt{\alpha_1}T_0) + c.c. \end{aligned} \tag{5.17}$$

In order to eliminate secular terms from Eq. 5.17, we must put

$$2i\sqrt{\alpha_i} D_2 A + \frac{9\alpha_1\alpha_3 - 10\alpha_2^2}{3\alpha_1} A^2\bar{A} = 0 \tag{5.18}$$

In solving equations having the form of Eq. 5.18, we find it convenient to write A in the polar form:

$$A = \frac{1}{2} a \ \exp(i\beta) \quad (5.19)$$

where a and β are real functions of T_2. Substituting Eq. 5.19 into Eq. 5.18 and separating the result into real and imaginary parts, we obtain

$$\sqrt{\alpha_1} a' = 0 \quad (5.20)$$

$$\sqrt{\alpha_1} a\beta' + \frac{10\alpha_2^2 - 9\alpha_1\alpha_3}{24\alpha_1} a^3 = 0 \quad (5.21)$$

where the prime denotes the derivate with respect to T_2. It follows that a = constant and hence that

$$\beta = \frac{9\alpha_1\alpha_3 - 10\alpha_2^2}{24\alpha_1\sqrt{\alpha_1}} a^2 T_2 + \beta_0 \quad (5.22)$$

where β_0 is a constant. Returning to Eq. 5.19, we find that

$$A = \frac{1}{2} a \ \exp\left(i \frac{9\alpha_1\alpha_3 - 10\alpha_2^2}{24\alpha_1\sqrt{\alpha_1}} \varepsilon^2 a^2 t + i\beta_0 \right) \quad (5.23)$$

where we used the fact that $T_2 = \varepsilon^2 t$.

Substituting u_1 and u_2 from Eqs. 5.8 and 5.12 into Eq. 5.4 and using Eq. 5.23, we obtain

$$u = \varepsilon a \cos(\omega t + \beta_0) - \frac{\varepsilon^2 a^2 \alpha_2}{6\alpha_1} [3 - \cos(2\omega t + 2\beta_0)] + 0(\varepsilon^3) \quad (5.24)$$

where

$$\omega = \sqrt{\alpha_1} \left[1 + \frac{9\alpha_1\alpha_3 - 10\alpha_2^2}{24\alpha_1^2} \varepsilon^2 a^2 \right] + 0(\varepsilon^3) \quad (5.25)$$

The constants a and β_0 can be determined from the initial conditions.

Using real form of the functions u_1 and u_2 given by Eqs. 5.9, 5.16 and substituting into Eq. 5.7 we obtain

$$\begin{aligned} D_0^2u_3 + \alpha_1 u_3 &= 2\sqrt{\alpha_1}D_2B(T_2)\sin[\sqrt{\alpha_1}T_0 + \varphi(T_2)] + \\ &+ \left[2\sqrt{\alpha_1}B(T_2)D_2\varphi(T_2) + \frac{5\alpha_2^2}{6\alpha_1}\beta^3(T_2) - \frac{3}{4}\alpha_3 B^3(T_2)\right] - \\ &- \left(\frac{\alpha_2^2}{6\alpha_1} + \frac{\alpha_3}{4}\right)B^3(T_2)\cos 3[\sqrt{\alpha_1}(T_0) + \varphi(T_2)] \end{aligned} \tag{5.26}$$

The secular terms from Eq. 5.26 can be eliminated if

$$D_2B(T_2) = 0 \tag{5.27}$$

$$2\sqrt{\alpha_1}B(T_2)D_2\varphi(T_2) + \frac{5\alpha_2^2}{6\alpha_1}B^3(T_2) - \frac{3}{4}\alpha_3 B^3(T_2) = 0 \tag{5.28}$$

which imply that

$$B(T_2) = B_0 \tag{5.29}$$

$$\varphi(T_2) = \frac{9\alpha_1\alpha_3 - 10\alpha_2^2}{24\alpha_1\sqrt{\alpha_1}}B_0^2T_2 + \varphi_0 \tag{5.30}$$

where β_0 and φ_0 are real constants.

Substituting Eqs. 5.9, 5.16, 5.29 and 5.30, into Eq. 5.4 we obtain an expression equivalent with Eq. 5.24:

$$u = \varepsilon B_0\cos(\Omega t + \varphi_0) - \frac{\varepsilon^2 B_0^2\alpha_2}{6\alpha_1}[3 - \cos(2\Omega t + 2\varphi_0)] + 0(\varepsilon^3) \tag{5.31}$$

where

$$\Omega = \sqrt{\alpha_1}\left[1 + \frac{9\alpha_1\alpha_3 - 10\alpha_2^2}{24\alpha_1^2}\varepsilon^2 a^2\right] + 0(\varepsilon^3)(=\omega) \tag{5.32}$$

Several variants of basic method of multiple scales have emerged in recent times. Such method of multiple scales has been computerized with parallelization strategies introduced for reasons of optimization [59]. The major texts [22] and [58] are the most authoritative on multiple scales in the area of vibration research. An alternative to the standard method [22, 34] is offered in [60] where time scales are rewritten as $T_0 = \Omega t$ and in general $T_n = \varepsilon^n \Omega t$. Also Ω^2 and Ω are independently so that $\Omega^2 = 1 + \varepsilon\delta_{11} + \varepsilon^2\delta_{12}$ and $\Omega = 1 + \varepsilon\delta_{21}$. For other appreciations, see [61, 62] and [63].

The structure of the power series for the dependent variable, the systematic use of timescales and the perturbation parameter ε are very significant. We consider, in what follows the motion of a pendulum with viscous damping having the equation

$$\ddot{\theta} + 2\bar{\mu}\dot{\theta} + \omega_0^2 f(\theta) = 0 \tag{5.33}$$

where $f(\theta) = \sin\theta \approx \theta - \frac{1}{6}\theta^3$ and the very light damping in the form $\mu = \varepsilon^2\mu$.

In [58] are present three alternative cases. In the first alternative, it is suppose that

$$\theta(t,\varepsilon) = \varepsilon\theta_1(T_0,T_1,T_2) + \varepsilon^2\theta_2(T_0,T_1,T_2) + \varepsilon^3\theta_3(T_0,T_1,T_2) + \ldots \tag{5.34}$$

The perturbation equations are

$$\begin{aligned} &D_0^2\theta_1 + \omega_0^2\theta_1 = 0 \\ &D_0^2\theta_2 + \omega_0^2\theta_2 = -2D_0D_1\theta_1 \\ &D_0^2\theta_3 + \omega_0^2\theta_3 = -2D_0D_1\theta_2 - 2D_0D_1\theta_1 - D_1^2\theta_1 - 2\mu D_0\theta_1 + \frac{\omega_0^2}{6}\theta_1^3 \end{aligned} \tag{5.35}$$

In the second alternative, a standard power series has the form

$$\theta(t,\varepsilon) = \theta_0(T_0,T_1,T_2) + \varepsilon\theta_1(T_0,T_1,T_2) + \varepsilon^2\theta_2(T_0,T_1,T_2) + \ldots \tag{5.36}$$

For the nonlinear term $f(\theta) = \theta - \varepsilon\gamma\theta^3$ with $\varepsilon\gamma$ replacing the 1/6 used previously, the perturbation equations become.

$$\begin{aligned} &D_0^2\theta_0 + \omega_0^2\theta_0 = 0 \\ &D_0^2\theta_1 + \omega_0^2\theta_1 = -2D_0D_1\theta_0 + \gamma\omega_0^2\theta_0^2 \\ &D_0^2\theta_2 + \omega_0^2\theta_2 = -2D_0D_1\theta_1 - 2D_0D_2\theta_0 - D_1^2\theta_0 - 2\mu D_0\theta_0 + 3\gamma\omega_0^2\theta_0^2\theta_1 \end{aligned} \tag{5.37}$$

In the last alternative, when $f(\theta) = \theta - \varepsilon^2\gamma\theta^3$ we find the following equations

$$\begin{aligned} &D_0^2\theta_0 + \omega_0^2\theta_0 = 0 \\ &D_0^2\theta_1 + \omega_0^2\theta_1 = -2D_0D_1\theta_0 \\ &D_0^2\theta_2 + \omega_0^2\theta_2 = -2D_0D_1\theta_1 - 2D_0D_2\theta_0 - D_1^2\theta_0 - 2\mu D_0\theta_0 + \gamma\omega_0^2\theta_0^3 \end{aligned} \tag{5.38}$$

By comparing the three sets of perturbation equation, taking Eqs. 5.35 as the base-line set, it can be seen that the standard power series together with $f(\theta) = \theta - \varepsilon\gamma\theta^3$ yields an identical lowest order perturbation equation except for the subscripting which in this context is irrelevant. However, at the intermediate perturbation level we find that the standard power series method together with the first-order, nonlinearity $\varepsilon\gamma\theta^3$ expansion leads to explicit presence of the cubic term at this order, as shown in Eq. 5.37_2 but also noting that this is not present in Eq. 5.35_2. At the highest perturbation level, it can be seen that both right-hand sides are structurally identical, with subscripts representing the same functionality within both sets of equations, except for the modified cubic term ${\omega_0}^2{\theta_1}^3/6$ in Eq. 5.35_3 and $3\gamma\omega_0^2\theta_0^2\theta_1$ in Eq. 5.37_3.

Apparently, Eqs. 5.38 are structurally identical to Eqs. 5.35 except for the subscripting of the θ_i ($i = 1,2,3$ and $0,1,2$ respectively) and also the coefficient of the cubic terms $\omega_0^2\theta^3/6$ in Eq. $5.35_3$and $\gamma\omega_0^2\theta_0^3$in Eq. 5.38_3.

For Eqs. 5.35 the solution can be written in the form

$$\theta = \varepsilon a_0 e^{-\bar{\mu}t}\cos\left[\omega_0 t - \frac{\varepsilon^2 a_0^2}{32\omega_0\bar{\mu}}e^{-2\bar{\mu}t} + \beta_0\right] + H.O.T \tag{5.39}$$

where a_0 and β_0 are constants of integration and H.O.T stands for higher order terms. The same solution procedure can be adopted for Eq. 5.38 and thus the first approximate solution is as follows:

$$\theta = a_0 e^{-\bar{\mu}t}\cos\left[\omega_0 t - \frac{3a_0^2\varepsilon^2\gamma}{16\omega_0\bar{\mu}}e^{-2\bar{\mu}t} + \beta_0\right] + H.O.T \tag{5.40}$$

We assumed that $1/6 \equiv \varepsilon^2\gamma$ and the second term in the brackets of Eq. 5.40 resembles that of Eq. 5.39 noting that the initial displacement is defined as εa_0 in Eq. 5.39 and a_0 in Eq. 5.40.

The previous, comparative, example shows that the form of the adopted power series and the ordering of terms can have a major bearing on the structure of the solution, with clear connotations for accuracy and physical relevance.

One way to deal with ordering is to have it on some sort of physical appreciation of the problem, leading to notions of "hard" ($F(t) = F_0\cos\Omega t$ and $F(t) = \varepsilon F_0\cos\Omega t$ as respective examples of this in the case of a single frequency harmonic oscillation) and "strong" or "weak" damping where damping ratio might be expressed to zero-th order ε for strong damping, or first or second order ε in the case of weaker damping.

5.1 Duffing Oscillator with Softening Nonlinearity

The Duffing oscillator with softening nonlinearity has been extensively studied in the context of a large variety of physical systems [22, 64–70] such as the oscillations of a pendulum, the rolling motion of a ship, or charge oscillations in super ionic conductors. In spite of its simplicity this system has been shown to exhibit complex behaviours in the presence of harmonic forcing and damping, including, among other features, bistability, period-doubling sequences, chaos and sometimes unbounded motions.

This above behaviours, apart from being interesting, must be considered because they can forecast catastrophic consequences for the physical system being modelled.

In what follows we study the response of one-degree-of-freedom systems with cubic nonlinearity to a fundamental parametric excitation ($\Omega = 1$), governed by a non-dimensional differential equation of the form

$$\ddot{u} + \varepsilon\mu\dot{u} + u - \varepsilon\alpha u^3 = \varepsilon f \cos \Omega t \tag{5.41}$$

where μ, α, f and Ω are positive constant parameters and $\varepsilon << 1$ (see also Sect. 4.2.3, Example 2).

We consider second-order expansion of the type

$$u(t, \varepsilon) = u_0(T_0, T_1, T_2) + \varepsilon u_1(T_0, T_1, T_2) + \varepsilon^2 u_2(T_0, T_1, T_2) \tag{5.42}$$

for the proximity of Ω to unity:

$$\Omega^2 = 1 + \varepsilon\sigma \tag{5.43}$$

where σ is a detuning parameter. The natural frequency, equal to unity, of the oscillators defined in Eq. 5.41 can be rewritten in terms of Ω by using Eq. 5.43. The result is the following

$$\ddot{u} + \Omega^2 u + \varepsilon\left(\mu\dot{u} - \sigma u - \alpha u^3 - f \cos \Omega t\right) = 0 \tag{5.44}$$

Substituting Eqs. 5.42, 5.1 and 5.2 into Eq. 5.44, and equating coefficients of like powers of ε, we obtain

$$D_0^2 u_0 + \Omega^2 u_0 = 0 \tag{5.45}$$

$$D_0^2 u_1 + \omega^2 u_1 = -2D_0 D_1 u_0 + \sigma u_0 - 2\mu D_0 u_0 + \alpha u_0^3 + f \cos \Omega t \tag{5.46}$$

$$\begin{aligned} D_0^2 u_2 + \omega^2 u_2 &= -2D_0 D_2 u_0 - 2D_0 D_1 u_1 - \\ &- D_1^2 u_0 - 2\mu D_0 u_1 - 2\mu D_1 u_0 + \sigma u_1 + 3\alpha u_0^2 u_1 \end{aligned} \tag{5.47}$$

In this first case we express the solution of Eq. 5.45 in the complex form

$$u_0(T_0, T_1, T_2) = A(T_1, T_2)\, \exp(i\Omega T_0) + c.c \tag{5.48}$$

where A is an unknown complex function of T_1 and T_2 at this level of approximation. It is determined by imposing solvability conditions at the next levels of approximation. Substituting Eq. 5.48 into Eq. 5.46 yields

$$\begin{aligned} D_0^2 u_1 + \omega^2 u_1 &= (\sigma A - 2iD_1 A - 2i\mu\Omega A)\exp(i\Omega T_0) + \alpha A^3 \exp(3i\Omega T_0) + \\ &+ 3\alpha A^2 \bar{A} \exp(i\Omega T_0) + \frac{1}{2} f \exp(i\Omega T_0) + c.c \end{aligned} \tag{5.49}$$

By imposing the solvability condition at the next levels of approximation, from Eq. 5.49 we obtain

$$2i\Omega D_1 A + (-\sigma + 2i\mu\Omega)A - 3\alpha A^2 \bar{A} - \frac{1}{2}f = 0 \tag{5.50}$$

If we propose to obtain only first-order approximate solution, then A can be considered to be a function of T_1 only. Therefore, a first-order approximation is obtained by expressing A in the polar form

$$A = \frac{1}{2}a(T_1)\exp[i\beta(T_1)] \tag{5.51}$$

where $a(T_1)$ and $\beta(T_1)$ are unknown amplitude and phase of the fundamental frequency, respectively.

Substituting Eq. 5.51 into Eq. 5.50 and separating real and imaginary parts, we find that

$$\dot{a} = -\mu a - \frac{f}{2\Omega}\sin\beta \tag{5.52}$$

$$\dot{\beta} = -\frac{\sigma}{2\Omega} - \frac{3\alpha}{8\Omega}a^2 - \frac{f}{2a\Omega}\cos\beta \tag{5.53}$$

The first-order solution then becomes

$$u(t) = a(\varepsilon t)\cos[\Omega t + \beta(\varepsilon t)] \tag{5.54}$$

where $a(\varepsilon t)$ and $\beta(\varepsilon t)$ are determined from Eqs. 5.52 and 5.53. An analytic approximate solution of Eqs. 5.52 and 5.53 can be obtained by averaging method. We obtain in this way:

$$\dot{a} = -\mu a \tag{5.55}$$

$$\dot{\beta} = -\frac{\sigma}{2\Omega} - \frac{3\alpha}{8\Omega}a^2 \tag{5.56}$$

The solution of Eqs. 5.55 and 5.56 are, respectively:

$$a(t) = a_o \exp(-\varepsilon\mu t) \tag{5.57}$$

$$\beta(t) = \beta_o - \frac{3\sigma}{2\Omega}t - \frac{3\alpha a_o^2}{16\mu\Omega}e^{-2\varepsilon\mu t} \tag{5.58}$$

where a_0 and β_0 are constants.

The first-order analytic approximate solution by means of Eqs. 5.51 and 5.58 becomes

$$u(t) = a_0 e^{-\varepsilon\mu t} \cos\left[\left(\Omega - \frac{\varepsilon\sigma}{2\Omega}\right)t + \frac{3\alpha a_0^2}{16\mu\Omega} e^{-2\varepsilon\mu t} + \beta_0\right] \tag{5.59}$$

The steady-state periodic motion corresponds to the solution of the system of equations, obtained from Eqs. 5.52 and 5.53 by conditions $\dot{a} = 0$ and $\dot{\beta} = 0$. We obtain

$$\mu a + \frac{f}{2\Omega} \sin\beta = 0 \tag{5.60}$$

$$\frac{\sigma}{2\Omega} + \frac{3\alpha}{9\Omega} a^2 + \frac{f}{2a\Omega} \cos\beta = 0 \tag{5.61}$$

Eliminating β from Eqs. 5.59 and 5.60 leads to the frequency-response equation (resonance curve)

$$4\mu^2 a^2 \Omega^2 + \left(\sigma a + \frac{3\alpha a^3}{4}\right)^2 - f^2 = 0 \tag{5.62}$$

The detuning parameter σ is obtained from Eq. 5.61 in the form

$$\sigma = -\frac{3\alpha a^2}{4} \pm \sqrt{\frac{f^2}{a^2} - 4\mu^2\Omega^2} \tag{5.63}$$

The second-order approximation requires elimination of the secular terms in Eqs. 5.46 and 5.47 simultaneously, but now A is considered to be in Eq. 5.51 a function of T_1 and T_2. With Eq. 5.50, the solution of Eq. 5.46 can be expressed as

$$u_1 = -\frac{\alpha A^3}{8\Omega^2} \exp(3i\Omega T_0) + c.c \tag{5.64}$$

Substituting Eqs. 5.48 and 5.64 into Eq. 5.47 yields

$$\begin{aligned} D_0^2 u_2 + \Omega^2 u_2 = -\left(D_1^2 A + 2i\Omega D_2 A + 2\mu D_1 A\right) \exp(i\Omega T_0) - \\ -\frac{3\alpha^2}{8\Omega^2} A^3 \bar{A}^2 \exp(i\Omega T_0) + H.O.T. + c.c \end{aligned} \tag{5.65}$$

Eliminating secular terms from Eq. 5.65, we obtain

$$D_1^2A + 2i\Omega D_2A + 2\mu D_1A + \frac{3\alpha^2}{8\Omega^2}A^3\bar{A}^2 = 0 \tag{5.66}$$

In order to solve Eqs. 5.50 and 5.66, we find it convenient to combine them into a single first-order ordinary-differential equation. For this, from Eq. 5.50 we obtain:

$$D_1^2A = D_1(D_1A) = \left(\mu + \frac{i\sigma}{2\Omega}\right)D_1A - \frac{3i\alpha}{2\Omega}D_1\left(A^2\bar{A}\right) \tag{5.67}$$

But:

$$\begin{aligned} D_1\left(A^2\bar{A}\right) &= 2A\bar{A}D_1A + A^2D_1\bar{A} = \\ &= -\left(3\mu + \frac{i\sigma}{2\Omega}\right)A^2\bar{A} - \frac{3i\alpha}{2\Omega}A^3\bar{A} + \left(A^2 - 2A\bar{A}\right)\frac{if}{4\Omega} \end{aligned} \tag{5.68}$$

and, therefore from Eqs. 5.68 and 5.67 we have

$$\begin{aligned} D_1^2A = &\left(-\frac{3i\alpha}{2\Omega^2} + \frac{6i\alpha\mu}{\Omega}\right)A^2\bar{A} - \frac{9\alpha^2}{4\Omega^2}A^3\bar{A}^2 + \left(\mu + \frac{i\sigma}{2\Omega}\right)^2A + \\ &+ \left[i\mu + \left(\frac{3iA^2}{2\Omega} - \frac{3\alpha A\bar{A}}{\Omega} - \frac{\sigma}{2\Omega}\right)\right]\frac{f}{4\Omega} \end{aligned} \tag{5.69}$$

Substituting Eqs. 5.69 and 5.50 into Eq. 5.66 yields

$$\begin{aligned} 2iD_2A &+ \left(\frac{3i\alpha\mu}{\Omega} - \frac{3\sigma\alpha}{2\Omega^2}\right)A^2\bar{A} - \frac{15\alpha^2}{8\Omega^2}A^3\bar{A}^2 - \left(\mu^2 + \frac{\sigma^2}{4\Omega^2}\right)A + \\ &+ \left(\frac{3\alpha A^2}{8\Omega^2} - \frac{3\alpha A\bar{A}}{4\Omega^2} - \frac{\sigma}{8\Omega^2} - \frac{i\mu}{4\Omega}\right)f = 0 \end{aligned} \tag{5.70}$$

From Eqs. 5.50 and 5.70 and taking account that $\varepsilon D_1A + \varepsilon^2D_2A = \frac{dA}{dt}$, we obtain

$$\begin{aligned} 2i\Omega\frac{dA}{dt} &+ \varepsilon\left[(2i\mu\Omega - \sigma)A - 3\alpha A^2\bar{A} - \frac{1}{2}f\right] + \\ &+ \varepsilon^2\left[\frac{3\alpha f}{8\Omega^2}A^2 + \left(\frac{3i\alpha\mu}{\Omega} - \frac{3\alpha\sigma}{2\Omega^2}\right)A^2\bar{A} - \left(\mu^2 + \frac{\sigma^2}{4\Omega^2}\right)A - \right. \\ &\left. - \frac{3\alpha f}{4\Omega^2}A\bar{A} - \frac{15\alpha^2}{8\Omega^2}A^3\bar{A}^2 - \frac{i\mu f}{4\Omega} - \frac{\sigma f}{8\Omega^2}\right] = 0 \end{aligned} \tag{5.71}$$

In this case, A is a function of t and is expressed in the polar form

$$A(t) = \frac{1}{2}a(t)\exp[i\beta(t)] \tag{5.72}$$

Substituting Eq. 5.72 into Eq. 5.71 and separating real and imaginary parts, we obtain

$$\dot{a} = -\varepsilon\mu a - \varepsilon^2\frac{3\alpha\mu}{8\Omega^2}a^3 + \varepsilon^2\frac{\mu f}{4\Omega^2}\cos\beta - \left(\frac{\varepsilon f}{2\Omega} + \frac{\varepsilon^2\sigma f}{8\Omega^3} + \frac{9\varepsilon^2\alpha f}{32\Omega^3}a^2\right)\sin\beta \tag{5.73}$$

$$\begin{aligned}\dot{\beta} = &-\frac{\varepsilon\sigma}{2\Omega} - \frac{\varepsilon^2\mu^2}{2\Omega} - \frac{\varepsilon^2\sigma^2}{8\Omega^3} - \left(\frac{3\varepsilon\alpha}{8\Omega} + \frac{3\varepsilon^2\alpha\sigma}{16\Omega^3}\right)a^2 - \frac{15\varepsilon^2\alpha^2a^4}{256\Omega^3} - \\ &-\left(\frac{\varepsilon f}{2a\Omega} + \frac{\varepsilon^2\sigma f}{8a\Omega^3} + \frac{3\varepsilon^2\alpha f a}{32\Omega^3}\right)\cos\beta - \frac{\varepsilon^2\mu f}{4a\Omega^2}\sin\beta\end{aligned} \tag{5.74}$$

Substituting Eqs. 5.48, 5.64 and 5.72 into Eq. 5.42, we find that the second-order approximation to the solution of Eq. 5.44 for the fundamental parametric excitation case ($\Omega \approx 1$) is

$$u(t) = a\cos(\Omega t + \beta) - \frac{\varepsilon\alpha a^3}{32\Omega^2}\cos(3\Omega t + 3\beta) + \ldots \tag{5.75}$$

where a and β are given by Eqs. 5.73 and 5.74.

By means of averaging method, from Eqs. 5.73 and 5.74 we obtain an analytic approximate solution in the form

$$\dot{a} = -\varepsilon\mu a\left(1 + \frac{3\varepsilon\alpha a^2}{8\Omega^2}\right) \tag{5.76}$$

$$\dot{\beta} = -\frac{3\sigma}{2\Omega} - \frac{\varepsilon^2\mu^2}{2\Omega} - \frac{\varepsilon^2\sigma^2}{8\Omega^3} - \left(\frac{3\varepsilon\alpha}{8\Omega} + \frac{3\varepsilon^2\alpha\sigma}{16\Omega^3}\right)a^2 - \frac{15\varepsilon^2\alpha^2a^4}{256\Omega^3} \tag{5.77}$$

The solution Eqs. 5.76 and 5.77 are obtained in the form

$$a(t) = \frac{a_0\exp(-\varepsilon\mu t)}{\sqrt{1 + \frac{3\varepsilon\alpha a_0^2}{8\Omega^2}[1 - \exp(-2\varepsilon\mu t)]}} \tag{5.78}$$

$$\begin{aligned}\beta(t) = \beta_0 = & -\left(\frac{3\sigma}{2\Omega}+\frac{\varepsilon^2\mu^2}{2\Omega}+\frac{\varepsilon^2\sigma^2}{8\Omega^3}\right)t+ \\ & +\left(\frac{4\Omega^3}{3\varepsilon^2\mu\alpha}+\frac{2\sigma\Omega}{3\varepsilon\alpha}-\frac{\Omega}{2\varepsilon\mu}-\frac{\sigma}{4\mu\Omega}\right)\ln\left\{1+\frac{3\varepsilon\alpha a_0^2}{8\Omega^2}[1-\exp(-2\varepsilon\mu t)]\right\}+ \\ & +\left(\frac{4\Omega^3}{3\varepsilon^2\mu\alpha}+\frac{2\sigma\Omega}{3\varepsilon\alpha}\right)(8\Omega^2+3\varepsilon\alpha a_0^2)[1-\exp(-2\varepsilon\mu t)]\{8\Omega^2+ \\ & +3\varepsilon\alpha a_0^2[1-\exp(-2\varepsilon\mu t)]\}^{-1}\end{aligned} \tag{5.79}$$

where a_0 and β_0 are constants of integrations.

For the steady-state periodic responses $\dot{a}=0$ and $\dot{\beta}=0$so that from Eqs. 5.73 and 5.74 we obtain

$$\begin{aligned}\cos\beta &= \frac{M}{8N} \\ \sin\beta &= \frac{\Omega P}{N}\end{aligned} \tag{5.80}$$

where

$$\begin{aligned}M = & \, 1024\varepsilon\mu^2 a\Omega^4 + 768\varepsilon^2\alpha\mu^2 a^3\Omega^2 - \\ & -\left(16\Omega^2+2\varepsilon\sigma+9\varepsilon\alpha a^2\right)\left(15\varepsilon\alpha^2 a^5+48\varepsilon\sigma a^3+96\alpha a^3\Omega^2+32\varepsilon\sigma^2 a+\right. \\ & \left.+128\varepsilon^2\mu^2 a\Omega^2+128\sigma a\Omega^2\right) \\ N = & \, 256\Omega^4+64\varepsilon^2\mu^2\Omega^4+128\varepsilon\sigma\Omega^2+192\varepsilon\alpha a^2\Omega^2+16\varepsilon^2\sigma^2+ \\ & +16\varepsilon^2\alpha\sigma a^2+27\varepsilon^2\alpha^2 a^4 \\ P = & \left(32\mu a\Omega^2+12\varepsilon\mu\alpha a^3\right)\left(16\Omega^2+4\varepsilon\sigma+3\varepsilon\alpha f a^2\right)+128\varepsilon\mu\sigma a\Omega^2+ \\ & +128\varepsilon^2\mu^3 a\Omega^2+32\varepsilon^2\sigma^2\mu a+96\varepsilon\mu\alpha a^3\Omega^2+48\varepsilon^2\mu\sigma a^3+15\varepsilon\alpha^2 a^5\end{aligned} \tag{5.81}$$

with conditions:

$$-1 \le \frac{M}{8N} \le 1, -1 \le \frac{\Omega \mathrm{P}}{N} \le 1 \tag{5.82}$$

The frequency-response curve of a system governed by Eq. 5.44 is obtained from Eq. 5.80, eliminating β and has the form

$$M^2+64\Omega^2P^2-64N^2=0 \tag{5.83}$$

where M, N and P are given by Eqs. 5.81.

Figure 5.1 shows the first-order approximation $u(t)$ given by Eq. 5.59, which is compared with the numerical results obtained from Eq. 5.44 for $\varepsilon = 0.01$, $\sigma = 0.1$,

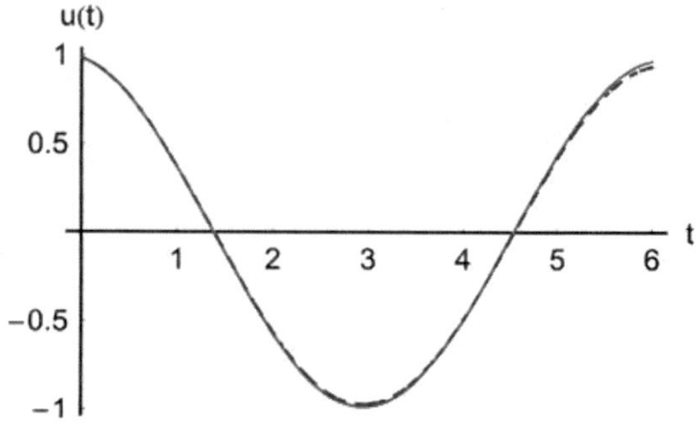

Fig. 5.1 Comparison between the approximate solution (5.59) and numerical solution of Eq. 5.44 for $\varepsilon = 0.01$, $\sigma = 0.1$, $\Omega = 1.0005$, $\mu = \alpha = 1$, $f = 0.1$: _____ numerical results, _ _ _ _ analytical results

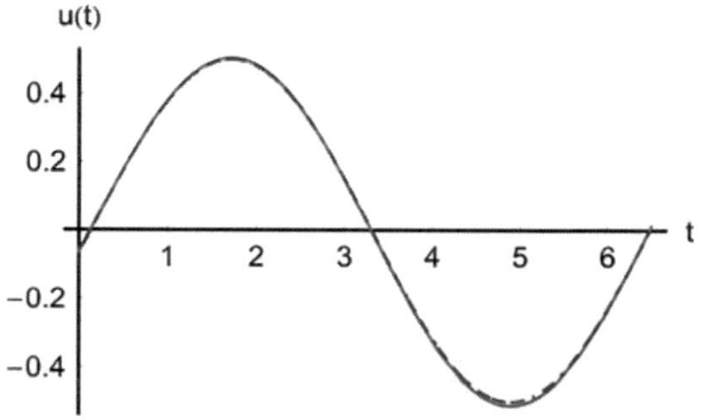

Fig. 5.2 Comparison between the approximate periodic solution and numerical solution of Eq. 5.44 for $\varepsilon = 0.01$, $\sigma = 0.00088$, $a = 1/2$, $\Omega = 0.099$, $\mu = 1$, $\alpha = 1.5$, $f = 1$: _____ numerical results, _ _ _ _ analytical results

$\Omega = 1.0005$, $\mu = \alpha = 1, f = 0.1$. Figure 5.2 shows the comparison between the first-order approximate periodic solution and numerical solution in the conditions $\varepsilon = 0.01$, $\sigma = 0.00088$, $a = 1/2$, $\Omega = 0.099$, $\mu = 1$, $\alpha = 1.5$, $f = 1$. One can observe that the first-order approximate solutions are in very good agreement with the numerical ones.

5.2 A Parametric System with Cubic Nonlinearity Coupled with a Lanchester Damper

In what follows, the two equations governing a parametric system with a single-degree-of-freedom coupled with a Lanchester damper are studied using the method of multiple scales for the case of fundamental parametric resonance.

Parametrically excited systems are quite frequently encountered in various applications. These are systems which, in contrast to those externally excited, have the excitation appearing as a time-dependent coefficient in the governing differential equation. These systems may exhibit large responses even when the excitation is small and not close to the natural frequency of the system.

The oscillatory-bare pendulum is a classical example of such systems and has been treated extensively [22]. Ships also experience this type of behaviour. Systems with one and two-degree-of-freedom including the effect of quadratic and cubic nonlinearities have been investigated in [71] and [72]. Non-stationary parametric oscillators with a single-degree-of-freedom near the principal parameter resonance is treated in [73] and the response of single-degree-of-freedom parametric system with cubic nonlinearity is studied near the principal parametric resonance in [74]. In [75] is investigated a non-linear oscillator having quadratic and cubic nonlinearities coupled with a Lanchester damper.

The governing equations of a parametric system with a cubic nonlinearity coupled with a Lanchester damper are [73, 74, 75]:

$$\ddot{u}_1 + u_1 + 2\varepsilon\mu(\dot{u}_1 - \dot{u}_2) + \varepsilon\alpha u_1^3 + 2\varepsilon F u_1 \cos\Omega t = 0 \tag{5.84}$$

$$\ddot{u}_2 + 2\gamma(\dot{u}_2 - \dot{u}_1) = 0 \tag{5.85}$$

where μ, α, F and γ are known parameters depending of the system and $\varepsilon << 1$.

Using the method of multiple scales for the case of the fundamental parametric resonance ($\Omega \approx 1$), one seeks an approximate solution in the form

$$u_1(t,\varepsilon) = u_{10}(T_0,T_1,T_2) + \varepsilon u_{11}(T_0,T_1,T_2) + \varepsilon^2 u_{12}(T_0,T_1,T_2) + \ldots \tag{5.86}$$

$$u_2(t,\varepsilon) = u_{20}(T_0,T_1,T_2) + \varepsilon u_{21}(T_0,T_1,T_2) + \varepsilon^2 u_{22}(T_0,T_1,T_2) + \ldots \tag{5.87}$$

We define the detuning parameter σ according to

$$\Omega^2 = 1 + \varepsilon\sigma \tag{5.88}$$

Using Eqs. 5.88, 5.84 and 5.85 become

$$\ddot{u}_1 + \Omega^2 u_1 + \varepsilon\left[-\sigma u_1 + 2\mu(\dot{u}_1 - \dot{u}_2) + \alpha u_1^3 + F u_1 \cos\Omega t\right] = 0 \tag{5.89}$$

$$\ddot{u}_2 + 2\gamma(\dot{u}_2 - \dot{u}_1) = 0 \tag{5.90}$$

Using Eqs. 5.1, 5.2, 5.86 and 5.87 in Eqs. 5.89 and 5.90 yields the following sets of equations:

$$D_0^2 u_{10} + \Omega^2 u_{10} = 0 \tag{5.91}$$

$$D_0^2 u_{20} + 2\gamma D_0 u_{20} = 2\gamma D_0 u_{10} \tag{5.92}$$

$$\begin{aligned} D_0^2 u_{11} + \Omega^2 u_{11} = & -2D_0D_1u_{10} + \sigma u_{10} - 2\mu(D_0u_{10} - D_0u_{20}) - \\ & -\alpha u_{10}^3 - 2Fu_{10}\cos\Omega T_0 \end{aligned} \tag{5.93}$$

$$D_0^2 u_{21} + 2\gamma D_0 u_{21} = 2\gamma(D_0u_{11} - D_1u_{20} + D_1u_{10}) - 2D_0D_1u_{20} \tag{5.94}$$

$$D_0^2u_{12} + \Omega^2u_{12} = -2D_0D_1u_{11} - \left(D_1^2 + 2D_0D_2\right)u_{10} + \sigma u_{11} - \\ -2\mu(D_1u_{11} - D_1u_{21}) - 3\alpha u_{10}^2u_{11} - 2Fu_{11}\cos\Omega t \tag{5.95}$$

$$D_0^2u_{22} + 2\gamma D_0u_{22} = -2D_0D_1u_{21} - \left(D_1^2 + 2D_0D_2\right)u_{20} + \\ +2\gamma(D_0u_{12} - D_1u_{21} + D_1u_{11} + D_2u_{10} - D_2u_{20}) \tag{5.96}$$

In this the second case we express the solutions of Eqs. 5.91 and 5.92 in the real form

$$u_{10} = A(T_1,T_2)\cos[\Omega T_0 + \varphi(T_1,T_2)] \tag{5.97}$$

$$u_{20} = \frac{4\gamma^2}{\Omega^2 + 4\gamma^2}A(T_1,T_2)\cos[\Omega T_0 + \varphi(T_1,T_2)] + \\ + \frac{2\gamma\Omega}{\Omega^2 + 4\gamma^2}A(T_1,T_2)\sin[\Omega T_0 + \varphi(T_1,T_2)] \tag{5.98}$$

Substituting Eqs. 5.97 and 5.98 into Eq. 5.93, expanding and arranging yields

$$D_0^2u_{11} + \Omega^2u_{11} = \\ = \left[2\Omega A(T_1,T_2)\frac{\partial\varphi(T_1,T_2)}{\partial T_1} + \sigma A(T_1,T_2) + \frac{4\mu\gamma\Omega^2}{\Omega^2 + 4\gamma^2}A(T_1,T_2) - \right. \\ \left. - \frac{3\alpha}{4}A^3(T_1,T_2)\right]\cos[\Omega T_0 + \varphi(T_1,T_2)] + \left[2\Omega\frac{\partial A(T_1,T_2)}{\partial T_1} + 2\mu\Omega A(T_1,T_2) - \right. \\ \left. - \frac{8\mu\gamma^2\Omega}{\Omega^2 + 4\gamma^2}A(T_1,T_2)\right]\sin[\Omega T_0 + \varphi(T_1,T_2)] - \frac{\alpha}{4}A^3(T_1,T_2)\cos[3\Omega T_0 + \\ + 3\varphi(T_1,T_2)] - FA(T_1,T_2)\cos[2\Omega T_0 + \varphi(T_1,T_2)] - FA(T_1,T_2)\cos[\varphi(T_1,T_2)] \tag{5.99}$$

For a uniformly valid expansion, the secular producing terms in Eq. 5.99 must be eliminated. Hence, the solvability conditions are

$$\frac{\partial\varphi(T_1,T_2)}{\partial T_1} + \frac{\sigma}{2\Omega} + \frac{2\mu\gamma\Omega}{\Omega^2 + 4\gamma^2} - \frac{3\alpha}{8\Omega}A^2(T_1,T_2) = 0 \tag{5.100}$$

$$\frac{\partial A(T_1,T_2)}{\partial T_1} + \frac{\mu\Omega^2}{\Omega^2 + 4\gamma^2}A(T_1,T_2) = 0 \tag{5.101}$$

The solution of Eq. 5.101 can be written in the form

$$A(T_1,T_2) = A(T_2)\exp\left(-\frac{\mu\Omega^2}{\Omega^2+4\gamma^2}T_1\right) \tag{5.102}$$

From Eqs. 5.100 and 5.102 we obtain

$$\begin{aligned}\varphi(T_1,T_2) = \varphi(T_2) - \frac{3\alpha(\Omega^2+4\gamma^2)}{16\mu\Omega^3}A^2(T_2)\exp\left(-\frac{2\mu\Omega^3}{\Omega^2+4\gamma^2}T_1\right) - \\ - \left(\frac{\sigma}{2\Omega}+\frac{2\mu\gamma\Omega}{\Omega^2+4\gamma^2}\right)T_1\end{aligned} \tag{5.103}$$

The solutions (5.97) and (5.98) by means of Eq. 5.102, become

$$u_{10} = A(T_2)\exp\left(-\frac{\mu\Omega^2}{\Omega^2+4\gamma^2}T_1\right)\cos[\Omega T_0+\varphi(T_1,T_2)] \tag{5.104}$$

$$\begin{aligned}u_{20} = \frac{4\gamma^2}{\Omega^2+4\gamma^2}A(T_2)\exp\left(-\frac{\mu\Omega^2}{\Omega^2+4\gamma^2}T_1\right)\cos[\Omega T_0+\varphi(T_1,T_2)] + \\ + \frac{2\gamma\Omega}{\Omega^2+4\gamma^2}A(T_2)\exp\left(-\frac{\mu\Omega^2}{\Omega^2+4\gamma^2}T_1\right)\sin[\Omega T_0+\varphi(T_1,T_2)]\end{aligned} \tag{5.105}$$

Substituting Eq. 5.102 into Eq. 5.99, we obtain the equation for u_{11}:

$$\begin{aligned}D_0^2u_{11}+\Omega^2u_{11} = -\frac{\alpha A^3(T_2)}{4}\exp\left(-\frac{3\mu\Omega^2}{\Omega^2+4\gamma^2}T_1\right)\cos[3\Omega T_0+3\varphi(T_1,T_2)] - \\ - FA(T_2)\exp\left(-\frac{\mu\Omega^2}{\Omega^2+4\gamma^2}T_1\right)\{\cos\varphi(T_1,T_2)+\cos[2\Omega T_0+\varphi(T_1,T_2)]\}\end{aligned} \tag{5.106}$$

The solution of Eq. 5.106 can be written in the form

$$\begin{aligned}u_{11} = \frac{\alpha A^3(T_2)}{32\Omega^2}\exp\left(-\frac{3\mu\Omega^2}{\Omega^2+4\gamma^2}T_1\right)\cos[3\Omega T_0+3\varphi(T_1,T_2)] + \\ + \frac{FA(T_2)}{3\Omega^2}\exp\left(-\frac{\mu\Omega^2}{\Omega^2+4\gamma^2}T_1\right)\{\cos[2\Omega T_0+\varphi(T_1,T_2)] - 3\cos[\varphi(T_1,T_2)]\}\end{aligned} \tag{5.107}$$

where the function $\phi(T_1,T_2)$ is given by Eq. 5.103.

Now, substituting Eqs. 5.97, 5.98 and 5.107 into Eq. 5.94 yields

$$
\begin{aligned}
D_0^2 u_{21} + 2\gamma D_0 u_{21} = & -\frac{3\alpha\gamma A^3(T_2)}{16\Omega}\exp\left(-\frac{3\mu\Omega^2}{\Omega^2+4\gamma^2}T_1\right)\sin[3\Omega T_0 + 3\varphi(T_1,T_2)]- \\
& -\frac{4\gamma F A(T_2)}{3\Omega}\exp\left(-\frac{\mu\Omega^2}{\Omega^2+4\gamma^2}T_1\right)\sin[2\Omega T_0+\varphi(T_1,T_2)]+ \\
& +\left\{\frac{4\Omega\gamma^2 A(T_2)}{(\Omega^2+4\gamma^2)^2}\exp\left(-\frac{\mu\Omega^2}{\Omega^2+4\gamma^2}T_1\right)\left[\frac{3\alpha A^2(T_2)}{8\Omega}\exp\left(-\frac{3\mu\Omega^2}{\Omega^2+4\gamma^2}T_1\right)-\frac{\sigma}{2\Omega}-\right.\right. \\
& \left.\left.-\frac{3\mu\gamma\Omega}{\Omega^2+4\gamma^2}\right]-\frac{2\mu\gamma\Omega^4 A(T_2)}{(\Omega^2+4\gamma^2)^2}\exp\left(-\frac{\mu\Omega^2}{\Omega^2+4\gamma^2}T_1\right)\right\}\cos[\Omega T_0+\varphi(T_1,T_2)]- \\
& -\left\{\frac{2\gamma\Omega^2 A(T_2)}{\Omega^2+4\gamma^2}\exp\left(-\frac{\mu\Omega^2}{\Omega^2+4\gamma^2}T_1\right)\left[\frac{3\alpha A^2(T_2)}{8\Omega}\exp\left(-\frac{3\mu\Omega^2}{\Omega^2+4\gamma^2}T_1\right)-\frac{\sigma}{2\Omega}-\right.\right. \\
& \left.\left.-\frac{2\mu\gamma\Omega}{\Omega^2+4\gamma^2}\right]+\frac{4\mu\gamma^2\Omega^3 A(T_2)}{(\Omega^2+4\gamma^2)^2}\exp\left(-\frac{\mu\Omega^2}{\Omega^2+4\gamma^2}T_1\right)\right\}\sin[\Omega T_0+\varphi(T_1,T_2)]
\end{aligned}
\tag{5.108}
$$

From Eq. 5.108 the solution has the form

$$
\begin{aligned}
u_{21}(T_0,T_1,T_2) = & -\frac{3\alpha\gamma A^3(T_2)}{16\Omega(9\Omega^2+4\gamma^2)}\exp\left(-\frac{3\mu\Omega^2}{\Omega^2+4\gamma^2}T_1\right)\sin[3\Omega T_0+ \\
& +3\varphi(T_1,T_2)] - \frac{\alpha\gamma^2 A^3(T_2)}{8\Omega^2(9\Omega^2+4\gamma^2)}\exp\left(-\frac{3\mu\Omega^2}{\Omega^2+4\gamma^2}T_1\right)\cos[3\Omega T_0+ \\
& +3\varphi(T_1,T_2)] + \frac{\gamma F A(T_2)}{3\Omega(\Omega^2+\gamma^2)}\exp\left(-\frac{\mu\Omega^2}{\Omega^2+4\gamma^2}T_1\right)\sin[2\Omega T_0+ \\
& +\varphi(T_1,T_2)] + \frac{\gamma^2 F A(T_2)}{3\Omega^2}\exp\left(-\frac{\mu\Omega^2}{\Omega^2+4\gamma^2}T_1\right)\cos[2\Omega T_0+ \\
& +\varphi(T_1,T_2)] + \frac{2\mu\gamma\Omega^3 A(T_2)}{\Omega^2+4\gamma^2}\exp\left(-\frac{\mu\Omega^2}{\Omega^2+4\gamma^2}T_1\right)\cos[\Omega T_0+\varphi(T_1,T_2)]+ \\
& +2\gamma\Omega A(T_2)\exp\left(-\frac{\mu\Omega^2}{\Omega^2+4\gamma^2}T_1\right)\left[\frac{3\alpha A^2(T_2)}{8\Omega}\exp\left(-\frac{3\mu\Omega^2}{\Omega^2+4\gamma^2}T_1\right)-\right. \\
& \left.-\frac{\sigma}{2\Omega}-\frac{2\mu\gamma\Omega}{\Omega^2+4\gamma^2}\right]\sin[\Omega T_0+\varphi(T_1,T_2)]
\end{aligned}
\tag{5.109}
$$

We remark that the terms of the form $K(T_2)T_0\sin[\Omega T_0+\varphi(T_1,T_2)]$and $K_2(T_2)T_0\cos[\Omega T_0+\varphi(T_1,T_2)]$not are resonant terms. Also, in Eqs. 5.95 and 5.96 do not exist resonant terms, such as we can write a second-order approximation to the solutions of Eqs. 5.84 and 5.85. In this case, we can consider $A(T_2)=A_0$ and $\varphi(T_2)=\varphi_0$, where A_0 and φ_0 are constants, because the terms in ε^2 are negligible.

In this way, we obtain the following expression for the second-order approximate solution for Eqs. 5.84 and 5.85, respectively.

$$\begin{aligned} u_1(t) = A_0 \exp\left(-\frac{\varepsilon\mu\Omega^2}{\Omega^2+4\gamma^2}t\right)\cos[\Omega t+\varphi(t)]+ \\ +\frac{\varepsilon\alpha A_0^3}{32\Omega^2}\exp\left(-\frac{3\varepsilon\mu\Omega^2}{\Omega^2+4\gamma^2}t\right)\cos[3\Omega t+3\varphi(t)]+ \\ +\frac{\varepsilon F A_0}{32\Omega^2}\exp\left(-\frac{\varepsilon\mu\Omega^2}{\Omega^2+4\gamma^2}t\right)\{\cos\varphi(t)+\cos[2\Omega t+\varphi(t)]\} \end{aligned} \quad (5.110)$$

$$\begin{aligned} u_2(t) = \frac{4\gamma^2 A_0}{\Omega^2+4\gamma^2}\exp\left(-\frac{\varepsilon\mu\Omega^2}{\Omega^2+4\gamma^2}t\right)\cos[\Omega t+\varphi(t)]+ \\ +\frac{2\gamma\Omega A_0}{\Omega^2+4\gamma^2}\exp\left(-\frac{\varepsilon\mu\Omega^2}{\Omega^2+4\gamma^2}t\right)\sin[\Omega t+\varphi(t)]- \\ -\frac{3\alpha\gamma A_0^3}{16\Omega(9\Omega^2+4\gamma^2)}\exp\left(-\frac{3\varepsilon\mu\Omega^2}{\Omega^2+4\gamma^2}t\right)\sin[3\Omega t+3\varphi(t)]- \\ -\frac{\alpha\gamma^2 A_0^3}{8\Omega^2(9\Omega^2+4\gamma^2)}\exp\left(-\frac{3\varepsilon\mu\Omega^2}{\Omega^2+4\gamma^2}t\right)\cos[3\Omega t+3\varphi(t)]+ \\ +\frac{\gamma F A_0}{3\Omega(\Omega^2+\gamma^2)}\exp\left(-\frac{\varepsilon\mu\Omega^2}{\Omega^2+4\gamma^2}t\right)\sin[2\Omega t+\varphi(t)]+ \\ +\frac{\gamma^2 F A_0}{3\Omega^2}\exp\left(-\frac{\varepsilon\mu\Omega^2}{\Omega^2+4\gamma^2}t\right)\cos[2\Omega t+\varphi(t)]+ \\ \frac{2\mu\gamma\Omega^3 A_0}{\Omega^2+4\gamma^2}\exp\left(-\frac{\varepsilon\mu\Omega^2}{\Omega^2+4\gamma^2}t\right)\cos[\Omega t+\varphi(t)]+ \\ +2\gamma\Omega A_0\exp\left(\frac{-\varepsilon\mu\Omega^2}{\Omega^2+4\gamma^2}t\right)\left[\frac{3\alpha A_0^2}{8\Omega}\exp\left(\frac{-3\varepsilon\mu\Omega^2}{\Omega^2+4\gamma^2}t\right)-\frac{\sigma}{2\Omega}-\frac{2\mu\gamma\Omega}{\Omega^2+4\gamma^2}\right] \end{aligned} \quad (5.111)$$

where

$$\begin{aligned} \varphi(t) = \varphi_0 - \frac{3\alpha(\Omega^2+4\gamma^2)}{16\mu\Omega^3}A_0^2\exp\left(-\frac{2\varepsilon\mu\Omega^2}{\Omega^2+4\gamma^2}t\right)- \\ -\left(\frac{\varepsilon\sigma}{2\Omega}+\frac{2\varepsilon\mu\gamma\Omega}{\Omega^2+4\gamma^2}\right)t \end{aligned} \quad (5.112)$$

The constants A_0 and φ_0 can be determined from the initial conditions for the first variable u_1 because in practical cases we are interested on the main mass (having the displacement u_1).

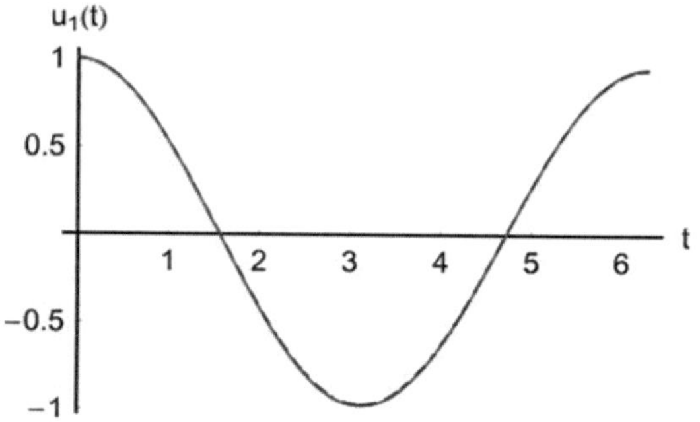

Fig. 5.3 Comparison between numerical solution and analytical solution (5.110) of Eq. 5.89 in the case $A_0 = \mu = 1$, $\gamma = \alpha = 0.1$, $\Omega = 1.002$, $\sigma = 0.2$, $\varepsilon = 0.01$: _______ numerical results, _ _ _ _ analytical results

If we are interested on the behaviour of the small mass, then from Eq. 5.92 it must be taken into account the initial conditions which refer to the variable u_2 [74, 75].

For Eq. 5.89, in the particular case when $A_0 = 1$, $\gamma = 0.1$, $\mu = 1$, $\alpha = 0.1$, $\Omega = 1.002$, $\sigma = 0.2$, $\varepsilon = 0.01$, $F = 0.2$, the comparison between the numerical results and approximate analytical results given by (5.110) is shown in Fig. 5.3.

From Fig. 5.3 one can observe that the analytical results perfectly match the numerical ones.

Chapter 6
The Optimal Homotopy Asymptotic Method

The homotopy continuation method was known as early as in the 1930s. This method was used by kinematicians in the 1960s in the US for solving mechanism synthesis problems [76]. The latest development was done by A.P. Morgan at General Motors [77]. We also have two important literature studies by Garcia [78] and Allgowar [79]. The continuation method gives a set of certain answers and some iteration processes to obtain our solutions more exactly.

Consider the following nonlinear algebraic equation

$$f(x) = 0 \tag{6.1}$$

for a given set of equations in n variables $x_1, x_2, \ldots, x_n$. We modify the Eq. 6.1 by omitting some of the terms and adding new ones until we have a new system of equations, the solutions to which may be easily guessed (known). We then deform the coefficients of the new system into the coefficients of the original system by a series of small increments to obtain our solutions. This is called homotopy continuation technique.

If we wish to find the solution in Eq. 6.1, we choose a new starting system or the auxiliary homotopy function

$$g(x) = 0 \tag{6.2}$$

The auxiliary homotopy function $g(x)$ must be known or controllable and easy to solve. Then, we define the homotopy continuation function as

$$H(x,t) = tf(x) + (1-t)g(x) = 0 \tag{6.3}$$

where t is an arbitrary parameter which varies from 0 to 1, i.e. $t \in [0,1]$. Therefore we have the following two boundary conditions

$$H(x,0) = g(x) \tag{6.4}$$

V. Marinca and N. Herisanu, *Nonlinear Dynamical Systems in Engineering*,
DOI 10.1007/978-3-642-22735-6_6,

$$H(x,1) = f(x) \tag{6.5}$$

This is the famous homotopy continuation method. It is also called the Bootstrap method or parameter-perturbation method, but these names did not become popular.

The perturbation procedure requires the existence of a small parameter which is not always the case. To eliminate these disadvantages, S.J. Liao [80–82], adopted the homotopy technique which is widely applied in differential topology for obtaining the approximate analytic solution of the second-order strong nonlinear differential equation. The embedded artificial parameter p $\in$ [0,1] is introduced and the homotopy transformation of the differential equation is done.

For a given nonlinear differential equation

$$N(u(x)) = 0, \quad x \in D \tag{6.6}$$

where N is a nonlinear operator and $u(x)$ is a unknown function, Liao constructed a one parameter family of equations in the embedding parameter p, called the zeroth-order deformation equation

$$(1-p)L[U(x;p) - u_0(x)] + pN[U(x,p)] = 0, \quad x \in D, \; p \in [0,1] \tag{6.7}$$

where L is an auxiliary linear operator and $u_0(x)$ is an initial guess. The homotopy provides us with larger freedom than the traditional non-perturbation method to choose both the auxiliary linear operator L and the initial guess. At $p = 0$ and $p = 1$, we have $U(x,0) = u_0(x)$and $U(x,1) = u(x)$, respectively. So, if the Taylor series

$$U(x,p) = \sum_{k=0}^{\infty} u_k(x) p^k \tag{6.8}$$

converge at $p = 1$, then we obtain the so-called homotopy-series solution

$$u(x) = \sum_{k=0}^{\infty} u_k(x) \tag{6.9}$$

which must satisfy the original Eq. 6.6. Here $u_k(x)$ is governed by a linear differential equation related to the auxiliary linear operator L and therefore is easy to solve. But this early homotopy-analysis method cannot always guarantee the convergence approximation series. To overcome this restriction, Liao introduced a non-zero auxiliary parameter $\hbar$ to construct a two-parameter family of equations, i.e. zeroth-order deformation equation

$$(1-p)[U(x;p) - u_0(x)] = \hbar p N[U(x;p)] \quad x \in D, \; p \in [0,1] \tag{6.10}$$

In this way, the homotopy-series solution Eq. 6.9 is dependent upon x and the auxiliary parameter $\hbar$. This auxiliary parameter $\hbar$ can adjust and control the convergence region and rate of homotopy-series solutions. Later, Liao introduced the zeroth-order deformation equation in a more general form

$$[1 - B(p))]L[U(x;p) - u_0(x)] = \hbar A(p)H(x)N[U(x;p)], \quad x \in D, \ p \in [0, 1] \tag{6.11}$$

where $A(p)$ and $B(p)$ are called deformation functions satisfying

$$A(0) = B(0) = 0, \quad A(1) = B(1) = 1 \tag{6.12}$$

and $H(x)$ is an auxiliary function.

The so-called rule of solution expression, rule of coefficient-ergodicity and rule of solution existence play important roles and greatly simplify the application of the homotopy analysis method. The convergence of the series Eq. 6.8 depends upon the auxiliary parameter $\hbar$, the auxiliary function H, the initial guess $u_0(x)$ and the auxiliary linear operator L. The combination of the convergence theorem and the freedom of the choice of the auxiliary parameter $\hbar$, the auxiliary function $H(x)$, the initial guess $u_0(x)$ and the auxiliary operator L establishes the cornerstone of the validity and flexibility of the homotopy analysis method.

For the zeroth-order deformation Eq. 6.11, Liao suggested to choose a proper value of $\hbar$ by plotting the so-called $\hbar$-curves, but the homotopy analysis method is a method for the time of computers: without high-performance computers and symbolic computation software such as Mathematica, Maple and so on, it is impossible to solve high-order deformation equations quickly so as to get approximations at high enough order. Without computer and symbolic computation software, it is also impossible to choose a proper value of the parameter $\hbar$. It is true that expressions given by the homotopy analysis method are often lengthy and thus can be hardly expressed on only one page. Note that one needs much more time to calculate a traditional "analytic" expression in a length of half page by means of a traditional computational tool such as slide rule [82]. This is an original concept of "analytic solution" unlike the traditional concept of analytic solution, which has been formed 100 years ago.

In a whole different manner, we proposed in [83–89] another similar technique called the optimal homotopy asymptotic method (OHAM). Instead of an infinite series, we need only a few terms, mostly two or three terms. Our method is applied successfully to obtain the solutions of currently important problems in dynamical systems and we have also shown its effectiveness, generalization and reliability.

The OHAM is not only useful for nonlinear differential equations but also it is useful for linear and nonlinear partial differential equations [90–100]. It is a powerful method for solving nonlinear problems without depending on small parameters, which shows its validity and potential for the solution of nonlinear problems in science and engineering applications.

6.1 Basic Idea of OHAM

We apply OHAM to the following differential equation

$$L(u(x)) + g(x) + N(u(x)) = 0, \quad B\left(u, \frac{du}{dx}\right) = 0, \quad x \in D \tag{6.13}$$

where L is a linear operator, x denotes independent variable, $u(x)$ is an unknown function, $g(x)$ is a known function, $N(u(x))$ is a nonlinear operator and B is a boundary operator.

By means of OHAM one first constructs a family of equation:

$$\begin{aligned} &(1-p)[L(\phi(x;p)) + g(x)] = H(x,p)[L(\phi(x;p)) + g(x) + N(\phi(x;p))] \\ &B\left(\phi(x,p), \frac{\partial \phi(x,p)}{\partial x}\right) = 0 \end{aligned} \tag{6.14}$$

where $p \in [0,1]$ is an embedding parameter, $H(x,p)$ is a nonzero auxiliary function for $p \neq 0$ and $H(x,0) = 0$, $\phi(x,p)$ is an unknown function, respectively. Obviously, when $p = 0$ and $p = 1$ it holds

$$\phi(x,0) = u_0(x), \quad \phi(x,1) = u(x) \tag{6.15}$$

respectively. Thus, as p increases from 0 to 1, the solution $\phi(x,p)$ varies from $u_0(x)$ to the solution $u(x)$, where $u_0(x)$ is obtained from Eq. 6.14 for $p = 0$:

$$L(u_0(x)) + g(x)) = 0, \quad B\left(u_0, \frac{du_0}{dx}\right) = 0 \tag{6.16}$$

We choose the auxiliary function $H(x,p)$ in the form

$$H(x,p) = ph_1(x, C_i) + p^2 h_2(x, C_j) + \ldots \tag{6.17}$$

where $h_k(x, C_n)$, $k = 1,2,\ldots$ are the auxiliary functions depending on x and on the constants C_n, $n = 1,2,\ldots$, which will be defined later.

Let us consider the solution of Eq. 6.14 in the form

$$\phi(x,p,C_k) = u_0(x) + \sum_{i \geq 1} u_i(x, C_k) p^i \quad k = 1, 2, \ldots \tag{6.18}$$

Now, substituting Eq. 6.18 into Eq. 6.14 and equating the coefficients of like powers of p, we obtain the governing equation of $u_0(x)$ given by Eq. 6.16, and the governing equation of $u_k(x)$, i.e.

$$L(u_1(x)) = h_1(x, C_i)N_0(u_0(x)), \quad B\left(u_1, \frac{du_1}{dx}\right) = 0 \tag{6.19}$$

$$L(u_k(x) - u_{k-1}(x)) = h_k N_0(u_0(x)) + \sum_{i=1}^{k-1} h_i[L(u_{k-i}(x)) +$$
$$+ N_{k-i}(u_0(x), u_1(x), ..., u_{k-i}(x))], \quad k = 2, 3, ... \quad B\left(u_k, \frac{du_k}{dt}\right) = 0 \tag{6.20}$$

where $N_m(u_0(x), u_1(x), ..., u_m(x))$is the coefficient of p^{m}, obtained expanding $N(\phi(x, p, C_i))$in series with respect to the embedding parameter p:

$$N(\phi(x, p, C_i)) = N_0(u_0(x)) + \sum_{m \geq 1} N_m(u_0, u_1, ..., u_m)p^m, \quad i = 1, 2, ... \tag{6.21}$$

where $\phi(x, p, C_i)$is given by Eq. 6.18.

It should be emphasized that u_{k} for $k \geq 0$ are governed by the linear Eqs. 6.16, 6.19 and 6.20 with the linear boundary conditions that come from the original problem, which can be easily solved. The convergence of the series Eq. 6.18 depends upon the auxiliary functions $h_i(x, C_n)$. There are many possibilities to choose the functions $h_i(x, C_n)$. The convergence of the solutions u_{k} and consequently the convergence of the solution $u(x)$ depend on the auxiliary function $h_i(x, C_n)$. Basically, the shape of $h_i(x, C_n)$must follow the terms appearing in Eq. 6.20. Therefore, we try to choose $h_i(x, C_n)$so that in Eq. 6.20, the product

$$h_i[L(u_{k-i}(x)) + N_{k-i}(u_0(x), u_1(x), ..., u_{k-i}(x))]$$

be of the same shape with other terms which appear into Eq. 6.20. For example $h_i(x, C_n)$could be chosen as exponential function, trigonometric function, polynomial function and so on, depending on the shape of the terms already present in the iteration given by Eq. 6.20.

If the series Eq. 6.18 is convergent at $p = 1$, one has

$$u(x, C_i) = u_0(x) + \sum_{k \geq 1} u_k(x, C_i), \quad i = 1, 2, ..., s \tag{6.22}$$

Generally speaking, the solution of Eq. 6.13 can be determined approximately in the form

$$\bar{u}(x, C_i) = u_0(x) + \sum_{k=1}^{m} u_k(x, C_i), \quad i = 1, 2, ...s \tag{6.23}$$

The Eq. 6.23 is named approximate solution of order m.

Now, substituting Eq. 6.23 into Eq. 6.13 it results the following residual

$$R(x, C_i) = L(\bar{u}(x, C_i)) + g(x) + N(\bar{u}(x, C_i)), \quad i = 1, 2, ...s \tag{6.24}$$

If $R(x, C_i) = 0$ then $\bar{u}(x, C_i)$ happens to be the exact solution. Generally such case will not arise for nonlinear problems. At this moment, the m th-order approximation given by Eq. 6.23 depends on the constants C_i, $i = 1,2,\ldots,s$ and these constants can be optimally identified via various methods, such as the least square method, the Galerkin method, the collocation method and so on. The first option always should be minimizing the square residual error:

$$J(C_1, C_2, ..., C_s) = \int_a^b R^2(x, C_1, C_2, ..., C_s)dx \tag{6.25}$$

where a and b are two values depending on the given problem. The unknown constants $C_1, C_2, \ldots, C_s$ can be identified from the conditions

$$\frac{\partial J}{\partial C_1} = \frac{\partial J}{\partial C_2} = ... = \frac{\partial J}{\partial C_s} = 0 \tag{6.26}$$

With these constants known (namely convergence-control constants), the approximate solution of order m (Eq. 6.23) is well-determined. Another convenient possibility to determine the optimal values of the constants C_i is to solve the system:

$$R(x_1, C_i) = R(x_2, C_i) = ... = R(x_s, C_i) = 0, \quad i = 1, 2, ..., s \tag{6.27}$$

where the residual R is given by Eq. 6.24.

Our procedure contains the auxiliary functions $h_i(x, C_n)$, which provides us with a simple way to adjust and control convergence of the solution. It is very important to properly choose the functions $h_i(x, C_n)$which appear in the m th-order approximation Eq. 6.23.

In particular, we consider a nonlinear differential equation of the form

$$\ddot{u}(t) + \omega^2 u(t) = f(u(t), \dot{u}(t), \ddot{u}(t)) \tag{6.28}$$

where the dot denotes derivative with respect to time, ω is a constant and f is in general a nonlinear term. The initial conditions are

$$u(0) = A, \quad \dot{u}(0) = 0 \tag{6.29}$$

where A is the amplitude of the oscillations. Note that it is unnecessary to assume the existence of any small or large parameter in Eq. 6.13 or Eq. 6.28.

Equation 6.28 describes a system oscillating with an unknown period T. We switch to a scalar time $\tau = \frac{2\pi t}{T} = \Omega t$. Under the transformations

$$\tau = \Omega t, \quad u(t) = Ax(\tau) \tag{6.30}$$

the original Eq. 6.28 becomes

$$\Omega^2 x''(\tau) + \omega^2 x(\tau) = \frac{1}{A} f(Ax(\tau), A\Omega x'(\tau), A\Omega^2 x''(\tau)) \tag{6.31}$$

and the initial conditions become

$$x(0) = 1, \quad x'(0) = 0 \tag{6.32}$$

where the prime denotes the derivative with respect to τ.

The family of Eqs. 6.14 can be written as

$$(1-p)L(\phi(\tau,p)) = H(\tau,p)[N(\phi(\tau,p),\Omega(\lambda,p))] \tag{6.33}$$

where L is a linear operator

$$L(\phi(\tau,p)) = \Omega_0^2 \left[\frac{\partial^2 \phi(\tau,p)}{\partial \tau^2} + \phi(\tau,p) \right] \tag{6.34}$$

while N is a nonlinear operator

$$\begin{aligned} N(\phi(\tau,p),\Omega(\lambda,p)) &= \Omega^2(\lambda,p)\frac{\partial^2 \phi(\tau,p)}{\partial \tau^2} + (\omega^2 + \lambda)\phi(\tau,p) \\ &- \frac{1}{A} f(A\phi(\tau,p), A\Omega(\lambda,p)\frac{\partial \phi(\tau,p)}{\partial \tau}, A\Omega^2(\lambda,p)\frac{\partial^2 \phi(\tau,p)}{\partial \tau^2}) - p\lambda\phi(\tau,p) \end{aligned} \tag{6.35}$$

where λ is an arbitrary unknown parameter and Ω_0 will be given later. From Eqs. 6.29 and 6.30 we obtain the initial conditions

$$\phi(0,p) = 1, \quad \left. \frac{\partial \phi(\tau,p)}{\partial \tau} \right|_{\tau=0} = 0 \tag{6.36}$$

In this case, when $p = 0$ and $p = 1$ it holds

$$\phi(\tau,0) = x_0(\tau), \quad \phi(\tau,1) = x(\tau), \quad \Omega(0) = \Omega_0, \Omega(1) = \Omega \tag{6.37}$$

where $x_0(\tau)$ is an initial approximation of $x(\tau)$. Therefore, as the embedding parameter p increases from 0 to 1, $\phi(\tau,p)$ varies from the initial approximation $x_0(\tau)$ to the solution $x(\tau)$, so does $\Omega(p)$ from the initial approximation Ω_0 to the exact frequency Ω.

Expanding $\phi(\tau, p)$and $\Omega(p)$in series with respect to the parameter p, one has respectively

$$\phi(\tau, p) = x_0(\tau) + px_1(\tau) + p^2x_2(\tau) + ... \tag{6.38}$$

$$\Omega(p, \lambda) = \Omega_0 + p\Omega_1 + p^2\Omega_2 + ... \tag{6.39}$$

If the initial approximation $x_0(\tau)$ and the auxiliary function $H(\tau, p)$are properly chosen so that the above series converges at $p = 1$, one has

$$x(\tau) = x_0(\tau) + x_1(\tau) + x_2(\tau) + ... \tag{6.40}$$

$$\Omega = \Omega_0 + \Omega_1 + \Omega_2 + ... \tag{6.41}$$

The series Eqs. 6.40 and 6.41 contain the auxiliary function $H(\tau, p)$. The results of the m th-order approximations are given by

$$\bar{x}(\tau) \approx x_0(\tau) + x_1(\tau) + ... + x_m(\tau) \tag{6.42}$$

$$\bar{\Omega} = \Omega_0 + \Omega_1 + ... + \Omega_{m-1} \tag{6.43}$$

If we substitute Eqs. 6.42 and 6.43 into Eq. 6.33 and if we equate to zero the coefficients of various powers of p we obtain the following linear equations

$$L(x_0) = 0, \quad x_0(0) = 1, \quad x_0'(0) = 0 \tag{6.44}$$

$$\begin{gathered} L(x_i) - L(x_{i-1}) = \sum_{j=1}^{i} h_j N_{i-j}(x_0, x_1, ..., x_{i-j}, \Omega_0, \Omega_1, ...\Omega_{i-j}, a, \lambda); \\ x_i(0) = x'_i(0) = 0, \quad i = 1, 2, ...m - 1 \end{gathered} \tag{6.45}$$

$$L(x_m) - L(x_{m-1}) = \sum_{j=1}^{m-1} h_j N_{m-1-j} + h_m N_0, \quad x_m(0) = x_m'(0) = 0 \tag{6.46}$$

where N_k are obtained from the equation

$$\begin{aligned} N(\phi, \Omega) = {} & N_0(x_0, \Omega_0, A, \lambda) + pN_1(x_0, x_1, \Omega_0, \Omega_1, A, \lambda) \\ & + p^2N_2(x_0, x_1, x_2, \Omega_0, \Omega_1, \Omega_2, A, \lambda) + ... \end{aligned} \tag{6.47}$$

Note that Ω_k can be determined avoiding the presence of secular terms in Eqs. 6.45 and 6.46.

The frequency Ω depends upon the arbitrary parameter λ and we apply the so-called "principle of minimal sensitivity" [101] in order to fix the value of λ. We do this imposing that

$$\frac{d\Omega}{d\lambda} = 0 \tag{6.48}$$

This principle of the minimal sensitivity appears for the first time in the quantum field theory, $\lambda\phi^4$ theory or quantum chromodynamics [101]. In its original formulation a Lagrangian density $\mathcal{L}$ which is not exactly solvable is interpolated with a solvable Lagrangian $\mathcal{L}_O(\lambda)$ depending upon one or more parameters λ:$\mathcal{L}\delta = \mathcal{L}_O(\lambda) + \delta(\mathcal{L}\text{-}\mathcal{L}_O(\lambda))$, δ being a parameter. We notice that the interpolation of the full Lagrangian with the solvable one, $\mathcal{L}_O(\lambda)$, brings an artificial dependence upon the arbitrary parameter λ. Such dependence, which would vanish if all perturbative orders were calculated, can be made weaker to a finite perturbative order, by requiring some physical observable P to be locally insensitive to λ, i.e. $\partial P/\partial\lambda = 0$.

In the above application this physical observable P is in fact the frequency Ω. This condition is known as the principle of minimal sensitivity and is normally seen to improve the convergence to the exact solution.

Unlike homotopy analysis method, in the proposed procedure (OHAM) the construction of the homotopy is quite different. In the frame of OHAM the linear operator L is well defined by Eqs. 6.16 or 6.34 and the initial approximation is rigorously determined from Eq. 6.44, while in the homotopy analysis method these ones are arbitrarily chosen. Instead of an infinite series (as is the case of the homotopy analysis method) the OHAM searches for only a few terms (mostly two or three terms). The way to ensure the convergence in OHAM is quite different and more rigorous. Unlike other homotopy procedures, OHAM ensure a very rapid convergence since it needs only two or three iterations for achieving a very accurate solution. This is in fact the true power of the method. OHAM does not need a recurrence formula as other homotopy procedures (such as homotopy analysis method) do. OHAM is an iterative procedure which often converges to the exact solution after only two or three iterations. Iterations are performed in a very simple manner by identifying some coefficients. OHAM does not need high-order approximations. OHAM does not use the rules established in the frame of homotopy analysis method, it is a self-sustained method which has no "open questions" as is the case of other homotopy procedures. OHAM does not need the restrictive conditions $A(1) = 1$, $B(1) = 1$ as the homotopy analysis does. The homotopy analysis method is a special case of OHAM when $H(\tau,p) = p\hbar$ where the parameter $\hbar$is chosen from the so-called "$\hbar$-curves". An important feature of the OHAM is that using Eqs. 6.26, 6.27 or 6.48 a minimization of errors is obtained. Finally, OHAM provides an analytic solution for complicated nonlinear problems expressed on only two rows, unlike other homotopy procedures which need few pages to express an analytic solution.

In the following, several examples are given to demonstrate the general validity and the great potential of the OHAM

6.2 Duffing Oscillator

We consider the well-known Duffing oscillator

$$\ddot{u} + \omega^2 u = f(u) \tag{6.49}$$

where $f(u) = -au^3$(from Eq. 6.28) and ω and a are real parameters. The initial conditions are

$$u(0) = A, \quad \dot{u}(0) = 0 \tag{6.50}$$

Under the transformations Eq. 6.30

$$\tau = \Omega t, \quad u(t) = Ax(\tau) \tag{6.51}$$

the original Eq. 6.49 becomes

$$\Omega^2 x''(\tau) + \omega^2 x(\tau) + aA^2 x^3(\tau) = 0 \tag{6.52}$$

where prime denotes the derivative with respect to τ.

We construct Eq. 6.33 in the form

$$(1-p)L(\phi(\tau,p)) = H(p,\tau)\left[\Omega^2\phi''(\tau,p) + \omega^2\phi(\tau,p) + aA^2\phi^3(\tau,p)\right][0,1] \tag{6.53}$$

where

$$\phi(\tau,p) = x_0(\tau) + px_1(\tau) + p^2 x_2(\tau) + \ldots \tag{6.54}$$

$$\Omega^2(p) = \omega_0^2 + \omega_1 p + \omega_2 p^2 + \ldots\ldots \tag{6.55}$$

while L is the following linear operator:

$$L\phi(\tau,p) = \omega_0^2\left[\frac{\partial^2\phi(\tau,p)}{\partial\tau^2} + \phi(\tau,p)\right] \tag{6.56}$$

The Eq. 6.44 becomes:

$$\omega_0^2(x_0'' + x_0) = 0,\ x_0(0) = 1,\ \text{x'}(0) = 0 \tag{6.57}$$

Equation 6.57 has the solution

$$x_0(\tau) = \cos\tau \tag{6.58}$$

From Eqs. 6.54, 6.55 and 6.53 and expanding the nonlinear operator Eq. 6.35 for $\lambda = 0$, because we consider only one iteration, we obtain:

$$N(\phi,\Omega) = N_0(x_0,\Omega_0,A) + pN_1(x_0,x_1,\Omega_0,\omega_1,A) + \ldots \tag{6.59}$$

where

$$N_0(x_0,\Omega_0,A) = \omega_0^2 x_0'' + \omega^2 x_0 + aA^2 x_0^3 \tag{6.60}$$

$$N_1(x_0,x_1,\Omega_0,\omega_1,A) = \omega_1 x_0'' + \omega_0^2 x_1'' + \omega^2 x_1 + 3aA^2 x_0 x_1 \tag{6.61}$$

If we choose

$$h_1(\tau,p) = C_1 + 2C_2\cos 2\tau + 2C_3\cos 4\tau \tag{6.62}$$

and substituting Eqs. 6.58, 6.50 and 6.62 into Eq. 6.45, for $i = 1$, we obtain:

$$\begin{aligned} \omega_0^2(x_1'' + x_1) &= (C_1 + 2C_2\cos 2\tau + \\ &+ 2C_3\cos 4\tau)\left[(\omega^2 - \omega_0^2 + \frac{3}{4}aA^2)\cos\tau + \frac{aA^2}{4}\cos 3\tau\right] \end{aligned} \tag{6.63}$$

We note that in the right-side of Eq. 6.63 only odd-order harmonics appear in the square brackets. By using simple manipulations, Eq. 6.63 becomes

$$\begin{aligned} \omega_0^2(x_1'' + x_1) = &\left[(C_1 + C_2)(\omega^2 - \omega_0^2 + \frac{3}{4}aA^2) + \frac{aA^2(C_2 + C_3)}{4}\right]\cos\tau + \\ &+ \left[(C_2 + C_3)(\omega^2 - \omega_0^2 + \frac{3}{4}aA^2) + \frac{C_1 aA^2}{4}\right]\cos 3\tau + \\ &+ \left[C_3(\omega^2 - \omega_0^2 + \frac{3}{4}aA^2) + \frac{C_2 aA^2}{4}\right]\cos 5\tau + \\ &+ \frac{C_3 aA^2}{4}\cos 7\tau,\ x_1(0) = x'_1(0) = 0 \end{aligned} \tag{6.64}$$

Avoiding the presence of a secular term needs:

$$\omega_0^2 = \omega^2 + \frac{3C_1 + 4C_2 + C_3}{4(C_1 + C_2)}aA^2 \tag{6.65}$$

With this requirement, the solution of Eq. 6.64 yields

$$\begin{aligned} x_1(\tau) = &\frac{aA^2(C_1^2 + C_1C_2 - C_2^2 - 2C_2C_3 - C_3^2)}{32\omega_0^2(C_1 + C_2)}(\cos\tau - \cos 3\tau) + \\ &+ \frac{aA^2(C_1C_2 + C_2^2 - C_2C_3 - C_3^2)}{96\omega_0^2(C_1 + C_2)}(\cos\tau - \cos 5\tau) + \frac{aA^2C_3}{192\omega_0^2}(\cos\tau - \cos 7\tau) \end{aligned} \tag{6.66}$$

For $m = 1$ into Eq. 6.42, we obtain the first-order approximate solution:

$$\bar{x}(\tau) = x_0(\tau) + x_1(\tau) \tag{6.67}$$

where x_0 and x_1 are given by Eqs. 6.58 and 6.66 respectively. Using the transformation Eqs. 6.51 and 6.67, the first-order approximate solution of Eq. 6.49 becomes:

$$\bar{u}(t) = \alpha\cos\Omega t + \beta\cos 3\Omega t + \gamma\cos 5\Omega t + \delta\cos 7\Omega t \tag{6.68}$$

where Ω is obtained from Eq. 6.55 and 6.65

$$\Omega^2 = \omega_0^2 = \omega^2 + \frac{(3C_1 + 4C_2 + C_3)}{4(C_1 + C_2)}aA^2 \tag{6.69}$$

and

$$\begin{aligned}
\alpha &= A + \frac{aA^3(C_1^2 + 4C_1C_2 + 6C_1C_3 - 11C_2C_3 - 4C_3^2 + 2C_2^2}{32\Omega^2(C_1 + C_2)} \\
\beta &= \frac{aA^3(C_2^2 + 2C_2C_3 + C_3^2 - C_1^2 - C_1C_2)}{32\Omega^2(C_1 + C_2)} \\
\gamma &= \frac{aA^3(C_2C_3 + C_3^2 - C_1C_2 - C_2^2)}{96\Omega^2(C_1 + C_2)}, \quad \delta = -\frac{aA^3C_3}{192\Omega^2}
\end{aligned} \tag{6.70}$$

The substitution of Eq. 6.68 into Eq. 6.49 leads to the residual of the form

$$R(t, C_1, C_2, C_3) = \ddot{\bar{u}} + \omega^2\bar{u} + a\bar{u}^3 \tag{6.71}$$

The constants C_1, C_2, C_3 can be determined by means of Eqs. 6.26.

6.2.1 Numerical Examples

In order to illustrate the remarkable accuracy of this method, we compare the approximate frequency given by Eq. 6.69 with the exact frequency known in [40]

$$\Omega_{ex} = \frac{\pi\sqrt{1 + aA^2}}{2}\left(\int_0^{\frac{\pi}{2}} \frac{d\theta}{\sqrt{1 - m^2\sin^2\theta}}\right)^{-1}, \quad m = \frac{aA^2}{2(1 + aA^2)} \tag{6.72}$$

For $A = 1$ and $\omega = 1$, and different values of a, we consider the following three cases:

Case 6.2.1. (a) $a = 1$. In this case from Eqs. 6.26 we obtain: $C_1 = -1.08262$; $C_2 = -0.417895; C_3 = 0.498748$

Case 6.2.1. (b) $a = 2$. The constants are: $C_1 = -1.05329$; $C_2 = 0.127917$; $C_3 = -0.0450043$

Case 6.2.1. (c) $a = 10$. It is obtained: $C_1 = -0.865779; C_2 = 0.167738$; $C_3 = -0.0794433$

For comparison, the exact frequency obtained by integrating Eq. 6.72 and the approximate frequency computed by Eq. 6.69 are listed in Table 6.1

Table 6.1 indicates that the formula Eq. 6.69 can give excellent approximate frequencies for different values of the parameter a, when $A = \omega = 1$.

Figures 6.1–6.3 show the comparison between the present solutions obtained using our procedure and the numerical integration results obtained by using a fourth-order Runge–Kutta method.

Table 6.1 Comparison of approximate frequencies with the corresponding exact frequency of the Duffing oscillator

a	Ω_{ex} Eq. 6.72	Ω Eq. 6.69
1	1.31778	1.317774307
2	1.56911	1.56907907
10	2.86664	2.866430119

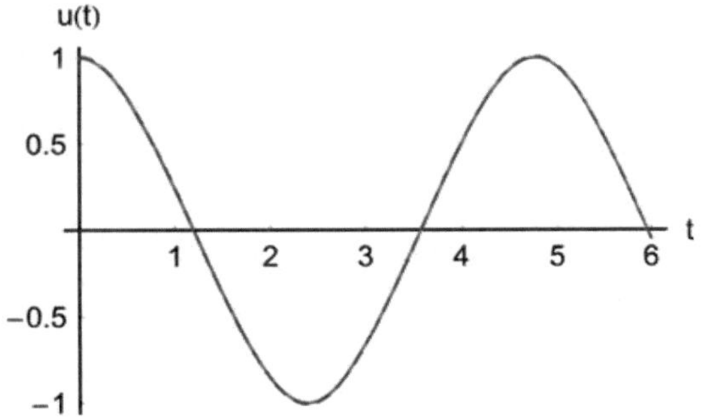

Fig. 6.1 Comparison between the solutions of Eq. 6.49 for $\omega = A = a = 1$: _____ numerical solution; _ _ _ approximate solution given by Eq. 6.68

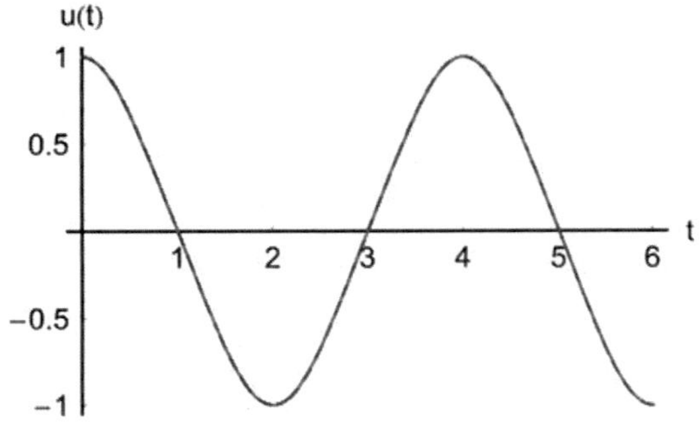

Fig. 6.2 Comparison between the solutions of Eq. 6.49 for $\omega = A = 1$, $a = 2$: _____ numerical solution; _ _ _ approximate solution given by Eq. 6.68

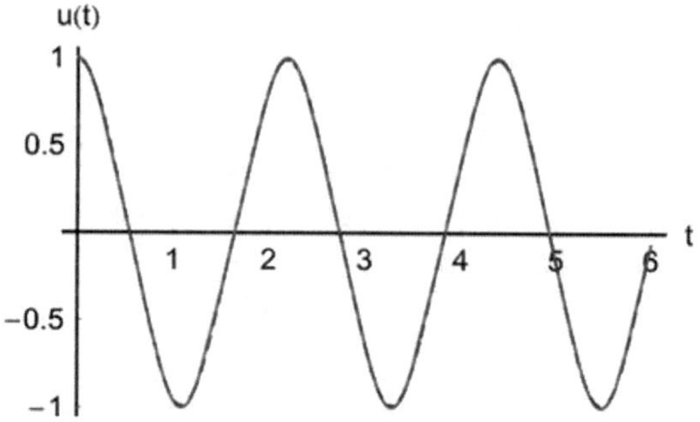

Fig. 6.3 Comparison between the solutions of Eq. 6.49 for $\omega = A = 1$, $a = 10$: _____ numerical solution; _ _ _ approximate solution given by Eq. 6.68

In summary, it can be seen that the solutions obtained by OHAM are nearly identical with that given by the numerical method. We find that the parameter a does not need to be small.

6.3 Thin Film Flow of a Fourth Grade Fluid Down a Vertical Cylinder

Consider a fourth grade fluid falling on the outside of an infinitely long vertical cylinder of a radius R. The flow is considered in thin, uniform, axisymmetric film with thickness δ, in contact with stationary air. In cylindrical coordinates, we have [83, 102–104]

$$\frac{\partial p}{\partial r} = (2\alpha_1 + \alpha_2)\frac{1}{r}\frac{d}{dr}\left[r\left(\frac{du}{dr}\right)^2\right] + \frac{4}{r}\left(\gamma_3 + \gamma_4 + \gamma_5 + \frac{\gamma_6}{2}\right)\frac{d}{dr}\left[r\left(\frac{du}{dr}\right)^4\right] \quad (6.73)$$

$$\frac{\partial p}{\partial z} = \frac{\mu}{r}\frac{d}{dr}\left(r\frac{du}{dr}\right) + \frac{2}{r}(\beta_2 + \beta_3)\frac{d}{dr}\left[r\left(\frac{du}{dr}\right)^3\right] + \rho g \quad (6.74)$$

where $p \neq p(z)$is pressure, α_1, α_2, β_2, β_3, γ_3, γ_4, γ_5 and γ_6 are known parameters and Eq. 6.74 further gives

$$r\frac{d^2u}{dr^2} + \frac{du}{dr} + \frac{2(\beta_2 + \beta_3)}{\mu}\left[3r\left(\frac{du}{dr}\right)^2\frac{d^2u}{dr^2} + \left(\frac{du}{dr}\right)^3\right] + \frac{\rho g}{\mu}r = 0 \quad (6.75)$$

The boundary conditions are:

$$u(R) = 0, \ \frac{du}{dr} = 0 \ at \ r = R + \delta \quad (6.76)$$

Defining

$$\eta = \frac{r}{R}, f = \frac{R}{v}u, \ k = \frac{\rho g R^3}{\mu^2}, \ b = \frac{\mu(\beta_2 + \beta_3)}{R^4}, \ d = 1 + \frac{\delta}{R}, \tag{6.77}$$

Equations 6.75 and 6.76 reduces to

$$\eta \frac{d^2 f}{d\eta^2} + \frac{df}{d\eta} + k\eta + 2b\left[\left(\frac{df}{d\eta}\right)^3 + 3\eta\left(\frac{df}{d\eta}\right)^2 \frac{d^2 f}{d\eta^2}\right] = 0 \tag{6.78}$$

$$f(1) = 0, f'(d) = 0 \tag{6.79}$$

In accordance with Eqs. 6.78 and 6.79, the linear operator is chosen as:

$$\mathrm{L}(\phi(\eta, p)) = \eta \frac{\partial^2 \phi(\eta, p)}{\partial \eta^2} + \frac{\partial \phi(\eta, p)}{\partial \eta} \tag{6.80}$$

and we define a nonlinear operator as

$$N(\phi(\eta, p)) = 2b\left[\left(\frac{\partial \phi(\eta, p)}{\partial \eta}\right)^3 + 3\eta\left(\frac{\partial \phi(\eta, p)}{\partial \eta}\right)^2 \frac{\partial^2 \phi(\eta, p)}{\partial \eta^2}\right] \tag{6.81}$$

and the function $g(\eta)$as

$$g(\eta) = k\eta \tag{6.82}$$

The initial conditions are

$$\phi(0, p) = 1, \ \frac{\partial \phi(d, p)}{\partial \eta} = 0 \tag{6.83}$$

Equation 6.16 can be written as

$$\eta f_0'' + f_0' + k\eta = 0 \tag{6.84}$$

$$f_0'(d) = 0 \tag{6.85}$$

It is obtained

$$f_0'(\eta) = \frac{k}{2}\left(\frac{d^2}{\eta} - \eta\right) \tag{6.86}$$

Equation 6.19 becomes

$$\eta f_1'' + f_1' = 2h_1 b(f_0'^3 + 3\eta f_0'^2 f_0'') \tag{6.87}$$

$$f_1'(d) = 0 \tag{6.88}$$

If we choose $h_1(\eta, C_i) = C_1$, where C_1 is a unknown constant, then the derivative of solution of Eq. 6.87 subject to condition Eq. 6.88 is given by

$$f_1'(\eta) = \frac{1}{4} C_1 b k^3 \left(\frac{d^2}{\eta} - \eta\right)^3 \tag{6.89}$$

Equation 6.20 for $k = 2$, reduces to

$$\begin{aligned} \eta f_2'' + f_2' = \eta f_1'' + f_1' + C_1\{\eta f_1'' + f_1' + 6b[f_0'^2 f_1' + \\ + \eta(f_0'^2 f_1'' + 2f_0' f_1' f_0'')]\} + 2bh_2(f_0'^3 + 3\eta f_0'^2 f_0'') \end{aligned} \tag{6.90}$$

$$f_2'(d) = 0 \tag{6.91}$$

Choosing $h_2(\tau, C_i) = C_2$, with C_2 constant and substituting Eqs. 6.86 and 6.89 into 6.90 and then integrating Eq. 6.90 with condition Eq. 6.91, we obtain the derivative of second-order solution:

$$f_2'(\eta) = \frac{1}{4}(C_1^2 + C_1 + C_2) b k^2 \left(\frac{d^2}{\eta} - \eta\right)^3 + \frac{3}{8} C_1 b^2 k^5 \left(\frac{d^2}{\eta} - \eta\right)^5 \tag{6.92}$$

Thus the derivative of solution up to second order is given by

$$\bar{f}'(\eta) = f_o' + f_1' + f_2' = \alpha\left(\frac{d^2}{\eta} - \eta\right) + \beta\left(\frac{d^2}{\eta} - \eta\right)^3 + \gamma\left(\frac{d^2}{\eta} - \eta\right)^5 \tag{6.93}$$

where

$$\alpha = \frac{k}{2}, \ \beta = \frac{1}{4}(C_1^2 + 2C_1 + C_2) b k^3, \ \gamma = \frac{3}{8} C_1 b^2 k^5 \tag{6.94}$$

Substituting the derivative of the solution up to second order Eq. 6.93 into Eq. 6.78, results in the residual:

$$R(\eta,C_1,C_2)=-2(\beta+2b\alpha^3)\left(\frac{d^2}{\eta}-\eta\right)^2\left(\frac{d^2}{\eta}+2\eta\right)-$$
$$-2(\gamma+6b\alpha^2A_3)\left(\frac{d^2}{\eta}-\eta\right)^4\left(2\frac{d^2}{\eta}+3\eta\right)-$$
$$-12b\alpha(\beta^2+\alpha\gamma)\left(\frac{d^2}{\eta}-\eta\right)^6\left(3\frac{d^2}{\eta}+4\eta\right)-4b\beta^3\left(\frac{d^2}{\eta}-\eta\right)^8\left(4\frac{d^2}{\eta}+5\eta\right)-$$
$$-12b\gamma(\beta^2+\alpha\gamma)\left(\frac{d^2}{\eta}-\eta\right)^{10}\left(5\frac{d^2}{\eta}+6\eta\right)-$$
$$-8b\beta\gamma\left(\frac{d^2}{\eta}-\eta\right)^{12}\left(7\frac{d^2}{\eta}+8\eta\right)-4b\alpha\beta\gamma\left(\frac{d^2}{\eta}-\eta\right)^8\left(23\frac{d^2}{\eta}+28\eta\right) \tag{6.95}$$

For $\eta \in \left[\frac{d}{3}, d\right]$, the functional Eq. 6.25 becomes:

$$J(C_1,C_2) = \int_{d/3}^{d} R^2(\eta,C_1,C_2)d\eta \tag{6.96}$$

The constants C_1 and C_2 result from the conditions:

$$\frac{\partial J}{\partial C_1} = 2\int_{d/3}^{d} R\frac{\partial R}{\partial C_1}d\eta = 0, \quad \frac{\partial J}{\partial C_2} = 2\int_{d/3}^{d} R\frac{\partial R}{\partial C_2}d\eta = 0 \tag{6.97}$$

In a particular case when $k = b = 1$, $d = 1.02$ (in Ref.[102], $b \geq 0.3$ is considered a parameter corresponding to strong nonlinearity), from Eqs. 6.97 are obtained:

$$C_1 = -0.0010842826666$$
$$C_2 = -0.0225161113357$$

From Eq. 6.93 the derivative of solution is obtained up to second order in the form:

$$\bar{f}'(\eta) = 0.5\left(\frac{1.0404}{\eta}-\eta\right) - 0.024683501\left(\frac{1.0404}{\eta}-\eta\right)^3 - 0.000406606\left(\frac{1.0404}{\eta}-\eta\right)^5 \tag{6.98}$$

By integrating Eq. 6.98 in the initial condition Eq. 6.79, we obtain the solution of second order for Eq. 6.78 in the form:

$$\bar{f}(\eta) = \frac{0.000123912}{\eta^4} + \frac{0.012707805}{\eta^2} - 0.286320453\eta^2 + 0.005642083\eta^4 + \\ + 0.000067767\eta^6 + 0.595775602\ln\eta + 0.2677778886,\ \eta \in \left[\frac{1.02}{3}, 1.02\right] \tag{6.99}$$

Figure 6.4 shows the comparison between the present solution obtained from formula Eq. 6.99 and the numerical integration results obtained from Eq. 6.78 by using the fourth-order Runge–Kutta method.

It can be seen from Fig. 6.4 that the solution obtained by the present method is quasi-identical with that given by the numerical method, demonstrating very good accuracy.

6.4 Damped Oscillator with Fractional-Order Restoring Force

In this section we consider the nonlinear oscillator with damping and fractional-order restoring force

$$\frac{d^2x}{dt^2} + 2k\frac{dx}{dt} + ax + bx|x|^{n-1} = 0 \tag{6.100}$$

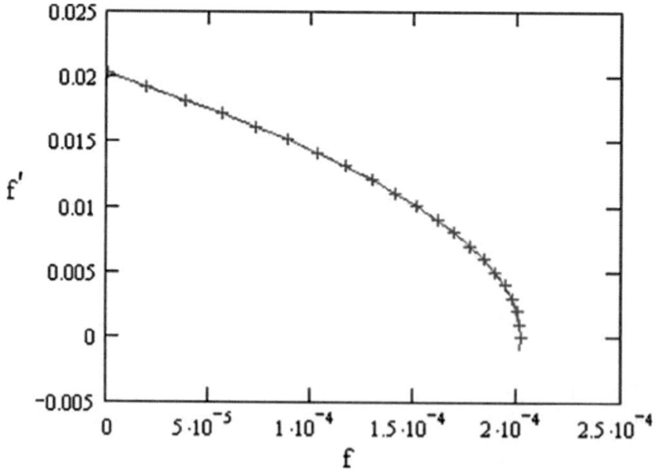

Fig. 6.4 Comparison between the results obtained using the proposed method and numerical method for Eq. 6.78: ____ numerical simulation; -+-+-+- approximate solution Eq. 6.99

where k, a and b are constants and $n > 0$.

The initial conditions are given by

$$x(0) = A, \quad \dot{x}(0) = 0$$

where the dot denotes derivatives with respect to time t.

Under the transformations

$$x = Ae^{-kt}y, \; \tau = kt \tag{6.101}$$

the original Eq. 6.100 becomes

$$y'' + (\alpha - 1)y + \beta e^{(1-n)\tau} y|y|^{n-1} = 0 \tag{6.102}$$

The initial conditions transform to

$$y(0) = 1, \; y'(0) = k, \tag{6.103}$$

where prime denotes the derivative with respect to τ and

$$\alpha = \frac{a}{k^2}, \; \beta = \frac{b}{k^2} A^{n-1} \tag{6.104}$$

We apply OHAM for Eqs. 6.102 and 6.103. The linear operator is

$$L(\phi(\tau, p)) = \frac{\partial^2 \phi(\tau, p)}{\partial \tau^2} + \gamma^2 \phi(\tau, p) \tag{6.105}$$

where γ is an arbitrary constant unknown at this moment.

We modify the family of Eqs. 6.14 in the following form

$$(1 - p)\left[L(\phi(\tau, p)) + g(\tau)\right] = \mathrm{H}(\tau, p) N(\phi(\tau, p)) \tag{6.106}$$

where the nonlinear operator is

$$N(\phi(\tau, p)) = \frac{\partial^2 \phi(\tau, p)}{\partial \tau^2} + (\alpha - 1)\phi(\tau, p) + \beta e^{(1-n)\tau} \phi(\tau, p)|\phi(\tau, p)|^{n-1} \tag{6.107}$$

The initial conditions become

$$\phi(0, p) = 1, \; \left.\frac{\partial \phi(\tau, p)}{\partial \tau}\right|_{\tau=0} = k \tag{6.108}$$

According to Eqs. 6.105 and 6.16, we choose

$$y_0(\tau) = e^{-\lambda\tau}\cos\gamma\tau,\ g(\tau) = e^{-\lambda\tau}(\lambda^2\cos\gamma\tau + 2\lambda\gamma\sin\gamma\tau) \tag{6.109}$$

where λ and γ are arbitrary unknown constants.

In order to obtain the first-order approximate solution of Eq. 6.102, the case $m = 1$ in Eq. 6.23 is considered. Therefore, from Eq. 6.19 the first-order problem is obtained:

$$L(y_1(\tau)) = h_1(\tau, C_i)(N_0(y_0(\tau)) \tag{6.110}$$

The operator $N_0(y_0)$ is

$$N_0(y_0(\tau)) = y_0'' + (\alpha - 1)y_0 + \beta e^{(1-n)\tau} y_0|y_0|^{n-1} \tag{6.111}$$

The last term of Eq. 6.111 can be written as

$$y_0|y_0|^{n-1} = |y_0|^n sign(y_0)$$

where $sign(y_0) = 1$ if $y_0 > 0$ and $sign(y_0) = -1$ if $y_0 \leq 0$.

Substituting Eq. 6.109$_1$ into this expression, the following Fourier series expansion is obtained

$$|y_0|^n \mathrm{sgn}(y_0) = e^{-n\lambda\tau}(a_{1n}\cos\gamma\tau + a_{3n}\cos 3\gamma\tau + a_{5n}\cos 5\gamma\tau + \ldots.) \tag{6.112}$$

where

$$a_{2j+1,n} = \frac{4}{\pi}\int_0^{\frac{\pi}{2}} (\cos u)^n \cos(2j+1)u du,,\ \mathrm{j} = 0, 1, 2, \ldots \tag{6.113}$$

In this way, from Eqs. 6.109, 6.112 and 6.111 one has

$$\begin{aligned} N_0(y_0) = {} & e^{-\lambda\tau}[(\lambda^2 + \alpha - 1)\cos\gamma\tau + (2\lambda - \gamma)\gamma\sin\gamma\tau] + \\ & + \beta e^{-(\lambda n + n - 1)\tau}(a_{1n}\cos\gamma\tau + a_{3n}\cos 3\gamma\tau + \ldots) \end{aligned} \tag{6.114}$$

It is assumed that

$$h_1(\tau, C_i) = C_1 + 2C_2\cos 2\gamma\tau$$

where C_1 and C_2 are unknown constants.

Substituting this equation and Eqs. 6.114 into 6.110, the equation in $y_1(\tau)$ is derived:

$$\begin{aligned} y''_1 + \gamma^2 y_1 = e^{-\lambda\tau}[(C_1 + C_2)(\lambda^2 + \alpha - 1)\cos\gamma\tau + \\ + (C_1 - C_2)(2\gamma\lambda - \gamma^2)\sin\gamma\tau + + C_2(\lambda^2 + \alpha - 1)\cos 3\gamma\tau + \\ + C_2(2\lambda\gamma - \gamma^2)\sin 3\gamma\tau] + \beta e^{-(\lambda n + n - 1)\tau}\{[a_{1n}C_1 + (a_{1n} + a_{3n})C_2]\cos\gamma\tau + \\ + [a_{3n}C_1 + (a_{1n} + a_{5n})C_2]\cos 3\gamma\tau + (a_{5n}C_1 + a_{3n}C_2)\cos 5\gamma\tau + \ldots.\} \end{aligned} \tag{6.115}$$

which has the solution

$$\begin{aligned} y_1(\tau) = B\cos\gamma\tau + D\sin\gamma\tau + e^{-\lambda\tau}(M_1\cos\gamma\tau + M_3\cos 3\gamma\tau + \\ + N_1\sin\gamma\tau + N_3\sin 3\gamma\tau) + e^{-(\lambda n + n - 1)\tau}(P_1\cos\gamma\tau + P_3\cos 3\gamma\tau + \\ + P_5\cos 5\gamma\tau + Q_1\sin\gamma\tau + Q_3\sin 3\gamma\tau + Q_5\sin 5\gamma\tau + \ldots.) \end{aligned} \tag{6.116}$$

where B and D are integration constants and

$$M_1 = \frac{C_1[(\alpha-1)\lambda + \lambda^3 + 4\lambda\gamma^2 - 2\gamma^3] + C_2[\lambda(\alpha-1) + \lambda^3 - 4\gamma^2\lambda + 2\gamma^3]}{\lambda(\lambda^2 + 4\gamma^2)}$$

$$M_3 = \frac{C_2[(\alpha-1)(\lambda^2 + \gamma^2 - 9\lambda\gamma) + \lambda^4 + 4\lambda^2\gamma^2 - 3\lambda^3\gamma - 3\lambda\gamma^3]}{(\lambda^2 + \gamma^2 - 9\lambda\gamma)^2 + 9(\lambda\gamma + \lambda^2)^2}$$

$$N_1 = \frac{C_1[2\gamma(1-\alpha) - \gamma^2\lambda] + C_2[2\gamma(1-\alpha) - 4\lambda^2\gamma + \gamma^2\lambda]}{\lambda(\lambda^2 + 4\gamma^2)^2}$$

$$N_3 = \frac{C_2[3(1-\alpha)(\lambda^2 + \lambda\gamma) - 3\lambda^4 - \lambda^3\gamma - 19\lambda^2\gamma^2 + 11\lambda\gamma^3 - \gamma^4]}{(\lambda^2 + \gamma^2 - 9\lambda\gamma)^2 + 9(\lambda\gamma + \lambda^2)^2}$$

$$P_1 = \frac{\beta[(\lambda n + n - 1)^2 + \gamma^2 - \lambda^2][a_{1n}C_1 + (a_{1n} + a_{3n})C_2]}{[(\lambda n + n - 1)^2 + \gamma^2 - \lambda^2]^2 + 4\lambda^2(\lambda n + n - 1)^2}$$

$$P_3 = \frac{\beta[(\lambda n + n - 1)^2 + \gamma^2 - 9\lambda^2][a_{3n}C_1 + (a_{1n} + a_{5n})C_2]}{[(\lambda n + n - 1)^2 + \gamma^2 - 9\lambda^2]^2 + 36\lambda^2(\lambda n + n - 1)^2}$$

$$P_5 = \frac{\beta[(\lambda n + n - 1)^2 + \gamma^2 - 25\lambda^2](a_{5n}C_1 + a_{3n}C_2)}{(\lambda n + n - 1)^2 + \gamma^2 - 25\lambda^2]^2 + 100\lambda^2(\lambda n + n - 1)^2}$$

$$Q_1 = \frac{-2\beta\lambda(\lambda n + n - 1)[a_{1n}C_1 + (a_{1n} + a_{3n})C_2]}{[(\lambda n + n - 1)^2 + \gamma^2 - \lambda^2]^2 + 4\lambda^2(\lambda n + n - 1)^2}$$

$$Q_3 = \frac{-6\beta\lambda(\lambda n + n - 1)[a_{3n}C_1 + (a_{1n} + a_{5n})C_2]}{[(\lambda n + n - 1)^2 + \gamma^2 - 9\lambda^2]^2 + 36\lambda^2(\lambda n + n - 1)^2}$$

$$Q_5 = \frac{-10\beta\lambda(\lambda n + n - 1)(a_{5n}C_1 + a_{3n}C_2)}{[(\lambda n + n - 1)^2 + \gamma^2 - 25\lambda^2]^2 + 100\lambda^2(\lambda n + n - 1)^2} \tag{6.117}$$

The first-order approximate solution is obtained from Eqs. 6.23, 6.109 and 6.116 and therefore

$$\bar{y}(\tau) = y_0(\tau) + y_1(\tau) \tag{6.118}$$

The constants B and D which appear in Eq. 6.116 can be determined from the initial conditions Eq. 6.103. The first-order approximate solution of Eq. 6.100 is obtained from Eqs. 6.118, 6.102 and 6.101 and can be written in the form

$$\begin{aligned}\bar{x}(t) = {} & Ae^{-kt}\{-(M_1 + M_3 + P_1 + P_3 + P_5)\cos\gamma kt + \\ & + [\frac{1+\lambda}{\gamma} + \frac{\lambda}{\gamma}(M_1 + M_3) + \frac{\lambda n + n - 1}{\gamma}(P_1 + P_3 + P_5) - (N_1 + 3N_3 + Q_1 + \\ & + 3Q_3 + 5Q_5)]\sin\gamma kt\} + Ae^{-(\lambda+1)kt}[(1 + M_1)\cos\gamma kt + M_3\cos 3\gamma kt + \\ & + N_1\sin\gamma kt + N_3\sin 3\gamma kt] + Ae^{-(\lambda+1)nkt}(P_1\cos\gamma kt + P_3\cos 3\gamma kt + \\ & + P_5\cos 5\gamma kt + Q_1\sin\gamma kt + Q_3\sin 3\gamma kt + Q_5\sin 5\gamma kt)\end{aligned} \tag{6.119}$$

where the coefficients M_{i}, N_{i}, P_{j}, Q_{j}, $i = 1{,}3;\ j = 1{,}3{,}5$ are given by Eqs. 6.117

The convergence of the solution $x(\tau)$ and, consequently, the convergence of the approximate solution $\bar{x}(\tau)$ depend on the auxiliary function $h_1(\tau, C_i)$. Basically, the shape of $h_1(\tau, C_i)$ (or $h_2(\tau, C_i)$,..., $h_m(\tau, C_i)$ in other nonlinear problems) must follow the terms appearing in Eq. 6.114 which are $\cos\gamma\tau$, $\sin\gamma\tau$, $\cos 3\gamma\tau$,... Therefore, $h_1(\tau, C_i)$ should be chosen so that in Eq. 6.110 the product $h_1(\tau, C_i)N_0(\tau)$ be of the same shape with the other terms from $\mathrm{N}_0(\tau)$ (a combination of functions $\cos\gamma\tau$, $\cos 3\gamma\tau$, $\sin\gamma\tau$). For example, one can choose

$$h_1(\tau, C_i') = C_1' + 2C_2'\cos 2\gamma\tau + 2C_3'\cos 4\gamma\tau$$

or

$$h_1(\tau, C_i'') = C_1'' + 2C_2''\cos 4\gamma\tau + 2C_3''\sin 6\gamma\tau$$

where $C_1', C_2', C_3', C_1'', C_2'', C_3''$ are unknown constants.

In other applications, such as those presented in [84–89], the functions $h_1(\tau, C_i)$, $h_2(\tau, C_j)$ or $h_m(\tau, C_k)$, can be chosen as exponential functions or polynomial functions, depending on the shape of the terms already existing in the specific iteration.

6.4.1 Numerical Examples

We illustrate the accuracy of this approach by comparing previously obtained approximate solutions with numerical integration results calculated by means of the fourth-order Runge–Kutta method.

Case 6.4.1.a

In the first case considered, it is assumed that $a = 1$, $b = 1$, $k = 0.1$, $n = 3$, $A = 0.5$. This corresponds to the oscillator with a linear-plus-cubic restoring force. It should be noted that the value of the parameter of nonlinearity b is not small.

From Eq. 6.113, one finds

$$a_{1,3} = 0.75, \quad a_{3,3} = 0.25, \quad a_{5,3} = 0$$

By means of the collocation method, one obtains

$$C_1 = -1.133641, \quad C_2 = 0.227603, \quad \lambda = 13.0497722, \quad \gamma = 10.244569$$

The first-order approximate solution becomes

$$\begin{aligned}\bar{x} = &\frac{1}{2}e^{-0.1t}(1.01717\cos 1.02446t - 0.063556\sin 1.02446t)+\\ &+\frac{1}{2}e^{-1.40498t}(-0.00176381\cos 1.02446t - 0.00857627\cos 3.07337t+\\ &+0.274502\sin 1.02446t - 0.0532868\sin 3.07337t)+\\ &+\frac{1}{2}e^{-4.21498t}(-0.00666211\cos 1.02446t - 0.0000715917\cos 3.07337t-\\ &-0.000100285\cos 5.12228t + 0.00439515\sin 1.02446t+\\ &+0.000868631\sin 3.07337t - 0.000218981\sin 5.12228t)\end{aligned} \tag{6.120}$$

This solution is plotted in Fig. 6.5 together with the numerical results and good agreement between them is found. These time solutions represent an oscillation

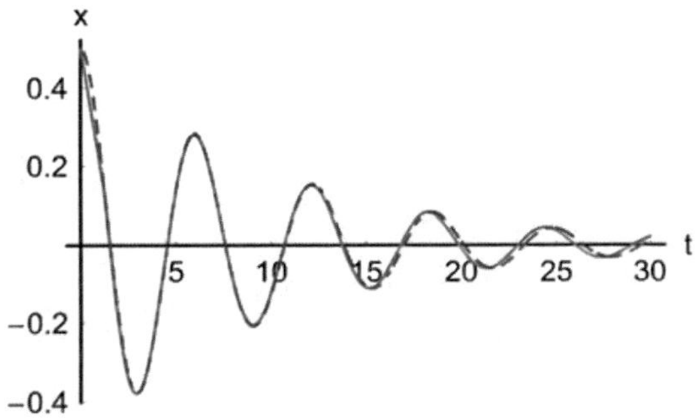

Fig. 6.5 Comparison between the approximate result of Eq. 6.100 (*solid line*) given by Eq. 6.120 and numerical results (*dashed line*) for $a = b = 1$, $A = 0.5$, $k = 0.1$, $n = 3$

with decreasing amplitude, approaching the zero equilibrium point, which corresponds to stable focus in the phase plane.

Case 6.4.1.b

For the second case we consider $a = 0, b = 1, k = 0.1, n = 3, A = 0.5$ and we obtain

$$C_1 = -0.405949, C_2 = 0.391642, \lambda = -0.227844, \gamma = 0.645668$$

and the first-order approximate solution is

$$\begin{aligned}\bar{x}(t) = & \frac{1}{2}e^{-0.1t}(-0.479616\cos 0.0645668t - 1.12506\sin 0.0645668t) + \\ & + \frac{1}{2}e^{-0.0772156t}(-0.861615\cos 0.0645668t + 0.374566\sin 0.0645668t - \\ & - 0.165853\cos 0.1937t - 0.183676\sin 0.1937t) + \\ & + \frac{1}{2}e^{-0.231647t}(0.960311\cos 0.0645668t + 0.274582\sin 0.0645668t + \\ & + 1.33225\cos 0.1937t + 0.214526\cos 0.322834t - \\ & - 1.42483\sin 0.1937t + 0.755112\sin 0.322834t)\end{aligned} \tag{6.121}$$

This analytical solution is plotted in Fig. 6.6 and agrees well with the numerical solution. This solution indicates that there is no oscillation in this case.

Case 6.4.1.c

In this case, if we consider $a = b = 1, k = 0.1, n = 3/2, A = 0.5$, then

$$a_{1,\frac{3}{2}} = 0.915311716, \quad a_{3,\frac{3}{2}} = 0.101701301, \quad a_{5,\frac{3}{2}} = -0.009663924$$

and

$$C_1 = 0.302498, C_2 = -0.257482, \lambda = 2.16821, \gamma = 12.3432$$

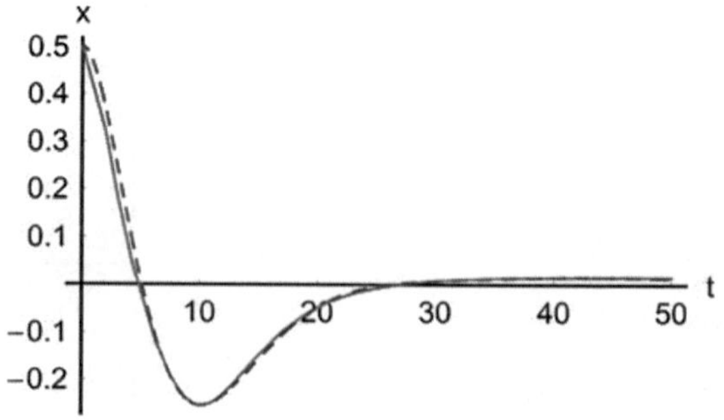

Fig. 6.6 Comparison between the approximate result of Eq. 6.100 (*solid line*) given by Eq. 6.121 and numerical results (*dashed line*) for $a = 0, b = 1, A = 0.5, k = 0.1, n = 3$

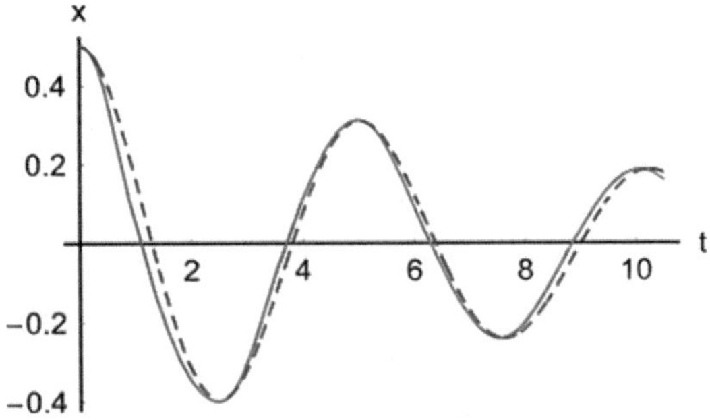

Fig. 6.7 Comparison between the approximate result of Eq. 6.100 (*solid line*) given by Eq. 6.122 and numerical results (*dashed line*) for $a = b = 1, A = 0.5, k = 0.1, n = 3/2$

The approximate solution is

$$\begin{aligned}\bar{x}(t) = & \frac{1}{2}e^{-0.1t}(1.013284\cos 1.23432t - 0.0524102\sin 1.23432t) + \\ & + \frac{1}{2}e^{-0.316821t}(-0.018425\cos 1.23432t - 0.176664\sin 1.23432t + \\ & + 0.109537\cos 3.70295t + 0.0243331\sin 3.70295t) + \\ & + \frac{1}{2}e^{-0.4752315t}(0.00649996\cos 1.23432t - -0.000653951\sin 1.23432t - \\ & - 0.098116\cos 3.70295t - 0.01277\cos 6.17159t + 0.02125\sin 6.17159t)\end{aligned} \tag{6.122}$$

The comparison between the approximate and numerical solutions shown in Fig. 6.7 indicates that they agree well for the first two cycles, for $t \in [0,12.6]$ and then the accuracy of the analytical solution deteriorates. In this case one can consider the motion interval by interval, for $t \in [12.6\text{–}21.8]$ and so on.

Figures 6.5–6.7 show a comparison between the first-order approximate analytical solutions of Eq. 6.100 and the numerical results. One can observe that the first-order approximate analytical results obtained through OHAM are near the numerical simulation results for various values of the parameters.

6.5 Nonlinear Equations Arising in Heat Transfer

6.5.1 Cooling of a Lumped System with Variable Specific Heat

Consider the cooling of a lumped system [84, 105–107]. Let the system have volume V, surface area A, density ρ, specific heat c and initial temperature T_i. At time $t = 0$, the system is exposed to a convective environment at temperature T_a with convective heat transfer coefficient h. Assume that the specific heat c is a linear function of temperature of the form

$$c = c_a[1 + \beta(T - T_a)] \tag{6.123}$$

where c_a is the specific heat, at temperature T_a and β is a constant. The cooling equation and the initial conditions are as follows:

$$\rho Vc\frac{dT}{dt} + hA(T - T_a) = 0,\ T_0 = T_i \tag{6.124}$$

where by using

$$u = \frac{T - T_a}{T_i - T_a},\ x = \frac{t(hA)}{\rho Vc_a},\ \varepsilon = \beta(T - T_a)$$

we have

$$(1 + \varepsilon u)\frac{du}{dx} + u = 0,\ u(0) = 1,\ x \in [0, \infty) \tag{6.125}$$

According to Eq. 6.13 we have:

$$L(u(x)) = u'(x) + u(x),\ g(x) = 0,\ N(u(x)) = \varepsilon u(x)u'(x),\ u(0) = 1 \tag{6.126}$$

where $' = \frac{d}{dx}$.

Equation 6.16 becomes

$$u_0'(x) + u_0(x) = 0, u_0(0) = 1 \tag{6.127}$$

from which we obtain

$$u_0(x) = e^{-x} \tag{6.128}$$

Substituting Eq. 6.128 into Eq. 6.19 we get

$$u_1'(x) + u_1(x) = -\varepsilon h_1(x, C_i)e^{-2x},\ u_1(0) = 0 \tag{6.129}$$

where $h_1(x, C_i) = C_1$(constant) and therefore

$$u_1(x) = \varepsilon C_1(e^{-2x} - e^{-x}) \tag{6.130}$$

For $k = 2$ and $k = 3$ into Eq. 6.20 we have respectively:

$$\begin{aligned} u_2'(x) + u_2(x) &= [(2\varepsilon^2 - \varepsilon)C_1^2 - \varepsilon C_1 - \varepsilon h_2(x, C_j)]\, e^{-2x} - 3\varepsilon^2 C_1^2 e^{-3x}, \\ u_2(0) &= 0 \end{aligned} \tag{6.131}$$

with the solution for $h_2(x, C_j) = C_2$(constant):

$$u_2(x) = \left[\left(\frac{\varepsilon^2}{2} - \varepsilon\right)C_1^2 - \varepsilon C_1 - \varepsilon C_2\right] e^{-x} + \\ + \left[(\varepsilon - 2\varepsilon^2)C_1^2 + \varepsilon C_1 + \varepsilon C_2\right] e^{-2x} + \frac{3}{2}\varepsilon^2 C_1^2 e^{-3x} \tag{6.132}$$

and

$$u_3'(x) + u_3(x) = \left[\left(-2\varepsilon^3 + 4\varepsilon^2 - \varepsilon\right) C_1^3 + (4\varepsilon^2 - 2\varepsilon)C_1^2 + \\ + (4\varepsilon^2 - 2\varepsilon)C_1C_2 - \varepsilon C_1 - \varepsilon C_2 - \varepsilon h_3(x, C_i)\right] e^{-2x} + \\ + \left[\left(9\varepsilon^3 - 6\varepsilon^2\right) C_1^3 - 6\varepsilon^2 C_1^2 - 6\varepsilon^2 C_1 C_2\right] e^{-3x} - 8\varepsilon^3 C_1^3 e^{-4x}, u_3(0) = 0 \tag{6.133}$$

which have the solution, if we consider $h_3(x, C_i) = C_3$(constant):

$$u_3(x) = -\left[(8\varepsilon^3 + 8\varepsilon^2 + \varepsilon)C_1^3 + (8\varepsilon^2 + 2\varepsilon)C_1^2 + (8\varepsilon^2 + 2\varepsilon)C_1C_2 + \\ + \varepsilon C_1 + \varepsilon C_2 + \varepsilon C_3\right]e^{-x} + \left[(2\varepsilon^3 - 4\varepsilon^2 + \varepsilon)C_1^3 + (2\varepsilon - 4\varepsilon^2)C_1^2 \\ + (2\varepsilon - 4\varepsilon^2)C_1C_2 + \varepsilon C_1 + \varepsilon C_2 + \varepsilon C_3\right] e^{-2x} + \\ + \left[(12\varepsilon^2 - 18\varepsilon^3)C_1^3 + 12\varepsilon^2 C_1^2 + 12\varepsilon^2 C_1 C_2\right] e^{-3x} + 24\varepsilon^3 C_1^3 e^{-4x} \tag{6.134}$$

The approximate solution of this third order is obtained from Eqs. 6.128, 6.129, 6.132, 6.134 and 6.23 for $m = 3$:

$$\bar{u}(x) = u_0(x) + u_1(x) + u_2(x) + u_3(x)$$

and we obtain:

$$\bar{u}(x) = A_1 e^{-x} + A_2 e^{-2x} + A_3 e^{-3x} + A_4 e^{-4x} \tag{6.135}$$

where

$$A_1 = 1 - 3\varepsilon C_1 - 2\varepsilon C_2 - \varepsilon C_3 - \left(3\varepsilon + \frac{15}{2}\varepsilon^2\right) C_1^2 - (2\varepsilon + 8\varepsilon^2)C_1C_2 - \\ - \left(\varepsilon + 8\varepsilon^2 + 8\varepsilon^3\right) C_1^3$$
$$A_2 = 3\varepsilon C_1 + 2\varepsilon C_2 + \varepsilon C_3 + (3\varepsilon - 6\varepsilon^2)C_1^2 + (2\varepsilon - 4\varepsilon^2)C_1C_2 + \\ + (\varepsilon - 4\varepsilon^2 + 2\varepsilon^3)C_1^3$$
$$A_3 = \frac{27}{2}\varepsilon^2 C_1^2 + 12\varepsilon^2 C_1 C_2 + (12\varepsilon^2 - 18\varepsilon^3)C_1^3$$
$$A_4 = 24\varepsilon^3 C_1^3 \tag{6.136}$$

The residual Eq. 6.24 for $m = 3$ is obtained in the form:

$$R(x, C_1, C_2, C_3) = (A_2 + \varepsilon A_1^2)e^{-2x} + (2A_3 + 3\varepsilon A_1 A_2)e^{-3x} + \\ + (3A_4 + 2\varepsilon A_2^2 + 4\varepsilon A_1 A_3)e^{-4x} + 5\varepsilon(A_1 A_4 + A_2 A_3)e^{-5x} + \\ + 3\varepsilon(A_3^2 + 2A_2 A_4)e^{-6x} + 7\varepsilon A_3 A_4 e^{-7x} + 4\varepsilon A_4^2 e^{-8x} \quad (6.137)$$

For $x_1 = 1, x_2 = 2, x_3 = 3$ into Eq. 6.137 we obtain the approximate solution of the third order for ε=1 and ε=2 respectively (see Eq. 6.27).

$$\bar{u}(x) = 2.666666\, e^{-x} - 4e^{-2x} + 3e^{-3x} - 0.6666666\, e^{-4x} \quad (6.138)$$

$$\bar{u}(x) = 4.460824085e^{-x} - 11.54898853e^{-2x} + \\ +14.65491942e^{-3x} - 6.566754975e^{-4x} \quad (6.139)$$

Figures 6.8 and 6.9 show the comparison between the present solution and the numerical integration results obtained by a fourth-order Runge–Kutta method, for cases ε=1 and ε=2, respectively.

It can be seen from these figures that the solution obtained by OHAM is nearly identical with that given by numerical simulation.

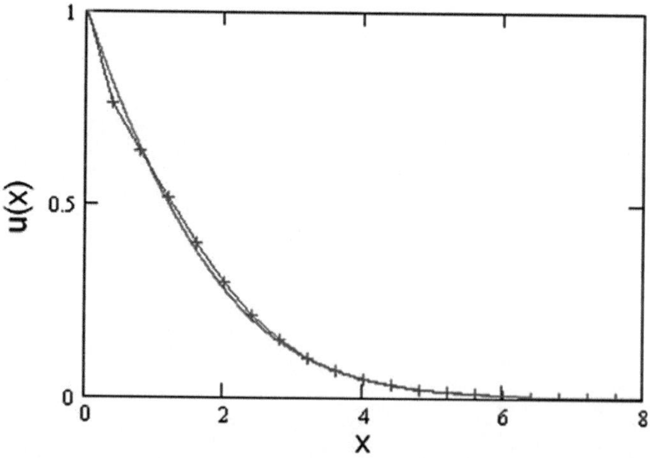

Fig. 6.8 Comparison between the results obtained by the present method Eq. 6.138 and the numerical results of Eq. 6.125 for $\varepsilon = 1$; -+-+-+-+- present solution ________ numerical solution

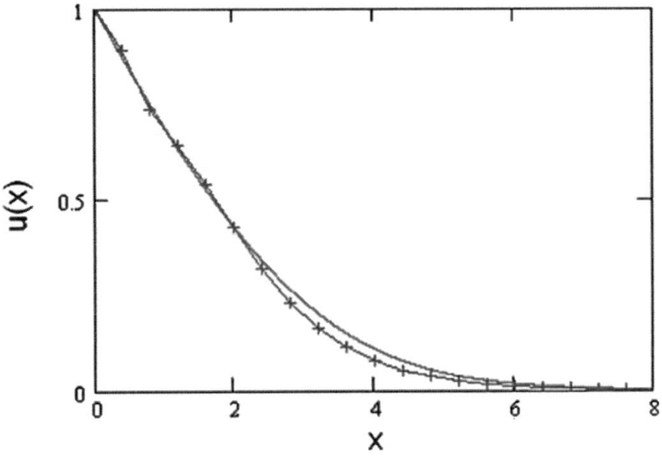

Fig. 6.9 Comparison between the results obtained by the present method Eq. 6.139 and the numerical results of Eq. 6.125 for $\varepsilon = 2$; -+-+-+-+- present solution; ________ numerical solution

6.5.2 *The Temperature Distribution Equation in a Thick Rectangular Fin Radiation to Free Space*

Now we will consider a nonlinear equation, the temperature distribution equation in a uniformly thick rectangular fin radiation to free space with non-linearity of high order [84, 105–107]:

$$u''(x) - \varepsilon u^4(x) = 0, \; u(1) = 1, \; u'(0) = 0, \; x \in [0, 1] \tag{6.140}$$

According to Eq. 6.13, we define the operators:

$$L(u(x)) = u''(x), \; N(u(x)) = -\varepsilon u^4(x), \; g(x) = 0, \tag{6.141}$$

In this case Eq. 6.16 becomes

$$u_0''(x) = 0, \; u(1) = 1, \; u'(0) = 0 \tag{6.142}$$

It is obtained:

$$u_0(x) = 1 \tag{6.143}$$

Equation 6.19 can be written as

$$u_1''(x) = -\varepsilon h_1 u_0''(x), \; u_1(1) = 0, \; u_1'(0) = 0 \tag{6.144}$$

with $h_1 = C_1$ (constant) and therefore the solution is

$$u_1(x) = -\frac{1}{2}\varepsilon C_1(x^2-1) \tag{6.145}$$

Equation 6.20 for $k=2$ and $h_2 = C_2$ (constant), becomes

$$\begin{aligned} u''_2(x) - u''_1(x) &= C_1[u''_1(x) - 4\varepsilon u_0^3(x)u_1(x)] + C_2[-\varepsilon u_0^4], \\ u_2(1) &= 0,\ u'_2(0) = 0 \end{aligned} \tag{6.146}$$

and we obtain the following solution:

$$\begin{aligned} u_2(x) = {} & \frac{1}{6}[(5\varepsilon^2+3\varepsilon)C_1^2 + 3\varepsilon C_1 + 3\varepsilon C_2] - \\ & -\frac{1}{2}[(2\varepsilon^2+\varepsilon)C_1^2 + \varepsilon C_1 + \varepsilon C_2]x^2 + \frac{1}{6}\varepsilon^2 C_1^2 x^4 \end{aligned} \tag{6.147}$$

Equation 6.20 for k $=3$ and $h_3 = C_3$ (constant) is

$$\begin{aligned} u''_3(x) - u''_2(x) = {} & C_1\left[u''_2(x) - \varepsilon(6u_0^2(x)\cdot u_1^2(x) + 4u_0^3(x)\cdot u_2(x))\right] + \\ & + C_2\left[u''_1(x) - 4\varepsilon u_0^3(x)u_1(x)\right] + C_3\left[-\varepsilon u_0^4(x)\right], u_3(1)=0,\ u'_3(0)=0 \end{aligned} \tag{6.148}$$

The solution of Eq. 6.148 becomes:

$$\begin{aligned} u_3(x) = {} & \frac{1}{180}\left[432\varepsilon^2 C_1^3 + 89\varepsilon^2 C_1^2 + 300\varepsilon^2 C_1 C_2\right] + \\ & + \frac{1}{24}[7\varepsilon C_1^3 + 14\varepsilon C_1^2 - 5\varepsilon C_1 C_2 + 7\varepsilon C_1 + 7\varepsilon C_2 - 12\varepsilon C_3] - \\ & - \frac{1}{12}\left[(36\varepsilon^2+3\varepsilon)C_1^3 + (7\varepsilon^2+6\varepsilon)C_1^2 + (24\varepsilon^2-3\varepsilon)C_1C_2 + \right. \\ & \left. + 3\varepsilon C_1 + 3\varepsilon C_2 - 6\varepsilon C_3\right]x^2 + \frac{1}{24}\left[(16\varepsilon^2-\varepsilon)C_1^3 + (2\varepsilon^2-2\varepsilon)C_1^2 + \right. \\ & \left. + (8\varepsilon^2-\varepsilon)C_1C_2 - \varepsilon C_1 - \varepsilon C_2\right]x^4 + \frac{1}{180}\left[\varepsilon^2 C_1^2 - 12\varepsilon^3 C_1^3\right]x^6 \end{aligned} \tag{6.149}$$

Finally, the approximate solution of the third-order of Eq. 6.140 is given by

$$\bar{u}(x) = B_0 + B_2x^2 + B_4x^4 + B_6x^6 \tag{6.150}$$

Where

$$B_0 = \frac{1}{180}\left[432\varepsilon^2 C_1^3 + 89\varepsilon^2 C_1^2 + 300\varepsilon^2 C_1 C_2\right] + \\ + \frac{1}{24}\left[7\varepsilon C_1^3 + 26\varepsilon C_1^2 - 5\varepsilon C_1 C_2 + 31\varepsilon C_1 + 31\varepsilon C_2 - 12\varepsilon C_3 + 24\right]$$

$$B_2 = -\frac{1}{12}\left[(36\varepsilon^2 + 3\varepsilon)C_1^3 + (19\varepsilon^2 + 12\varepsilon)C_1^2 \\ + (24\varepsilon^2 - 3\varepsilon)C_1 C_2 + 15\varepsilon C_1 + 15\varepsilon C_2 - 6\varepsilon C_3\right]$$

$$B_4 = \frac{1}{24}\left[(16\varepsilon^2 - \varepsilon)C_1^3 + (6\varepsilon^2 - 2\varepsilon)C_1^2 + (8\varepsilon^2 - \varepsilon)C_1 C_2 - \varepsilon C_1 - \varepsilon C_2\right]$$

$$B_6 = \frac{1}{180}\left[\varepsilon^2 C_1^2 - 12\varepsilon^3 C_1^3\right] \tag{6.151}$$

From Eq. 6.24 for $x_1 = \frac{1}{3}, x_2 = \frac{1}{2}, x_3 = \frac{3}{4}$ we obtain the following approximate solution of the third order for different values of ε:

$$\varepsilon = 1: \ \mathrm{u}(x) = 0.775052997 + 0.204645365x^2 + \\ +0.020278343x^4 + 0.000023293x^6 \tag{6.152}$$

$$\varepsilon = 2: \ \mathrm{u}(x) = 0.698299431 + 0.282644054x^2 + \\ + 0.019027065x^4 + 0.000029291x^6 \tag{6.153}$$

$$\varepsilon = 3: \ \mathrm{u}(x) = 0.654819681 + 0.325418252x^2 + \\ + 0.019710533x^4 + 0.000051532x^6 \tag{6.154}$$

Figures 6.10–6.12 show the comparison between the results obtained by the present procedure and the numerical integration results. It can be seen that the obtained results are nearly identical with the results obtained numerically using a fourth-order Runge–Kutta method, for the cases $\varepsilon = 1$, $\varepsilon = 2$ and $\varepsilon = 3$.

6.6 Blasius' Problem

Blasius equation is one of the basic equations of fluid dynamics. Blasius' equation describes the velocity profile of the fluid in the boundary layer theory on a half-infinite interval. A broad class of analytical solutions method and numerical solutions methods were used to handle this problem. The Blasius equation is the

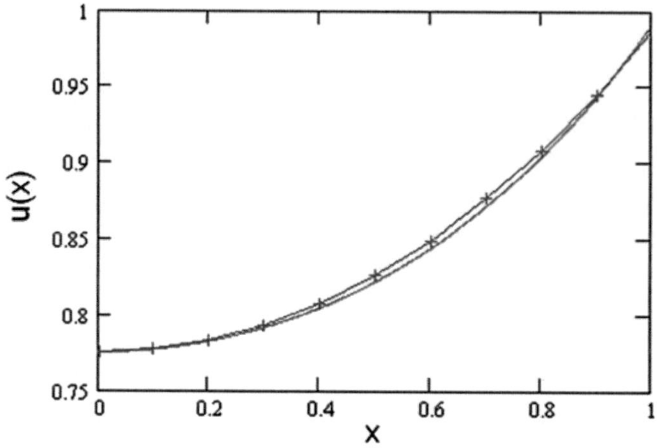

Fig. 6.10 Comparison between the results obtained by the present method Eq. 6.152 and the numerical results of Eq. 6.140 for $\varepsilon = 1$; +-+-+-+- OHAM; _______ numerical solution

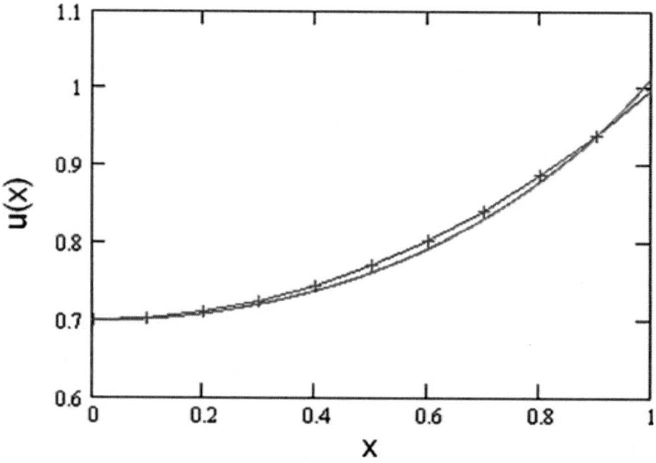

Fig. 6.11 Comparison between the results obtained by the present method Eq. 6.153 and the numerical results of Eq. 6.140 for $\varepsilon = 2$; +-+-+-+- OHAM; _______ numerical method

mother of all boundary layer equations in fluid mechanics. Two forms of Blasius' equation appear in the fluid mechanics theory, where each is subject to specific physical conditions. For instance, consider the two-dimensional laminar viscous flow governed by a nonlinear ordinary differential equation [81, 108–112]

$$u'''(\eta) + \frac{1}{2}u(\eta)u''(\eta) = 0, \quad \eta \in [0, \infty) \tag{6.155}$$

subject to the boundary conditions

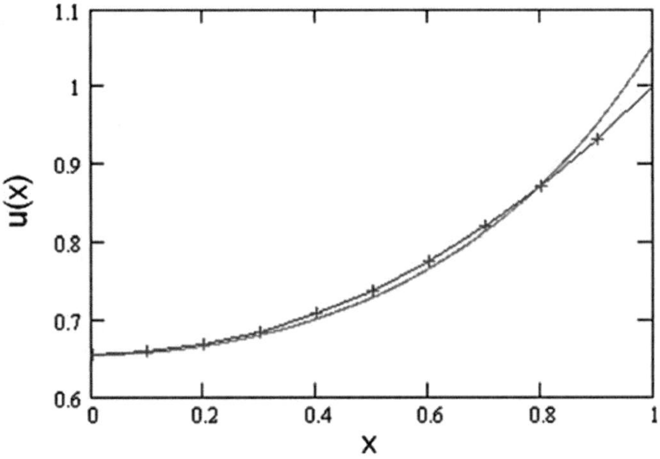

Fig. 6.12 Comparison between the results obtained by the present method Eq. 6.154 and the numerical results of Eq. 6.140 for $\varepsilon = 3$; +-+-+-+- OHAM; _______ numerical method

$$u(0) = 0, \quad u'(0) = 0, \quad u'(\infty) = 1 \tag{6.156a}$$

or

$$u(0) = 0, \quad u'(0) = 1, \quad u'(\infty) = 0 \tag{6.156b}$$

It is obvious that the differential equations are the same, but differ in boundary conditions. For small η [81] and by using perturbation method, a solution of Blasius' equation was obtained in the form

$$u(\eta) = \sum_{k=0}^{\infty} \left(-\frac{1}{2}\right)^k \frac{A_k \sigma^{k+1}}{(3k+2)!} \eta^{3k+2} \tag{6.157}$$

where

$$A_0 = A_1 = 1, \quad A_k = \sum_{i=0}^{k-1} \binom{3k-1}{3i} A_i A_{k-i-1} \quad (k \geq 2) \tag{6.158}$$

Note that the expression Eq. 6.157 is not closed, because $\sigma = u''(0)$ is unknown. By means of matching two different approximations at a proper point, Blasius obtained the numerical result $\sigma = 0.332$. In 1938, Howarth [113] gained a more accurate value $\sigma = 0.33206$ by means of a numerical technique. However, by means of $\sigma = 0.33206$, $u(\eta)$ given by Eq. 6.157 is valid in a small region $0 \leq \eta \leq 5.69$.

By the homotopy analysis method [81], Liao obtained a similar series solution, which is valid in the whole region [0, ∞]. The solution reads

$$u(\eta) = \lim_{m\to\infty} \sum_{k=0}^{+\infty} \left[\left(-\frac{1}{2}\right)^k \frac{A_k \sigma^{k+1}}{(3k+2)!} \eta^{3k+2} \right] \Phi_{m,n}(h), \ \eta \in [0,\infty), \ h \in (-2,0) \tag{6.159}$$

where

$$\Phi_{m,n}(h) = \begin{cases} 0 \quad n > m \\ (-h)^n \sum_{k=0}^{m-n} \begin{pmatrix} m \\ m-n-k \end{pmatrix} (n+k-1) h^k 1 \leq n \leq m \\ 1 \quad n \leq 0 \end{cases} \tag{6.160}$$

Many of the mathematical methods employed in nonlinear problems may be successfully tested on the Blasius equation, which seems to be very simple, but what appears to be a simple problem is really not that simple, it is very difficult to search for an approximate analytical solution. So, it has appeared ever-increasing interest of scientists and engineers in searching for analytical solution of the Blasius equation. An explicit, totally analytic approximate solution for Blasius equation, by means of Homotopy Analysis Method is obtained by Liao [114]:

$$u(\eta) = \eta + \lim_{M\to\infty} \left[\sum_{m=0}^{M} b_0^{m,0} + \sum_{n=1}^{M+1} \exp(-n\lambda\eta) \left(\sum_{m=n-1}^{M} \sum_{k=0}^{2(m-n+1)} b_k^{m,n} \eta^k \right) \right] \tag{6.161}$$

where the coefficients $b_k^{m,n}$ contain the auxiliary parameter h and the so-called spatial scale parameter λ.

J.H. He [115] proposed a perturbation technique coupled with an iteration technique. Comparison with Howarth's numerical solution reveals that the proposed method leads to the approximate value $\sigma = 0.3296$ with 0.73% accuracy.

The variational iteration method is applied for a reliable treatment by Wazwaz [111]. Abbasbandy [112] applied in his work an efficient modification of the standard Adomian decomposition method.

In the following, we apply OHAM to Blasius equation. We note that there are many possibilities to choose the auxiliary function $H(\eta,p)$, initial guess $u_0(\eta)$and linear operator L. Using the initial conditions Eq. 6.156a, we choose the initial guess u_0 as

$$u_0(\eta) = \eta - \frac{1}{k} + \frac{1}{k}e^{-k\eta} \tag{6.162}$$

where k is a positive unknown constant. Then, from Eqs. 6.156, 6.16 and 6.162 we choose the linear operator in the form:

$$L(f(\eta)) = u''' + ku'' \tag{6.163}$$

The nonlinear operator N given by Eqs. 6.21 and 6.155 becomes:

$$\begin{aligned} N(\eta, p) = u_0^{'''} + \frac{1}{2}u_0 u_0^{''} + p[u_1^{'''} + \frac{1}{2}(u_0 u_1^{''} + u_1 u_0^{''})] + \\ + p^2[u_2^{'''} + \frac{1}{2}(u_0 u_2^{''} + u_1 u_1^{''} + u_2 u_0^{''})] + \ldots \end{aligned} \tag{6.164}$$

From Eq. 6.164 we obtain

$$N_0(u_0) = u_0^{'''} + \frac{1}{2}u_0 u_0^{''} \tag{6.165}$$

$$N_1(u_0, u_1) = u_1^{'''} + \frac{1}{2}(u_0 u_1^{''} + u_1 u_0^{''}) \tag{6.166}$$

$$N_2(u_0, u_1, u_2) = u_2^{'''} + \frac{1}{2}(u_0 u_2^{''} + u_1 u_1^{''} + u_2 u_0^{''}) \tag{6.167}$$

Substituting Eq. 6.162 into Eq. 6.165, it is obtained:

$$N_0(u_0) = (\frac{1}{2}k\eta - k^2 - \frac{1}{2})e^{-k\eta} + \frac{1}{2}e^{-2k\eta} \tag{6.168}$$

Case 1. If we choose $m = 1$ into Eq. 6.23 and using Eq. 6.168, we consider the auxiliary function $h_1(\eta, p)$ in the form

$$h_1(\eta, p) = C_1 + C_2 e^{-k\eta} + C_3 e^{-2k\eta} \tag{6.169}$$

where C_1, C_2 and C_3 are unknown constants at this moment. In this case, Eq. 6.19 becomes:

$$\begin{aligned} & u_1^{'''} + k u_1^{''} = (C_1 + C_2 e^{-k\eta} + C_3 e^{-2k\eta})[(\frac{1}{2}k\eta - k^2 - \frac{1}{2})e^{-k\eta} + \frac{1}{2}e^{-2k\eta}] \\ & u_1(0) = u_1^{'}(0) = u_1^{'}(\infty) = 0 \end{aligned} \tag{6.170}$$

The solution of Eq. 6.170 becomes:

$$u_1(\eta) = \frac{C_1}{k} - \frac{5C_1}{8k^3} + \frac{C_2}{4k} - \frac{C_2}{18k^3} + \frac{C_3}{9k} - \frac{11C_3}{864k^3} + \\ + \left[\frac{C_1}{4k}\eta^2 + \left(\frac{C_1}{2k^2} - C_1\right)\eta + \frac{3C_1}{4k^3} - \frac{C_1}{k} + \frac{5C_2}{24k^3} - \frac{C_2}{2k} + \\ + \frac{C_3}{36k^3} - \frac{C_3}{6k}\right]e^{-k\eta} + \left[-\frac{C_2}{8k^2}\eta - \frac{C_1}{8k^3} + \frac{C_2}{4k} - \frac{C_2}{8k^3}\right]e^{-2k\eta} + \\ + \left[-\frac{C_3}{36k^2}\eta - \frac{C_2}{36k^3} + \frac{C_3}{18k} - \frac{C_3}{216k^3}\right]e^{-3k\eta} - \frac{C_3}{96k^3}e^{-4k\eta} \tag{6.171}$$

The first approximate solution Eq. 6.23 is

$$\bar{u}(\eta) = u_0(\eta) + u_1(\eta) \tag{6.172}$$

Substituting Eqs. 6.162 and 6.171 into 6.172 we obtain:

$$\bar{u}(\eta) = \eta - \frac{1}{k} + \frac{C_1}{k} - \frac{5C_1}{8k^3} + \frac{C_2}{4k} - \frac{C_2}{18k^3} + \frac{C_3}{9k} - \frac{11C_3}{864k^3} + \\ + \left[\frac{C_1}{4k}\eta^3 + \left(\frac{C_1}{2k^2} - C_1\right)\eta + \frac{3C_1}{4k^3} + \frac{1 - C_1}{k} + \frac{5C_2}{24k^3} - \frac{C_2}{2k} + \\ + \frac{C_3}{36k^3} - \frac{C_3}{6k}\right]e^{-k\eta} + \left[-\frac{C_2}{8k^2}\eta + \frac{C_2}{4k} - \frac{C_1}{8k^3} - \frac{C_2}{8k^3}\right]e^{-2k\eta} + \\ + \left[-\frac{C_3}{36k^2}\eta - \frac{C_2}{36k^3} + \frac{C_3}{18k} - \frac{C_3}{216k^3}\right]e^{-3k\eta} - \frac{C_3}{96k^3}e^{-4k\eta} \tag{6.173}$$

The residual Eq. 6.24 becomes in this case:

$$R(\eta, C_1, C_2, C_3, k) = \bar{u}'''(\eta) + \frac{1}{2}\bar{u}(\eta)\bar{u}''(\eta) \tag{6.174}$$

From the Eqs. 6.26, which become

$$\frac{\partial J}{\partial C_i} = \frac{\partial J}{\partial k} = 0, i = 1, 2, 3$$

we obtain:

$$C_1 = 0.117625, \; C_2 = -5.47261, \; C_3 = 5.20066, \; k = 0.453827 \tag{6.175}$$

From the Eq. 6.173 we obtain:

$$u''(0) = 0.330626 \tag{6.176}$$

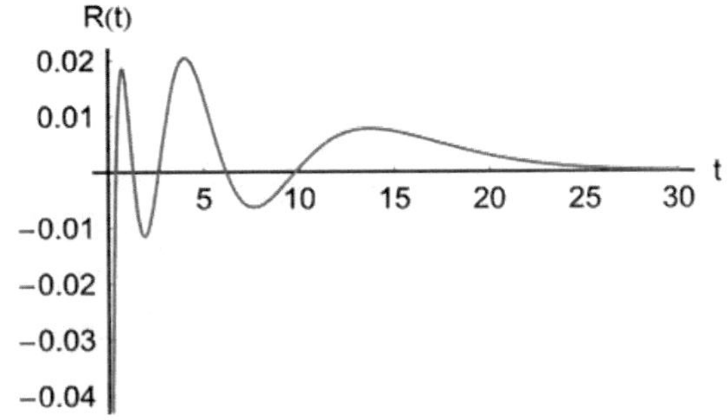

Fig. 6.13 Residual of Eq. 6.173

From Fig. 6.13 and. Eq. 6.176 we can observe that the first approximate solution Eq. 6.173 is in good agreement with the exact solution

Case 2. If we choose $m = 2$ and the auxiliary function of the form:

$$H(p,\eta) = pD_1 + p^2(D_2\eta^2 + D_3\eta + D_4 + D_5e^{-k\eta}) \tag{6.177}$$

where $D_1 \ldots .D_5$ are unknown constants, Eq. 6.19 becomes in this case:

$$u_1''' + ku_1'' = D_1\left[\left(\frac{k}{2}\eta - k^2 - \frac{1}{2}\right)e^{-k\eta} + \frac{1}{2}e^{-2k\eta}\right] \tag{6.178}$$

and the solution is given by

$$\begin{aligned} u_1(\eta) = &\frac{D_1}{k} - \frac{5D_1}{8k^3} + \\ &+ \left[\frac{D_1}{4k}\eta^2 + \left(\frac{D_1}{2k^2} - D_1\right)\eta + \frac{3D_1}{4k^3} - \frac{D_1}{k}\right]e^{-k\eta} - \frac{D_1}{8k^3}e^{-2k\eta} \end{aligned} \tag{6.179}$$

Equation 6.20 for $k = 2$ can be written as

$$\begin{aligned} u_2''' + ku_2'' - (u_1'' + ku_1'') = D_1\left[u_1''' + \frac{1}{2}\left(u_0u_1'' + u_1u_0''\right)\right] + \\ + \left(D_2\eta^2 + D_3\eta + D_4 + D_5e^{-k\eta}\right)\left(u_0'' + ku_0''\right) \end{aligned} \tag{6.180}$$

Substituting Eqs. 6.162 and 6.179 into Eq. 6.180, it is obtained:

$$u_2''' + ku_2'' = \left[\frac{1}{8}k(D_1^2+4D_2)\eta^3 + \left(\frac{1}{2}kD_3 - k^2D_2 - \frac{1}{2}D_2 - \frac{3}{8}D_1^2 - \right.\right.$$
$$\left. - \frac{3}{4}k^2D_1^2\right)\eta^2 + \left(\frac{1}{2}kD_1 + 2kD_1^2 + k^3D_1^2 + \frac{3}{8}\frac{D_1^2}{k} + \frac{1}{2}kD_4 - k^2D_3 - \right.$$
$$\left. -\frac{1}{2}D_3\right)\eta - k^2D_1 - \frac{1}{2}D_1 - \frac{3}{4}D_1^2 - 2k^2D_1^2 - \frac{7D_1^2}{16k^2} - k^2D_4 -$$
$$\left. -\frac{1}{2}D_4\right]e^{-k\eta} + \left[\frac{1}{4}(D_1^2+2D_2)\eta^2 + \left(\frac{1}{2}kD_5 + \frac{1}{2}D_3 - kD_1^2 - \frac{D_1^2}{4k}\right)\eta + \right.$$
$$\left. + \frac{1}{2}D_1 + D_1^2 + \frac{3D_1^2}{4k^2} + \frac{1}{2}D_4 - k^2D_5 - \frac{1}{2}D_5\right]e^{-2k\eta} + \left(\frac{1}{2}D_5 - \frac{5D_1^2}{16k^2}\right)e^{-3k\eta} \tag{6.181}$$

In the initial conditions

$$f_2(0) = f_2'(0) = f_2'(\infty) = 0, \tag{6.182}$$

Equation 6.181 has the solution

$$f_2(\eta) = \frac{8k^2-5}{8k^3}D_1 + \frac{96k^2-147}{16k^5}D_2 + \frac{16k^2-17}{8k^4}D_3 + \frac{8k^2-5}{8k^3}D_4 +$$
$$+ \frac{9k^2-2}{36k^3}D_5 + \frac{360k^2-359}{288k^5}D_1^2 + \left[\frac{D_1^2+4D_2}{32k}\eta^4 + \left(\frac{5-2k^2}{6k^2}D_2 + \right.\right.$$
$$\left. + \frac{D_3}{6k} + \frac{1-2k^2}{8k^2}\right)\eta^3 + \left(\frac{D_1}{4k} + \frac{7-4k^2}{2k^3}D_2 + \frac{3-2k^2}{4k^2}D_3 + \frac{D_4}{4k}\right.$$
$$\left. + \frac{8k^4-8k^2+9}{16k^3}D_1^2\right)\eta^2 + \left(\frac{1-2k^2}{2k^2}D_1 + \frac{9-6k^2}{k^4}D_2 + \frac{2-2k^2}{k^3}D_3 + \right.$$
$$\left. + \frac{1-2k^2}{2k^2}D_4 + \frac{17-20k^2}{16k^4}D_1^2\right)\eta + \frac{3-4k^2}{4k^3}D_1 + \frac{79-48k^2}{8k^5}D_2 +$$
$$\left. + \frac{19-16k^2}{8k^4}D_3 + \frac{3-4k^2}{4k^3}D_4 + \frac{5-12k^2}{24k^3}D_5 + \frac{157-144k^2}{96k^5}D_1^2\right]e^{-k\eta} +$$
$$+ \left[-\frac{D_1^2+2D_2}{16k^3}\eta^2 + \left(-\frac{D_2}{2k^4} - \frac{D_3}{8k^3} - \frac{D_5}{8k^2} + \frac{4k^2-3}{16k^4}D_1^2\right)\eta - \frac{D_1}{8k^3} - \right.$$
$$\left. - \frac{11D_2}{16k^5} - \frac{D_3}{4k^4} - \frac{D_4}{8k^3} + \frac{2k^2-1}{8k^3}D_5 + \frac{8k^2-13}{32k^5}D_1^2\right]e^{-2k\eta} +$$
$$+ \frac{5D_1^2 - 8k^2D_5}{288k^5}e^{-3k\eta} \tag{6.183}$$

The second approximate solution Eq. 6.23 is

$$\bar{u}(\eta) = u_0(\eta) + u_1(\eta) + u_2(\eta) \tag{6.184}$$

Substituting Eqs. 6.162, 6.179 and 6.183 into Eq. 6.184 we have:

$$\begin{aligned}\bar{u}(\eta) &= \eta + A_0 + \left(A_{14}\eta^4 + A_{13}\eta^3 + A_{12}\eta^2 + A_{11}\eta + A_{10}\right)e^{-k\eta} + \\ &+ \left(A_{22}\eta^2 + A_{21}\eta + A_{20}\right)e^{-2k\eta} + A_{30}e^{-3k\eta}\end{aligned} \tag{6.185}$$

where

$$\begin{aligned}
A_0 &= -\frac{1}{k} + \frac{8k^2-5}{4k^3}D_1 + \frac{96k^2-147}{16k^5}D_2 + \frac{16k^2-17}{8k^4}D_3 + \\
&+ \frac{8k^2-5}{8k^3}D_4 + \frac{9k^2-2}{36k^3}D_5 + \frac{360k^2-359}{288k^5}D_1^2 \\
A_{14} &= \frac{D_1^2+4D_2}{32k}, \quad A_{13} = \frac{5-2k^2}{6k^2}D_2 + \frac{D_3}{6k} + \frac{1-2k^2}{8k^2}D_1^2 \\
A_{12} &= \frac{D_1}{2k} + \frac{7-4k^2}{2k^3}D_2 + \frac{3-2k^2}{4k^2}D_3 + \frac{D_4}{4k} + \frac{8k^4-8k^2+9}{16k^3}D_1^2 \\
A_{11} &= \frac{1-2k^2}{k^2}D_1 + \frac{9-6k^2}{k^4}D_2 + \frac{2-2k^2}{k^3}D_3 + \frac{1-2k^2}{2k^2}D_4 + \frac{17-20k^2}{16k^4}D_1^2 \\
A_{10} &= \frac{1}{k} + \frac{3-4k^2}{2k^3}D_1 + \frac{79-48k^2}{8k^5}D_2 + \frac{19-16k^2}{8k^4}D_3 + \\
&+ \frac{3-4k^2}{4k^3}D_4 + \frac{5-12k^2}{24k^3}D_5 + \frac{157-144k^2}{96k^5}D_1^2 \\
A_{22} &= -\frac{D_1^2+2D_2}{16k^3} \\
A_{21} &= -\frac{D_2}{2k^4} - \frac{D_3}{8k^3} - \frac{D_5}{8k^2} + \frac{4k^2-3}{16k^4}D_1^2 \\
A_{20} &= -\frac{D_1}{4k^3} - \frac{11D_2}{16k^5} - \frac{D_3}{4k^4} - \frac{D_4}{8k^3} + \frac{2k^2-1}{8k^3}D_5 + \frac{8k^2-13}{32k^5}D_1^2 \\
A_{30} &= \frac{5D_1^2-8k^2D_5}{288k^5}
\end{aligned} \tag{6.186}$$

In this case, the residual Eq. 6.24 is

$$R(\eta, D_1, ..., D_5, k) = \bar{u}'''(\eta) + \frac{1}{2}\bar{u}(\eta)\bar{u}''(\eta) \tag{6.187}$$

Equations 6.26 become:

$$\frac{\partial J}{\partial D_i} = \frac{\partial J}{\partial k} = 0, \quad i = 1, 2, ..., 5 \tag{6.188}$$

From Eq. 6.188 the constants D_i and k are determined as:

$$\begin{aligned} &D_1 = -2.036463964;\ D_2 = -0.874158223;\ D_3 = 4.585694066; \\ &D_4 = -3.825618257;\ D_5 = 0.319068718;\ k = 0.948929591 \end{aligned} \tag{6.189}$$

From Eq. 6.188 we obtain

$$\bar{u}''(0) = 0.330667 \tag{6.190}$$

The second analytic approximate solution agrees very well with Howarth's numerical results [113], as shown in Table 6.2 and in Fig. 6.14. Clearly, the higher the order of approximation is, the better the approximate results will be.

Table 6.2 Comparison of the analytic approximate solutions $\bar{u}(\eta)$in cases m = 1 and m = 2 with Howarth's numerical results [113]

η	$\bar{u}(\eta)$given by Eq. 6.173	$\bar{u}(\eta)$given by Eq. 6.185	Numerical results [113]
0.4	0.02562	0.02676	0.0266
2	0.62234	0.65177	0.6500
3.2	1.50473	1.57388	1.5690
4	2.21506	2.31028	2.3058
5	3.16978	3.28834	3.2833
6	4.15538	4.28703	4.2797
7	5.14907	5.28814	5.2793
8	6.14257	6.28756	6.2793
10	8.12469	8.28358	8.2793

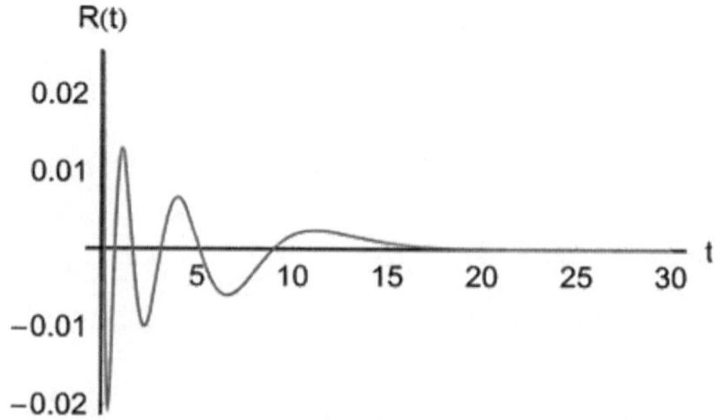

Fig. 6.14 Residual of Eq. 6.185

It is clear that OHAM provides us with a simple way to optimally control and adjust the convergence of the solution and can give good approximations in few terms.

6.7 Oscillations of a Uniform Cantilever Beam Carrying an Intermediate Lumped Mass and Rotary Inertia

The problems related to large-amplitude oscillations of non-linear engineering structures have received considerable attention in the past years [116, 117]. Engineering structures undergoing large-amplitude oscillations often involve discretizing the structure, when free vibration analysis is performed, and results in a temporal problem having inertia and static non-linearities. Such problems are not amenable to exact treatment because of their complexity and approximate techniques must be resorted to [118].

The problem of large-amplitude vibration of a uniform cantilever beam approached in this section is of practical interest because many engineering structures can be modelled as a slender, flexible cantilever beam carrying a lumped mass with rotary inertia at an intermediate point along its span hence they experience large-amplitude vibration.

Often linearization techniques are employed in order to approximate such non-linear problems. One may question the accuracy of using such a linear mode method, which are a frequently used method in the analysis of non-linear continuous systems to approximate the large amplitude non-linear behaviour [118].

In this section we consider a clamped beam at the base, free at the tip, which caries a lumped mass and rotary inertia at an arbitrary intermediate point along its span. The beam is considered to be uniform of constant length and mass per unit length and the thickness of this conservative beam is assumed to be small compared to the length so that the effects of rotary inertia and shearing deformation will be ignored. Moreover, the beam is assumed to be inextensible, which implies that the length of beam's neutral axis remains constant during the motion. These assumptions are the same as those used in references [117, 118]. In these conditions one can derive the discrete, single-mode, of order three nonlinearities, beam temporal problem [118]:

$$\frac{d^2u}{dt^2} + u + \alpha u^2 \frac{d^2u}{dt^2} + \alpha u \left(\frac{du}{dt}\right)^2 + \beta u^3 = 0 \tag{6.191}$$

subject to the initial conditions

$$u(0) = A, \quad \frac{du}{dt}(0) = 0 \tag{6.192}$$

This system, where α and β are modal constants which result from the discretization procedure [118], describes the large-amplitude free vibrations of the considered slender inextensible cantilever beam, which is assumed undergoing planar flexural vibrations. The third and fourth-terms in Eq. 6.191 represent inertia-type cubic non-linearity arising from the inextensibility assumption. The last term is a static-type cubic non-linearity associated with the potential energy stored in bending.

In order to analytically solve this problem, we introduce a new independent variable

$$\tau = \Omega t \tag{6.193}$$

then Eqs. 6.191 and 6.192 become

$$u'' + k^2 u + \alpha u^2 u'' + \alpha u u'^2 + \beta k^2 u^3 = 0 \tag{6.194}$$

and

$$u(0) = A, \quad u'(0) = 0 \tag{6.195}$$

where the prime denotes differentiation with respect to τ and $k = \Omega^{-1}$. The new independent variable τ is chosen in such a way that the solution of Eq. 6.194, which satisfies the assigned initial conditions in Eq. 6.195 is a periodic function of τ, of period 2π. The period of the corresponding non-linear oscillation is given by $T = 2\pi/\Omega$. Here, both periodic solution $u(\tau)$ and frequency Ω depend on A.

Under the transformation

$$u(\tau) = Ax(\tau) \tag{6.196}$$

Equation 6.194 becomes

$$x'' + k^2 x + a x^2 x'' + a x x'^2 + b k^2 x^3 = 0 \tag{6.197}$$

where $a = \alpha A^2$, $b = \beta A^2$ and the initial conditions become

$$x(0) = 1, \quad x'(0) = 0 \tag{6.198}$$

From Eqs. 6.16, 6.19 and 6.20 ($m = 2$), we obtain the following equations:

$$x_0'' + x_0 = 0, \quad x_0(0) = 1, \quad x_0'(0) = 0 \tag{6.199}$$

$$\begin{aligned} &x_1'' + x_1 = h_1(\tau, C_i)[x_0'' + k_0^2 x_0 + a x_0^2 x_0'' + a x_0 x_0' 2 + b k_0^2 x_0^3] \\ &x_1(0) = x_1'(0) = 0 \end{aligned} \tag{6.200}$$

$$\begin{aligned} x_2'' + x_2 - (x_1'' + x_1) = h_1(\tau, C_i)[x_1'' + k_0^2 x_1 + 2k_0 k_1 x_0 + \\ + a(2x_0 x_1 x_0'' + x_0^2 x_1'' + 2x_0 x'_0 x_1' + x_0'^2 x_1) + \\ + b(2k_0 k_1 x_0^3 + 3k_0^2 x_0^2 x_1)] + h_2(\tau, C_i)[x_0'' + k_0^2 x_0 + \\ + a(x_0^2 x_0'' + x_0 x_0'^2) + bk_0^2 x_0^3], x_2(0) = x'_2(0) = 0 \end{aligned} \tag{6.201}$$

where

$$k = k_0 + k_1 + k_2 + \ldots \tag{6.202}$$

Equation 6.199 has the solution

$$x_0(\tau) = \cos\tau \tag{6.203}$$

If this result is substituted into Eq. 6.200 and assuming $h_1(\tau,C_i) = C_1$, we obtain the following equation

$$\begin{aligned} x''_1 + x_1 + C_1\left(k_0^2 - 1 + \frac{3bk_0^2 - 2a}{4}\right)\cos\tau + \frac{C_1(2a + ab - b)}{3b + 4}\cos 3\tau = 0, \\ x_1(0) = x'_1(0) = 0 \end{aligned} \tag{6.204}$$

where C_1 is an unknown constant at this moment. Avoiding the presence of a secular term needs:

$$k_0^2 = \frac{2a + 4}{3b + 4} \tag{6.205}$$

with this requirement, the solution of Eq. 6.204 is

$$x_1(\tau) = \frac{C_1(2a + ab - b)}{8(3b + 4)}(\cos 3\tau - \cos\tau) \tag{6.206}$$

If we substitute Eqs. 6.203, 6.205 and 6.206 into Eq. 6.201, we obtain the equation in x_2:

$$x''_2 + x_2 = \frac{C_1(2a+ab-b)}{3b+4}\cos 3\tau + C_1\left[\frac{-C_1(2a+ab-b)(2a+3ab+3b)}{8(3b+4)^2} + \right.$$
$$\left. + \frac{k_0 k_1}{2}(3b+4)\right]\cos\tau + C_1\left[\frac{bk_0k_1}{2} - \frac{C_1(2a+ab-b)(3a+8)}{8(3b+4)}\right]\cos 3\tau +$$
$$+\left[\frac{3C_1^2(2a+ab-b)(b-4ab-6a)}{8(3b+4)^2}\right]\cos 5\tau +$$
$$+ h_2(\tau, C_i)\left[\frac{b-2a-ab}{3b+4}\right]\cos 3\tau,\ x_2(0) = x'_2(0) = 0 \tag{6.207}$$

No secular terms in $x_2(\tau)$ requires that

$$k_1 = \frac{C_1(2a+ab-b)(2a+3ab+3b)}{4k_0(3b+4)^3} \tag{6.208}$$

From Eqs. 6.201, 6.205 and 6.208 we obtain

$$k = \sqrt{\frac{2a+4}{3b+4}} + \frac{C_1(2a+ab-b)(2a+3ab+3b)}{4(3b+4)^3}\sqrt{\frac{3b+4}{2a+4}} \tag{6.209}$$

and therefore

$$\Omega = k^{-1} = \left(\sqrt{\frac{2a+4}{3b+4}} + \frac{C_1(2a+ab-b)(2a+3ab+3b)}{4(3b+4)^3}\sqrt{\frac{3b+4}{2a+4}}\right)^{-1} \tag{6.210}$$

Substituting Eq. 6.208 into Eq. 6.207 and considering C_2, C_3 and C_4 unknown constants in $h_2(\tau) = C_2 + 2C_3\cos 2\tau + 2C_4\cos 4\tau$, we obtain $C_4 = -C_3$ and

$$x_2(\tau) = \left[\frac{(C_1+C_2)(b-2a-ab)}{8(3b+4)} - \frac{C_1^2(2a+ab-b)(3a+8)}{64(3b+4)} + \right.$$
$$\left. + \frac{bC_1^2(2a+ab-b)(2a+3ab+3b)}{64(3b+4)^3}\right](\cos\tau - \cos 3\tau) +$$
$$+ \left[\frac{C_1^2(2a+ab-b)(b-4ab-6a)}{64(3b+4)^2} + \frac{C_3(b-2a-ab)}{24(3b+4)}\right](\cos\tau - \cos 5\tau) -$$
$$- \frac{C_3(b-ab-2a)}{48(3b+4)}(\cos\tau - \cos 7\tau) \tag{6.211}$$

The second-order approximate solution is

$$\bar{x}(\tau) = x_0(\tau) + x_1(\tau) + x_2(\tau)$$

where x_0, x_1 and x_2 are given by Eqs. 6.203, 6.206 and 6.211 respectively.

Using the transformations Eqs. 6.193 and 6.196, the second order approximate solution of Eq. 6.194 becomes:

$$\bar{u}(t) = B\cos\Omega t + C\cos 3\Omega t + D\cos 5\Omega t + E\cos 7\Omega t \tag{6.212}$$

where B, C, D, E and Ω are given respectively by:

$$B = A - \frac{F(12C_1 + 6C_2 + C_3)}{48(3\beta A^2 + 4)} - \frac{C_1^2 F(18\alpha\beta A^5 + 33\beta^2 A^4 + 52\alpha\beta A^4 + 36\alpha A^2 + 94\beta A^2 + 64)}{32(3\beta A^2 + 4)^3}$$

$$C = \frac{(2C_1 + C_2)F}{8(3\beta A^2 + 4)} + \frac{C_1^2 F(24\alpha\beta^2 A^5 + 70\alpha\beta A^4 + 69\beta^2 A^4 + 48\alpha A^2 + 192\beta A^2 + 128)}{64(3\beta A^2 + 4)^3}$$

$$D = \frac{C_1^2 F(6\alpha A^2 + 4\alpha\beta A^4 - \beta A^2)}{64(3\beta A^2 + 4)^2} + \frac{C_3(2\alpha A^2 + \alpha\beta A^4 - \beta A^2)}{24(3\beta A^2 + 4)}$$

$$E = \frac{C_3(\beta A^2 - 2\alpha A^2 - \alpha\beta A^4)}{48(3bA^2 + 4)}, \quad F = 2\alpha A^3 + \alpha\beta A^5 - \beta A^3 \tag{6.213}$$

$$\Omega = \left(\sqrt{\frac{2\alpha A^2 + 4}{3\beta A^2 + 4}} + \frac{C_1 F(2\alpha A + 3\alpha\beta A^3 + 3\beta A)}{4(3\beta A^2 + 4)^3}\sqrt{\frac{3\beta A^2 + 4}{2\alpha A^2 + 4}}\right)^{-1} \tag{6.214}$$

The constants C_i, $i = 1,2,3,4$ will be obtained using the least square method.

In this analysis, periodic solutions are analyzed for the cantilever beam under study. Beside the role of the large amplitude A, a special role is played by the modal constants α and β, which depends on the inertia parameters of the attached inertia element with mass M and rotary inertia J. We do not approach the simplest cases when the modal constants α, β are small (0.1 or 0.2), because in these cases it is easy to achieve accurate periodic solutions even for large amplitudes using known procedures. Difficulties appear when these modal constants become larger [119] and the oscillator experiences large amplitudes. Here, the meaning of "large" implies the fact that the peak amplitude reach a value where the non-linear terms are of an order comparable to that of the linear ones. More specific, the amplitude may be of the order of beam length.

We illustrate the accuracy of our procedure for large modal constants and large amplitudes comparing the obtained approximate analytical solutions with the numerical integration results obtained using a fourth-order Runge–Kutta method. We will also compare these results with published results [119].

We further consider larger values for the modal constants α and β (1 or 2) and we also consider large values of the initial amplitude A (5 or 10). In order to prove the accuracy of the obtained results, two examples are analysed.

Example (a) For the modal constants $\alpha = 1$, $\beta = 1$, and the initial amplitude $A = 5$, following the procedure described above it is obtained the convergence-control constants:

$$C_1 = -0.102968962;\ C_2 = -0.00561304;\ C_3 = 0.03843892;\ C_4 = -0.03843892$$

and consequently the approximate periodic solution becomes:

$$\begin{aligned}\bar{u}(t) &= 5.317402609\cos\Omega t - 0.576807191\cos 3\Omega t + \\ &+ 0.292349339\cos 5\Omega t - 0.032944754\cos 7\Omega t\end{aligned} \tag{6.215}$$

where $\Omega = 1.342143172$, while for the same modal constants α and β, when the initial amplitude A raises to $A = 10$, we obtain the constants:

$$\begin{aligned}&C_1 = -0.029698817;\ C_2 = -0.002408906;\\ &C_3 = 0.001571857;\ C_4 = -0.001571857\end{aligned}$$

and the approximate periodic solution in this case will be:

$$\begin{aligned}\bar{u}(t) &= 10.686783311\cos\Omega t - 1.307660795\cos 3\Omega t + \\ &+ 0.631757266\cos 5\Omega t - 0.010879766\cos 7\Omega t\end{aligned} \tag{6.216}$$

where $\Omega = 1.382367422$.

Example (b) For the modal constants $\alpha = 2$, $\beta = 2$, and the initial amplitude $A = 5$, following the same procedure we obtain the convergence-control constants:

$$\begin{aligned}&C_1 = -0.056226544;\ C_2 = -0.003304042;\\ &C_3 = -0.073429731;\ C_4 = 0.073429731\end{aligned}$$

and consequently the approximate periodic solution is in this case:

$$\begin{aligned}\bar{u}(t) &= 5.47307199\cos\Omega t - 0.618622196\cos 3\Omega t + \\ &+ 0.018895842\cos 5\Omega t + 0.126654363\cos 7\Omega t\end{aligned} \tag{6.217}$$

where $\Omega = 1.374317516$, while for the same modal constants α and β, but for a larger initial amplitude $A = 10$, we obtain the constants:

$$C_1 = -0.015351906;\ C_2 = -0.001411507;$$
$$C_3 = -0.000041392;\ C_4 = 0.000041392$$

and the approximate periodic solution in this case:

$$\bar{u}(t) = 10.69151238\cos\Omega t - 1.344253519\cos 3\Omega t + \\ +0.652167205\cos 5\Omega t + 0.000014221\cos 7\Omega t \tag{6.218}$$

where $\Omega = 1.400493712$.

Figures 6.15 and 6.16 present a comparison of the obtained analytical solutions Eqs. 6.215 and 6.216 with numerical ones and also with known results [119] for the modal constants $\alpha = \beta = 1$ and initial amplitudes $A = 5$ and $A = 10$, respectively,

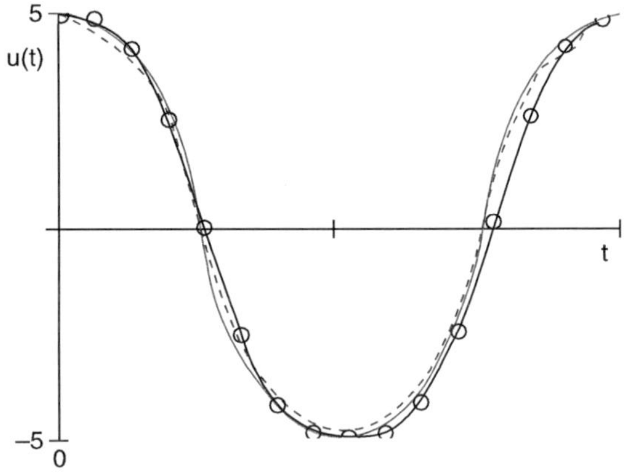

Fig. 6.15 Comparison of the results for $\alpha = \beta = 1$ and initial amplitudes $A = 5$ for Eq. 6.194: ______ numerical solution; _ _ _ _ OHAM Eq. 6.215; -o-o-o-o results from [119]

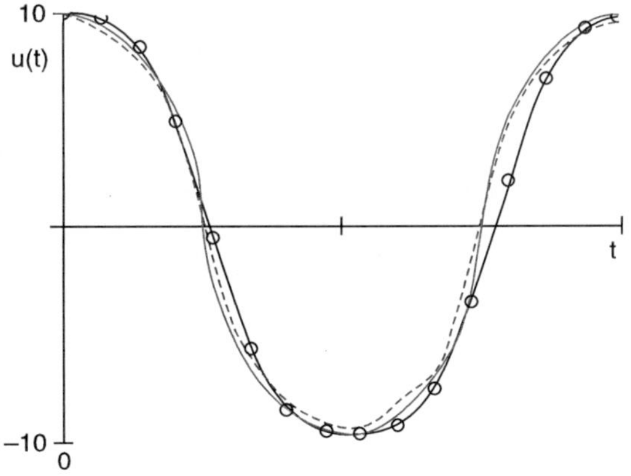

Fig. 6.16 Comparison of the results for $\alpha = \beta = 1$ and initial amplitudes $A = 10$ for Eq. 6.194: ______ numerical solution; _ _ _ _ OHAM Eq. 6.216; -o-o-o-o results from [119]

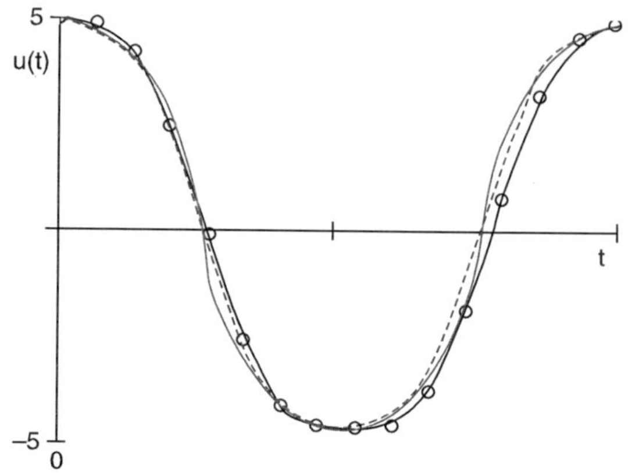

Fig. 6.17 Comparison of the results for $\alpha = \beta = 2$ and initial amplitudes $A = 5$ for Eq. 6.194: ______ numerical solution; _ _ _ _ OHAM Eq. 6.217; -o-o-o-o results from [119]

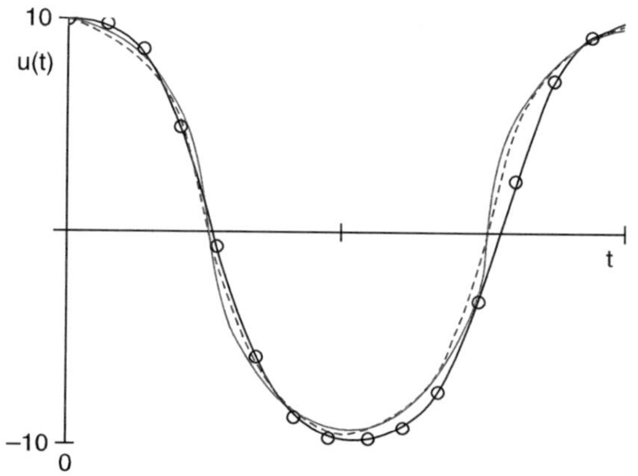

Fig. 6.18 Comparison of the results for $\alpha = \beta = 2$ and initial amplitudes $A = 10$ for Eq. 6.194: ______ numerical solution; _ _ _ _ OHAM Eq. 6.218; -o-o-o-o results from [119]

for a period of the motion. Similar comparisons are presented in Figures 6.6.17 and 6.6.18 for the modal constants α=β=2.

It can be seen from the above Figures that the solutions obtained by the proposed procedure is nearly identical with the numerical solutions obtained using a fourth-order Runge–Kutta method. Moreover, the analytical solutions obtained by our procedure proved to be more accurate than other known results obtained by combining the linearization of the governing equation with the method of harmonic balance [51], especially for large values of the modal constants α and β. In the case of large modal constants $\alpha = \beta = 2$ and very large amplitude $A = 10$, Fig. 6.18 indicates an evident discrepancy of numerical solutions and solutions from [51], while the analytic solution obtained in this paper is still valid.

6.8 Oscillations of an Electrical Machine

Electrical machines are widely used in engineering applications and industry due to their reliability. They are dynamical systems encountering dynamical phenomena which can be detrimental to the system. From engineering point of view it is very important to predict the nonlinear dynamic behaviour of complex dynamical systems, such as the electrical machines. This is a significant stage in the design process, before the machine is exploited in real conditions, avoiding in this way undesired dynamical phenomena which could damage the system. Basically, the electric machines share the same dynamical problems with classical rotor systems, having specific sources of excitation, which lead to nonlinear vibration occurrence.

The main sources of dynamic problems are the unbalanced forces of the rotor [120], bad bearings or nonlinear bearings [121], mechanical looseness, misalignments, other electrical and mechanical faults which generate nonlinear vibration in the system. These problems are usually solved by numerical simulations, experimental investigations or by analytical developments [85, 122].

In what follows, the investigated electrical machine is considered to be supported by nonlinear bearings and the assumption made in development of the mathematical model is that these bearings are characterised by nonlinear stiffness of Duffing type. In the same time, the entire dynamical system is subjected to a parametric excitation caused by an axial thrust and a forcing excitation caused by an unbalanced force of the rotor, which is obviously harmonically shaped. In these conditions, the dynamical behaviour of the investigated electrical machine will be governed by the following second-order strongly nonlinear differential equation:

$$m\ddot{u} + k_1(1 - q\sin\omega_2 t)u + k_2u^3 = f\sin\omega_1 t \tag{6.219}$$

with the initial conditions

$$u(0) = A, \quad \dot{u}(0) = 0 \tag{6.220}$$

which can be written in the more convenient form:

$$\ddot{u} + \omega^2 u - \alpha u\sin\omega_2 t + \beta u^3 - \gamma\sin\omega_1 t = 0 \tag{6.221}$$

where $\omega^2 = \frac{k_1}{m}$, $\alpha = \frac{k_1 q}{m}$, $\beta = \frac{k_2}{m}$, $\gamma = \frac{f}{m}$, the dot denotes derivative with respect to time and A is the amplitude of the oscillations. Note that it is unnecessary to assume the existence of any small or large parameter in Eq. 6.221.

The Eq. 6.221 describes a system oscillating with an unknown period T. We switch to a scalar time$\tau = 2\pi t/T = \Omega t$. Under the transformations

$$\begin{aligned} \tau &= \Omega t \\ u(t) &= Ax(\tau) \end{aligned} \tag{6.222}$$

the original Eq. 6.221 becomes

$$\Omega^2 x'' + \omega^2 x - \alpha x \sin\frac{\omega_2}{\Omega}\tau + \beta A^2 x^3 - \frac{\gamma}{A}\sin\frac{\omega_1}{\Omega}\tau = 0 \tag{6.223}$$

with the initial conditions

$$x(0) = 1, \quad x'(0) = 0 \tag{6.224}$$

where the prime denotes the derivative with respect to τ.

By the homotopy technique, we construct a family of Eqs. 6.33 with linear operator L given by Eq. 6.34 and nonlinear operator N in the form

$$\begin{aligned} N[\phi(\tau,p),\Omega(\lambda,p)] &= \Omega^2(p)\frac{\partial^2\phi(\tau,p)}{\partial\tau^2} + (\omega^2+\lambda)\phi(\tau,p) - \alpha\phi(\tau,p)\sin\frac{\omega_2}{\Omega}\tau \\ &+ \beta A^2\phi^3(\tau,p) - \frac{\gamma}{A}\sin\frac{\omega_1}{\Omega}\tau - p\lambda\phi(\tau,p) \end{aligned} \tag{6.225}$$

Form Eq. 6.44 it is obtained the following solution:

$$x_0(\tau) = \cos\tau \tag{6.226}$$

From Eq. 6.225 we obtain:

$$N_0(x_0,\Omega_0,\lambda) = \Omega_0^2 x''_0 + (\omega^2+\lambda)x_0 - \alpha x_0 \sin\frac{\omega_2}{\Omega}\tau + \beta A^2 x_0^3 - \frac{\gamma}{A}\sin\frac{\omega_1}{\Omega}\tau \tag{6.227}$$

Substituting Eq. 6.226 into Eq. 6.227 it is obtained:

$$\begin{aligned} N_0(x_0,\Omega_0,\lambda) &= \left(-\Omega_0^2 + \omega^2 + \lambda + \frac{3}{4}\beta A^2\right)\cos\tau + \frac{\beta A^2}{4}\cos 3\tau - \\ &- \frac{1}{2}\alpha\left[\sin\left(\frac{\omega_2}{\Omega}+1\right)\tau + \sin\left(\frac{\omega_2}{\Omega}-1\right)\tau\right] - \frac{\gamma}{A}\sin\frac{\omega_1}{\Omega}\tau \end{aligned} \tag{6.228}$$

If we choose

$$h_1(\tau, C_i) = C_1 + 2C_2\cos 2\tau + 2C_3\cos 4\tau \tag{6.229}$$

then Eq. 6.45 for $i = 1$ becomes:

$$\begin{aligned}\Omega_0^2(x''_1 + x_1) = {} & (C_1 + C_2)(-\Omega_0^2 + \omega^2 + \lambda\frac{3}{4}\beta A^2)\cos\tau + \\ & + \frac{1}{4}C_1\beta A^2\cos 3\tau + [\frac{1}{4}C_2\beta A^2 - C_2(-\Omega^2 + \omega^2 + \lambda + \frac{3}{4}\beta A^2)]\cos 5\tau - \\ & - \frac{1}{4}C_2\beta A^2\cos 7\tau - \frac{1}{2}(C_1 + C_2)\alpha[\sin\left(\frac{\omega_2}{\Omega} + 1\right)\tau + \sin\left(\frac{\omega_2}{\Omega} - 1\right)\tau] + \\ & + \frac{1}{2}C_2\alpha[\sin\left(\frac{\omega_2}{\Omega} + 5\right)\tau + \sin\left(\frac{\omega_2}{\Omega} - 5\right)\tau] - C_1\frac{\gamma}{A}\sin\frac{\omega_1}{\Omega}\tau - \\ & - C_2\frac{\gamma}{A}[\sin\left(\frac{\omega_1}{\Omega} + 2\right)\tau + \sin\left(\frac{\omega_1}{\Omega} - 2\right)\tau - \sin\left(\frac{\omega_1}{\Omega} + 4\right)\tau - \sin\left(\frac{\omega_1}{\Omega} - 4\right)\tau]\end{aligned} \tag{6.230}$$

Avoiding the presence of a secular term into Eq. 6.230 needs:

$$\Omega_0^2 = \omega^2 + \lambda + \frac{3}{4}\beta A^2 \tag{6.231}$$

With this requirement, the solution of Eq. 6.230 is

$$x_1(\tau) = M\cos\tau + N\cos 3\tau + P\cos 5\tau + Q\cos 7\tau + R\sin\tau \tag{6.232}$$

where

$$\begin{aligned} M = {} & \frac{C_1\beta A^2}{32\Omega_0^2} + \frac{C_2\beta A^2}{192\Omega_0^2},\ N = -\frac{C_1\beta A^2}{32\Omega_0^2},\ P = -\frac{C_2\beta A^2}{96\Omega_0^2},\ Q = \frac{C_2\beta A^2}{192\Omega_0^2}, \\ R = {} & \frac{(C_1 + C_2)\alpha\Omega(2\Omega^2 - \omega_2^2)}{\Omega_0^2(\omega_2^2 - 4\Omega^2)\omega_2} + \frac{C_2\alpha\Omega\omega_2(\omega_2^2 - 26\Omega^2)}{\Omega_0^2(576\Omega^4 - 52\Omega^2\omega_2^2 + \omega_2^4)} + \\ & + \frac{C_1\gamma\Omega\omega_1}{\Omega_0^2(\Omega^2 - \omega_1^2)A} + \frac{2C_2\gamma\Omega\omega_1(5\Omega^2 - \omega_1^2)}{A\Omega_0^2(9\Omega^4 - 10\Omega^2\omega_1^2 + \omega_1^4)} + \\ & + \frac{2C_2\gamma\Omega\omega_1(\omega_1^2 - 17\Omega^2)}{\Omega_0^2(225\Omega^4 - 34\Omega^2\omega_1^2 + \omega_1^4)A}\end{aligned} \tag{6.233}$$

Substituting Eqs. 6.226 and 6.232 into Eq. 6.45 for $i = 2$, we obtain the following equation:

$$\begin{aligned}\Omega_0^2(x''_2 + x_2) = {} & \cos\tau[\frac{3}{4}(C_1 + C_2)\beta A^2(2M + N) - \\ & - A(C_1 + C_2)(2\Omega_0\Omega_1 + \lambda) - \frac{(C_1 + C_2)^2\alpha^2\Omega^2 A}{2\Omega_0^2(\omega_2^2 - 4\Omega^2)} - 25C_2P\Omega_0^2 - \\ & - C_2(\omega^2 + \lambda)P - \frac{3}{4}C_2\beta A^2(N + Q + 2P)] + \frac{3}{4}\beta A^2 C_2 R\sin\tau + H.O.T.\end{aligned} \tag{6.234}$$

No secular term in $x_2(\tau)$ requires that

$$\Omega_1 = \frac{(3C_1 + C_2)\beta^2 A^4}{256\Omega_0^3} - \frac{\lambda}{2\Omega_0} + \frac{(C_1 + C_2)\alpha^2\Omega^2}{4\Omega_0^3(4\Omega^2 - \omega_2^2)} - \\ - \frac{25C_2^2\beta A^2}{192\Omega_0(C_1 + C_2)} + \frac{(\omega^2 + \lambda)C_2^2\beta A^2}{192\Omega_0^3(C_1 + C_2)} + \frac{3(2C_1 + C_2)\beta^2 A^4 C_2}{512\Omega_0^3(C_1 + C_2)} \tag{6.235}$$

$$R = 0 \tag{6.236}$$

From Eqs. 6.231 and 6.235 we obtain the frequency in the form:

$$\Omega = \Omega_0 + \Omega_1 \tag{6.237}$$

The parameter λ can be determined applying the "principle of minimal sensitivity". From Eq. 6.48, we obtain the following condition:

$$\lambda\Omega_0^2(C_1 + C_2) - \frac{3\beta^2 A^4(6C_1^2 + 14C_1C_2 + 5C_2^2)}{256} - \\ - \frac{3(C_1 + C_2)^2\alpha^2\Omega^2}{2(4\Omega^2 - \omega_2^2)} + \frac{C_2^2\beta A^2(29\Omega_0^2 - 3\omega^2 - 3\lambda)}{96} = 0 \tag{6.238}$$

By means of Eqs. 6.235 and 6.238, Eq. 6.237 becomes:

$$\Omega = \Omega_0 - \frac{\lambda}{3\Omega_0} - \frac{23C_2^2\beta A^2}{288\Omega_0(C_1 + C_2)} \tag{6.239}$$

The Eq. 6.236 can be written as:

$$\frac{(C_1 + C_2)\alpha A(2\Omega^2 - \omega_2^2)}{\omega_2(\omega_2^2 - 4\Omega^2)} + \frac{C_2\alpha A\omega_2(\omega_2^2 - 26\Omega^2)}{576\Omega^4 - 52\Omega^2\omega_2^2 + \omega_2^4} + \frac{C_1\gamma\omega_1}{\Omega^2 - \omega_1^2} + \\ + \frac{2C_2\gamma\omega_1(5\Omega^2 - \omega_1^2)}{9\Omega^4 - 10\Omega^2\omega_1^2 + \omega_1^4} + \frac{2C_2\gamma\omega_1(\omega_1^2 - 17\Omega^2)}{225\Omega^4 - 34\Omega^2\omega_1^2 + \omega_1^4} = 0 \tag{6.240}$$

The first-order approximate solution is

$$\bar{x}(\tau) = x_0(\tau) + x_1(\tau)$$

or by means of Eqs. 6.226, 6.232 and 6.222:

$$\bar{u}(t) = \left(A + \frac{C_1\beta A^3}{32\Omega_0^2} + \frac{C_2\beta A^3}{192\Omega_0^2}\right)\cos\Omega t - \frac{C_1\beta A^3}{32\Omega_0^2}\cos 3\Omega t -$$
$$-\frac{C_2\beta A^3}{96\Omega_0^2}\cos 5\Omega t + \frac{C_2\beta A^3}{192\Omega_0^2}\cos 7\Omega t + \frac{(C_1+C_2)\alpha A\Omega^2}{2\Omega_0^2\omega_2(\omega_2+2\Omega)}\sin(\omega_2+\Omega)t +$$
$$\frac{(C_1+C_2)\alpha A\Omega^2}{2\Omega_0^2\omega_2(\omega_2-2\Omega)}\sin(\omega_2-\Omega)t - \frac{C_2\alpha A\Omega^2}{2\Omega_0^2(24\Omega^2+10\Omega\omega_2+\omega_2^2)}\sin(\omega_2 +$$
$$+5\Omega)t - \frac{C_2\alpha A\Omega^2}{2\Omega_0^2(24\Omega^2-10\Omega\omega_2+\omega_2^2)}\sin(\omega_2-5\Omega)t -$$
$$-\frac{C_1\gamma\Omega^2}{\Omega_0^2(\Omega^2-\omega_1^2)}\sin\omega_1 t + \frac{C_2\gamma\Omega^2}{\Omega_0^2(3\Omega^2+4\Omega\omega_1+\omega_1^2)}\sin(\omega_1+2\Omega)t +$$
$$+\frac{C_2\gamma\Omega^2}{\Omega_0^2(3\Omega^2-4\Omega\omega_1+\omega_1^2)}\sin(\omega_1-2\Omega)t -$$
$$\frac{C_2\gamma\Omega^2}{\Omega_0^2(15\Omega^2+8\Omega\omega_1+\omega_1^2)}\sin(\omega_1+4\Omega)t -$$
$$-\frac{C_2\gamma\Omega^2}{\Omega_0^2(15\Omega^2-8\Omega\omega_1+\omega_1^2)}\sin(\omega_1-4\Omega)t \tag{6.241}$$

The constants λ, Ω_0, Ω, C_1 and C_2 can be determined from Eqs. 6.231, 6.238, 6.239, 6.240 and by means of the residual, which reads:

$$R(t,\lambda,\Omega_0,\Omega,C_1,C_2) = \ddot{\bar{u}} + \omega^2\bar{u} - \alpha\bar{u}\sin\omega_2 t + \beta\bar{u}^3 - \gamma\sin\omega_1 t$$

The last condition can be written with collocation method:

$$R(\frac{\pi}{6},\lambda,\Omega_0,\Omega,C_1,C_2) = 0 \tag{6.242}$$

Finally, five equations with five unknowns are obtained. In the case when $\omega_1 = 1.1$, $\omega_2 = 1.5$, $\omega = 1.58$, $\alpha = 2.75$, $\beta = 12.5$, $\gamma = 0.2$, we obtain:

$$\lambda = -1.112020191,\ \Omega_0 = 3.282055275,\ \Omega = 3.394160431,$$
$$\mathrm{C1} = -0.000729317,\ \mathrm{C2} = 0.000714849$$

Figure 6.19 shows the comparison between the approximate solution and the numerical solution obtained by a fourth-order Runge–Kutta method.

It can be seen that the solution obtained by our procedure is nearly identical with that given by the numerical method.

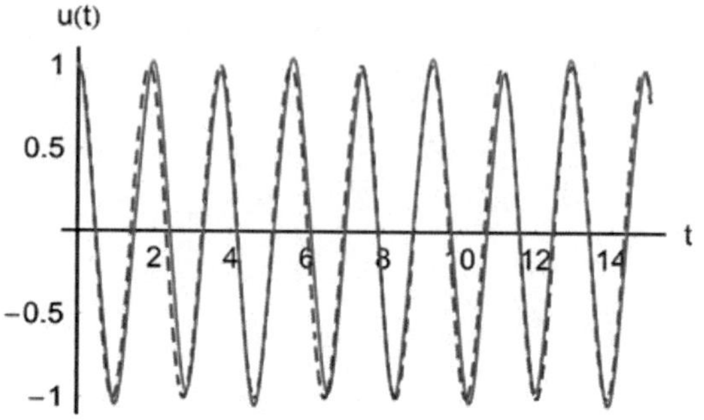

Fig. 6.19 Comparison between analytical and numerical solutions of Eq. 6.219: _______ numerical results; _ _ _ _ _ approximate results Eq. 6.241

6.9 Oscillations of a Mass Attached to a Stretched Elastic Wire

A mass attached to the centre of a stretched elastic wire, in dimensionless form, has the equation of motion [123, 124]:

$$\ddot{u} + u - \frac{\alpha u}{\sqrt{1+u^2}} = 0, \quad 0 < \alpha \leq 1 \tag{6.243}$$

with the initial conditions

$$u(0) = A, \quad \dot{u}(0) = 0 \tag{6.244}$$

Equation 6.243 is an example of a conservative nonlinear oscillatory system having an irrational elastic item.

Under the transformations

$$\begin{aligned} \tau &= \Omega t \\ u(t) &= Ax(\tau) \end{aligned} \tag{6.245}$$

Equations 6.243 and 6.2444 become

$$\Omega^2 x'' + x - \frac{\alpha x}{\sqrt{1+Ax^2}} = 0 \tag{6.246}$$

respectively

$$x(0) = 1, \quad x'(0) = 0 \tag{6.247}$$

where dot denotes differentiation with respect to t and prime denotes differentiation with respect to τ.

In this case, the operators Eqs. 6.34 and 6.35 are respectively:

$$L(\phi(u,p)) = \Omega_0^2[\phi''(\tau,p) + \phi(\tau,p)] \tag{6.248}$$

$$N[\phi(\tau,p))] = \Omega^2(p)\phi''(\tau,p) + (1+\lambda)\phi(\tau,p) - \frac{\alpha\phi(\tau,p)}{\sqrt{1+A^2\phi^2(\tau,p)}} - \\ - p\lambda\phi(\tau,p) \tag{6.249}$$

where ϕ and Ω are given by Eqs. 6.38, 6.39 respectively and λ is an unknown parameter. We obtain the following three equations from Eqs. 6.44, 6.45, 6.46 and 6.47, $(m = 2)$:

$$\Omega_0^2(x_0'' + x_0) = 0, \quad x_0(0) = 1, \quad x'(0) = 0 \tag{6.250}$$

$$\Omega_0^2(x''_1 + x_1) - \Omega_0^2(x''_0 + x_0) - h_1(\tau, C_i)[\Omega_0^2 x''_0 + (1+\lambda)x_0 - \\ - \frac{\alpha x_0}{\sqrt{1+A^2x_0^2}}] = 0, \qquad x_1(0) = 0, \quad x'_1(0) = 0 \tag{6.251}$$

$$\Omega_0^2(x''_2 + x_2) - \Omega_0^2(x''_1 + x_1) - h_1(\tau, C_i)[2\Omega_0\Omega_1 x''_0 + \Omega_0^2 x''_1 + (1+\lambda)x_1 - \\ - \frac{\alpha x_1}{(1+A^2x_0^2)^{3/2}} - \lambda x_0] - h_2(\tau, C_i)[\Omega_0^2 x''_0 + (1+\lambda)x_0 - \\ - \frac{\alpha x_0}{\sqrt{1+A^2x_0^2}}] = 0, \qquad x_2(0) = 0, x'_2(0) = 0 \tag{6.252}$$

Equation 6.250 has the following solution:

$$x_0(\tau) = \cos\tau \tag{6.253}$$

If this result is substituted into Eq. 6.251, and if we choose $h_1(\tau, C_i) = C_1$ we obtain the following equation:

$$\Omega_0^2(x''_1 + x_1) - C_1[(1 - \Omega_0^2 + \lambda - a_1\alpha)\cos\tau - \lambda a_3\cos 3\tau - \\ - \lambda a_5\cos 5\tau - \lambda a_7\cos 7\tau - \lambda a_9\cos 9\tau \ldots] = 0 \tag{6.254}$$

where the term $\cos\tau(1 + A^2\cos^2\tau)^{-1/2}$was expanded into a Fourier series:

$$\frac{\cos\tau}{\sqrt{1+A^2\cos^2\tau}} = \sum_{k=0}^{\infty} a_{2k+1}\cos(2k+1)\tau \tag{6.255}$$

and

$$a_{2k+1} = \frac{4}{\pi}\int_0^{\frac{\pi}{2}} \frac{\cos x \cos(2k+1)x}{\sqrt{1+A^2\cos^2 x}}dx \tag{6.256}$$

and C_1 is an unknown constant in this moment. Avoiding the presence of a secular term needs:

$$\Omega_0^2 = 1 + \lambda - a_1\alpha \tag{6.257}$$

With this requirement, the solution of Eq. 6.254 is

$$\begin{aligned} x_1(\tau) = & \frac{C_1\lambda a_3}{8\omega_0^2}(\cos 3\tau - \cos\tau) + \frac{C_1\lambda a_5}{24\omega_0^2}(\cos 5\tau - \cos\tau) + \\ & + \frac{C_1\lambda a_7}{48\omega_0^2}(\cos 7\tau - \cos\tau) + \frac{C_1\lambda a_9}{80\omega_0^2}(\cos 9\tau - \cos\tau) \end{aligned} \tag{6.258}$$

If we substitute Eqs. 6.253, 6.257 and 6.258 into Eq. 6.252 and if we expand the term $\frac{x_1}{(1+A^2x_0^2)^{3/2}}$ into a Fourier series:

$$\frac{x_1(\tau)}{\sqrt{1+A^2x_0^2}^3} = \frac{C_1\lambda}{8\Omega_0^2}\left[\sum_{k=0}^{\infty} b_{2k+1}\cos(2k+1)\tau\right] \tag{6.259}$$

where

$$\begin{aligned} b_{2k+1} = & \frac{2}{\pi}\int_0^{\frac{\pi}{2}} \frac{f(x)\cos(2k+1)}{(1+A^2\cos^2 x)^{3/2}}dx, \\ f(x) = & a_3(\cos 3x - \cos x) + \frac{a_5}{3}(\cos 5x - \cos x) + \\ & + \frac{a_7}{6}(\cos 7x - \cos x) + \frac{a_9}{10}(\cos 9x - \cos x) \end{aligned} \tag{6.260}$$

it is obtained the following equation in x_2:

$$\begin{aligned} \Omega_0^2(x''_2 + x_2) = & -\left[2\Omega_0\Omega_1 + \frac{C_1 a_1 \alpha^2}{8\Omega_0^2}(a_3 + \frac{a_5}{3} + \frac{a_7}{6} + \frac{a_9}{10}) + \right. \\ & \left. + \lambda + b_1\right]\frac{C_1\alpha^2}{8\Omega_0^2}\cos\tau + H.O..T. \end{aligned} \tag{6.261}$$

No secular term in $x_2(\tau)$ requires that

$$\Omega_1 = -\frac{\lambda}{2\omega_0} - \frac{C_1\alpha^2 K}{16\omega_0^2} \tag{6.262}$$

where

$$K = b_1 + a_1\left(a_3 + \frac{a_5}{3} + \frac{a_7}{6} + \frac{a_9}{10}\right) \tag{6.263}$$

From Eqs. 6.41 and 6.262, we obtain the frequency in the form:

$$\Omega = \Omega_0 - \frac{\lambda}{2\Omega_0} - \frac{C_1\alpha^2 K}{16\Omega_0^2} \tag{6.264}$$

where Ω_0 is given by Eq. 6.257.

The parameter λ can be determined applying the "principle of minimal sensitivity". From Eq. 6.48 we obtain:

$$\lambda = -\frac{3C_1\alpha^2 K}{8\omega_0^2} \tag{6.265}$$

This result is substituted into Eqs. 6.264 and 6.257 and we have:

$$\Omega = \Omega_0 + \frac{C_1\alpha^2 K}{8\Omega_0^3}, \quad \Omega_0^2 = \frac{1-a_1\alpha}{2} + \sqrt{\left(\frac{1-a_1\alpha}{2}\right)^2 - \frac{3}{8}C_1\alpha^2 K} \tag{6.266}$$

Substituting Eqs. 6.253, 6.258 and 6.266 into Eq. 6.252, we obtain:

$$\Omega_0^2(x''_2 + x_2) = \sum_{k=1}^{4}\Big[-(C_1 + h_2(\tau, C_i))\alpha a_{2k+1} + \\ + \frac{C_1^2\alpha a_{2k+1}}{8\Omega_0^2}\left(1 - (2k+1)^2\Omega_0^2 - \frac{3C_1\alpha^2 K}{8\Omega_0^2}\right) - \frac{C_1^2\alpha^2 b_{2k+1}}{8\Omega_0^2}\Big]\cos(2k+1)\tau \tag{6.267}$$

where $h_2(\tau,C_i) = C_2$ and C_2 is an unknown constant.

The solution $x_2(\tau)$ is obtained from Eq. 6.267 and becomes:

$$x_2(\tau) = \sum_{k=1}^{4}\frac{1}{4k(k+1)\Omega_0^2}\Big[(C_1 + C_2)\alpha a_{2k+1} + \frac{C_1^2\alpha a_{2k+1}}{4k(k+1)\Omega_0^4}\times \\ \times \left(1 - (2k+1)^2\Omega_0^2 - 3\frac{C_1\alpha^2 K}{8\Omega_0^2}\right) - \frac{C_1^2\alpha^2 b_{2k+1}}{4k(k+1)\Omega_0^4}\Big][\cos(2k+1)\tau - \cos\tau] \tag{6.268}$$

The second order approximate solution is

$$\bar{x}(\tau) = x_0(\tau) + x_1(\tau) + x_2(\tau) \tag{6.269}$$

where $x_0(\tau)$, $x_1(\tau)$, $x_2(\tau)$, are given by Eqs. 6.253, 6.258, 6.268 respectively. Using the transformations Eq. 6.245 the second order approximate solution of Eq. 6.243 becomes:

$$\bar{u}(t) = A\cos\Omega t + \sum_{k=1}^{4} D_{2k+1}[\cos(2k+1)\omega t - \cos\omega t] \tag{6.270}$$

where ω is given by Eq. 6.266 and

$$D_{2k+1} = \frac{(2C_1 + C_2)A\alpha a_{2k+1}}{4k(k+1)\Omega_0^2} + \left[(2k+1)^2\Omega_0^2 + \right.$$
$$\left. + \frac{3C_1\alpha^2 K}{8\Omega_0^2} - 1\right] \frac{C_1^2 A\alpha a_{2k+1}}{16k^2(k+1)^2\Omega_0^4} + \frac{C_1^2 A\alpha^2 b_{2k+1}}{32k(k+1)\Omega_0^4}, \quad k = 1,2,3,4 \tag{6.271}$$

The substitution of Eq. 6.270 into Eq. 6.243 results in a residual, which reads:

$$R(t, C_1, C_2) = \ddot{\bar{u}} + \bar{u} - \frac{\alpha\bar{u}}{\sqrt{1+\bar{u}^2}} \tag{6.272}$$

As numerical examples we consider the following four cases

Case (a) In the case when $A = 1$, $\alpha = \frac{1}{2}$, we obtain

$$C_1 = -0.585950249; C_2 = -4.748101012; \ \Omega = 0.786318306$$

and the second-order approximate solution is

$$\bar{u}(t) = 0.984079503\cos\Omega t + 0.017504901\cos 3\Omega t - \\ - 0.001584404\cos 5\Omega t \tag{6.273}$$

Figures 6.20–6.23 show the comparison between the present solution and the numerical integration results obtained by a fourth-order Runge–Kutta method:
Case (b) If we consider $A = 1$, $\alpha = 3/4$, it is obtained:

$$C_1 = -0.260422222; C_2 = -5.877201998; \Omega = 0.653805614$$

$$\bar{u}(t) = 0.968131548\cos\Omega t + 0.035536685\cos 3\Omega t - \\ - 0.003668233\cos 5\Omega t \tag{6.274}$$

Case (c) For $A = 10$, $\alpha = 1/2$, we obtain

$$C_1 = -1.436783545; C_2 = -2.611409879; \Omega = 0.96805811$$

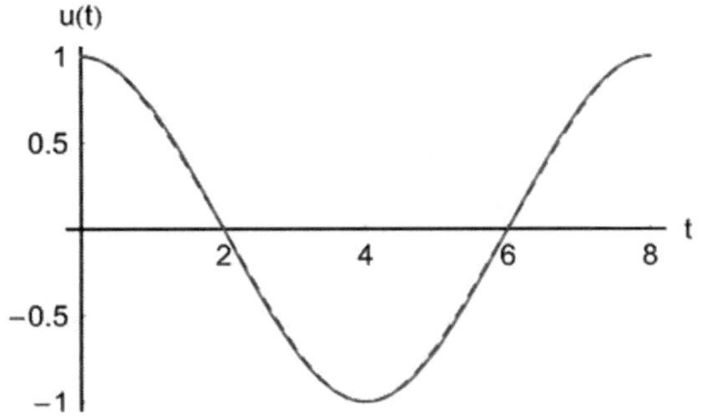

Fig. 6.20 Comparison between analytical and numerical solutions of Eq. 6.243: _______ numerical results; _ _ _ _ _ approximate results Eq. 6.273 for A = 1, α = 1/2

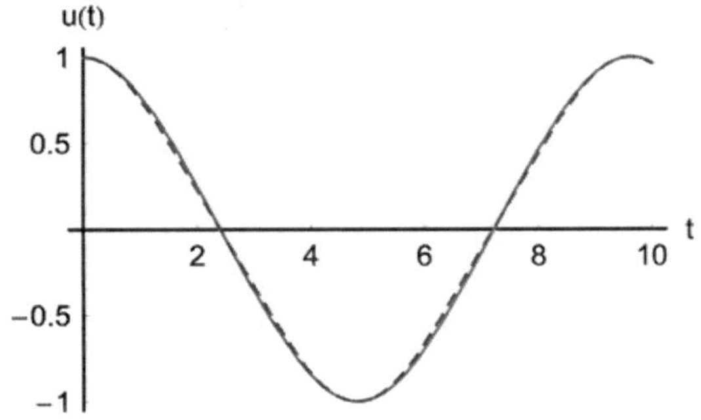

Fig. 6.21 Comparison between analytical and numerical solutions of Eq. 6.243: _______ numerical results; _ _ _ _ _ approximate results Eq. 6.274 for A = 1, α = 3/4

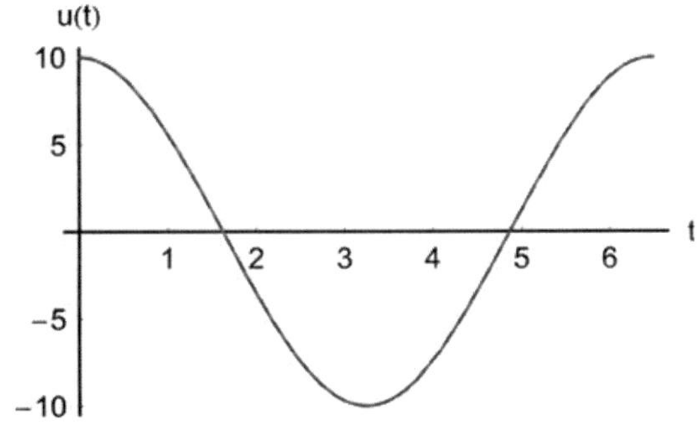

Fig. 6.22 Comparison between analytical and numerical solutions of Eq. 6.243: _______ numerical results; _ _ _ _ _ approximate results Eq. 6.275 for A = 10, α = 1/2

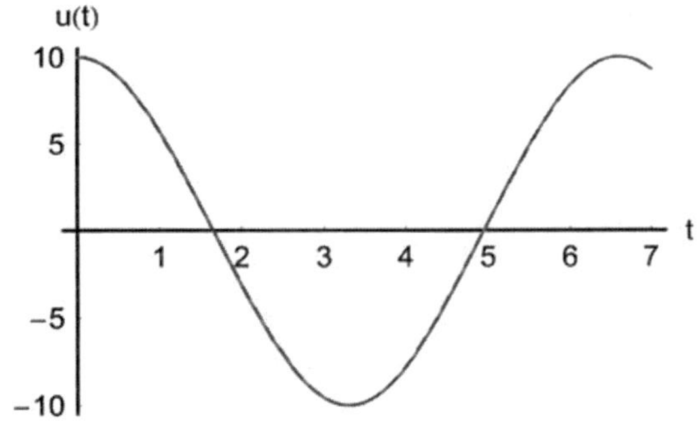

Fig. 6.23 Comparison between analytical and numerical solutions of Eq. 6.243: _______ numerical results; _ _ _ _ _ approximate results Eq. 6.276 for A = 10, α = 3/4

$$\bar{u}(t) = 9.973914015\cos\Omega t + 0.041993608\cos 3\Omega t - \\ - 0.015907625\cos 5\Omega t \tag{6.275}$$

Case (d) In the last case, we consider $A = 10$, $\alpha = 3/4$ and therefore:

$$C_1 = -0.638635455; C_2 = -1.123039001; \Omega = 0.951737427$$

$$\bar{u}(t) = 9.934493076\cos\Omega t + 0.079937932\cos 3\Omega t - \\ - 0.014430988\cos 5\Omega t \tag{6.276}$$

From Figures 6.20–6.23 it can be seen that the solution obtained by our procedure is nearly identical with that given by the numerical method.

6.10 Nonlinear Oscillator with Discontinuities

Consider the following antisymetric constant force oscillator with discontinuities

$$\ddot{u} + sign(u) = 0 \tag{6.277}$$

with the initial conditions:

$$u(0) = A, \quad \dot{u}(0) = 0 \tag{6.278}$$

The function $sign(u)$is defined as

$$sign(u) = \begin{cases} 1 & if \quad u > 0 \\ -1 & \text{if} \quad \text{u} \leq 0 \end{cases} \tag{6.279}$$

There exists no small parameter in the Eq. 6.277 and therefore the traditional perturbation methods cannot be applied directly.

Under the transformations

$$\tau = \omega t, \\ u(t) = Ax(\tau) \tag{6.280}$$

the original Eq. 6.277 becomes

$$\Omega^2 u''(\tau) + signu = 0 \tag{6.281}$$

where the prime denotes the derivative with respect to τ.

For Eq. 6.281, the nonlinear operator Eq. 6.35 is given by the equation

$$N(\phi(\tau,p),\Omega(\lambda,p)) = \Omega^2(\lambda,p)\phi''(\tau,p)+ \\ + sign\phi(\tau,p) + \lambda\phi(\tau,p) - p\lambda\phi(\tau,p) \tag{6.282}$$

Equation 6.44 can be written as:

$$\Omega_0^2(x_0'' + x_0) = 0,\ x_0(0) = 1,\ x'_0(0) = 0 \tag{6.283}$$

and has the solution

$$x_0(\tau) = \cos\tau \tag{6.284}$$

If we note $f(x) = sign(x)$, where x is given by Eq. 6.42, then

$$f(x) = f(x_0) + f'(x_0)(x_1 + x_2 + ...) + \frac{1}{2}f''(x_0)(x_1 + x_2 + ...)^2 + ... \tag{6.285}$$

but $f'(x_0) = f''(x_0) = ... = 0$ and therefore we obtain:

$$sign(x) = sign(x_0) = sign(\cos\tau) \tag{6.286}$$

The first term in Eq. 6.47 is given by

$$N_0(x_0, \Omega_0, \lambda) = \Omega_0^2 x_0'' + sign(x_0) + \lambda x_0 \tag{6.287}$$

For $i = 1$ into Eq. 6.45 we obtain the equation in x_1:

$$\Omega_0^2(x''_1 + x_1) - \Omega_0^2(x''_0 + x_0) - h_1(\tau, C_i)\left[\Omega_0^2 x''_0 + sign(x_0) + \lambda x_0\right] = 0, \\ x_1(0) = x'_1(0) = 0 \tag{6.288}$$

Substituting Eq. 6.284 into Eq. 6.288, using $h_1 = C_1$ (constant) and the identity

$$sign(\cos\tau) = \frac{4}{\pi}\left(\cos\tau - \frac{1}{3}\cos 3\tau + \frac{1}{5}\cos 5\tau - \frac{1}{7}\cos 7\tau + ...\right) \tag{6.289}$$

we obtain the following equation:

$$\Omega_0^2(x''_1 + x_1) = C_1\left[\left(\frac{4}{\pi} + \lambda - \Omega_0^2\right)\cos\tau + \\ + \frac{4}{\pi}\left(-\frac{1}{3}\cos 3\tau + \frac{1}{5}\cos 5\tau - \frac{1}{7}\cos 7\tau + \frac{1}{9}\cos 9\tau - \frac{1}{11}\cos 11\tau + ...\right)\right] \tag{6.290}$$

Avoiding the presence of a secular term needs:

$$\Omega_0^2 = \frac{4}{\pi} + \lambda \tag{6.291}$$

With this requirement, the solution of Eq. 6.290 becomes:

$$x_1(\tau) = \frac{C_1}{6\pi\Omega_0^2}\left[(\cos 3\tau - \cos\tau) - \frac{1}{5}(\cos 5\tau - \cos\tau) + \right.$$
$$\left. + \frac{1}{14}(\cos 7\tau - \cos\tau) - \frac{1}{30}(\cos 9\tau - \cos\tau) + \frac{1}{55}(\cos 11\tau - \cos\tau) + \ldots\right] \tag{6.292}$$

For $m = 2$ into Eq. 6.46 and choosing

$$\mathrm{h}_2(\tau, C_i) = \mathrm{C}_2 + \mathrm{C}_3\cos 2\tau + \mathrm{C}_4\cos 4\tau \tag{6.293}$$

where C_2, C_3, C_4 are constants, the equation in x_2 has the form:

$$\begin{aligned}\Omega_0^2(x''_2 + x_2) - \Omega_0^2(x''_1 + x_1) - C_1(2\Omega_0\Omega_1 x''_0 + \Omega_0^2 x''_1 + \lambda x_1 - \lambda x_0) - \\ -(C_2 + C_3\cos 2\tau + C_4\cos 4\tau)(\Omega_0^2 x''_0 + sign(x_0) + \lambda x_0) = 0\end{aligned} \tag{6.294}$$

Substituting Eqs. 6.284, 6.290 and 6.292 into Eq. 6.294, the equation in x_2 becomes:

$$\begin{aligned}\Omega_0^2(x''_2 + x_2) = &\left[-2\Omega_0\Omega_1 - \lambda + \frac{C_1}{6\pi}\left(1 - \frac{\lambda}{\Omega_0^2}\right)\left(1 - \frac{1}{5} + \frac{1}{14} - \frac{1}{30} + \frac{1}{55}\right) + \right.\\
&\left. + \frac{4}{\pi}\left(\frac{C_3}{6} + \frac{C_4}{15}\right)\right]\cos\tau - \left[\left(\frac{\lambda}{\Omega_0^2} - 9\right)\frac{C_1^2}{6\pi} - \frac{4}{\pi}\left(\frac{C_1 + C_2}{3} - \frac{C_3}{10} + \frac{C_4}{14}\right)\right]\times\\
&\times\cos 3\tau - \left[\left(25 - \frac{\lambda}{\Omega_0^2}\right)\frac{C_1^2}{30\pi} + \frac{4}{\pi}\left(\frac{C_1 + C_2}{5} - \frac{5C_3}{21} + \frac{C_4}{18}\right)\right]\cos 5\tau -\\
&- \left[\left(\frac{\lambda}{\Omega_0^2} - 49\right)\frac{C_1^2}{84\pi} - \frac{4}{\pi}\left(\frac{C_1 + C_2}{7} - \frac{7C_3}{45} + \frac{7C_4}{33}\right)\right]\cos 7\tau -\\
&- \left[\left(81 - \frac{\lambda}{\Omega_0^2}\right)\frac{C_1^2}{180\pi} + \frac{4}{\pi}\left(\frac{C_1 + C_2}{9} - \frac{9C_3}{77} + \frac{C_4}{10}\right)\right]\cos 9\tau -\\
&- \left[\left(\frac{\lambda}{\Omega_0^2} - 121\right)\frac{C_1^2}{330\pi} - \frac{4}{\pi}\left(\frac{C_1 + C_2}{11} - \frac{C_3}{18} + \frac{C_4}{14}\right)\right]\cos 11\tau\end{aligned} \tag{6.295}$$

No secular term in $x_2(\tau)$ requires that:

$$\Omega_1 = -\frac{\lambda}{2\Omega_0} + \frac{989C_1}{13860\pi\Omega_0}\left(1 - \frac{\lambda}{\Omega_0^2}\right),\ C_4 = -\frac{5}{2}C_3 \qquad (6.296)$$

From Eqs. 6.41 and 6.296 we obtain the frequency in the form:

$$\Omega = \Omega_0 - \frac{\lambda}{2\Omega_0} + \frac{989C_1}{13860\pi\Omega_0}\left(1 - \frac{\lambda}{\Omega_0^2}\right) \qquad (6.297)$$

where Ω_0 is given by Eq. 6.291.

The parameter λ can be determined applying the "principle of minimal sensitivity". From Eq. 6.48 we obtain:

$$\lambda = \frac{1}{\pi}\left(\sqrt{4 + \frac{1978C_1}{1155}} - 2\right) \qquad (6.298)$$

This result is substituted into Eqs. 6.297 and 6.291 and we have:

$$\Omega = \frac{8 + 2\sqrt{4 + \frac{1978C_1}{1155}}}{3\sqrt{\pi\left(2 + \sqrt{4 + \frac{1978C_1}{1155}}\right)}},\ \Omega_0^2 = \frac{1}{\pi}\left(2 + \sqrt{4 + \frac{1978C_1}{1155}}\right) \qquad (6.299)$$

Substituting Eqs. 6.284, 6.290, 6.292 and 6.296 into Eq. 6.295 and solving this equation we obtain:

$$\begin{aligned}
x_2(\tau) = &\left[\frac{C_1^2}{48\pi\Omega_0^2}\left(9 - \frac{\lambda}{\omega_0^2}\right) + \frac{1}{2\pi\Omega_0^2}\left(\frac{C_1 + C_2}{3} - \frac{39C_3}{140}\right)\right](\cos 3\tau - \cos\tau) +\\
&+ \left[\frac{C_1^2}{720\pi\Omega_0^2}\left(\frac{\lambda}{\omega_0^2} - 25\right) - \frac{1}{6\pi\Omega_0^2}\left(\frac{C_1 + C_2}{5} - \frac{95C_3}{252}\right)\right](\cos 5\tau - \cos\tau) +\\
&+ \left[\frac{C_1^2}{4032\pi\Omega_0^2}\left(49 - \frac{\lambda}{\Omega_0^2}\right) + \frac{1}{12\pi\Omega_0^2}\left(\frac{C_1 + C_2}{7} - \frac{679C_3}{990}\right)\right](\cos 7\tau - \cos\tau) +\\
&+ \left[\frac{C_1^2}{14400\pi\Omega_0^2}\left(\frac{\lambda}{\Omega_0^2} - 81\right) - \frac{1}{20\pi\Omega_0^2}\left(\frac{C_1 + C_2}{9} - \frac{113C_3}{308}\right)\right](\cos 9\tau - \cos\tau) +\\
&+ \left[\frac{C_1^2}{39600\pi\Omega_0^2}\left(121 - \frac{\lambda}{\Omega_0^2}\right) + \frac{1}{30\pi\Omega_0^2}\left(\frac{C_1 + C_2}{11} - \frac{59C_3}{252}\right)\right](\cos 11\tau - \cos\tau)
\end{aligned} \qquad (6.300)$$

The second-order approximate solution is:

$$\bar{x}(\tau) = x_0(\tau) + x_1(\tau) + x_2(\tau) \tag{6.301}$$

Using Eqs. 6.280, 6.284, 6.292 and 6.300, the second-order approximate solution of Eq. 6.277 becomes:

$$\begin{aligned}
\bar{u}(t) = {} & A\cos\Omega t + A\left[\frac{C_1^2}{48\pi\Omega_0^2}\left(9-\frac{\lambda}{\Omega_0^2}\right)+\frac{1}{2\pi\Omega_0^2}\left(\frac{2C_1+C_2}{3}-\right.\right.\\
& \left.\left.-\frac{39C_3}{140}\right)\right](\cos 3\Omega t-\cos\Omega t)+A\left[\frac{C_1^2}{720\pi\Omega_0^2}\left(\frac{\lambda}{\Omega_0^2}-25\right)-\right.\\
& \left.-\frac{1}{6\pi\Omega_0^2}\left(\frac{2C_1+C_2}{5}-\frac{95C_3}{252}\right)\right](\cos 5\Omega t-\cos\Omega t)\\
& +A\left[\frac{C_1^2}{4032\pi\Omega_0^2}\times\left(49-\frac{\lambda}{\Omega_0^2}\right)+\frac{1}{12\pi\Omega_0^2}\left(\frac{2C_1+C_2}{7}-\frac{679C_3}{990}\right)\right](\cos 7\Omega t-\cos\Omega t)+\\
& +A\left[\frac{C_1^2}{14400\pi\Omega_0^2}\left(\frac{\lambda}{\Omega_0^2}-81\right)-\frac{1}{20\pi\Omega_0^2}\left(\frac{2C_1+C_2}{9}-\right.\right.\\
& \left.\left.-\frac{113C_3}{308}\right)\right](\cos 9\Omega t-\cos\Omega t)+A\left[\frac{C_1^2}{39600\pi\Omega_0^2}\left(121-\frac{\lambda}{\Omega_0^2}\right)+\right.\\
& \left.+\frac{1}{30\pi\Omega_0^2}\left(\frac{2C_1+C_2}{11}-\frac{59C_3}{252}\right)\right](\cos 11\Omega t-\cos\Omega t)
\end{aligned} \tag{6.302}$$

where λ, Ω and Ω_0 are given by Eqs. 6.298 and 6.299 respectively.

The substitution of Eq. 6.302 into Eq. 6.277 results in a residual, which reads:

$$R(t, C_1, C_2 C_3) = \ddot{\bar{u}} + sign(\bar{u}) \tag{6.303}$$

As numerical examples we consider two cases

Case (a) In the case $A = 1$, from Eqs. 6.26 we obtain:

$$C_1 = -0.815614211,\ \mathrm{C}_2 = -0.351951443,\ \mathrm{C}_3 = -0.598628474,$$
$$\mathrm{C}_4 = 1.496571185$$

From Eqs. 6.298 and 6.299 we obtain

$$\lambda = -0.123043117, \Omega_0 = 1.07247211, \Omega = 1.110714944$$

The second-order approximate solution Eq. 6.302 becomes:

$$\bar{u}(t) = 1.030088562\cos\Omega t - 0.033468588\cos 3\Omega t + 0.00146585\cos 5\Omega t + \\ +0.009978953\cos 7\Omega t - 0.008820567\cos 9\Omega t + 0.001398542\cos 11\Omega t \tag{6.304}$$

(a) In the last case $A = 10$, we obtain:

$C_1 = -0.815614211, C_2 = 5.287986428, C_3 = 3.774015449, C_4 = -9.4350386225,$
$\lambda = -0.0123043117, \Omega_0 = 0.339145459, \Omega = 0.351238905$

In this case, the second-order approximate solution becomes:

$$\bar{u}(t) = 10.30035768\cos\Omega t - 0.401767921\cos 3\Omega t + 0.1745208\cos 5\Omega t - \\ -0.103175516\cos 7\Omega t + 0.026982622\cos 9\Omega t - 0.00930889\cos 11\Omega t \tag{6.305}$$

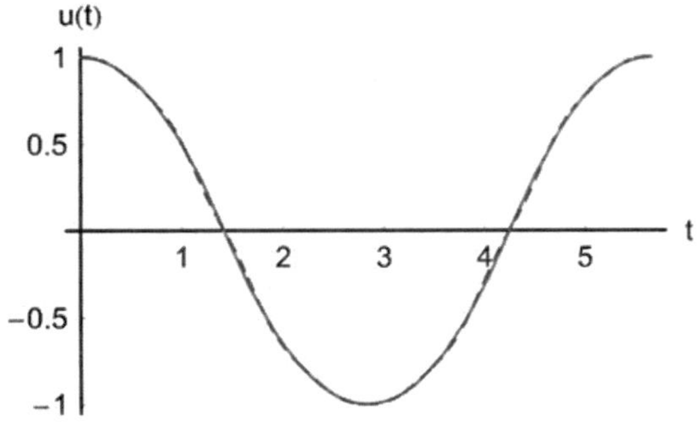

Fig. 6.24 Comparison between the approximate solution Eq. 6.304 and numerical solution of Eq. 6.277 in case A = 1: ______numerical solution; _ _ _ _ _ approximate solution

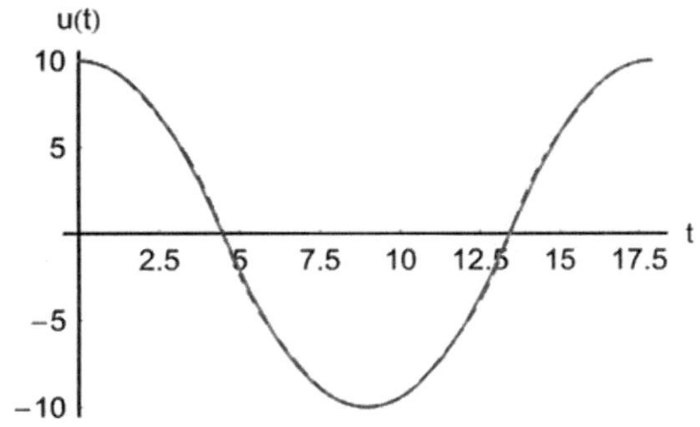

Fig. 6.25 Comparison between the approximate solution Eq. 6.305 and numerical solution of Eq. 6.277 in case A = 10: ______numerical solution; _ _ _ _ _ approximate solution

Figures 6.24 and 6.25 show the comparison between the present solutions and the numerical integration results obtained by a fourth-order Runge–Kutta method.

It is clear that the solutions obtained by OHAM are nearly identical with the solutions given by the numerical method. Additionally, we remark that the exact values of the frequencies are $\Omega_{ex} = 1.110720735$ for the case (a) ($A = 1$) and $\Omega_{ex} = 0.351240736$ in case (b) ($A = 10$), which means that very good approximations are found also for the frequencies.

6.11 Nonlinear Jerk Equations

The nonlinear jerk equations involving the third temporal derivative of displacement have been widely studied [125–127]. As well as originally being of some interest in mechanics, non-linear jerk equations are finding increasing importance also in the study of chaos.

Jerk appears in some structures exhibiting rotating and translating motions, such as robots and machine-tools structures. From a practical perspective, excessive jerk arising at some machine-tools leads to excitation of vibrations in components in the machine assembly, accelerated wear in the transmission and bearing elements, noisy operations and large contouring errors at discontinuities (such as corners) in the machining path. Also in the case of robots, limiting jerk (defined as the time derivative of the acceleration of the manipulator joints) is very important because high jerk values can wear out of the robot structure, and heavily excite its resonance frequencies. Vibrations induced by non-smooth trajectories can damage the robot actuators, and introduce large errors while the robot is performing tasks such as trajectory tracking. Moreover, low-jerk trajectories can be executed more rapidly and accurately.

Jerk equations, though not nearly as common as acceleration (or force) equation $\ddot{u} = f(u, \dot{u})$are therefore of direct physical interest. Moreover, simple forms of the jerk function

$$\dddot{u} + f(u, \dot{u}, \ddot{u}) = 0 \tag{6.306}$$

which lead to perhaps the simplest manifestation of chaos have found in [128].

Gottlieb [125] has explored the flexibility of applying the method of harmonic balance to achieve analytical approximations of periodic solutions to nonlinear jerk equations. Consequent restrictions on the jerk equations amenable to harmonic balance solution are that only problems which have zero initial acceleration and parity and time-reversal invariant (all terms have the same space-parity of reflective behaviour under the transformation $u \to -u$ and time-parity of reflective behaviour under the transformation $t \to -t$) can be considered. This situation of taking off with a constant velocity initially is a feasible condition and it depends on the actual physical meaning of the dependent variable u and the interpretation of the governing equation [128]. Wu et al. [126] proposed an improved harmonic balance method for determining the periodic solutions of nonlinear jerk equations. Ma et al. [127] applied homotopy perturbation method to the jerk equations.

Following Gottlieb [125], the most general jerk function with invariance of the time-reversal and space-reversal and which has only cubic non-linearities may be written as

$$\dddot{u} + \alpha\dot{u}^3 + \beta u^2\dot{u} + \gamma u\dot{u}\ddot{u} + \delta\dot{u}\ddot{u}^2 + \lambda\dot{u} = 0 \tag{6.307}$$

where the parameters α, β, γ, δ and λ are constants and the dot denotes derivative with respect to time. The corresponding initial conditions are

$$u(0) = 0, \dot{u}(0) = A, \ddot{u}(0) = 0 \tag{6.308}$$

Here, at least one of α, β, γ and δ should be non-zero. In addition, if $\delta = 0$, we require $\gamma \neq$ -2α such that the jerk equation is simply not the time-derivative of an acceleration equation.

A new independent variable $\tau = \omega t$ and a new dependent variable $u = \frac{a}{\omega}x$ are introduced. Thus, Eqs. 6.307 and 6.308 can be written as

$$\Omega^2(x''' + \delta A^2 x'x'') + \Omega(\alpha A x'^3 + \gamma A^2 x x' x'' + \lambda x') + \beta A^2 x^2 x' = 0 \tag{6.309}$$

$$x(0) = 0, \quad x'(0) = 1, \quad x''(0) = 0 \tag{6.310}$$

where dot denotes differentiation with respect to τ and $\Omega = \omega^2$. The new independent variable is chosen such that the solution of Eqs. 6.309 and 6.310 is a periodic function of τ of period 2π. The corresponding period of the non-linear jerk equation is given by $T = \frac{2\pi}{\sqrt{\Omega}}$.

The linear operator is defined by

$$L[\phi(\tau,p)] = \Omega_0^2\left[\frac{\partial^3\phi(\tau,p)}{\partial\tau^3} + \phi(\tau,p)\right] \tag{6.311}$$

and $g(\tau) = 0$, while the non-linear operator is defined by

$$\begin{aligned} N[\phi(\tau,p),\Omega(p)] = {} & \Omega^2(p)\left[\frac{\partial^3\phi(\tau,p)}{\partial\tau^3} + \delta A^2\frac{\partial\phi(\tau,p)}{\partial\tau}\frac{\partial^2\phi(\tau,p)}{\partial\tau^2}\right] + \\ & + \Omega(p)\left[\alpha A\left(\frac{\partial\phi(\tau,p)}{\partial\tau}\right)^3 + \gamma A^2\phi(\tau,p)\frac{\partial\phi(\tau,p)}{\partial\tau}\frac{\partial^2\phi(\tau,p)}{\partial\tau^2} + \lambda\frac{\partial\phi(\tau,p)}{\partial\tau}\right] + \\ & + \beta A^2\phi^2(\tau,p)\frac{\partial\phi(\tau,p)}{\partial\tau} \end{aligned} \tag{6.312}$$

From Eqs. 6.310 we obtain the initial conditions:

$$\phi(0,p)=0,\quad \frac{\partial\phi(0,p)}{\partial\tau}=1,\quad \frac{\partial^2(0,p)}{\partial\tau^2}=0, \tag{6.313}$$

From Eqs. 6.44, 6.45 and 6.46, we obtain the following three equations ($m=2$):

$$L[(x_0(\tau)]=0,\ x_0(0)=0,\ x'(0)=1,\ x''(0)=0 \tag{6.314}$$

$$\begin{aligned}L[x_1(\tau)]&=h_1(\tau,C_i)[\Omega_0^2(x'''_0+\delta A^2x'_0x''_02)+\Omega_0(\alpha A^2x'_03+\\&+\gamma A^2x_0x'_0x''_0+\lambda x'_0)+\beta A^2x_0^2x'_0],\quad x_1(0)=x'_1(0)=x''_1(0)=0\end{aligned} \tag{6.315}$$

$$\begin{aligned}L[x_2(\tau)-x_1(\tau)]=&h_1(\tau,C_i)\{\Omega_0^2[x'''_1+\delta A^2(x'''_02x'_1+2x'_0x''_0x''_1)]+\\&+2\Omega_0\Omega_1(x'''_0+\delta A^2x'_0x''^2_0)+\Omega_0[3\alpha A^2x'_02x'_1+\gamma A^2(x_0x'_0x''_1+\\&+x_0x''_0x'_1+x'_0x''_0x_1)+\lambda x'_1]+\Omega_1(\alpha A^2x'^3_0+\gamma A^2x_0x'_0x''_0+\lambda x'_0)+\\&+\beta A^2(x_0^2x'_1+2x_0x'_0x_1)\}+h_2(\tau,C_j)[\Omega_0^2(x'''_0+\\&+\delta A^2x'_0x''^2_0)+\Omega_0(\alpha A^2x'^3_0+\gamma A^2x_0x'_0x''_0+\lambda x'_0)+\beta A^2x_0^2x'_0],\\&x_2(0)=x'_2(0)=x''_2(0)=0\end{aligned} \tag{6.316}$$

Equation 6.314 has the following solution

$$x_0(\tau)=\sin\tau \tag{6.317}$$

If this result is substituted into Eq. 6.315, we obtain the following equation

$$\begin{aligned}\Omega_0^2(x'''_1+x'_1)=h_1(\tau,C_i)\Big\{&\frac{1}{4}[\Omega_0^2(\delta A^2-4)+\Omega_0(3\alpha A^2-\gamma A^2+4\lambda)+\\&+\beta A^2]\cos\tau+\frac{A^2}{4}[-\delta\Omega_0^2+\Omega_0(\alpha+\gamma)-\beta]\cos 3\tau\Big\}\end{aligned} \tag{6.318}$$

Avoiding the presence of a secular term needs

$$\Omega_0^2(\delta A^2-4)+\Omega_0(3\alpha A^2-\gamma A^2+4\lambda)+\beta A^2=0 \tag{6.319}$$

With this requirement, Eq. 6.318 becomes

$$\begin{aligned}x'''_1+x'_1&=\frac{A^2}{4\Omega_0^2}h_1(\tau,C_i)[-\delta\Omega_0^2+\Omega_0(\alpha+\gamma)-\beta]\cos 3\tau,\\x_1(0)&=x'_1(0)=x''_1(0)=0\end{aligned} \tag{6.320}$$

We choose for the auxiliary function h_1 the simplest form $h_1(\tau,C_i)=C_1$. The solution of Eq. 6.320 is

$$x_1(\tau) = S(3\sin\tau - \sin 3\tau) \tag{6.321}$$

where

$$S = \frac{a^2 C_1}{96\Omega_0^2}[\delta\Omega_0^2 - (\alpha+\gamma)\Omega_0 + \beta] \tag{6.322}$$

If we substitute Eqs. 6.317 and 6.321 into Eq. 6.316, we obtain the equation in the variable x_2:

$$\begin{aligned} L(x_{2)} - L(x_1) = \Bigg[& -\frac{3S\Omega_0^2}{2}(\delta A^2 + 2) - 2\Omega_0\Omega_1\left(1 - \frac{1}{4}\delta A^2\right) + \frac{S\Omega_0}{2}(9\alpha A^2 - \\ & -\gamma A^2 + 6\lambda) + \frac{\Omega_1}{4}(3\alpha A^2 - \gamma A^2 + 4\lambda) + 4\beta A^2 S\Bigg] \cos\tau + H.O..T. \end{aligned} \tag{6.323}$$

No secular terms in $x_2(\tau)$ requires that

$$\Omega_1 = \frac{2S[3\Omega_0^2(\delta A^2 + 2) - \Omega_0(9\alpha A^2 - \gamma A^2 + 6\lambda) - 8\beta A^2]}{3\alpha A^2 - \gamma A^2 + 4\lambda - 2\Omega_0(4 - \delta A^2)} \tag{6.324}$$

From Eq. 6.43 for $m = 2$, we obtain the frequency in the form

$$\Omega = \Omega_0 + \Omega_1 \tag{6.325}$$

where Ω_0 is given by Eq. 6.319 and $\Omega_1$is given by Eq. 6.324. Now, Eq. 6.323 (with $h_2(\tau, C_i) = C_2 = \text{constant}$) becomes

$$\begin{aligned} x'''_2 + x'_2 = \Bigg\{ & C_1\left[S\left(27 - \frac{15}{4}\delta A^2\right) - \frac{\delta A^2\Omega_1}{2\Omega_0} + \frac{S}{4\Omega_0}(15\gamma A^2 - 9\alpha A^2 - \right. \\ & \left. - 12\lambda) + \frac{(3\alpha+\gamma)A^2\Omega_1}{4\Omega_0^2} - \frac{27A^2\beta S}{4\Omega_0^2}\right] - 24S(1 + C_2)\Bigg\} \cos 3\tau + \\ & + \frac{a^2 S}{4}\left(21\delta - \frac{9\alpha + 13\gamma}{\Omega_0} + \frac{11\beta}{\Omega_0^2}\right)\cos 5\tau \qquad x_2(0) = x'_2(0) = x''_2(0) \end{aligned} \tag{6.326}$$

The solution of Eq. 6.326 becomes

$$x_2(\tau) = \frac{M}{24}(3\sin\tau - \sin 3\tau) + \frac{N}{120}(5\sin\tau - \sin 5\tau) \tag{6.327}$$

where

$$M = C_1\left[S\left(27 - \frac{15}{4}\delta A^2\right) - \frac{\delta A^2\Omega_1}{2\Omega_0} + \frac{S}{4\Omega_0}(15\gamma A^2 - 9\alpha A^2 - 12\lambda) + \right.$$
$$\left. + \frac{(3\alpha+\gamma)A^2\Omega_1}{4\Omega_0^2} - \frac{27A^2\beta S}{4\Omega_0^2}\right] - 24S(1+C_2)$$
$$N = \frac{A^2 S}{4}\left(21\delta - \frac{9\alpha + 13\gamma}{\Omega_0} + \frac{11\beta}{\Omega_0^2}\right) \tag{6.328}$$

The second-order approximate solution will be

$$\bar{x}(\tau) = x_0(\tau) + x_1(\tau) + x_2(\tau) \tag{6.329}$$

where x_0, x_1 and x_2 are given by Eqs. 6.317, 6.321 and 6.327, respectively, such that

$$\bar{u}(t) = \frac{A}{\sqrt{\Omega_0+\Omega_1}}\left[\left(1 + 3S + \frac{M}{8} + \frac{N}{24}\right)\sin\sqrt{\Omega_0+\Omega_1}t - \right.$$
$$\left. -\left(S + \frac{M}{24}\right)\sin 3\sqrt{\Omega_0+\Omega_1}t - \frac{N}{120}\sin 5\sqrt{\Omega_0+\Omega_1}t\right] \tag{6.330}$$

The constants C_1 and C_2 are determined from Eqs. 6.26, where J is given by

$$J = \int_0^T \left[\dddot{\bar{u}} + \alpha\dot{\bar{u}}^3 + \beta\bar{u}^2\dot{\bar{u}} + \gamma\bar{u}\dot{\bar{u}}\ddot{\bar{u}} + \delta\bar{u}\ddot{\bar{u}}^2 + \lambda\bar{u}\right]^2 dt \tag{6.331}$$

As a numerical example, by setting $\alpha = \beta = 1$, $\delta = \gamma = \lambda = 0$ (jerk function containing velocity-cubed and velocity times displacements-squared) in Eq. 6.307, the governing equation will be

$$\dddot{u} + \dot{u}(\dot{u}^2 + u^2) = 0 \tag{6.332}$$

subject to the initial conditions

$$u(0) = 0, \quad \dot{u}(0) = A, \quad \ddot{u}(0) = 0 \tag{6.333}$$

In what follows we analyze three distinct cases.

(a) The case A = 2
From Eqs. 6.319, 6.324 and 6.26 we obtain

$$\Omega_0 = 3.3022775638, \quad \Omega_1 = -0.104229926C_1$$

and from Eqs. 6.322 and 6.328 we obtain

$$S = -0.008795939C_1; \; M = C_1(0.071616608C_1 + 0.211102527C_2); \\ N = 0.015098903C_1$$

From Eqs. 6.26 we obtain

$$C_1 = 0.922185833, \quad C_2 = -0.10910221$$

and therefore, from Eq. 6.325 it is obtained

$$\omega = \sqrt{\Omega} = 1.790713901$$

The second-order approximate solution Eq. 6.330 becomes

$$\bar{u}(t) = 1.095880068 \sin \omega t + 0.007213632 \sin 3\omega t - 0.000129594 \times \sin 5\omega t \tag{6.334}$$

(b) The case $A = 5$
In the same manner as in the above case, we obtain

$$\Omega_0 = 19.07760919, C_1 = -0.101794266, C_2 = 10.84917301, \omega = 4.278311302$$

$$\bar{u}(t) = 1.125588397 \sin \omega t + 0.014303567 \sin 3\omega t + 0.000030279 \times \sin 5\omega t \tag{6.335}$$

(c) The case $A = 20$
In this last case, it is obtained

$$\Omega_0 = 300.3329638, C_1 = 1.028399265, C_2 = 0.49151121, \omega = 16.95546134$$

$$\bar{u}(t) = 1.148209795 \sin \omega t + 0.011145664 \sin 3\omega t - 0.000417163 \times \sin 5\omega t \tag{6.336}$$

It is easy to verify the accuracy of the obtained solution if we graphically compare the analytical results with the numerical simulation results. Figures 6.26–6.28 show the comparison between the present solutions and the numerical integration results obtained by a fourth-order Runge–Kutta method.

6.12 The Motion of a Particle on a Rotating Parabola

The motion of a particle on a rotating parabola is mentioned by Nayfeh and Mook in [22] and [88]:

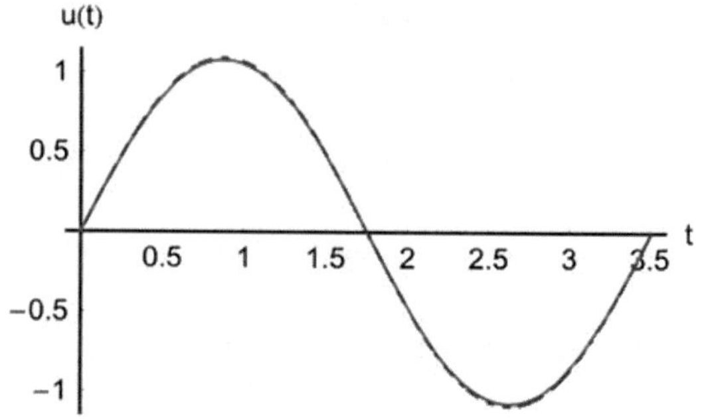

Fig. 6.26 Comparison between the approximate and numerical results of Eq. 6.307 in case (a), $A = 2$ _________numerical solution; _ _ _ _ _ analytical solution, Eq. 6.334

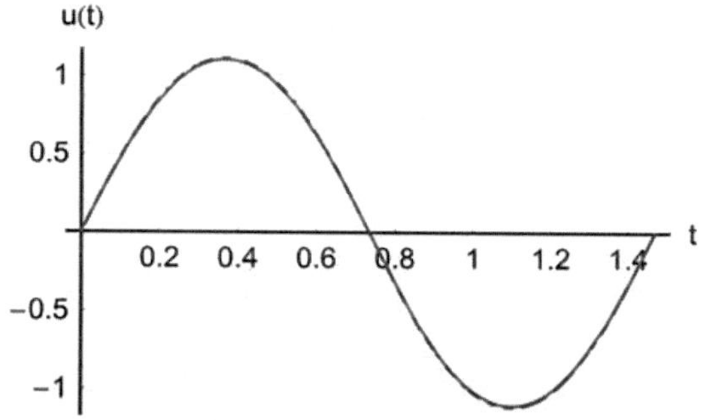

Fig. 6.27 Comparison between the approximate and numerical results of Eq. 6.307 in case (b), A = 5 _________numerical solution, _ _ _ _ _ analytical solution, Eq. 6.335

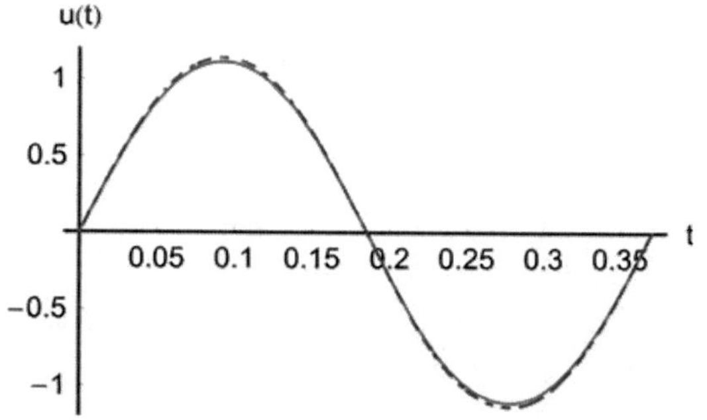

Fig. 6.28 Comparison between the approximate and numerical results of Eq. 6.307 in case (c), A = 20 _________numerical solution, _ _ _ _ _ analytical solution, Eq. 6.336

$$\left(1+4q^2u^2\right)\frac{\mathrm{d}^2u}{\mathrm{d}t^2}+\Lambda u+4q^2\left(\frac{\mathrm{du}}{\mathrm{d}t}\right)^2u=0 \tag{6.337}$$

with the boundary conditions:

$$u(0)=A,\frac{\mathrm{du}}{\mathrm{d}t}(0)=0 \tag{6.338}$$

where q and A are known constants and need not to be small.

Under the transformations Eqs. 6.30, 6.337 and 6.338 become:

$$\Omega^2 x'' + \omega_0^2 x + 4q^2A^2\Omega^2(x^2x'' + xx'^2) = 0 \tag{6.339}$$

respectively

$$x(0) = 1, \quad \mathrm{x}'(0) = 0 \tag{6.340}$$

where $\Lambda = \omega_0^2$ and $' = \frac{\mathrm{d}}{\mathrm{d}\tau}$.

The operators Eqs. 6.34 and 6.35 are respectively:

$$L(\phi(\tau,p)) = \Omega_0^2[\phi''(\tau,p) + \phi(\tau,p)] \tag{6.341}$$

$$\begin{aligned} N[\phi(\tau,p),\Omega(\tau,p)] &= \Omega^2(p)\phi''(\tau,p) + (\omega_0^2 + \lambda)\phi(\tau,p) + \\ &+ 4q^2A^2\Omega^2\left[\phi^2(\tau,p)\phi''(\tau,p) + \phi(\tau,p)\phi'^2(\tau,p)\right] - p\lambda\phi(\tau,p) \end{aligned} \tag{6.342}$$

where ϕ and Ω are given by Eqs. 6.38 and 6.39 respectively, and λ is an unknown parameter. From Eqs.6.44, 6.45 and 6.46, ($m = 2$), we obtain the following three equations:

$$\Omega_0^2(x_0'' + x_0) = 0, \ \mathrm{x}_0(0) = 1, \quad \mathrm{x'}(0) = 0 \tag{6.343}$$

$$\begin{aligned} &\Omega_0^2(x''_1 + x_1) - \Omega_0^2(x''_0 + x_0) - h_1(\tau, C_i)\left[\Omega_0^2 x''_0 + (\omega_0^2 + \lambda)x_0 + \right. \\ &\left. + 4q^2A^2\Omega_0^2(x_0x''_0 + x'_0{}^2 x_0)\right] = 0, \ \mathrm{x}_1(0) = x'_1(0) = 0 \end{aligned} \tag{6.344}$$

$$\begin{aligned} &\Omega_0^2(x''_2 + x_2) - \Omega_0^2(x''_1 + x_1) - h_1(\tau, C_i)\{2\Omega_0\Omega_1 x''_0 + \Omega_0^2 x''_1 + (\omega_0^2 + \lambda)x_1 + \\ &\quad + 4q^2A^2[\Omega_0^2(2x_0x''_0x_1 + x_0^2x''_1 + 2x_0x'_0x'_1 + 2x_0x'_0x'_1 + x'_0{}^2 x_1) + \\ &\quad + 2\Omega_0\Omega_1(x_0^2x''_0 + x'_0{}^2 x_0)] - \lambda x_0\} - h_2(\tau, C_j)[\Omega_0^2 x''_0 + (\omega_0^2 + \lambda)x_0 + \\ &\quad + 4q^2A^2\Omega_0^2(x_0^2x''_0 + x'_0{}^2 x_0)] = 0, \quad \mathrm{x}_2(0) = x'_2(0) = 0 \end{aligned} \tag{6.345}$$

Equation 6.343 has the following solution:

$$x_0(\tau) = \cos\tau \tag{6.346}$$

If this result is substituted into Eq. 6.344 and assuming that $h_I = C_I =$ constant, we obtain the following equation:

$$\Omega_0^2(x''_1 + x_1) - C_1[(\omega_0^2 + \lambda - \Omega_0^2 - 2q^2A^2\Omega_0^2)\cos\tau - \\ -2q^2A^2\Omega_0^2\cos 3\tau] = 0, \quad x_1(0) = x'_1(0) = 0 \tag{6.347}$$

where C_1 is an unknown constant at this moment. Avoiding the presence of a secular term needs:

$$\Omega_0^2 = \frac{\omega_0^2 + \lambda}{1 + 2q^2A^2} \tag{6.348}$$

With this requirement, the solution of Eq. 6.347 is:

$$x_1(\tau) = \frac{1}{4}C_1q^2A^2(\cos 3\tau - \cos\tau) \tag{6.349}$$

If we substitute Eqs. 6.346, 6.348 and 6.349 into Eq. 6.345, we obtain the equation in x_2:

$$\begin{aligned} &\Omega_0^2(x''_2 + x_2) + \frac{2C_1q^2A^2(\omega_0^2 + \lambda)}{1 + 2q^2A^2}\cos 3\tau + \\ &+ C_1\left\{\left[\frac{C_1q^4A^4(\omega_0^2 + \lambda)}{2(1 + 2q^2A^2)} + 2\Omega_0\Omega_1(1 + 2q^2A^2) + \lambda\right]\cos\tau + \right. \\ &+ \left[\frac{(\omega_0^2 + \lambda)C_1q^2A^2(3q^2a^2 + 16)}{2(1 + 2q^2A^2)} + 2\Omega_0\Omega_1q^2A^2\right]\cos 3\tau + \\ &\left. + \frac{9C_1q^4A^4(\omega_0^2 + \lambda)}{2(1 + 2q^2A^2)}\cos 5\tau\right\} + h_2(\tau)\left[\frac{2q^2A^2(\omega_0^2 + \lambda)}{1 + 2q^2A^2}\cos 3\tau\right] = 0, \\ &x_2(0) = x'_2(0) = 0 \end{aligned} \tag{6.350}$$

No secular term in $x_2(\tau)$ requires that

$$2\Omega_0\Omega_1 = -\frac{\lambda}{1 + 2q^2a^2} - \frac{C_1q^4A^4(\omega_0^2 + \lambda)}{2(1 + 2q^2A^2)^2} \tag{6.351}$$

From Eqs. 6.351 and 6.43, we obtain the frequency in the form:

$$\Omega = \Omega_0 - \frac{\lambda}{\Omega_0(1 + 2q^2A^2)} - \frac{C_1q^4A^4\Omega_0}{4(1 + 2q^2A^2)} \tag{6.352}$$

where Ω_0 is given by Eq. 6.348.

The parameter λ can be determined applying the "principle of minimal sensitivity" and thus we obtain

$$\lambda = \frac{C_1\omega_0 q^4A^4}{2 + 4q^2A^2 - C_1q^4A^4} \tag{6.353}$$

This result is substituted into Eq. 6.352 and we have:

$$\Omega = \frac{\omega_0}{1 + 2q^2A^2}\sqrt{1 + 2q^2A^2 - \frac{1}{2}C_1q^4A^4} \tag{6.354}$$

Substituting Eqs. 6.352, 6.353 and 6.354 into Eq. 6.350, we obtain:

$$x''_2 + x_2 + 2C_1q^2A^2\cos 3\tau + \frac{C_1^2q^2A^2(5q^4A^4 + 7q^2A^2 + 2)}{1 + 2q^2A^2}\cos 3\tau +$$
$$+\frac{9}{2}C_1^2q^4A^4\cos 5\tau + 2h_2(\tau, C_j)q^2A^2\cos 3\tau = 0, \qquad x_2(0) = x'_2(0) = 0 \tag{6.355}$$

As we shown, there are many possibilities to choose the function $h_2(\tau,C_j)$. The convergence of the solution $x_2(\tau)$ and consequently the convergence of the approximate solution $\bar{x}(\tau)$ depend on the auxiliary function $h_2(\tau,C_j)$. Basically, the shape of $h_2(\tau,C_j)$ must follow the terms appearing in Eq. 6.350, which are $\cos\tau$, $\cos 3\tau$, $\cos 5\tau$ (odd-order harmonics). Therefore we try to choose $h_2(\tau,C_j)$ so that in Eq. 6.350 the product

$$h_2\left[\frac{2q^2A^2(\omega_0^2 + \lambda)}{1 + 2q^2A^2}\cos 3\tau\right]$$

be of the same shape with the other terms (a combination of functions $\cos\tau$, $\cos 3\tau$, $\cos 5\tau \ldots$).

All three cases presented in this section demonstrate the importance of the function $h_2(\tau,C_j)$ on the accuracy of the solution. In the same time, a larger number of constants in $h_2(\tau,C_j)$ lead to a better accuracy of the results. If the error obtained using a certain $h_2(\tau,C_j)$ is unsatisfactory, one can choose other shapes for this function.

We will consider three cases:

Case (A): We consider the function h_2 of the form:

$$h_2(\tau, C_j) = C'_2 \tag{6.356}$$

where $C'_2$is a constant.

Substituting Eq. 6.356 into Eq. 6.355, we obtain the equation in x_2:

$$x_2'' + x_2 + \left[2(C_1 + C_2')q^2A^2 + \frac{C_1^2q^2A^2(5q^4A^4 + 7q^2A^2 + 2)}{1 + 2q^2A^2}\right]\cos 3\tau + \\ + \frac{9}{2}C_1^2q^4A^4\cos 5\tau = 0, \quad x_2(0) = x'_2(0) = 0 \tag{6.357}$$

The solution of Eq. 6.357 becomes:

$$x_2(\tau) = \left[\frac{C_1 + C'_2}{4} + \frac{C_1^2q^2a^2(5q^4A^4 + 7q^2A^2 + 2)}{8(1 + 2q^2A^2)}\right](\cos 3\tau - \cos\tau) + \\ + \frac{3}{16}C_1^2q^4A^4(\cos 5\tau - \cos\tau) \tag{6.358}$$

The second-order approximate solution is

$$\bar{x}(\tau) = x_0(\tau) + x_1(\tau) + x_2(\tau)$$

where x_0, x_1 and x_2 are given by Eqs. 6.346, 6.349 and 6.358. Using the transformations Eq. 6.30, the second-order approximate solution of Eq. 6.337 becomes

$$\bar{u}(t) = M\cos\Omega t + N\cos 3\Omega t + P\cos 5\Omega t \tag{6.359}$$

where Ω is given by Eq. 6.354 and

$$M = A - \frac{2C_1 + C_2'}{4}q^2A^3 - \frac{C_1^2q^2A^3(16q^4A^4 + 17q^2A^2 + 4)}{16(1 + 2q^2A^2)} \\ N = \frac{2C_1 + C_2'}{4}q^2A^3 + \frac{C_1^2q^2A^3(5q^4A^4 + 7q^2A^2 + 2)}{8(1 + 2q^2A^2)} \\ P = \frac{3}{16}C_1^2q^4A^5 \tag{6.360}$$

Case (B): We consider the function $h_2(\tau, C_j)$ if the form

$$h_2(\tau, C_j) = C_2 + C_3\cos 2\tau + C_4\cos 4\tau \tag{6.361}$$

where C_2, C_3 and C_4 are constants.

Substituting Eq. 6.361 into Eq. 6.355 and avoiding the presence of a secular term, we obtain:

$$C_4 = -C_3 \tag{6.362}$$

respectively:

$$x_2'' + x_2 + \left[2(C_1 + C_2)q^2A^2 + \frac{C_1^2q^2A^2(5q^4A^4 + 7q^2A^2 + 2)}{1 + 2q^2A^2}\right]\cos 3\tau +$$
$$+ \left[\frac{9}{2}C_1^2q^4A^4 + C_3q^2A^2\right]\cos 5\tau - C_3q^2A^2\cos 7\tau = 0$$
$$x_2(0) = x'_2(0) = 0 \qquad (6.363)$$

With these requirements, the solution of Eq. 6.363 becomes:

$$x_2(\tau) = \left[\frac{C_1 + C_2}{4}q^2A^2 + \frac{C_1^2q^2A^2(5q^4A^4 + 7q^2A^2 + 2)}{8(1 + 2q^2A^2)}\right](\cos 3\tau -$$
$$- \cos\tau) + \left[\frac{3}{16}C_1^2q^4A^4 + \frac{1}{24}C_3q^2A^2\right](\cos 5\tau - \cos\tau) -$$
$$- \frac{1}{48}C_3q^2A^2(\cos 7\tau - \cos\tau) \qquad (6.364)$$

The second-order approximate solution in this case is:

$$\bar{u} = \bar{M}\cos\Omega t + \bar{N}\cos 3\Omega t + \bar{P}\cos 5\Omega t + \bar{Q}\cos 7\Omega t \qquad (6.365)$$

where Ω is given by Eq. 6.354 and the coefficients are:

$$\bar{M} = A - \frac{2C_1 + C_2}{4}q^2A^3 -$$
$$- \frac{C_1^2q^2A^3(16q^4A^4 + 17q^2A^2 + 4)}{16(1 + 2q^2A^2)} - \frac{1}{48}C_3q^2A^3$$
$$\bar{N} = \frac{2C_1 + C_2}{4}q^2A^3 + \frac{C_1^2q^2A^3(5q^4A^4 + 7q^2A^2 + 2)}{8(1 + 2q^2A^2)}$$
$$\bar{P} = \frac{3}{16}C_1^2q^4A^5 + \frac{1}{24}C_3q^2A^3$$
$$\bar{Q} = -\frac{1}{48}C_3q^2A^3 \qquad (6.366)$$

Case (C): We consider the function $h_2(\tau, C_j)$ of the form:

$$h_2(\tau, C_j) = C_2^* + C_3^*\cos 2\tau + C_4^*\cos 4\tau + C_5^*\cos 6\tau + C_6^*\cos 8\tau \qquad (6.367)$$

where C_2^*, C_3^*, C_4^*, C_5^* and $C_6^*$are constants.

Substituting Eq. 6.367 into Eq. 6.355, we obtain:

$$C_4^* = -C_3^* \qquad (6.368)$$

and then:

$$x_2(\tau) = \left[\frac{2C_1 + 2C_2^* + C_5^*}{8} q^2A^2 + \right.$$
$$\left. + \frac{C_1^2 q^2A^2(5q^4A^4 + 7q^2A^2 + 2)}{8(1 + 2q^2A^2)}\right](\cos 3\tau - \cos\tau) +$$
$$+ \frac{1}{24}(C_3^* + C_6^*)q^2A^2\Big](\cos 5\tau - \cos\,\tau) + \frac{1}{48}C_3^* q^2A^2(\cos\,\tau -$$
$$- \cos 7\tau) + \frac{1}{80}C_5^* q^2A^2(\cos 9\tau - \cos\tau) + \frac{C_6^*}{120}q^2A^2(\cos 11\tau - \cos\tau) \quad (6.369)$$

The second-order approximate solution of Eq. 6.337 becomes:

$$\bar{u}(t) = M^*\cos\Omega t + N^*\cos 3\Omega t + P^*\cos 5\Omega t + \\ + Q^*\cos 7\Omega t + R^*\cos 9\Omega t + S^*\cos 11\Omega t \quad (6.370)$$

where

$$M^* = A - \frac{q^2A^3}{240}(120C_1 + 60C_2^* + 5C_3^* + 33C_4^* + 12C_5^*) -$$
$$- \frac{C_1^2 q^2A^3(16q^4A^4 + 17q^2A^2 + 4)}{16(1 + 2q^2A^2)}$$
$$N^* = \frac{4C_1 + 2C_2^* + C_4^*}{8}q^2A^3 + \frac{C_1^2 q^2A^3(5q^4A^4 + 7q^2A^2 + 2)}{8(1 + 2q^2A^2)}$$
$$P^* = \frac{C_3^* + C_5^*}{24}q^2A^3 + \frac{3}{16}C_1^2 q^4A^5$$
$$Q^* = -\frac{1}{48}C_3^2A^3$$
$$R^* = \frac{1}{80}C_4^* q^2A^3$$
$$S^* = \frac{1}{120}C_5^* q^2A^3 \quad (6.371)$$

We will show through six numerical examples that the error of the solutions decreases when the number of terms in the auxiliary function *H(τ,p)* increases. In Eqs. 6.337 and 6.338, we consider $\Lambda = \omega_0 = 1, A = 1$and two cases for q in every of the cases A, B and C. The constants C_i are obtained using the least square method.

(a) For $q = 1$ in the case A, it is obtained:

$$\mathrm{C}_1 = -0.401483291,\ \mathrm{C'}_2 = -0.065781508$$

The second-order approximate solution Eq. 6.359 becomes in this case:

$$\bar{u}(t) = 1.092937297\cos\Omega t - 0.123160203\cos 3\Omega t + 0.030222906\cos 5\Omega t \quad (6.372)$$

where Ω is obtained from Eq. 6.354: $\Omega = 0.596353888$.
(b) For $q = 1$ in the case B, it is obtained

$$C_1 = -0.398431527;\ C_2 = -0.052485317;\ C_3 = 0.0341786762$$

$$\begin{aligned}\bar{u}(t) &= 1.089257032\cos\Omega t - 0.119734278\cos 3\Omega t + \\ &\quad + 0.031189301\cos 5\Omega t - 0.00712055\cos 7\Omega t \end{aligned} \tag{6.373}$$

where $\Omega = 0.596211722$.
(c) For $q = 1$ in the case C, we obtain the following results

$$\begin{aligned} &C_1 = -0.395753003;\ C_2^* = -0.24453992;\ C_3^* = 0.396618201; \\ &C_4^* = -0.396618201;\ C_5^* = 0.534493194;\ C_6^* = -0.497490133 \end{aligned}$$

$$\begin{aligned}\bar{u}(t) &= 1.08140204\cos\Omega t - 0.100837908\cos 3\Omega t + 0.025163334\cos 5\Omega t - \\ &\quad - 0.008262879\cos 7\Omega t + 0.006681164\cos 9\Omega t - 0.004145751\cos 11\Omega t \end{aligned} \tag{6.374}$$

where $\Omega = 0.596087918$.
(d) For $q = 2$ in the case A we obtain

$$C_1 = -0.167434521,\ C_2' = -0.02382096$$

$$\bar{u}(t) = 1.081827345\cos\Omega t - 0.165930301\cos 3\Omega t + 0.084102956\cos 5\Omega t \tag{6.375}$$

where $\Omega = 0.357278398$.
(e) For $q = 2$ in the case B it is obtained

$$\begin{aligned} C_1 &= -0.164357411; C_2 = 0.017447955; C_3 = -0.073610524 \\ \bar{u}(t) &= 1.071279364\cos\Omega t - 0.146185231\cos 3\Omega t + \\ &\quad + 0.068771659\cos 5\Omega t + 0.00613421\cos 7\Omega t \end{aligned} \tag{6.376}$$

where $\Omega = 0.356852829$.
(f) For $q = 2$ in the case C it is obtained

$$\begin{aligned} &C_1 = -0.16124603;\ C_2^* = -0.103086948;\ C_3^* = 0.254864679; \\ &C_4^* = -0.254864679;\ C_5^* = 0.236002415;\ C_6^* = -0.345719606 \end{aligned}$$

$$\bar{u}(t) = 1.1067914118\cos\Omega t - 0.148687186\cos 3\Omega t + 0.062858358\cos 5\Omega t - \\ -0.021238723\cos 7\Omega t + 0.01180012\cos 9\Omega t - 0.011523986\cos 11\Omega t \tag{6.377}$$

where $\Omega = 0.356422004$.

It is easy to verify the accuracy of the obtained solutions if we graphically compare these analytical solutions with the numerical ones. Figures 6.29–6.34 show the comparison between the present solutions and the numerical integration results obtained by a fourth-order Runge–Kutta method.

It can be seen from Figures 6.29–6.34 that the solutions obtained by OHAM are very accurate being nearly identical with the solutions obtained by a fourth-order Runge–Kutta method. Moreover, the analytical solutions obtained by our procedure prove to be more accurate along with an increased number of terms in the auxiliary function $H(\tau,p)$.

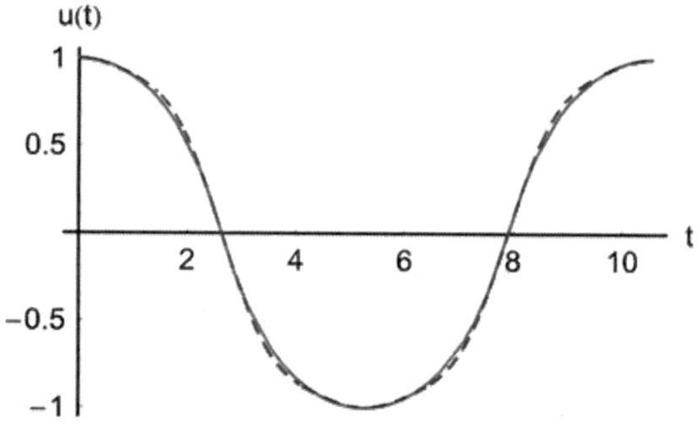

Fig. 6.29 Comparison between the approximate and numerical results of Eq. 6.337 in case (a), for $A = \omega_0 = a = q = 1$: ________numerical solution; _ _ _ _ _ approximate solution Eq. 6.372

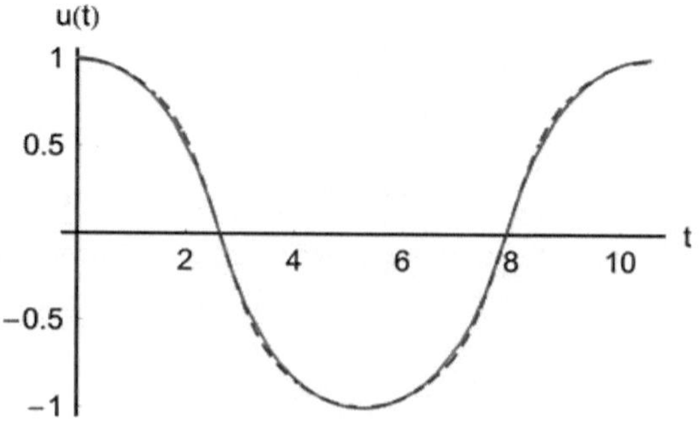

Fig. 6.30 Comparison between the approximate and numerical results of Eq. 6.337 in case (b), for $A = \omega_0 = a = q = 1$: ______numerical solution, _ _ _ _ approximate solution Eq. 6.373

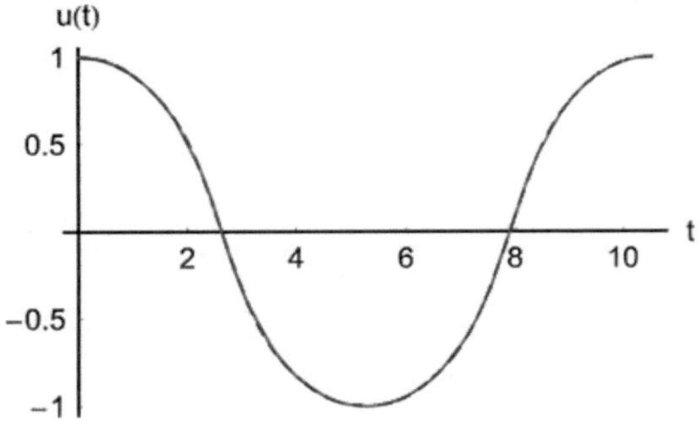

Fig. 6.31 Comparison between the approximate and numerical results of Eq. 6.337 in case (c), for $A = \omega_0 = a = q = 1$: ______ numerical solution; _ _ _ _ approximate solution Eq. 6.374

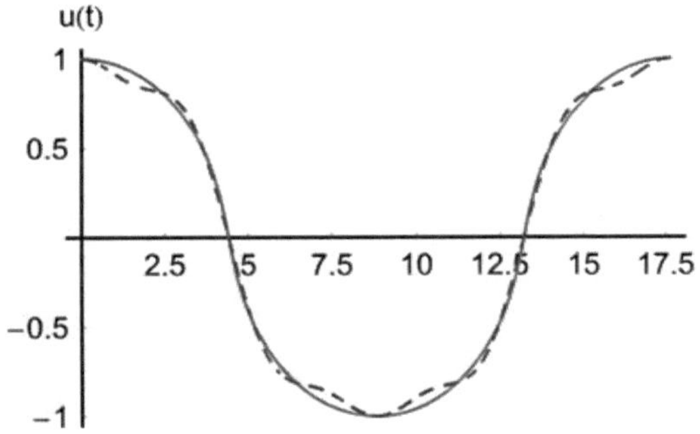

Fig. 6.32 Comparison between the approximate and numerical results of Eq. 6.337 in case (d), for $A = \omega_0 = a = 1, q = 2$: _________numerical solution; _ _ _ _ _ approximate solution Eq. 6.375

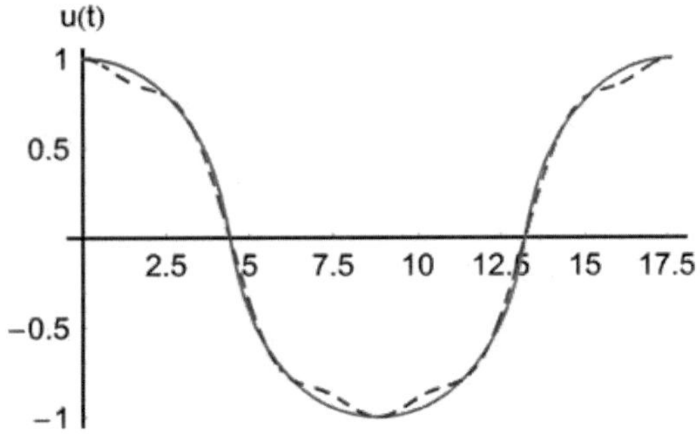

Fig. 6.33 Comparison between the approximate and numerical results of Eq. 6.337 in case (e), for $A = \omega_0 = a = 1, q = 2$: _________numerical solution, _ _ _ _ _ approximate solution Eq. 6.376

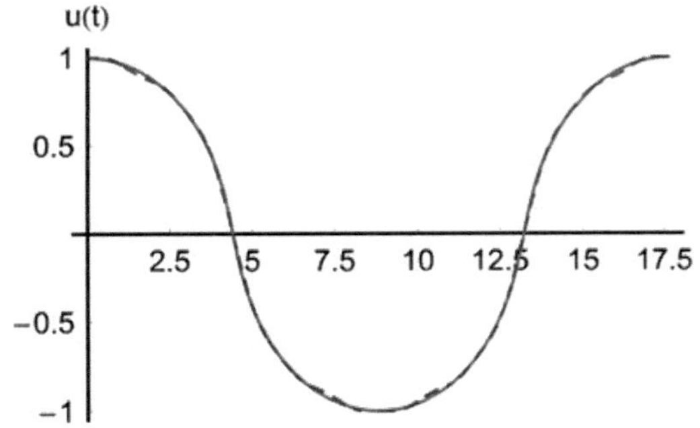

Fig. 6.34 Comparison between the approximate and numerical results of Eq. 6.377 in case (f), for $A = \omega_0 = a = 1, q = 2$: _________numerical solution, _ _ _ _ _ approximate solution Eq. 6.377

6.13 Nonlinear Oscillator with Discontinuities and Fractional-Power Restoring Force

This kind of nonlinear oscillator was studied up to now using different methodologies [87, 129–133]. We consider an ordinary differential equation a with single-term positive-power nonlinear oscillator with fractional-power restoring force:

$$\ddot{u} + sign(u)|u|^{\alpha} = 0, \quad \alpha > 0 \tag{6.378}$$

with initial conditions

$$u(0) = A, \quad \underline{u}(0) = 0 \tag{6.379}$$

With a new independent variable $\tau = \Omega t$, Eq. 6.378 can be written as

$$\Omega^2 u'' + sign(u)|u|^{\alpha} = 0 \tag{6.380}$$

with initial conditions

$$u(0) = A,\ \mathrm{u'}(0) = 0 \tag{6.381}$$

For Eq. 6.380, the linear operator Eq. 6.34 is defined by

$$L(\phi(\tau,p)) = \Omega_0^2 \left[\frac{\partial^2 \phi(\tau,p)}{\partial t^2} + \phi(\tau,p) \right] \tag{6.382}$$

and the nonlinear operator Eq. 6.35 is given by the equation:

$$\begin{aligned} &N[\phi(\tau,p),\Omega(\lambda,p)] = \\ &= \Omega^2(\lambda,p)\phi''(\tau,p) + \lambda\phi(\tau,p) + sign(\phi(\tau,p))|\phi(\tau,p)|^{\alpha} - p\lambda\phi(\tau,p) \end{aligned} \tag{6.383}$$

Equation 6.44 can be written as

$$\Omega_0^2(u_0'' + u_0) = 0, \quad \mathrm{u}_0(0) = A,\ \mathrm{u'}(0) = 0 \tag{6.384}$$

and has the solution

$$u_0(\tau) = A\cos\tau \tag{6.385}$$

If $f(u(\tau)) = sign(u(\tau))|u(\tau)|^{\alpha}$, where $u = u_0 + pu_1 + p^2u_2 + \ldots$, then in the following we have taken into account the identity:

$$\begin{aligned} f(u) &= f(u_0 + pu_1 + p^2u_2 + \ldots) = \\ &= f(u_0) + pu_1 f'(u_0) + p^2[u_2 f'(u_0) + \frac{1}{2}u_1^2 f''(u_0)] + 0(p^3) \end{aligned} \quad (6.386)$$

where for example

$$f'(u_0) = \alpha|u_0|^{\alpha-1} \quad (6.387)$$

Taking into account Eq. 6.386, the first term in Eq. 6.383 is given by:

$$N_0(u_0, \Omega_0, \lambda) = \Omega_0^2 u_0'' + \lambda u_0 + sign(u_0)|u_0|^\alpha \quad (6.388)$$

For $i = 1$ into Eq. 6.45, we obtain the equation in u_1:

$$\begin{aligned} &\Omega_0^2(u_1'' + u_1) - \Omega_0^2(u''_0 + u_0) - h_1(\tau, C_i)[\Omega_0^2 u_0'' + \lambda u_0 + sign(u_0)|u_0|^\alpha] = 0, \\ &u_1(0) = u_1'(0) = 0 \end{aligned} \quad (6.389)$$

Using Eq. 6.385, we obtain the following Fourier series expansions:

$$sign(u_0)|u_0|^\alpha = A^\alpha(a_{1\alpha}\cos\tau + a_{3\alpha}\cos 3\tau + \ldots) \quad (6.390)$$

where

$$a_{2k+1\alpha} = \frac{4}{\pi}\int_0^{\frac{\pi}{2}} (\cos\tau)^\alpha \cos(2k+1)\tau d\tau, \quad k = 0, 1, 2, \ldots \quad (6.391)$$

Considering $h_1(\tau, C_i) = C_1$ (constant) and substituting Eqs. 6.385 and 6.390 into Eq. 6.389, we have:

$$\begin{aligned} &\Omega_0^2(u_1'' + u_1) - C_1(-A\Omega_0^2 + \lambda A + a_{1\alpha}A^\alpha)\cos\tau - C_1 A^\alpha a_{3\alpha}\cos 3\tau - \\ &- C_1 A^\alpha a_{5\alpha}\cos 5\tau - C_1 A^\alpha a_{7\alpha}\cos 7\tau - \ldots = 0 \end{aligned} \quad (6.392)$$

No secular terms in u_1 requires eliminating contributions proportional to $\cos\tau$ in Eq. 6.392:

$$\Omega_0^2 = \lambda + a_{1\alpha}A^{\alpha-1}, \quad \lambda \geq 0 \quad (6.393)$$

The solution of Eq. 6.392 can be written

$$u_1(\tau) = \frac{C_1 a_{3\alpha} A^\alpha}{8\Omega_0^2}(\cos\tau - \cos 3\tau) + \frac{C_1 a_{5\alpha} A^\alpha}{24\Omega_0^2}(\cos\tau - \cos 5\tau) + \\ + \frac{C_1 a_{7\alpha} A^\alpha}{48\Omega_0^2}(\cos\tau - \cos 7\tau) + \ldots.. \tag{6.394}$$

For $m = 2$ into Eq. 6.46 and if we consider the simplest case $h_2(\tau, C_i) = C_2$ (constant), then the equation in u_2 has the form:

$$\Omega_0^2(u_2'' + u_2) - \Omega_0^2(u''_1 + u_1) - C_1[\Omega_0^2 u''_1 + 2\Omega_0\Omega_1 u''_0 + \lambda(u_1 - u_0) + \\ + \alpha|u_0|^{\alpha-1}u_1] - C_2[\Omega_0^2 u''_0 + \lambda u_0 + sign(u_0)|u_0|^\alpha] = 0, \\ u_2(0) = u'_2(0) = 0 \tag{6.395}$$

Having in view Eqs. 6.385, 6.395 and 6.394, we can write the identities:

$$|\alpha| = \alpha sign(\alpha) \tag{6.396}$$

$$\frac{\cos(2k+1)\tau}{\cos\tau} = \\ = 2\cos 2k\tau - 2\cos 2(k-1)\tau + 2\cos 2(k-2)\tau + \ldots. + (-1)^k \cos\tau \tag{6.397}$$

$$|u_0|^{\alpha-1} = A^{\alpha-1}[a_{1\alpha} - a_{3\alpha} + a_{5\alpha} - a_{7\alpha} + \ldots + \\ + 2(a_{3\alpha} - a_{5\alpha} + a_{7\alpha} + \ldots)\cos 2\tau + 2(a_{5\alpha} - a_{7\alpha} + \ldots)\cos 4\tau + \ldots.] \tag{6.398}$$

Substitution of Eqs. 6.385, 6.390, 6.394, 6.396 and 6.397 into Eq. 6.395 yields:

$$\Omega_0^2(u''_2 + u_2) = \left[2AK_1\frac{\lambda - \Omega_0^2}{\Omega_0^2} - 2A\Omega_0\Omega_1 - \lambda A + \frac{C_1^2\alpha A^{2\alpha-1}}{48\Omega_0^2}\beta_1 + \right. \\ \left. + \frac{2AK_2}{\Omega_0^2}\right]\cos\tau + \left[\frac{C_1^2 A^\alpha a_{3\alpha}}{8}\left(9 - \frac{\lambda}{\Omega_0^2}\right) + \frac{\alpha C_1^2 A^{2\alpha-1}\beta_3}{48\Omega_0^2} + \right. \\ \left. + (C_1 + C_2)A^\alpha a_{3\alpha}\right]\cos 3\tau + \left[\frac{C_1^2 A^\alpha a_{5k}}{24}\left(25 - \frac{\lambda}{\Omega_0^2}\right) + \right. \\ \left. + \frac{\alpha C_1^2 A^{2\alpha-1}\beta_5}{48\Omega_0^2} + (C_1 + C_2)A^\alpha a_{5\alpha}\right]\cos 5\tau + \\ + \left[\frac{C_1^2 A^\alpha a_{7\alpha}}{48}\left(49 - \frac{\lambda}{\Omega_0^2}\right) + \frac{\alpha C_1^2 A^{2\alpha-1}\beta_7}{48\Omega_0^2} + (C_1 + C_2)A^\alpha a_{7\alpha}\right]\cos 7\tau + \ldots. \tag{6.399}$$

where

$$K_1 = \frac{C_1 A^{\alpha-1}(6a_{3\alpha} + 2a_{5\alpha} + a_{7\alpha})}{96}$$

$$K_2 = \frac{C_1 A^{2\alpha-2}(6a_{1\alpha}a_{3\alpha} - 6a_{3\alpha}^2 + 2a_{1\alpha}a_{5\alpha} - 2a_{5\alpha}^2 + a_{1\alpha}a_{7\alpha} - a_{7\alpha}^2)}{96}$$

$$\beta_1 = 6a_{1\alpha}a_{3\alpha} + 2a_{1\alpha}a_{5\alpha} + a_{1\alpha}a_{7\alpha} - 6a_{3\alpha}^2 - 6a_{5\alpha}^2 + 6a_{5\alpha}a_{7\alpha}$$

$$\beta_3 = -6a_{1\alpha}a_{3\alpha} + 12a_{3\alpha}^2 - 6a_{3\alpha}a_{5\alpha} + 2a_{5\alpha}^2 + 7a_{3\alpha}a_{7\alpha} - 3a_{5\alpha}a_{7\alpha} + a_{7\alpha}^2$$

$$\beta_5 = -6a_{3\alpha}^2 + 14a_{3\alpha}a_{5\alpha} - 2a_{1\alpha}a_{5\alpha} - 6a_{3\alpha}a_{7\alpha} + 2a_{5\alpha}a_{7\alpha} - 2a_{7\alpha}^2$$

$$\beta_7 = -8a_{3\alpha}a_{5\alpha} + 2a_{5\alpha}^2 + 5a_{3\alpha}a_{7\alpha} - a_{5\alpha}a_{7\alpha} - 3a_{7\alpha}^2 \tag{6.400}$$

The secular term in the solution of u_2 can be eliminated from Eq. 6.399 if

$$\Omega_1 = \frac{K_1(\lambda - \Omega_0^2)}{\Omega_0^3} - \frac{\lambda}{2\Omega_0} + \frac{K_2}{\Omega_0^3} + \frac{C_1^2 \alpha A^{2\alpha-2}}{96\Omega_0^3}\beta_1 \tag{6.401}$$

From Eqs. 6.43 and 6.401 we obtain the frequency in the form:

$$\Omega = \Omega_0 + \frac{K_1(\lambda - \Omega_0^2)}{\Omega_0^3} - \frac{\lambda}{2\Omega_0} + \frac{K_2}{\Omega_0^3} + \frac{C_1^2 \alpha A^{2\alpha-2}}{96\Omega_0^3}\beta_1 \tag{6.402}$$

where Ω_0 is given by Eq. 6.393.

The parameter λ can be determined applying the "principle of minimal sensitivity". From Eq. 6.48 we obtain:

$$\lambda = -\frac{1}{2}a_{1\alpha}A^{\alpha-1} + \sqrt{\frac{1}{4}a_{1\alpha}^2 A^{2\alpha} + 6K_2 - 6K_1 a_{1\alpha}A^{\alpha} + \frac{C_1^2 \alpha A^{2\alpha-2}}{16}\beta_1} \tag{6.403}$$

From Eqs. 6.403 and 6.393 it follows that

$$\Omega_0^2 = \frac{1}{2}a_{1\alpha}A^{\alpha-1} + \sqrt{\frac{1}{4}a_{1\alpha}^2 A^{2\alpha} + 6K_2 - 6K_1 a_{1\alpha}A^{\alpha} + \frac{C_1^2 \alpha A^{2\alpha-2}}{16}\beta_1} \tag{6.404}$$

where K_1, K_2 and β_1 are given by Eq. 6.400.

Now, we can write the solution of Eq. 6.399 in the form

$$u_2(\tau) = \left[\frac{C_1^2 A^\alpha a_{3\alpha}}{64\Omega_0^2}\left(9 - \frac{\lambda}{\Omega_0^2}\right) + \frac{\alpha C_1^2 A^{2\alpha-1}\beta_3}{384\Omega_0^4} + \right.$$
$$\left. + \frac{(C_1 + C_2)A^\alpha a_{3\alpha}}{8\Omega_0^2}\right](\cos\tau - \cos 3\tau) + \left[\frac{C_1^2 A^\alpha a_{5\alpha}}{576\Omega_0^2}\left(25 - \frac{\lambda}{\Omega_0^2}\right) + \right.$$
$$\left. + \frac{\alpha C_1^2 A^{2\alpha-1}\beta_5}{1152\Omega_0^4} + \frac{(C_1 + C_2)A^\alpha a_{5\alpha}}{24\Omega_0^2}\right](\cos\tau - \cos 5\tau) +$$
$$+ \left[\frac{C_1^2 A^\alpha a_{7\alpha}}{2304\Omega_0^2}\left(49 - \frac{\lambda}{\Omega_0^2}\right) + \frac{\alpha C_1^2 A^{2\alpha-1}\beta_7}{2304\Omega_0^4} + \frac{(C_1 + C_2)A^\alpha a_{7\alpha}}{48\Omega_0^2}\right](\cos\tau -$$
$$\cos 7\tau) \tag{6.405}$$

In order to determine the second-order approximate solution it is necessary to substitute Eqs. 6.385, 6.394 and 6.405 into the equation:

$$\bar{u}(\tau) = u_0(\tau) + u_1(\tau) + u_2(\tau) \tag{6.406}$$

By means of the transformation $\tau = \Omega t$, the second-order approximate solution of Eq. 6.378 is:

$$\bar{u}(t) = A\cos\Omega t + \left[\frac{C_1^2 A^\alpha a_{3\alpha}}{64\Omega_0^2}\left(9 - \frac{\lambda}{\Omega_0^2}\right) + \frac{\alpha C_1^2 A^{2\alpha-1}\beta_3}{384\Omega_0^4} + \right.$$
$$\left. + \frac{(2C_1 + C_2)A^\alpha a_{3\alpha}}{8\Omega_0^2}\right](\cos\Omega t - \cos 3\Omega t) + \left[\frac{C_1^2 A^\alpha a_{5\alpha}}{576\Omega_0^2}\left(25 - \frac{\lambda}{\Omega_0^2}\right) + \right.$$
$$\left. + \frac{\alpha C_1^2 A^{2\alpha-1}\beta_5}{1152\Omega_0^4} + \frac{(2C_1 + C_2)A^\alpha a_{5\alpha}}{24\Omega_0^2}\right](\cos\Omega t - \cos 5\Omega t) +$$
$$+ \left[\frac{C_1^2 A^\alpha a_{7\alpha}}{2304\Omega_0^2}\left(49 - \frac{\lambda}{\Omega_0^2}\right) + \frac{\alpha C_1^2 A^{2\alpha-1}\beta_7}{2304\Omega_0^4} + \right.$$
$$\left. + \frac{(2C_1 + C_2)A^\alpha a_{7\alpha}}{48\Omega_0^2}\right](\cos\Omega t - \cos 7\Omega t) \tag{6.407}$$

where $\beta_1, \beta_3, \beta_5, \beta_7$ are given by Eq. 6.400 and Ω, λ, Ω_0 depends on the constants C_1 and C_2, which will be optimally determined following Eqs. 6.26.

We illustrate the accuracy of the OHAM by comparing previously obtained approximate solutions with the numerical integration results obtained by means of a fourth-order Runge–Kutta method.

In the first case we consider $\alpha = \frac{3}{5}$ and from Eq. 6.391 we obtain:

$$a_{1\frac{3}{5}} = 1.0872931957;\ a_{3\frac{3}{5}} = -\frac{1}{9}a_{1\frac{3}{5}};\ a_{5\frac{3}{5}} = \frac{1}{21}a_{1\frac{3}{5}};\ a_{7\frac{3}{5}} = -\frac{11}{399}a_{1\frac{3}{5}}$$

(a) For $A = 1$, applying the conditions Eq. 6.26 we obtain

$$C_1 = -0.768214231; C_2 = -3.562815287; \Omega = 1.040588306;$$
$$\frac{\lambda}{\Omega_0^2} = -0.01253803$$

The exact frequency in the case $A = 1$ is $\Omega_{ex} = 1.04075$ [131] and therefore the relative error between the approximate and the exact frequency is 0.016%.

The second-order approximate solution Eq. 6.407 becomes:

$$\bar{u}(t) = 1.00699391 \cos \Omega t - 0.013465301 \cos 3\Omega t + \\ +0.009104865 \cos 5\Omega t - 0.002633483 \cos 7\Omega t \tag{6.408}$$

(b) For $A = 5$, we obtain the following expressions:

$$C_1 = -0.768214231;\ C_2 = -4.21476475;\ \Omega = 0.75419724;\ \frac{\lambda}{\Omega_0^2} = -0.012538032$$

The exact frequency in this case is $\Omega_{ex} = 0.754314435$ [131] and therefore the relative error between the approximate and the exact frequency is 0.015%.

The second-order approximate solution Eq. 6.407 becomes in this case:

$$\bar{u}(t) = 5.076158387 \cos \Omega t - 0.113168423 \cos 3\Omega t + \\ +0.052073171 \cos 5\Omega t - 0.015063135 \cos 7\Omega t \tag{6.409}$$

In the second case we consider$\alpha = \frac{1}{3}$. From Eq. 6.391 it is obtained:

$$a_{1\frac{1}{3}} = 1.1595952669;\ a_{3\frac{1}{3}} = -\frac{1}{5}a_{1\frac{1}{3}};\ a_{5\frac{1}{3}} = \frac{1}{10}a_{1\frac{1}{3}};\ a_{7\frac{1}{3}} = -\frac{7}{110}a_{1\frac{1}{3}}$$

(c) For $A = 1$, we find

$$C_1 = -0.812457981;\ C_2 = 0.040087495;\ \Omega = 1.07000511;\ \frac{\lambda}{\Omega_0^2} = -0.040045851$$

The exact frequency for $A = 1$ is $\Omega_{ex} = 1.07045$ [131] and therefore the relative error between the approximate and the exact frequency is 0.044%.

The second-order approximate solution Eq. 6.407 becomes:

$$\bar{u}(t) = 1.033439236 \cos \Omega t - 0.02397233 \cos 3\Omega t - \\ -0.008055218 \cos 5\Omega t - 0.001411688 \cos 7\Omega t \tag{6.410}$$

(d) For $A = 5$, we obtain the values:

$$C_1 = -0.812457981; C_2 = -0.68943007; \Omega = 0.62574278; \frac{\lambda}{\Omega_0^2} = -0.040045851$$

The exact frequency is $\Omega_{ex} = 0.626002957$ [131] and therefore the relative error between the approximate and the exact frequency is 0.042%.
The second-order approximate solution Eq. 6.407 becomes in this case:

$$\bar{u}(t) = 5.014642562 \cos \Omega t - 0.008561393 \cos 3\Omega t - \\ -0.005486728 \cos 5\Omega t - 0.000594441 \cos 7\Omega t \tag{6.411}$$

Figures 6.35–6.38 show a comparison between the present analytical solutions and the numerical integration results obtained using a fourth-order Runge–Kutta method.

One can observe that the second-order approximate analytical results obtained through OHAM are almost identical with the numerical simulation results in all

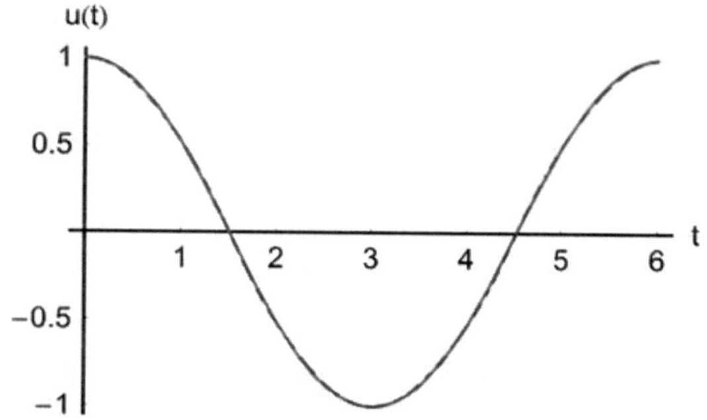

Fig. 6.35 Comparison between the approximate solution Eq. 6.408 and numerical results of Eq. 6.378 in the case $\alpha = 3/5$, $A = 1$: ________numerical solution; _ _ _ _ _ approximate solution

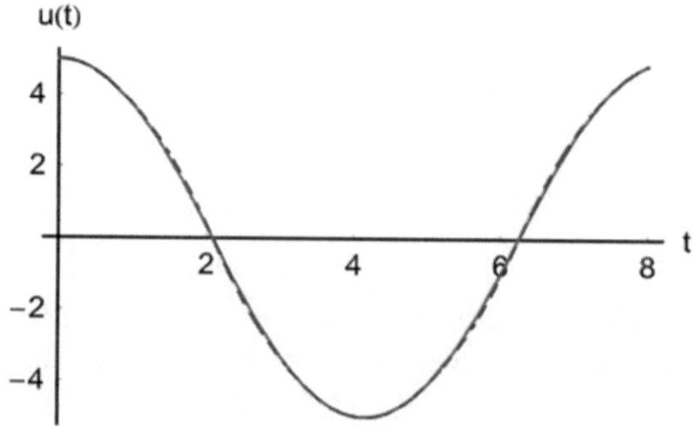

Fig. 6.36 Comparison between the approximate solution Eq. 6.409 and numerical results for Eq. 6.378 in the case $\alpha = 3/5$, $A = 5$: ______numerical solution; _ _ _ approximate solution

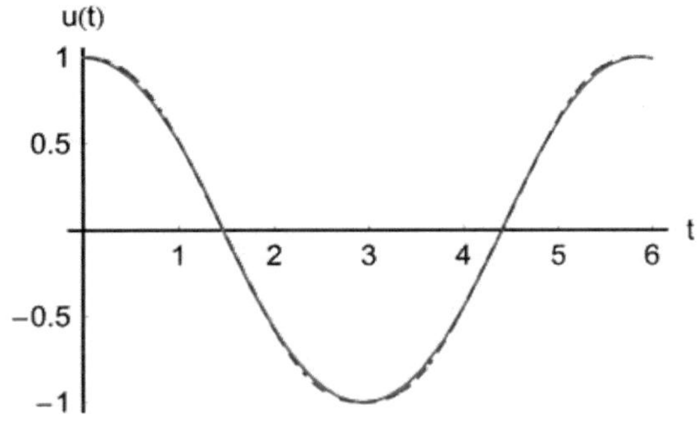

Fig. 6.37 Comparison between the approximate solution Eq. 6.410 and numerical results for Eq. 6.378 in the case $\alpha = 1/3$, A = 1: ______numerical solution; _ _ _ _ approximate solution

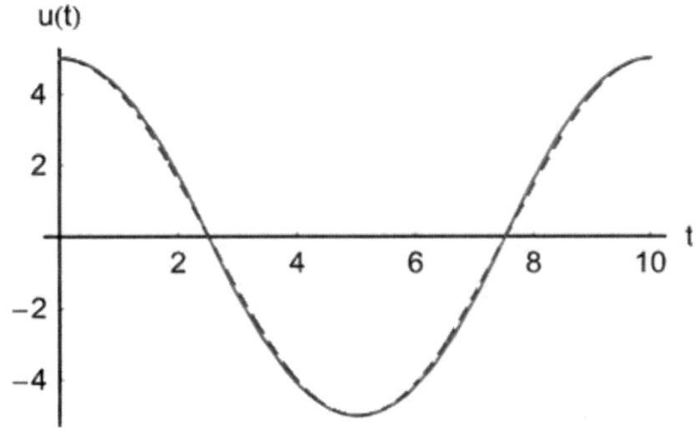

Fig. 6.38 Comparison between the approximate solution Eq. 6.411 and numerical results for Eq. 6.378 in the case $\alpha = 1/3$, A = 5: _____numerical solution; _ _ _ approximate solution

considered cases for various values of the parameters α and A. Moreover, the relative error between the approximate and the exact frequency presented in [131] varies between 0.015% and 0.044%, which proves the accuracy of the method.

6.14 Oscillations of a Flexible Cantilever Beam with Support Motion

A masless cantilever beam with a lumped mass attached to its free end while being excited harmonically at the base has received widespread attention in connection with interest in applications such as mast antennas, towers, flexible robot manipulators, and space structures. When beam support undergoes motion, the beam might be subject to external or parametric excitation. Since the parametric resonance is considered, the nonlinearities begin to affect the motion, and hence they cannot be ignored.

Many investigations have considered the influence of a tip mass on the behaviour of transverse vibration of a beam clamped at the other end. The method which Timoshenko used to solve a special case of a problem for the longitudinal vibration of a rod with a mass attached to its end is not easily applicable to a more general case of transverse vibrations of beams [134]. Chen has investigated the problem as a purely mathematical one [135]. He used the method of separation of variables and determined the eigenfunctions of the resulting ordinary differential equation, but he neither considered the non-linearities of the system nor the excitations at the base. Esmailzadeh et al. [136] used the application of fixed theorems (Schauder) which is a topological character and they guarantee the existence of at least one solution.

In what follows, we consider oscillations of a flexible cantilever beam with suport motion [136] governed by the ordinary differential equation

$$\ddot{u} + \omega^2 u + u^2 \ddot{u} + u\dot{u}^2 + a\cos t = 0 \tag{6.412}$$

where ω and a are known parameters.

Initial conditions are:

$$u(0) = A, \quad \dot{u}(0) = 0$$

Under the transformation:

$$\tau = \Omega t \tag{6.413}$$

the original Eq. 6.412 becomes

$$\Omega^2 u'' + \omega^2 u + \Omega^2(u^2 u'' + uu'^2) + a\cos\frac{\tau}{\Omega} = 0 \tag{6.414}$$

with initial conditions

$$u(0) = A, \quad u'(0) = 0 \tag{6.415}$$

where prime denote derivative with respect to τ

The linear operator Eq. 6.34 is given by

$$L[\phi(\tau,p)] = \Omega_0^2\left[\frac{\partial^2\phi(\tau,p)}{\partial\tau^2} + \phi(\tau,p)\right] \tag{6.416}$$

while the nonlinear operator Eq. 6.35 is defined by

$$\begin{aligned} N[\phi(\tau,p),\Omega(p)] &= \Omega^2(p)\frac{\partial^2\phi(\tau,p)}{\partial\tau^2} + \omega^2\phi(\tau,p) + \\ &+ \Omega^2\left[\phi^2(\tau,p)\frac{\partial^2\phi(\tau,p)}{\partial\partial\tau^2} + \phi(\tau,p)\left(\frac{\partial\phi(\tau,p)}{\partial\tau}\right)^2\right] + a\cos\frac{\tau}{\Omega} \end{aligned} \tag{6.417}$$

Because we consider only one iteration it follows that $\lambda = 0$.
The Eq. 6.44 becomes

$$\Omega_0^2(u''_0 + u_0) = 0, \qquad u_0(0) = A, \; u'_0(0) = 0 \tag{6.418}$$

and has the solution

$$u_0(\tau) = A\cos\tau \tag{6.419}$$

The first term in Eq. 6.47 is

$$N_0 = \Omega_0^2 u_0'' + \omega^2 u_0 + \Omega^2(u_0^2 u_0'' + u_0 u_0' 2) + a\cos\frac{\tau}{\Omega} \tag{6.420}$$

For $m = 1$ into Eq. 6.46, supposing that

$$h_1(\tau, C_i) = C_1 + 2C_2\cos 2\tau + 2C_3\cos 4\tau \tag{6.421}$$

where C_1, C_2 and C_3 are constants, then substituting Eq. 6.419 into Eq. 6.420 we obtain

$$\begin{aligned} h_1 N_0(u_0, \Omega_0) = &\{(C_1+C_2)\left[\omega^2 A - \Omega_0^2\left(A + \frac{A^3}{2}\right)\right] - \frac{1}{2}(C_2 + \\ &+C_3)\Omega_0^2 A^3\}\cos\tau + \{(C_2+C_3)\left[\omega^2 A - \Omega_0^2\left(A+\frac{A^3}{2}\right)\right] - \frac{1}{2}\Omega_0^2 A^3 C_1 - \\ &-\Omega_0^2\left(A+\frac{A^3}{2}\right)\}\cos 3\tau + \{C_3[\omega^2 A - \Omega_0^2\left(A+\frac{A^3}{2}\right)] - \frac{1}{2}\Omega_0^2 A^3 C_2\}\cos 5\tau - \\ &-\frac{1}{2}\Omega_0^2 A^3 C_3 \cos 7\tau - aC_1\cos\frac{\tau}{\Omega} - aC_2\left[\cos\left(2+\frac{1}{\Omega}\right)\tau + \cos\left(2-\frac{1}{\Omega}\right)\tau\right] - \\ &-aC_3\left[\cos\left(4+\frac{1}{\Omega}\right)\tau + \cos\left(4-\frac{1}{\Omega}\right)\tau\right] \end{aligned} \tag{6.422}$$

The elimination of the secular term requires

$$\Omega^2 = \Omega_0^2 = \frac{\omega^2}{1 + \frac{1}{2}A^2\left(1 + \frac{C_2+C_3}{C_1+C_2}\right)} \tag{6.423}$$

Equation 6.46 can be written for $m = 1$ in the form:

$$\Omega_0^2(u''_1 + u_1) + \frac{A^3}{2}\left[C_1 - \frac{(C_2+C_3)^2}{C_1+C_2}\right]\cos 3\tau + \frac{A^2}{2}\left[C_2 - .\right.$$
$$\left. - \frac{C_3(C_2+C_3)}{C_1+C_2}\right]\cos 5\tau + \frac{A^3}{2}C_3\cos 7\tau + aC_1\cos\frac{\tau}{\Omega} +$$
$$+ aC_2\left[\cos\left(2+\frac{1}{\Omega}\right)\tau + \cos\left(2-\frac{1}{\Omega}\right)\tau\right] +$$
$$+ aC_3\left[\cos\left(4+\frac{1}{\Omega}\right)\tau + \cos\left(4-\frac{1}{\Omega}\right)\tau\right] = 0,\ \mathrm{u}_1(0) = u'_1(0) = 0 \quad (6.424)$$

From Eq. 6.424 we obtain the solution

$$u_1(\tau) = \frac{A^3}{16}\left[C_1 - \frac{(C_2+C_3)^2}{C_1+C_2}\right](\cos 3\tau - \cos\tau) + \frac{A^3}{48}\left[C_2 - .\right.$$
$$\left. - \frac{C_3(C_2+C_3)}{C_1+C_2}\right](\cos 5\tau - \cos\tau) + \frac{A^3C_3}{96}(\cos 7\tau - \cos\tau) -$$
$$- \frac{a\Omega^2 C_1}{\Omega_0^2(\Omega^2-1)}\left(\cos\frac{\tau}{\Omega} - \cos\tau\right) + \frac{a\Omega^2 C_2}{\Omega_0^2(3\Omega^2+4\Omega+1)}\left[\cos\left(2+\frac{1}{\Omega}\right)\tau - \right.$$
$$\left. - \cos\tau\right] + \frac{a\Omega^2 C_2}{\Omega_0^2(3\Omega^2-4\Omega+1)}\left[\cos\left(2-\frac{1}{\Omega}\right)\tau - \cos\tau\right] +$$
$$+ \frac{a\Omega^2 C_3}{\Omega_0^2(15\Omega^2+8\Omega+1)}\left[\cos(4+\frac{1}{\Omega}\right)\tau - \cos\tau\right] +$$
$$+ \frac{a\Omega^2 C_3}{\Omega_0^2(15\Omega^2-8\Omega+1)}\left[\cos\left(4+\frac{1}{\Omega}\right)\tau - \cos\tau\right]$$

(6.425)

The first-order approximate solution Eq. 6.42 becomes:

$$\bar{u}(\tau) = u_0(\tau) + u_1(\tau) \quad (6.426)$$

Substituting Eqs. 6.413, 6.419 and 6.425 into Eq. 6.426, we obtain:

$$\bar{u}(t) = \left[A + \frac{A^3(-6C_1^2 - 8C_1C_2 - C_1C_3 + 5C_2C_3 + 4C_3^2)}{48(C_1 + C_2)} + \frac{aC_1}{\Omega^2 - 1} - \right.$$
$$\left. - \frac{2a(3\Omega^2 + 1)C_2}{9\Omega^4 - 10\Omega^2 + 1} - \frac{2a(15\Omega^2 + 1)C_3}{225\Omega^4 - 34\Omega^2 + 1}\right] \cos \Omega t + \frac{A^3}{16}\left[C_1 - .\right.$$
$$\left. - \frac{(C_2 + C_3)^2}{C_1 + C_2}\right] \cos 3\Omega t + \frac{A^3}{48}\left[C_2 - \frac{C_3(C_2 + C_3)}{C_1 + C_2}\right] \cos 5\Omega t +$$
$$+ \frac{A^3C_3}{96} \cos 7\Omega t - \frac{aC_1}{\Omega^2 - 1} \cos t + \frac{aC_2}{3\Omega^2 + 4\Omega + 1} \cos(2\Omega + 1)t +$$
$$+ \frac{aC_3}{3\Omega^2 - 4\Omega + 1} \cos(2\Omega - 1)t + \frac{aC_3}{15\Omega^2 + 8\Omega + 1} \cos(4\Omega + 1)t +$$
$$+ \frac{aC_3}{15\Omega^2 - 8\Omega + 1} \cos(4\Omega - 1)t \tag{6.427}$$

where Ω is given by Eq. 6.423 and the constants C_1, C_2, C_3 are determined from Eq. 6.26.

In order to show the validity of the OHAM, Eq. 6.412 has been numerically solved for the following characteristics:$a = 0.12337$, $\omega^2 = 1.709678$.

(a) In the first case $A = 0.3$, we obtain using the least square method:

$$C_1 = 0.9027885273, C_2 = -0.0865822331,$$
$$C_3 = -0.2465694380, \Omega = 1.673124$$

The first order approximate solution Eq. 6.427 becomes:

$$\begin{aligned}\bar{u}(t) = &- 0.0618116 \text{ cost} + 0.365437 \text{ cos1.6738 t}- \\ &- 0.0039414 \text{ cos2.3477 t} - 0.00066340 \text{ cos4.3477 t}+ \\ &+ 0.00129398 \text{ cos5.0216 t} - 0.00102639 \text{ cos5.6955 t}- \\ &- 0.00053916 \text{ cos7.6955 t} - 0.00010531 \text{ cos8.3694 t}- \\ &- 0.00006934 \text{ cos11.717 t}\end{aligned} \tag{6.428}$$

(b) In the second case $A = 0.4$, we obtain:

$$C_1 = 0.9236350757, C_2 = -0.1344892332,$$
$$C_3 = 0.2051665021, \quad \Omega = 1.644324$$

The first order approximate solution becomes:

$$\bar{u}(t) = -0.067173\cos t + 0.46621\cos 1.64206t - 0.0065819\cos 2.2841t - \\ -0.00105969\cos 4.2841t + 0.00366922\cos 4.9261t + \\ +0.00089411\cos 5.5682t + 0.00046373\cos 7.5682t - \\ -0.00020381\cos 8.2102t + 0.00013677\cos 11.4944t \tag{6.429}$$

(c) In the third case $A = 0.5$, we obtain:

$$C1 = 0.9529501648798903, C2 = -0.16063357719568372, \\ C3 = 0.3717062754599205, \Omega = 1.604512$$

The first order approximate solution becomes:

$$u(t) = -0.0743925\cos t + 0.567464\cos 1.6063t - 0.00855804\cos 2.2126t - \\ -0.00130666\cos 4.2126t + 0.00700563\cos 4.8190t + \\ +0.00170763\cos 5.4253t + 0.00087254\cos 7.42537t - \\ -0.00067618\cos 8.03172t + 0.00048399\cos 11.2444t \tag{6.430}$$

Figures 6.39–6.41 show the comparison between the present solutions and the numerical integration results obtained by a fourth-order Runge–Kutta method.

It is clear that the solutions obtained by OHAM are nearly identical with the solutions obtained through a fourth-order Runge–Kutta method. Additionally we remark that the exact values of the frequencies are $\Omega_{ex} = 1.673881823$ in the case $A = 0.3$, $\Omega_{ex} = 1.642056622$ for $A = 0.4$ and $\Omega_{ex} = 1.606343438$ for $A = 0.5$, which means that very good approximations were found also for the frequencies.

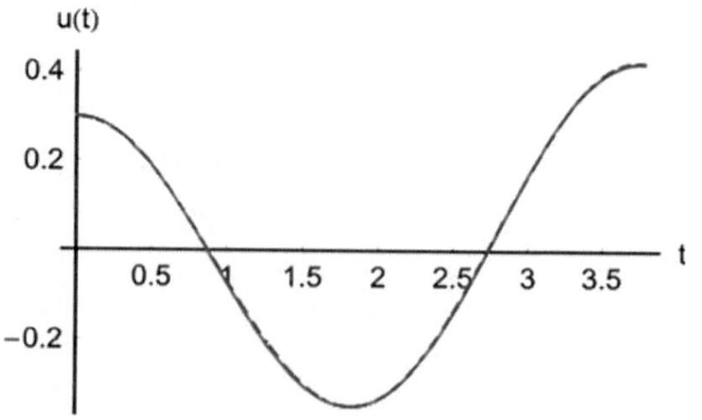

Fig. 6.39 Comparison between the approximate solution Eq. 6.428 and the numerical solution for Eq. 6.412 in case $A = 0.3$, $a = 0.12337$, $\omega^2 = 1.709678$: _______ numerical solution; _ _ _ _ _ approximate solution

6.15 The Jeffery-Hamel Flow Problem

We consider a system of cylindrical polar coordinates (r,θ,z) with a steady two-dimensional flow of an incompressible conducting viscous fluid from a source or sink at channel walls lying in planes, with angle 2α, as shown in Fig. 6.42.

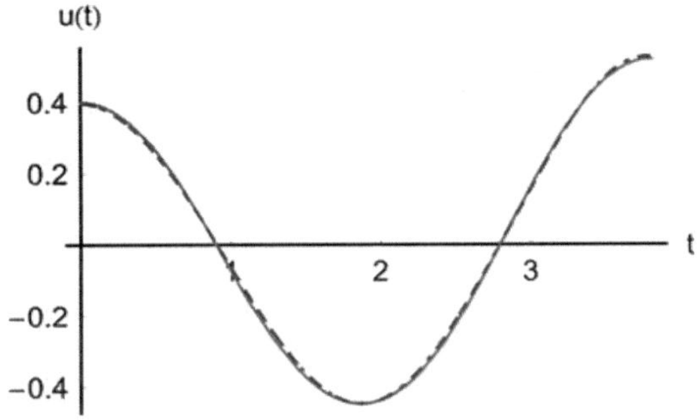

Fig. 6.40 Comparison between the approximate solution Eq. 6.429 and the numerical solution for Eq. 6.412 in case A = 0.4, a = 0.12337, $\omega^2 = 1.709678$: _______ numerical solution; _ _ _ _ _ approximate solution

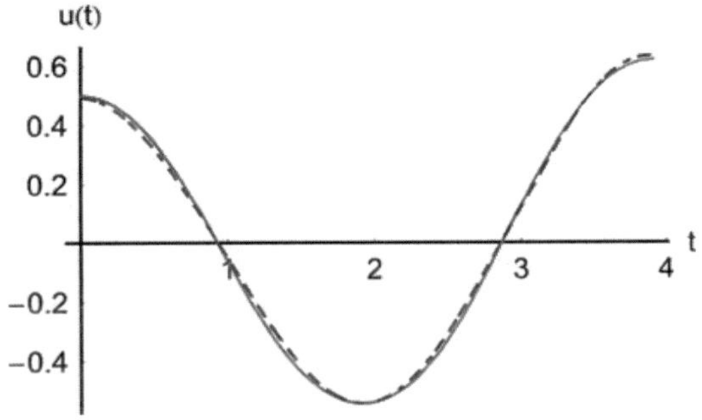

Fig. 6.41 Comparison between the approximate solution Eq. 6.430 and the numerical solution for Eq. 6.412 in case A = 0.5, a = 0.12337, $\omega^2 = 1.709678$: _______ numerical solution; _ _ _ _ _ approximate solution

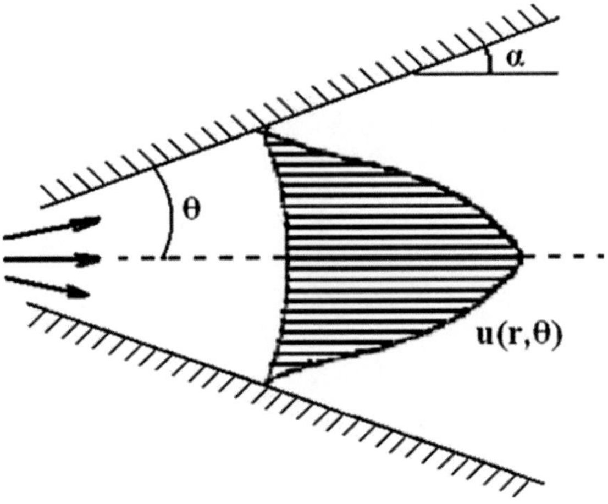

Fig. 6.42 Geometry of the Jeffery-Hamel flow problem

Assuming that the velocity is only along the radial direction and depends on r and θ, $V(u(r,\theta),0)$ using the continuity and the Navier–Stokes equations in polar coordinates, the governing equations are [137]

$$\frac{\rho}{r}\frac{\partial}{\partial r}(ru(r,\theta)) = 0 \tag{6.431}$$

$$u(r,\theta)\frac{\partial u(r,\theta)}{\partial r} = -\frac{1}{\rho}\frac{\partial p}{\partial r} + \\ + \nu\left[\frac{\partial^2 u(r,\theta)}{\partial r^2} + \frac{1}{r}\frac{\partial u(r,\theta)}{\partial r} + \frac{1}{r^2}\frac{\partial^2 u(r,\theta)}{\partial \theta^2} - \frac{u(r,\theta)}{r^2}\right] \tag{6.432}$$

$$-\frac{1}{\rho r}\frac{\partial p}{\partial \rho} + \frac{2\nu}{r^2}\frac{\partial u(r,\theta)}{\partial \theta} = 0 \tag{6.433}$$

where ρ is the fluid density, p is the pressure and ν is the kinematic viscosity. From Eq. 6.431 and using dimensionless parameters we get:

$$f(\theta) = ru(r,\theta) \tag{6.434}$$

$$F(x) = \frac{f(\theta)}{f_{\max}}, \quad x = \frac{\theta}{\alpha} \tag{6.435}$$

Substituting Eq. 6.435 into Eqs. 6.432 and 6.433 and eliminating the pressure, we obtain an ordinary differential equation for the normalized function profile $F(x)$:

$$F'''(x) + 2\alpha \mathrm{Re} F(x)F'(x) + 4\alpha^2 F'(x) = 0 \tag{6.436}$$

where prime denotes derivative with respect to x and the Reynolds number is

$$\mathrm{Re} = \frac{\alpha f_{\max}}{\nu} = \frac{u_{\max} r\alpha}{\nu}\left(\begin{array}{l} \textit{divergent channel}: \ \alpha > 0, \ u_{\max} > 0 \\ \textit{convergent channel}: \ \alpha < 0, \ u_{\max} < 0 \end{array}\right) \tag{6.437}$$

and u_{mx} is the maximum velocity at the centre of the channel.

The boundary conditions for Eq. 6.436 are

$$F(0) = 1, \quad F'(0) = 0, \quad F(1) = 0 \tag{6.438}$$

We choose $g(x) = 0$ into Eq. 6.13 and the linear operator

$$L(\phi(x,p)) = \frac{\partial^3 \phi(x,p)}{\partial x^3} \tag{6.439}$$

The nonlinear operator is

$$N(\phi(x,p)) = 2\alpha \text{Re}\phi(x,p)\frac{\partial \phi(x,p)}{\partial x} + 4\alpha^2 \frac{\partial \phi(x,p)}{\partial x} \tag{6.440}$$

and the boundary conditions are

$$\phi(0,p) = 1, \quad \frac{\partial \phi(0,p)}{\partial x} = 0, \quad \phi(1,p) = 0 \tag{6.441}$$

Equation 6.16 becomes

$$\begin{aligned} &F'''_0(x) = 0 \\ &F_0(0) = 1, \quad F'_0(0) = 0, \quad F_0(1) = 0 \end{aligned} \tag{6.442}$$

It is obtained that

$$F_0(x) = 1 - x^2 \tag{6.443}$$

From Eqs. 6.440 and 6.21, we obtain the following expression

$$N_0(x) = F_0'''(x) + 2\alpha \text{Re} F_0(x) F_0'(x) + 4\alpha^2 F_0'(x) \tag{6.444}$$

Note that substituting Eq. 6.443 into Eq. 6.444 it follows that

$$N_0(x) = 4\alpha \text{Re} x^3 - 4(\alpha \text{Re} + 2\alpha^2)x \tag{6.445}$$

There are many possibilities to choose the functions $h_i(x,C_i)$, $i = 1,2,\ldots$ The convergence of the solutions F_i, $i = 1,2,\ldots m$ and consequently the convergence of the approximate solution $\bar{F}(x)$ given by Eq. 6.23 depend on the auxiliary functions $h_i(x,C_i)$. Basically, the shape of $h_i(C_i)$ should follow the terms appearing in Eq. 6.445 which are polynomial functions. We consider the following cases ($m = 2$) in Eq. 6.23:

Case 6.15.a

If h_1 is of the form

$$h_1(x, C_i) = C_1 \tag{6.446}$$

where C_1 is a unknown constant at this moment, then Eq. 6.20 for $k = 1$ becomes:

$$F_1'''(x) - F_0'''(x) - h_1(x, C_i)N_0(x) = 0 \tag{6.447}$$

Substituting Eqs. 6.443, 6.445 and 6.446 into Eq. 6.447, we obtain the equation in F_1:

$$F_1'''(x) - 4C_1\alpha\text{Re}x^3 - 4C_1(\alpha\text{Re} + 2\alpha^2)x = 0,\ F_1(0) = F_1'(0) = F_1(1) = 0 \tag{6.448}$$

The solution of Eq. 6.448 is given by

$$F_1(x) = \frac{C_1\alpha\text{Re}}{30}x^6 - \frac{C_1(\alpha\text{Re} + 2\alpha^2)}{6}x^4 + \frac{2\alpha\text{Re} + 5\alpha^2}{15}C_1x^2 \tag{6.449}$$

Equation 6.20 for $m = 2$, can be written in the form

$$F_2'''(x) - F_1'''(x) - C_1N_1(F_0, F_1) - h_2(x, C_i)N_0(F_0) = 0 \tag{6.450}$$

where N_1 is obtained from Eq. 6.21:

$$N_1(F_0, F_1) = F_1''' + 2\alpha\text{Re}(F_0F_1' + F_0'F_1) + 4\alpha^2F_1' \tag{6.451}$$

If we consider

$$h_2(x, C_i) = C_2 \tag{6.452}$$

where C_2 is an unknown constant, then from Eqs. 6.443, 6.449, 6.450, 6.451 and 6.452 we obtain the following equation in F_2:

$$\begin{aligned} F'''_2 &+ \frac{8\alpha^2\text{Re}^2C_1^2}{15}x^7 - \frac{12\alpha^2\text{Re}^2 + 24\alpha^2\text{Re}}{5}C_1^2x^5 - \Big[4\,\alpha\text{Re}.(2C_1 + C_2) + \\ &+ \frac{60\alpha\text{Re} + 36\alpha^2\text{Re}^2 + 40\alpha^3\text{Re} - 16\alpha^4}{15}C_1^2\Big]x^3 + \Big[4(\alpha\text{Re} + 2\alpha^2)(C_1 + C_2).+ \\ &+ \frac{60\alpha\text{Re} + 120\alpha^2 - 8\alpha^2\text{Re}^2 - 36\alpha^3\text{Re} - 40\alpha^4}{15}C_1^2\Big]x = 0 \end{aligned} \tag{6.453}$$

So, the solution of Eq. 6.453 is given by

$$F_2(x) = -\frac{\alpha^2\mathrm{Re}^2C_1^2}{1350}x^{10} + \frac{\alpha^2\mathrm{Re}^2 + 2\alpha^2\mathrm{Re}}{140}C_1^2x^8 + \\ + \left[\frac{\alpha\mathrm{Re}(2C_1+C_2)}{30} + \frac{15\alpha\mathrm{Re} + 9\alpha^2\mathrm{Re}^2 + 10\alpha^3\mathrm{Re} - 4\alpha^4}{450}C_1^2\right]x^6 + \\ + \left[-\frac{(\alpha\mathrm{Re}+2\alpha^2)(C_1+C_2)}{6} + \right. \\ \left. + \frac{2\alpha^2\mathrm{Re}^2 + 9\alpha^3\mathrm{Re} + 10\alpha^4 - 15\alpha\mathrm{Re} - 3\alpha^2}{90}C_1^2\right]x^4 + \\ + \left[\frac{2\alpha\mathrm{Re}+5\alpha^2}{15}(C_1+C_2) + \right. \\ \left. + \frac{2520\alpha\mathrm{Re} - 1932\alpha^4 + 360\alpha^2 - 919\alpha^2\mathrm{Re}^2 - 2580\alpha^3\mathrm{Re}}{18900}\right]x^2 \tag{6.454}$$

The second-order approximate solution ($m = 2$) is obtained from the Eq. 6.23

$$\bar{F}(x) = F_0(x) + F_1(x) + F_2(x) \tag{6.455}$$

where F_0, F_1 and F_2 are given by Eqs. 6.443, 6.449 and 6.454 respectively.

Case 6.15.b

In this case we consider

$$\begin{aligned} h_1(x, C_i) &= C_1 \\ h_2(x, C_i) &= C_2x + C_3 \end{aligned} \tag{6.456}$$

where C_1, C_2 and C_3 are unknown constants. It is clear that the function F_1 is given by Eq. 6.449. Equation 6.20 becomes for $k = 2$

$$F'''_2 + \frac{8\alpha^2\mathrm{Re}^2C_1^2}{15}x^7 - \frac{12\alpha^2\mathrm{Re}^2 + 24\alpha^2\mathrm{Re}}{5}C_1^2x^5 - 4\alpha\mathrm{Re}C_3x^4 - \\ - \left[4\alpha\mathrm{Re}(C_1+C_2) + \frac{60\alpha\mathrm{Re} - 36\alpha^2\mathrm{Re}^2 - 120\alpha^3\mathrm{Re} - 80\alpha^4}{15}C_1^2\right]x^3 + \\ + 4(\alpha\mathrm{Re}+2\alpha^2)C_3x^2 + \left[4\left(\alpha\mathrm{Re}+2\alpha^2\right)(C_1+C_2) + \right. \\ \left. + \frac{60\alpha\mathrm{Re} + 12\alpha^2 - 8\alpha^2\mathrm{Re}^2 - 36\alpha^3\mathrm{Re} - 40\alpha^4}{15}C_1^2\right]x = 0, \\ F_2(0) = F'_2(0) = F_2(1) = 0 \tag{6.457}$$

and has the solution

$$F_2(x) = -\frac{\alpha^2\mathrm{Re}^2C_1^2}{1350}x^{10} + \frac{\alpha^2\mathrm{Re}^2 + 2\alpha^2\mathrm{Re}}{140}C_1^2x^8 + \frac{2\alpha\mathrm{Re}}{105}C_3x^7 + \\ + \left[\frac{\alpha\mathrm{Re}(2C_1+C_2)}{30} + \frac{15\alpha\mathrm{Re} + 9\alpha^2\mathrm{Re}^2 + 10\alpha^3\mathrm{Re} - 4\alpha^4}{450}C_1^2\right]x^6 + \\ + \frac{\alpha\mathrm{Re} + 2\alpha^2}{15}C_3x^5 + \left[-\frac{(\alpha\mathrm{Re}+2\alpha^2)(C_1+C_2)}{6} + \right. \\ \left. + \frac{2\alpha^2\mathrm{Re}^2 + 9\alpha^3\mathrm{Re} + 10\alpha^4 - 15\alpha\mathrm{Re} - 3\alpha^2}{90}C_1^2\right]x^4 + \\ + \left[\frac{2\alpha\mathrm{Re} + 5\alpha^2}{15}(C_1+C_2) + \frac{5\alpha\mathrm{Re} + 14\alpha^2}{105}C_3 + \right. \\ \left. + \frac{2520\alpha\mathrm{Re} - 1932\alpha^4 + 360\alpha^2 - 919\alpha^2\mathrm{Re} - 2580\alpha^3\mathrm{Re}}{18900}C_1^2\right]x^2 \quad (6.458)$$

The second-order approximate solution becomes

$$\bar{F}(x) = F_0(x) + F_1(x) + F_2(x) \quad (6.459)$$

where F_0, F_1 and F_2 are given by Eqs. 6.443, 6.449 and 6.458 respectively

Case 6.15.c

In the third case we consider

$$h_1(x) = C_1 + C_2x \\ h_2(x) = C_3 + C_4x + C_5x^2 \quad (6.460)$$

Equation 6.19 can be written as

$$F'''_1 - 4\alpha\mathrm{Re}C_2x^4 - 4\alpha\mathrm{Re}C_1x^3 + 4(\alpha\mathrm{Re}+2\alpha^2)C_2x^2 + \\ + 4(\alpha\mathrm{Re}+2\alpha^2)C_1x = 0, \quad F_1(0) = F'_1(0) = F_1(1) = 0 \quad (6.461)$$

From Eq. 6.461 we have

$$F_1(x) = \frac{2\alpha\mathrm{Re}C_2}{105}x^7 + \frac{\alpha\mathrm{Re}C_1}{30}x^6 - \frac{(\alpha\mathrm{Re}+2\alpha^2)C_2}{15}x^5 - \\ - \frac{(\alpha\mathrm{Re}+2\alpha^2)C_1}{6}x^4 + \left[\frac{(\alpha\mathrm{Re}+2\alpha^2)(5C_1+2C_2)}{30} - \frac{\alpha\mathrm{Re}(7C_1+4C_2)}{210}\right]x^2 \quad (6.462)$$

Equation 6.20 for $k = 2$ becomes

$$
\begin{aligned}
&F'''_2(x) + \frac{12\alpha^2\mathrm{Re}^2}{35}C_2^2x^9 + \frac{92\alpha^2\mathrm{Re}^2C_1C_2}{105}x^8 + \\
&+\left(\frac{8\alpha^2\mathrm{Re}^2C_1^2}{15} - \frac{6\alpha^2\mathrm{Re}^2 + 12\alpha^3\mathrm{Re}}{5}C_2^2\right)x^7 - \\
&-\frac{18\alpha^2\mathrm{Re}^2 + 36\alpha^3\mathrm{Re}}{5}C_1C_2x^6 - \left[4\alpha\mathrm{Re}C_5 + \frac{12\alpha^2\mathrm{Re}^2 + 24\alpha^3\mathrm{Re}}{5}C_1^2 + \right.\\
&\left.+\frac{12\alpha\mathrm{Re} - 2\alpha^2\mathrm{Re}^2 - 8\alpha^3\mathrm{Re} - 8\alpha^4}{3}C_2^2\right]x^5 - \\
&-\left[4\alpha\mathrm{Re}(C_2 + C_4) + \frac{120\alpha\mathrm{Re} - 46\alpha^2\mathrm{Re}^2 - 160\alpha^3\mathrm{Re} - 120\alpha^4}{15}C_1C_2 - \right.\\
&\left.-\frac{40\alpha^2\mathrm{Re}^2 + 112\alpha^3\mathrm{Re}}{105}C_2^2\right]x^4 - \left[4\alpha\mathrm{Re}(C_1 + C_3) - 4(\alpha\mathrm{Re} + 2\alpha^2)C_3 + \right.\\
&+\frac{60\alpha\mathrm{Re} - 36\alpha^2\mathrm{Re}^2 - 120\alpha^3\mathrm{Re} - 80\alpha^4}{15}C_1^2 - \\
&\left.-\frac{40\alpha^2\mathrm{Re}^2 + 112\alpha^3\mathrm{Re}}{105}C_1C_2 - 4(\alpha\mathrm{Re} + 2\alpha^2)C_2^2\right]x^3 + \\
&+[4(\alpha\mathrm{Re} + 2\alpha^2)(C_2 + C_4) + \\
&+\frac{120\alpha\mathrm{Re} + 240\alpha^2 - 8\alpha^2\mathrm{Re}^2 - 36\alpha^3\mathrm{Re} - 40\alpha^4}{15}C_1C_2 - \\
&\left.-\frac{20\alpha^2\mathrm{Re}^2 + 96\alpha^3\mathrm{Re} + 112\alpha^4}{105}C_2^2\right]x^2 + [4(\alpha\mathrm{Re} + 2\alpha^2)(C_1 + C_3) + \\
&+\frac{60\alpha\mathrm{Re} + 120\alpha^2 - 8\alpha^2\mathrm{Re}^2 - 36\alpha^3\mathrm{Re} - 40\alpha^4}{15}C_1^2 - \\
&\left.-\frac{20\alpha^2\mathrm{Re}^2 + 96\alpha^3\mathrm{Re} + 112\alpha^4}{105}C_1C_2\right]x = 0, \\
&F_2(0) = F'_2(0) = F_2(1) = 0
\end{aligned}
\tag{6.463}
$$

The solution of Eq. 6.463 is given by

$$
\begin{aligned}
F_2(x) = & -\frac{\alpha^2\mathrm{Re}^2C_2^2}{3850}x^{12} - \frac{46\alpha^2\mathrm{Re}^2C_1C_2}{51975}x^{11} + \\
& + \left[-\frac{\alpha^2\mathrm{Re}^2C_1^2}{1350} + \frac{\alpha^2\mathrm{Re}^2 + 2\alpha^3\mathrm{Re}}{600}C_2^2\right]x^{10} + \\
& + \frac{\alpha^2\mathrm{Re}^2 + 2\alpha^3\mathrm{Re}}{420}C_1C_2x^9 + \left[\frac{\alpha\mathrm{Re}C_5}{84} + \frac{\alpha^2\mathrm{Re}^2 + 2\alpha^3\mathrm{Re}}{140}C_1^2 + \right. \\
& \left. + \frac{6\alpha\mathrm{Re} - \alpha^2\mathrm{Re}^2 - 4\alpha^3\mathrm{Re} - 4\alpha^4}{504}C_2^2\right]x^8 + \\
& + \left[\frac{2\alpha\mathrm{Re}(C_2 + C_4)}{105} + \frac{60\mathrm{Re} - 23\alpha^2\mathrm{Re}^2 - 80\alpha^3\mathrm{Re} - 60\alpha^4}{1575}C_1C_2 - \right. \\
& \left. - \frac{20\alpha^2\mathrm{Re}^2 + 56\alpha^3\mathrm{Re}}{11025}C_2^2\right]x^7 + \left[\frac{\alpha\mathrm{Re}(C_1 + C_3)}{30} - \frac{(\alpha\mathrm{Re} + 2\alpha^2)C_5}{30} + \right. \\
& + \frac{15\alpha\mathrm{Re} - 9\alpha^2\mathrm{Re}^2 - 30\alpha^3\mathrm{Re} - 20\alpha^4}{450}C_1^2 - \frac{5\alpha^2\mathrm{Re}^2 + 14\alpha^3\mathrm{Re}}{1575}C_1C_2 - \\
& \left. - \frac{\alpha\mathrm{Re} + 2\alpha^2}{30}C_2^2\right]x^6 - \left[\frac{(\alpha\mathrm{Re} + 2\alpha^2)(C_2 + C_4)}{15} + \right. \\
& + \frac{30\alpha\mathrm{Re} + 60\alpha^2 - 2\alpha^2\mathrm{Re}^2 - 9\alpha^3\mathrm{Re} - 10\alpha^4}{225}C_1C_2 - \\
& \left. - \frac{5\alpha^2\mathrm{Re}^2 + 24\alpha^3\mathrm{Re} + 28\alpha^4}{1575}C_2^2\right]x^5 - \left[\frac{(\alpha\mathrm{Re} + 2\alpha^2)(C_1 + C_3)}{6} + \right. \\
& + \frac{15\alpha\mathrm{Re} + 30\alpha^2 - 2\alpha^2\mathrm{Re}^2 - 9\alpha^3\mathrm{Re} - 10\alpha^4}{90}C_1^2 - \\
& \left. - \frac{5\alpha^2\mathrm{Re}^2 + 24\alpha^3\mathrm{Re} + 28\alpha^4}{630}C_1C_2\right]x^4 + \left[\frac{(2\alpha\mathrm{Re} + 5\alpha^2)(C_1 + C_3)}{15} + \right. \\
& + \frac{(5\alpha\mathrm{Re} + 14\alpha^2)(C_2 + C_4)}{105} + \frac{9\alpha\mathrm{Re} + 28\alpha^2}{420}C_5 + \\
& + \frac{2520\alpha\mathrm{Re} - 1260\alpha^4 + 6300\alpha^2 - 163\alpha^2\mathrm{Re}^2 - 900\alpha^3\mathrm{Re}}{18900}C_1^2 + \\
& + \frac{19800\alpha\mathrm{Re} + 55440\alpha^2 - 1433\alpha^2\mathrm{Re}^2 - 8514\alpha^3\mathrm{Re} - 10560\alpha^4}{207900}C_1C_2 - \\
& \left. - \frac{4774\alpha^4 - 32340\alpha^2 - 10395\alpha\mathrm{Re} + 2695\alpha^3\mathrm{Re} + 380\alpha^2\mathrm{Re}^2}{485100}C_2^2\right]x^2
\end{aligned}
\tag{6.464}
$$

The second-order approximate solution is

$$
\bar{F}(x) = F_0(x) + F_1(x) + F_2(x) \tag{6.465}
$$

where F_0, F_1, and F_2 are given by Eqs. 6.443, 6.462 and 6.464 respectively.

Case 6.15.d
In the last case, we consider

$$K_1(x) = C_1 + C_2 x$$
$$K_2(x) = C_3 + C_4 x + C_5 x^2 + C_6 x^3 + C_7 x^4 + C_8 x^5 + C_9 x^6 \tag{6.466}$$

The solution of $F_1(x)$ is given by Eq. 6.462. On the other hand, Eq. 6.20 for $k = 2$ has the solution

$$\begin{aligned}
F_2(x) = &\left(\frac{\alpha\mathrm{Re}}{330}C_9 - \frac{\alpha^2\mathrm{Re}^2 C_2^2}{3850}\right)x^{12} + \left(\frac{2\alpha\mathrm{Re}}{495}C_8 - \frac{46\alpha^2\mathrm{Re}^2}{51975}C_1C_2\right)x^{11} + \\
&+ \left[\frac{\alpha\mathrm{Re}C_7 - (\mathrm{Re} + 2\alpha^2)C_9}{180} - \frac{\alpha^2\mathrm{Re}^2}{1350}C_1^2 + \frac{\alpha^2\mathrm{Re}^2 + 2\alpha^3\mathrm{Re}}{600}C_2^2\right]x^{10} + \\
&+ \left[\frac{\alpha\mathrm{Re}C_6 - (\alpha\mathrm{Re} + 2\alpha^2)C_8}{126} + \frac{\alpha^2\mathrm{Re}^2 + 2\alpha^3\mathrm{Re}}{420}C_1C_2\right]x^9 + \\
&+ \left[\frac{\alpha\mathrm{Re}C_5 - (\alpha\mathrm{Re} + 2\alpha^2)C_7}{84} + \frac{\alpha^2\mathrm{Re}^2 + 2\alpha^3\mathrm{Re}}{140}C_1^2 + \right. \\
&\left. + \frac{6\alpha\mathrm{Re} - \alpha^2\mathrm{Re}^2 - 4\alpha^3\mathrm{Re} - 4\alpha^4}{504}C_2^2\right]x^8 + \left[\frac{2\alpha\mathrm{Re}C_4 - 2(\alpha\mathrm{Re} + 2\alpha^2)C_6}{105} + \right. \\
&+ \frac{2\alpha\mathrm{Re}C_2}{105} + \frac{60\alpha\mathrm{Re} - 23\alpha^2\mathrm{Re}^2 - 80\alpha^3\mathrm{Re} - 60\alpha^4}{1575}C_1C_2 - \\
&\left. - \frac{20\alpha^2\mathrm{Re}^2 + 56\alpha^3\mathrm{Re}}{11025}C_2^2\right]x^7 + \left[\frac{\alpha\mathrm{Re}(C_1 + C_3) - (\alpha\mathrm{Re} + 2\alpha^2)C_5}{30} + \right. \\
&+ \frac{15\alpha\mathrm{Re} - 9\alpha^2\mathrm{Re}^2 - 30\alpha^3\mathrm{Re} - 20\alpha^4}{450}C_1^2 - \frac{5\alpha^2\mathrm{Re}^2 + 14\alpha^3\mathrm{Re}}{1575}C_1C_2 - \\
&\left. - \frac{(\alpha\mathrm{Re} + 2\alpha^2)C_2^2}{30}\right]x^6 + \left[\frac{2\alpha^2\mathrm{Re}^2 + 9\alpha^3\mathrm{Re} + 10\alpha^4 - 30\alpha\mathrm{Re} - 60\alpha^2}{225}C_1C_2 + \right. \\
&\left. + \frac{5\alpha^2\mathrm{Re}^2 + 24\alpha^3\mathrm{Re} + 28\alpha^4}{1575}C_2^2 - \frac{(\alpha\mathrm{Re} + 2\alpha^2)(C_2 + C_4)}{15}\right]x^5 + \\
&+ \left[\frac{2\alpha^2\mathrm{Re}^2 + 9\alpha^3\mathrm{Re} + 10\alpha^4 - 15\alpha\mathrm{Re} - 30\alpha^2}{90}C_1^2 + \right. \\
&\left. + \frac{5\alpha^2\mathrm{Re}^2 + 24\alpha^3\mathrm{Re} + 28\alpha^4}{630}C_1C_2 - \frac{(\alpha\mathrm{Re} + 2\alpha^2)(C_1 + C_3)}{6}\right]x^4 + \\
&+ \left[\frac{5\alpha\mathrm{Re} + 14\alpha^2}{105}(C_2 + C_4) + \frac{2\alpha\mathrm{Re} + 5\alpha^2}{15}C_1 + \frac{2\alpha\mathrm{Re} + 15\alpha^2}{15}C_3 + \right. \\
&+ \frac{9\alpha\mathrm{Re} + 28\alpha^2}{420}C_5 + \frac{7\alpha\mathrm{Re} + 24\alpha^2}{630}C_6 + \frac{4\alpha\mathrm{Re} + 15\alpha^2}{630}C_7 + \\
&+ \frac{27\alpha\mathrm{Re} + 110\alpha^2}{6930}C_8 + \frac{5\alpha\mathrm{Re} + 22\alpha^2}{1980}C_9 +
\end{aligned}$$

$$+\frac{2520\alpha\mathrm{Re}-1260\alpha^4+6300\alpha^2-163\alpha^2\mathrm{Re}^2-900\alpha^3\mathrm{Re}}{18900}C_1^2+$$
$$+\frac{10395\alpha\mathrm{Re}-380\alpha^2\mathrm{Re}^2+32340\alpha^2-2695\alpha^3\mathrm{Re}-4774\alpha^4}{485100}C_2^2+$$
$$\left.+\frac{(19800\alpha\mathrm{Re}-1443\alpha^2\mathrm{Re}^2+55440\alpha^2-8514\alpha^3\mathrm{Re}-10560\alpha^4)C_1C_2}{207900}\right]x^2 \tag{6.467}$$

The second-order approximate solution in this case is given by

$$\bar{F}(x)=F_0(x)+F_1(x)+F_2(x) \tag{6.468}$$

where F_0, F_1 and F_2 are given by Eqs. 6.443, 6.462 and 6.467 respectively.

6.15.1 Numerical Examples

In what follows we will show that the error of the solution decreases when the number of terms in the auxiliary function $h(x,p)$ increases.

Example 6.4.15.a. For Re $=50$ and $\alpha=5$ in the case **6.15.a** it is obtained two solutions for the constants C_1 and C_2:

(a) $C_1=0.017506079 \quad C_2=-0.047286881$
(b) $C_1=-0.017506079\, C_2=0.022737435$

but the second-order approximate solution Eq. 6.455 is the same in both cases:

$$\begin{aligned}\bar{F}(x)\approx{}&1-1.767845893x^2+1.236877181x^4-0.619019693x^6+\\&+0.164176501x^8-0.014188096x^{10}\end{aligned} \tag{6.469}$$

Example 6.15.1.b. For Re $=50$ and $\alpha=5$ in the case **6.15.b**, we obtain

(a) $C_1=0.007977212 \quad C_2=-0.041469373\; C_3=0.022761122$
(b) $C_1=-0.007977212\, C_2=-0.009560525\; C_3=0.022761122$

The second-order approximate solution Eq. 6.459 becomes

$$\begin{aligned}\bar{F}(x)\approx{}&1-1.769527092x^2+1.40514047x^4-0.45522244x^5-\\&-0.319921792x^6+0.108386295x^7+0.03409066x^8-0.002946101x^{10}\end{aligned} \tag{6.470}$$

Example 6.15.1.c. For Re $=50$ and $\alpha=5$ in the case **6.15.c**, the second-order approximate solution can be written in the form:

$$\bar{F}(x) \approx 1 - 1.769777647x^2 + 1.295738478x^4 - 0.010406134x^5 - \\ - 0.871743272x^6 + 0.178832164x^7 + 0.299266794x^8 - 0.090606944x^9 - \\ - 0.040442155x^{10} + 0.00935599x^{11} - 0.0002178x^{12} \tag{6.471}$$

Example 6.15.1.d. For Re $= 80$ and $\alpha = -5$ in the case 6.4.d, the second-order approximate solution becomes

$$\bar{F}(x) \approx 1 - 0.399291819x^2 - 0.461970063x^4 - 0.014703786x^5 - \\ - 0.12415397x^6 - 0.07325724x^7 - 0.08278982x^8 + 0.45101379x^9 - \\ - 0.648015234x^{10} + 0.47466473x^{11} - 0.121496588x^{12} \tag{6.472}$$

It is easy to verify the accuracy of the obtained solutions if we compare these analytical solutions with the numerical ones or with results obtained by other procedures.

It can be seen from Tables 6.3, 6.4, 6.5 and 6.6 that the analytical solutions of Jeffery-Hamel flows obtained by OHAM are very accurate.

The examples presented in this section lead to the very important conclusion that the accuracy of the obtained results is growing along with increasing the number of constants in the auxiliary function.

Some other methods such as DTM, HPM or HAM give a good accuracy, but OHAM is by far the best method delivering faster convergence and better accuracy. In this procedure, iterations are performed in a very simple manner by identifying some coefficients and therefore very good approximations are obtained in few terms.

The Optimal Homotopy Asymptotic Method is employed to propose new analytic approximate solutions for some nonlinear dynamical systems. This procedure

Table 6.3 The results of the second-order approximate solutions Eqs. 6.469, 6.470 and 6.471 and numerical solution of $F(x)$ for Re $= 50$, $\alpha = 5$

x	$\bar{F}(x)$, Eq. 6.469	$\bar{F}(x)$, Eq. 6.470	$\bar{F}(x)$, Eq. 6.471	Numerical solution
0	1	1	1	1
0.1	0.98244611	0.982440382	0.982430842	0.98243124
0.2	0.931225969	0.931302469	0.931225959	0.93122597
0.3	0.850471997	0.850810709	0.850611445	0.85061063
0.4	0.746379315	0.747074996	0.746790784	0.74679081
0.5	0.626298626	0.627192084	0.626947253	0.62694818
0.6	0.497665923	0.498340984	0.498235028	0.49823446
0.7	0.366966345	0.366966353	0.366970088	0.36696635
0.8	0.238952034	0.238148782	0.238142322	0.23812375
0.9	0.116313019	0.115260361	0.115219025	0.11515193
1	0	0	0	0

Table 6.4 Comparison between the OHAM and numerical solutions for Re = 50 and $\alpha = 5$ (error $= |x_{num} - x_{app}|$)

x	Error of the solution Eq. 6.469	Error of the solution Eq. 6.470	Error of the solution Eq. 6.471
00	0	0	0
0.1	0.00001487	0.000009142	0.000000398
0.2	0.000000001	0.000076499	0.000000011
0.3	0.000138633	0.0002	0.000000815
0.4	0.000411495	0.000284186	0.000000026
0.5	0.000649554	0.000243904	0.000000927
0.6	0.000568537	0.000106524	0.000000568
0.7	0.000000005	0.000000003	0.000003738
0.8	0.000828284	0.000024962	0.000018572
0.9	0.0001161089	0.000108431	0.000067095
1	0	0	0

Table 6.5 Comparison between differential transformation method (DTM) [137], homotopy perturbation method (HPM) [137], homotopy analysis method [137] and OHAM-Eq. 6.472 for Re = 80, $\alpha = -5$

x	$\bar{F}(x)$ (DTM)	$\bar{F}(x)$ (HPM)	$\bar{F}(x)$ (HAM)	$\bar{F}(x)$ (OHAM)	Numerical
00	1	1	1	1	1
0.1	0.9959603887	0.9960671874	0.9995960242	0.995960605	0.9959606278
0.2	0.9832745481	0.9836959424	0.9832755258	0.983275548	0.9832755383
0.3	0.9601775551	0.9610758773	0.9601798911	0.960179914	0.96017991139
0.4	0.9235170706	0.9249245156	0.9235215737	0.923521643	0.9235215894
0.5	0.8684511349	0.8701997697	0.8684588997	0.868458963	0.86845887772
0.6	0.7880785402	0.7898325937	0.7880910186	0.788090923	0.78809092032
0.7	0.673248448	0.6745334968	0.6731437690	0.673143633	0.6731436346
0.8	0.5119644061	0.5128373095	0.5119909939	0.511991107	0.5119910891
0.9	0.2915280122	0.2918936991	0.2915580178	0.291558742	0.29155874261
1	0	0	−0.000001149	0	0

Table 6.6 Comparison between OHAM results and numerical solutions [137] for Re = 80, $\alpha = -5$

x	$\bar{F}(x)$, Eq. 6.472	Numerical	Error
00	1	1	0
0.1	0.995960605	0.9959606278	0.000000022
0.2	0.983275548	0.9832755383	0.000000009
0.3	0.960179914	0.96017991139	0.000000002
0.4	0.923521643	0.9235215894	0.000000053
0.5	0.868458963	0.86845887772	0.000000085
0.6	0.788090923	0.78809092032	0.000000002
0.7	0.673143633	0.6731436346	0.000000001
0.8	0.511991107	0.5119910891	0.000000017
0.9	0.291558742	0.29155874261	0.000000006
1	0	0	0

is valid even if the nonlinear differential equations contain any small or large parameters.

In the presented construction of homotopy appear some distinctive concepts such as: the parameter λ (involved in the case of nonlinear oscillations) which is determined using the principle of minimal sensitivity, the auxiliary function $H(\tau,p)$, the operator L and several convergence-control constants $C_1, C_2, \ldots$..., which ensure a fast convergence of the solutions.

The examples presented in this section lead to the conclusion that the accuracy of the obtained results is growing along with increasing the number of convergence-control constants in the auxiliary function. The OHAM provides us with a simple and rigorous way to control and adjust the convergence of a solution through the auxiliary functions $h_i(x,C_i)$ involving several convergence-control constants C_i which are optimally determined. OHAM is an iterative procedure and iterations are performed in a very simple manner by identifying some coefficients and therefore very good approximations are obtained in few terms. Actually, the capital strength of OHAM is its fast convergence, which proves that this method is very efficient in practice.

Chapter 7
The Optimal Homotopy Perturbation Method

7.1 Homotopy Perturbation Method

An effective and convenient mathematical tool for nonlinear differential equations is the homotopy perturbation method, a combination of the classical perturbation method and the homotopy technique. This method, proposed by J.H. He in 1998 [30, 40, 115, 138, 139], does not require a small parameter in the equation in contrast to the traditional perturbation methods. Like the homotopy analysis method (HAM) and optimal homotopy asymptotic method (OHAM), an embedding parameter p ∈ [0,1] is employed. The homotopy perturbation method takes the full advantage of the traditional perturbation methods and the homotopy techniques. In this method the perturbation equation can be easily constructed by homotopy in topology and the initial approximation can be also freely selected. Consequently, the construction of the homotopy plays an important role to solve a nonlinear problem and therefore is problem dependent. Application of the homotopy perturbation method can be found in Refs. [140–145].

To explain the basic idea of the homotopy perturbation method, for solving nonlinear differential equation, we consider the following equation:

$$A(u) - f(r) = 0, \qquad r \in \Omega \tag{7.1}$$

subject to the boundary condition

$$B\left(u, \frac{\partial u}{\partial n}\right) = 0, \qquad r \in \Gamma \tag{7.2}$$

where A is a general differential operator, B a boundary operator, $f(r)$ is a known analytical function, $\Gamma$is the boundary of the domain Ω and $\partial/\partial n$denotes differentiation along the normal drawn outwards from the domain Ω. The operator A can generally be divided into two parts, a linear part L and a nonlinear part N. Eq. 7.1 can therefore be written as follows

V. Marinca and N. Herisanu, *Nonlinear Dynamical Systems in Engineering*,
DOI 10.1007/978-3-642-22735-6_7,

$$L(u) + N(u) - f(r) = 0 \tag{7.3}$$

We construct a homotopy $u(r,p) : \Omega \times [0,1] \to \Re$ which satisfies

$$H(v,p) = (1-p)[L(v) - L(u_0) - f(r)] + pA(v) = 0, \quad p \in [0,1],\ r \in \Omega \tag{7.4}$$

which is equivalent to

$$H(v,p) = L(v) - L(u_0) - f(r) + p[L(u_0) + N(v)] = 0 \tag{7.5}$$

where $p \in [0,1]$ is an embedding parameter and u_0 is an initial guess of the solution of Eq. 7.1 which satisfies the boundary conditions. It follows from Eqs. 7.4 and 7.5 that

$$\begin{aligned} H(v,0) &= L(v) - L(u_0) - f(r) = 0 \\ H(v,1) &= A(u) = 0 \end{aligned} \tag{7.6}$$

Thus, the changing process of p from zero to unity is just that of $u(r,p)$ from $v_0(r)$ to $v(r)$. In topology, this is called deformation and $L(v)$-$L(u_0)$ and $A(v)$-$f(r)$ are called homotopic. So, it is quite right to assume that the solutions of Eqs. 7.4 or 7.5 can be expressed as:

$$v = v_0 + pv_1 + p^2 v_2 + \ldots \tag{7.7}$$

The approximate solution of Eq. 7.1 can be readily obtained:

$$u = \lim_{p \to 1} (v_0 + pv_1 + p^2 v_2 + \ldots) = v_0 + v_1 + v_2 + \ldots. \tag{7.8}$$

The convergence of series (7.8) has been proved by J.H. He in the paper [30].

We illustrate the basic procedure of this method solving the following example [34]:

$$\ddot{u} + \frac{u}{1 + \varepsilon u^2} = 0, \quad t \in \Omega \tag{7.9}$$

with the initial conditions

$$u(0) = A, \quad \dot{u}(0) = 0 \tag{7.10}$$

We construct a homotopy which satisfies

$$(1-p)[L(v) - L(u_0)] + p[(1 + \varepsilon v^2)v'' + v] = 0 \tag{7.11}$$

The linear operator is

$$Lv = \ddot{v} + v \tag{7.12}$$

We assume that the initial approximation of Eq. 7.9 is of the form [138]

$$v_0 = u_0 = A\cos\alpha t \tag{7.13}$$

where $\alpha = \alpha(\varepsilon, A)$ is a non-zero unknown constant with $\alpha(0, A) = 1$.

Substituting Eq. 7.7 into Eq. 7.11 and equating the terms with identical powers of p, we have

$$L(v_0) - L(u_0) = 0, \quad v_0(0) = A, \quad \dot{v}_0(0) = 0 \tag{7.14}$$

$$L(v_1) - L(u_0) + L(u_0) + (1 + \varepsilon v_0^2)\ddot{v}_0 + v_0, \quad v_1(0) = \dot{v}_1(0) = 0 \tag{7.15}$$

Setting $v_0 = u_0 = A\cos\alpha t$, the unknown α can be determined by the Galerkin method

$$\int_o^{\frac{\pi}{2}} \sin\alpha t[(1 + \varepsilon u_0^2)\ddot{u}_0 + u_0]dt = 0 \tag{7.16}$$

The unknown α therefore can be identified

$$\alpha = \frac{1}{\sqrt{1 + \frac{3}{4}\varepsilon A^2}} \tag{7.17}$$

As a result, from Eq. 7.15 we obtain

$$\ddot{v}_1 + v_1 - \alpha^2 \frac{\varepsilon A^3}{4} \cos 3\alpha t = 0 \tag{7.18}$$

and therefore

$$v_1(t) = -\frac{\alpha^2 \varepsilon A^3}{4(9\alpha^2 - 1)}(\cos 3\alpha t - \cos t) \tag{7.19}$$

where α is defined in Eq. 7.17.

The period of the solution can be expressed as follows

$$T = 2\pi\sqrt{1 + \frac{3}{4}\varepsilon A^2} \tag{7.20}$$

while the first-order approximation is given by the expression

$$\bar{u}(t) = v_0(t) + v_1(t) = A\cos\alpha t - \frac{\alpha A^3}{32 - 3\varepsilon A^2}(\cos 3\alpha t - \cos\alpha t) \tag{7.21}$$

The approximate solution and its period obtained by the traditional perturbation method read:

$$u(t) = A\cos\left(1 - \frac{3}{8}\varepsilon A^2\right)t \tag{7.22}$$

$$T_{per} = \frac{2\pi}{1 - \frac{3}{8}\varepsilon A^2} \tag{7.23}$$

It is also interesting to point out that Eqs. 7.22 and 7.23 are valid only for small εA^2, while Eqs. 7.21 and 7.20 are valid for a very large region $0 \le \varepsilon A^2 < \infty$, furthermore the approximations obtained by the proposed method are highly accurate.

7.2 Modified Homotopy Perturbation Method

In this section we will apply the homotopy technique in a completely different way as in J.H. He's papers.

Let us further consider the damped, forced oscillations of a nonconservative nonlinear system governed by the equation [146]

$$\ddot{u}(t) + \omega_0^2 u(t) = F(\omega t, u, \dot{u}, \ldots, \overset{(i)}{u}) \tag{7.24}$$

where u(t) is a dimensionless variable, $\overset{(i)}{u}(t) = \frac{d^i u(t)}{dt^i}, i = 1, 2\ldots$, F is a nonlinear analytic function with the period T in the first variable, ω_0 and ω are constants. The initial conditions are

$$u(0) = A, \quad \dot{u}(0) = 0 \tag{7.25}$$

We present the modified homotopy perturbation method for solving nonlinear problems of the form Eq. 7.24, some ideas and improvements which point towards relatively new and interesting applications of this method. The periodic solutions obtained by this method are valid not only for small parameters, but also for very large parameters. This method sometimes leads to results according to the standard Lindstedt-Poincaré method or the harmonic balance method.

To explain the basic idea of the modified homotopy perturbation method, for Eq. 7.24 we construct a one-parameter family of equations

$$\frac{\partial^2 U(t;p)}{\partial t^2} + \Lambda(p)U(t;p) = pF\left(\omega t, U(t;p), \frac{\partial U(t;p)}{\partial t}, \ldots, \frac{\partial^i U(t;p)}{\partial t^i}\right), \qquad (7.26)$$
$$t \in [0,\infty), p \in [0,1], i = 1,2...$$

where $U(t;p)$ is an analytical function of both t and p, such as $U(t,0) = u_0(t)$ and $u_0(t)$ is the solution of Eq. 7.26 for $p = 0$, $U(t,1) = u(t)$ and $u(t)$ is exactly the solution that we want to know. As the well-known expanding parameter p varies from zero to one, $U(t,p)$ varies continuously from $u_0(t)$ to $u(t)$ and $\Lambda(p)$ varies from $\Lambda(0) = \Omega^2$ to $\Lambda(1) = \omega_0^2$, where Ω is the angular frequency of the system (7.24). The continuous deformations of $U(t;\ p)$ and $\Lambda(p)$ are completely governed by Eq. 7.26. Whether or not Eq. 7.24 contains small or large parameters this is not important at all for the validity of the modified homotopy method, because the only assumptions made in Eq. 7.24 are that F should be analytical and with the period T in the first variable.

The initial conditions become

$$U(0,p) = A, \quad \left.\frac{\partial U(t,p)}{\partial t}\right|_{t=0} = 0 \qquad (7.27)$$

By Taylor's formula, we have:

$$U(t;p) = u_0(t) + pu_1(t) + p^2 u_2(t) + \ldots. \qquad (7.28)$$

$$\Lambda(p) = \Omega^2 + p\Omega_1 + p\Omega_2 + \ldots. \qquad (7.29)$$

Setting $p = 1$, we obtain

$$u(t) = u_0 + u_1 + u_2 + \ldots \qquad (7.30)$$

$$\omega_0^2 = \Omega^2 + \Omega_1 + \Omega_2 + \ldots \qquad (7.31)$$

Substituting Eqs. 7.28 and 7.29 into Eq. 7.26 and equating the terms with identical powers of p, and taking into consideration that

$$\begin{aligned} F\left(\omega t, U(t;p), \ldots, \frac{\partial^i U(t;p)}{\partial t_i}\right) &= F_0\left(\Omega t, u_0, \ldots, \overset{(i)}{u_0}\right) + \\ + pF_1\left(\Omega t, u_0, u_1, \ldots, \overset{(i)}{u_0}, \overset{(i)}{u_1}\right) &+ p^2 F_2\left(\Omega t, u_0, u_1, u_2, \ldots, \overset{(i)}{u_0}, \overset{(i)}{u_1}, \overset{(i)}{u_2}\right) + \ldots \end{aligned} \qquad (7.32)$$

we have

$$\ddot{u}_0 + \Omega^2 u_0 = 0, \quad u_0(0) = A, \quad \dot{u}_0(0) = 0 \qquad (7.33)$$

$$\ddot{u}_1 + \Omega^2 u_1 = -\Omega_1 u_0 + F_0(\Omega t, u_0, \dot{u}_0, ..., \overset{(i)}{u_0}), \quad u_1(0) = \dot{u}_1(0) = 0 \tag{7.34}$$

$$\ddot{u}_2 + \Omega^2 u_2 = -\Omega_2 u_0 - \Omega_1 u_1 + F_1(\Omega t, u_0, u_1, ..., \overset{(i)}{u_0}, \overset{(i)}{u_1}), \quad u_2(0) = \dot{u}_2(0) = 0 \tag{7.35}$$

and so on.

The parameters Ω_i, $i = 1, 2, \ldots$ can be determined avoiding the presence of secular terms in Eqs. 7.34, 7.35 and so on.

We illustrate the basic evaluation procedure of this method by the following examples:

Example (a)

We consider the same Eq. 7.9 and thus $F(u, \ddot{u}) = -\varepsilon u^2 \ddot{u}$. From Eq. 7.33 we have

$$u_0(t) = A \cos \Omega t \tag{7.36}$$

Substituting (7.36) into Eq. 7.34, results in:

$$\ddot{u}_1 + \Omega^2 u_1 = (-\Omega_1 A + \frac{3}{4}\varepsilon A^3 \Omega^2) + \frac{1}{4}\varepsilon A^3 \Omega^2 \cos^3 \Omega t, \quad u_1(0) = \dot{u}_1(0) = 0 \tag{7.37}$$

Avoiding the presence of a secular term in Eq. 7.37, needs:

$$\Omega_1 = \frac{3}{4}\varepsilon A^2 \Omega^2 \tag{7.38}$$

The solution of Eq. 7.37, can be written as:

$$u_1(t) = \frac{1}{32}\varepsilon A^3 (\cos \Omega t - \cos 3\Omega t) \tag{7.39}$$

Substitution of Eqs. 7.36, 7.38 and 7.39 into Eq. 7.35 yields:

$$\begin{aligned} \ddot{u}_2 + \Omega^2 u_2 = & \left(-\Omega_2 A - \frac{5\varepsilon^2 A^5 \Omega^2}{128}\right) \cos \Omega t - \frac{\varepsilon^2 A^5 \Omega^2}{8} \cos 3\Omega t - \\ & - \frac{11\varepsilon^2 A^5 \Omega^2}{128} \cos 5\Omega t, \qquad u_2(0) = \dot{u}_2(0) = 0 \end{aligned} \tag{7.40}$$

The elimination of secular term in Eq. 7.40 requires:

$$\Omega_2 = -\frac{5}{128}\varepsilon^2 A^4 \Omega^2 \tag{7.41}$$

Solving Eq. 7.40, we obtain

$$u_2(t) = \frac{\varepsilon^2 A^5}{64}(\cos 3\Omega t - \cos \Omega t) + \frac{11\varepsilon^2 A^5}{3072}(\cos 5\Omega t - \cos \Omega t) \tag{7.42}$$

Substituting Eqs. 7.38 and 7.41 into Eq. 7.31, we have ($\omega_0 = 1$):

$$1 = \Omega^2 + \frac{3}{4}\varepsilon A^2\Omega^2 - \frac{5}{128}\varepsilon^2 A^4\Omega^2 + 0(\varepsilon^3) \tag{7.43}$$

From Eq. 7.43 we obtain

$$\Omega^2 = \frac{1}{1 + \frac{3}{4}\varepsilon A^2 - \frac{5}{128}\varepsilon^2 A^4} \tag{7.44}$$

The approximate period obtained from Eq. 7.44 is

$$T_{app} = 2\pi\sqrt{1 + \frac{3}{4}\varepsilon A^2 - \frac{5}{128}\varepsilon^2 A^4} \tag{7.45}$$

The formula (7.45) works well for small $\varepsilon(0<\varepsilon<<1)$ but breaks down quickly when ε becomes large. Here, we wish to develop uniformly valid expansions for Ω^2 and $u(t)$ for large values of ε, using a newly defined expansion parameter $\eta(\varepsilon, A)$from Eq. 7.44 as follows (see Sect. 2.1.3):

$$\eta(\varepsilon, A) = \frac{\frac{3}{4}\varepsilon A^2}{1 + \frac{3}{4}\varepsilon A^2} \tag{7.46}$$

This relation is quickly convergent regardless of the magnitude of εA^2, since $\eta<1$ for all εA^2. In terms of η the original parameter ε is given by the formula

$$\varepsilon = \frac{\eta}{\frac{3}{4}(1-\eta)A^2} \tag{7.47}$$

Equation 7.9 can be rewritten as

$$\ddot{u} + u = \eta\left(\ddot{u} + u - \frac{4u^2\ddot{u}}{3A^2}\right) \tag{7.48}$$

Equations 7.33, 7.34 and 7.35 are respectively:

$$\ddot{u}_0 + \bar{\Omega}^2 u_0 = 0, \ \mathrm{u}_0(0) = A, \ \underline{\mathrm{u}}_0(0) = 0 \tag{7.49}$$

$$\ddot{u}_1 + \bar{\Omega}^2 u_1 = -\bar{\Omega}_1 u_0 + \eta(\ddot{u}_0 + u_0 - \frac{4u_0^2\ddot{u}_0}{3A^2}); u_1(0) = \dot{u}_1(0) = 0 \tag{7.50}$$

$$\ddot{u}_2 + \bar{\Omega}^2 u_2 = -\bar{\Omega}_2 u_0 - \bar{\Omega}_1 u_1 + \eta(\ddot{u}_1 + u_1 + \frac{8u_0 u_1 \ddot{u}_0 + 4u_0^2 \ddot{u}_1}{3A^2}),$$
$$u_2(0) = \dot{u}_2(0) = 0 \tag{7.51}$$

where we used Eq. 7.29 in the form $\bar{\Lambda}(p) = \bar{\Omega}^2 + p\bar{\Omega}_1 + p^2\bar{\Omega}_2 + \ldots$.

Equation 7.49 has the solution

$$u_0(t) = A\cos\bar{\Omega}t \tag{7.52}$$

Substituting Eq. 7.52 into Eq. 7.50, by simple manipulations, we have

$$\ddot{u}_1 + \bar{\Omega}^2 u_1 = (-\bar{\Omega}_1 + \eta)\cos\bar{\Omega}t + \frac{1}{3}\eta A\bar{\Omega}^2 \cos 3\bar{\Omega}t,$$
$$u_1(0) = \dot{u}_1(0) = 0 \tag{7.53}$$

In order to ensure that no secular term appears in Eq. 7.53, the resonance must be avoided. To do so, coefficient of $\cos\bar{\Omega}t$ must be zero, i.e.

$$\bar{\Omega}_1 = \eta \tag{7.54}$$

The solution of Eq. 7.53 becomes

$$u_1(t) = \frac{1}{24}\eta A\left(\cos\bar{\Omega}t - \cos 3\bar{\Omega}t\right) \tag{7.55}$$

Substituting Eqs. 7.52 and 7.55 into Eq. 7.51 we obtain:

$$\ddot{u}_2 + \bar{\Omega}^2 u_2 = (-A\bar{\Omega}_2 - \frac{5}{72}\bar{\Omega}^2\eta^2 A)\cos\bar{\Omega}t + H.O.T. \tag{7.56}$$

No secular terms in Eq. 7.56, means that

$$\bar{\Omega}_2 = -\frac{5\bar{\Omega}^2\eta^2}{72} \tag{7.57}$$

Now, substituting Eqs. 7.54 and 7.57 into Eq. 7.31, we obtain:

$$1 = \bar{\Omega}^2 + \eta - \frac{5\bar{\Omega}^2}{72}\eta^2 \tag{7.58}$$

and therefore

$$\bar{\Omega}^2 = \frac{1-\eta}{1-\frac{5}{72}\eta^2} \tag{7.59}$$

Substituting Eq. 7.46 into Eq. 7.59 yields

$$\bar{\Omega}^2 = \frac{96\varepsilon A^2 + 128}{67\varepsilon^2 A^4 + 192\varepsilon A^2 + 128} \tag{7.60}$$

The approximate period obtained from Eq. 7.60 is

$$T^*_{app} = 2\pi\sqrt{\frac{67\varepsilon^2 A^4 + 192\varepsilon A^2 + 128}{96\varepsilon A^2 + 128}} \tag{7.61}$$

In order to illustrate the remarkable accuracy of the obtained results, we compare the approximate period T_{app} given by Eq. 7.20 with the exact one [147]:

$$T_{ex} = 4\sqrt{\varepsilon}\int_0^A \frac{du}{\sqrt{\ln(1+\varepsilon A^2) - \ln(1+\varepsilon u^2)}} \tag{7.62}$$

In case $\varepsilon A^2 \to \infty$, we have

$$\lim_{\varepsilon A^2 \to \infty} \frac{T_{ex}}{T_{app}} = \frac{2\sqrt{2\pi\varepsilon}A}{2\pi\sqrt{\frac{3}{4}\varepsilon A^2}} \approx 0,9213177 \tag{7.63}$$

Therefore, for any values of ε, it can be easily proved that the maximal relative error in the case of the homotopy perturbation method is less than 7.87%.

On the other hand, the approximate period T^*_{app} given by Eq. 7.61 is compared with the exact one and thus

$$\lim_{\varepsilon A^2 \to \infty} \frac{T_{ex}}{T^*_{app}} = \frac{2\sqrt{2\pi\varepsilon}A}{2\pi\sqrt{\frac{67}{96}\varepsilon A^2}} \approx 0,955076 \tag{7.64}$$

In the case of the modified homotopy perturbation method the maximal relative error is less than 4.5%.

We observe that in the frame of the modified homotopy perturbation method, the term $u_0(t)$ results from Eq. 7.33 while in the frame of the homotopy perturbation method this term is supposed to be in the form (7.13). Moreover, it is obvious that the methods differ each other by the construction of the homotopy, by the construction of the frequency given by Eq. 7.31 and by the construction of the solutions, which depends on the frequency Ω that is unknown a priori. One can observe that both methods provide approximate results that come very close to each other for small values of ε.

Example (b): Let us consider the well-known Duffing equation

$$\ddot{u} + u + au^3 = 0 \tag{7.65}$$

with the initial conditions

$$u(0) = A, \quad \dot{u}(0) = 0 \tag{7.66}$$

By the same manipulations as in the example (a), we obtain

$$\Omega = \sqrt{\frac{1}{2}(1 + \frac{3}{4}aA^2) + \frac{1}{2}\sqrt{1 + \frac{3}{2}aA^2 + \frac{15}{32}a^2A^4}} \tag{7.67}$$

$$u(t) = A\cos\Omega t + \frac{aA^3}{32\Omega^2}(\cos 3\Omega t - \cos\Omega t) + \frac{a^2A^5}{1024\Omega^2}(\cos 5\Omega t - \cos\Omega t) \tag{7.68}$$

For comparison, the exact frequency obtained by Eq. 7.62 and the approximate frequency computed by Eq. 7.67 are listed in table 7.1.

We also have

$$\lim_{aA^2\to\infty}\frac{\Omega_{app}}{\Omega_{ex}} = \frac{\sqrt{6+\sqrt{30}}}{2\pi}\int_0^{\frac{\pi}{2}}\frac{dx}{\sqrt{1-0.5\sin^2 x}} = 0,999699 \tag{7.69}$$

Note that the accuracy of the frequency Ω given by Eq. 7.67 is not strongly dependent upon the values of aA^2 because it is uniformly valid for any possible values of aA^2. Equation 7.69 shows that the formula (7.67) can give an excellent approximate frequency for both small and large values of oscillation amplitude. Therefore, for any value of $a > 0$ it can be easily proved that the maximal relative error of the frequency (7.67) is less than 0.03%. Without any cumbersome procedure, we can readily obtain the second or higher order approximation with high accuracy. Convergence and error study of the above mentioned examples is a

Table 7.1 Comparison between the approximate frequency (7.67) and the exact frequency for Eq. 7.65

aA^2	Ω, Eq. 7.67	Ω, Eq. 7.62
0.2	1.07200	1.07200
0.4	1.138906	1.13891
0.6	1.201731	1.20173
0.8	1.2611777	1.26118
1	1.3177644	1.31778
2	1.5690506	1.56911
5	2.1501774	2.15042
10	2.86612768	2.86664
100	8.5310997	8.53359
1,000	26.802504	26.8107
10,000	84.7013205	84.7245

further need and it is clear that many other modifications of the homotopy perturbation method can be made.

Example (c)

For the forced Duffing oscillator

$$\ddot{u} + \omega^2 u + \varepsilon\alpha u^3 = \varepsilon k \cos \Omega t \tag{7.70}$$

with ω, α, k, constants, $\omega \approx \Omega$ (primary resonance) and $0 < \varepsilon << 1$, the Eqs. 7.33, 7.34 and 7.35 become

$$\ddot{u}_0 + \Omega^2 u_0 = 0, \quad u_0(0) = A,\ \dot{u}_0(0) = 0, \tag{7.71}$$

$$\ddot{u}_1 + \Omega^2 u_1 + \Omega_1 u_0 + \varepsilon\alpha u_0^3 - \varepsilon k \cos \Omega t = 0, \quad u_1(0) = \dot{u}_1(0) = 0 \tag{7.72}$$

$$\ddot{u}_2 + \Omega^2 u_2 + \Omega_1 u_1 + \Omega_2 u_0 + 3\varepsilon\alpha u_0^2 u_1 = 0, \quad u_2(0) = \dot{u}_2(0) = 0 \tag{7.73}$$

From Eq. 7.71 we obtain

$$u_0(t) = A \cos \Omega t \tag{7.74}$$

Substituting Eq. 7.74 into Eq. 7.72 yields

$$\ddot{u}_1 + \Omega^2 u_1 + (A\Omega_1 + \frac{3}{4}\varepsilon\alpha A^3 - \varepsilon k) \cos \Omega t + \frac{1}{4}\varepsilon\alpha A^3 \cos 3\Omega t = 0 \tag{7.75}$$

Avoiding the presence of secular term in Eq. 7.75 needs

$$\Omega_1 = \frac{\varepsilon k}{A} - \frac{3}{4}\varepsilon\alpha A^2 \tag{7.76}$$

The solution of Eq. 7.75 become

$$u_1(t) = \frac{\varepsilon\alpha A^3}{32\Omega^2}(\cos 3\Omega t - \cos \Omega t) \tag{7.77}$$

From Eqs. 7.74, 7.76, 7.77 and 7.73 we obtain

$$\ddot{u}_2 + \Omega^2 u_2 + \left(\Omega_2 A - \frac{\alpha\varepsilon A^3 \Omega_1}{32\Omega^2} - \frac{3\varepsilon^2\alpha^2 A^5}{64\Omega^2}\right) \cos \Omega t + H.O.T. = 0 \tag{7.78}$$

The elimination of secular term in Eq. 7.78 requires

$$\Omega_2 = \frac{\varepsilon^2 \alpha k A}{32\Omega^2} + \frac{3\varepsilon^2\alpha^2 A^4}{128\Omega^2} \tag{7.79}$$

The first-order approximation to the solution of Eq. 7.70 for the primary resonant case $\Omega \approx \omega$ is

$$u(t) = A\cos\Omega t + \frac{\varepsilon\alpha A^3}{32\Omega^2}(\cos 3\Omega t - \cos\Omega t) \tag{7.80}$$

where the frequency Ω is obtained from Eqs. 7.76, 7.79 and 7.29 ($\Lambda(0) = \Omega^2$):

$$\Omega^2 = \frac{\omega^2}{2} - \varepsilon\left(\frac{k}{2A} - \frac{3}{8}\alpha A^2\right) + \\ + \frac{1}{2}\sqrt{\left[\omega^2 - a\left(\frac{k}{A} - \frac{3}{4}\alpha A^2\right)^2\right]^2 - \frac{1}{8}\varepsilon^2\left(\alpha k A + \frac{3}{4}\alpha^2 A^4\right)} \tag{7.81}$$

In order to show the validity of the modified homotopy perturbation method, the differential Eq. 7.70 has been numerically solved for the following characteristics: $\alpha = \omega = A = k = 1$. Figure 7.1–7.4 show the comparison between the numerical solutions obtained using a fourth-order Runge–Kutta method and the approximate solution (7.80) for different values of ε.

The displacement curves obtained by the modified homotopy perturbation method are quasi-identical to those obtained via numerical simulation for $\varepsilon \leq 0.3$. For moderate values of ε, e.g. $\varepsilon = 0.5$, the obtained results are satisfactory.

Example (d)

Consider the Van der Pol oscillator in the form

$$\ddot{u} + u = \varepsilon(1 - u^2)\dot{u} \tag{7.82}$$

and assume that $\varepsilon > 0$ is small. The initial conditions are

$$u(0) = A, \quad \dot{u}(0) = 0 \tag{7.83}$$

Equation 7.33 is

$$\ddot{u}_0 + \Omega^2 u_0 = 0, \quad u_0(0) = A, u_0(0) = 0 \tag{7.84}$$

and has the solution

$$u_0(t) = A\cos\Omega t \tag{7.85}$$

Equation 7.34 by means of Eq. 7.85 can be written in the form:

$$\ddot{u}_1 + \Omega^2 u_1 + \Omega_1 A\cos\Omega t + \varepsilon A\Omega\left(1 - \frac{A^2}{4}\right)\sin\Omega t - \\ - \frac{1}{4}A^3\Omega\sin 3\Omega t = 0, \quad u_1(0) = \dot{u}_1(0) = 0 \tag{7.86}$$

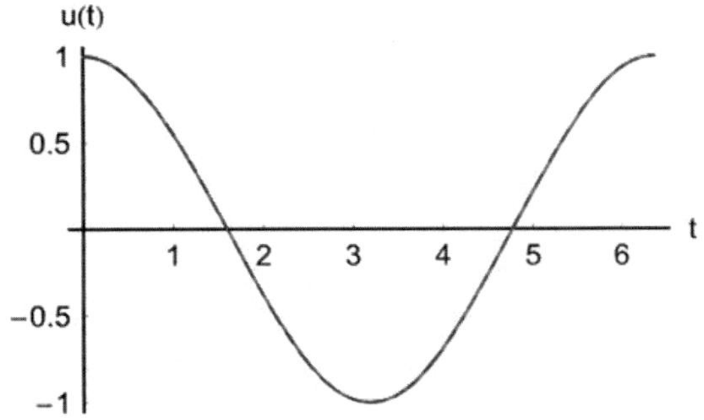

Fig. 7.1 Comparison between the numerical solution of Eq. 7.70 _____ and the approximate solution _ _ _ _ _ (7.80) for $\varepsilon = 0.1$ and $\alpha = \omega = A = k = 1$

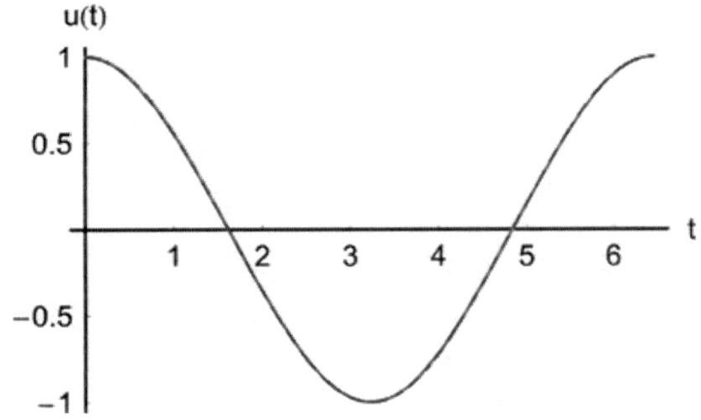

Fig. 7.2 Comparison between the numerical solution of Eq. 7.70 _____ and the approximate solution _ _ _ _ _ (7.80) for $\varepsilon = 0.2$ and $\alpha = \omega = A = k = 1$

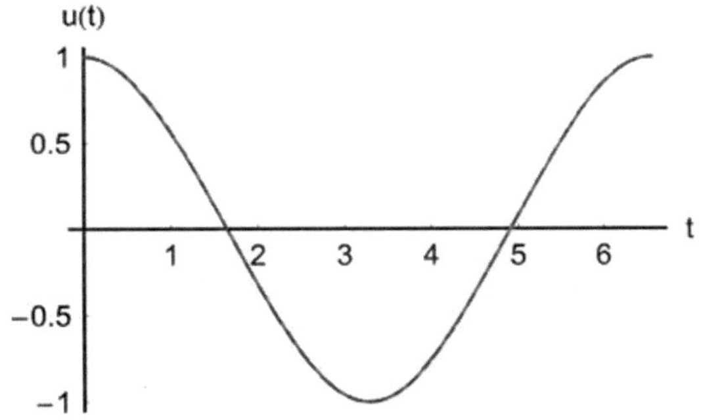

Fig. 7.3 Comparison between the numerical solution of Eq. 7.70 _____ and the approximate solution _ _ _ _ _ (7.80) for $\varepsilon = 0.3$ and $\alpha = \omega = A = k = 1$

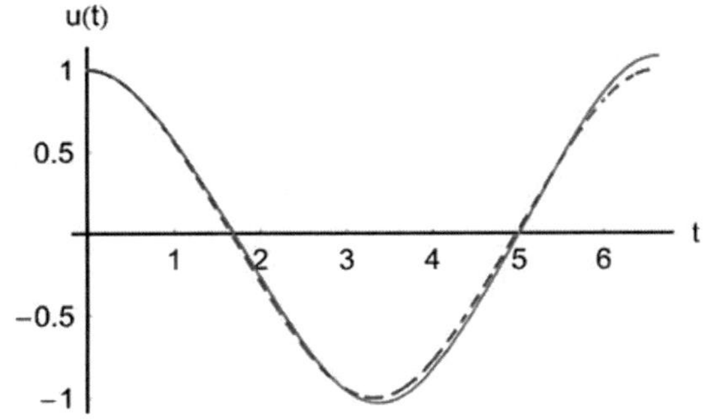

Fig. 7.4 Comparison between the numerical solution of Eq. 7.70 _____ and the approximate solution _ _ _ _ _ (7.80) for $\varepsilon = 0.5$ and $\alpha = \omega = A = k = 1$

From the elimination of secular terms in Eq. 7.86 we obtain

$$\Omega_1 = 0, \quad A = 2 \tag{7.87}$$

and therefore, the solution of Eq. 7.86 becomes

$$u_1(t) = -\frac{\varepsilon}{4\Omega}(3\sin\Omega t - \sin 3\Omega t) \tag{7.88}$$

Substituting Eqs. 7.85, 7.87 and 7.88 into Eq. 7.35 we obtain

$$\ddot{u}_2 + \Omega^2 u_2 + \left(2\Omega_2 - \frac{\varepsilon^2}{4}\right)\cos\Omega t + \frac{3}{2}\varepsilon^2\cos 3\Omega t - \frac{5}{4}\varepsilon^2\cos 5\Omega t = 0 \tag{7.89}$$

Avoiding the presence of secular term in Eq. 7.89 needs

$$\Omega_2 = \frac{\varepsilon^2}{8} \tag{7.90}$$

and therefore, from Eqs. 7.87, 7.90 and 7.31 it is obtained

$$\Omega^2 = 1 - \frac{\varepsilon^2}{8} \tag{7.91}$$

Substituting Eqs. 7.85 and 7.88 into Eq. 7.30, we find the first-order approximate solution in the form

$$u(t) = 2\cos\sqrt{1-\frac{\varepsilon^2}{8}}t + \frac{\varepsilon}{\sqrt{1-\frac{\varepsilon^2}{8}}}\left[3\sin\left(\sqrt{1-\frac{\varepsilon^2}{8}}\right)t - \sin\left(3\sqrt{1-\frac{\varepsilon^2}{8}}\right)t\right] \tag{7.92}$$

Example (e) (Generalized Van der Pol equation)

In this last example, we consider nonlinear oscillations governed by a generalized Van der Pol equation of the form [148]:

$$\ddot{u} + u = \varepsilon(1 - u^{2n})\dot{u} \tag{7.93}$$

with the initial conditions:

$$u(0) = A, \quad \dot{u}(0) = 0 \tag{7.94}$$

where n is any positive integer and the parameter ε is small. For $n = 1$, the Eq. 7.93 reduces to the traditional Van der Pol equation.

The Eq. 7.33 is

$$\ddot{u}_0 + \Omega^2 u_0 = 0 \quad u_0(0) = A, \quad \dot{u}_0 = 0 \tag{7.95}$$

The solution of Eq. 7.95 becomes

$$u_0 = A\cos\Omega t \tag{7.96}$$

In view of Eq. 7.34 we need the identity [149]

$$\cos^{2n}\Omega t = \frac{1}{2^{2n}}\left[\binom{2n}{n} + 2\sum_{k=1}^{n}\binom{2n}{n-k}\cos 2k\Omega t\right] \tag{7.97}$$

where

$$\binom{n}{k} = \frac{n!}{k!(n-k)!}, \binom{n}{0} = 1, \quad k! = 1\cdot 2\cdot 3 \ldots k$$

So, it can be verified that

$$\begin{aligned}\sin\Omega t\cos^{2n}\Omega t &= \frac{1}{2^{2n}}\sum_{k=0}^{n}\binom{2n}{n-k}\frac{2k+1}{n+k+1}\sin(2k+1)\Omega t\\ \sin\Omega t\cos^{2n-1}\Omega t &= \frac{1}{n2^{2n-1}}\sum_{k=1}^{n}\binom{2n}{n-k}k\sin 2k\Omega t\end{aligned} \tag{7.98}$$

Equation 7.34 can be written by means of Eq. 7.96 in the form

$$\ddot{u}_1 + \Omega^2 u_1 = -A\Omega_1\cos\Omega t - \varepsilon A\Omega(1 - A^{2n}\cos^{2n}\Omega t)\sin\Omega t, \quad u_1(0) = \dot{u}_1(0) = 0$$

The last equation, with the aid of the identity (7.98_1), becomes:

$$\begin{aligned}\ddot{u}_1 + \Omega^2 u_1 &= -A\Omega_1\cos\Omega t + \varepsilon A\Omega\left[\left(\frac{A}{2}\right)^{2n}\binom{2n}{n}\frac{1}{n+1} - 1\right]\sin\Omega t +\\ &+ \varepsilon\Omega\left(\frac{A}{2}\right)^{2n}\sum_{k=1}^{n}\binom{2n}{n-k}\frac{2k+1}{n+k+1}\sin(2k+1)\Omega t\end{aligned} \tag{7.99}$$

Avoiding the presence of a secular term in the Eq. 7.99 needs

$$\Omega_1 = 0, \quad A = \left[\frac{4n!(n+1)!}{(2n)!}\right]^{\frac{1}{2n}} \tag{7.100}$$

and thus, from Eq. 7.99 we obtain the following result

$$\begin{aligned}u_1(t) = &-\frac{\varepsilon}{2\Omega}\left(\frac{A}{2}\right)^{2n+1}\sum_{k=1}^{n}\binom{2n}{n-k}\frac{2k+1}{k(k+1)(n+k+1)}\times\\ &\times[\sin(2k+1)\Omega t - (2k+1)\sin\Omega t]\end{aligned} \tag{7.101}$$

Substituting Eqs. 7.95 and 7.101 into Eq. 7.35 yields

$$\ddot{u}_2+\Omega^2 u_2=\frac{\Omega_1\varepsilon}{2\Omega}\left(\frac{A}{2}\right)^{2n+1}\sum_{k=1}^{n}\binom{2n}{n-k}\frac{2k+1}{k(k+1)(n+k+1)}\times$$
$$\times\left[\sin(2k+1)\Omega t-(2k+1)\sin\Omega t\right]-\Omega_2 A\cos\Omega t-$$
$$-\frac{\varepsilon}{2}(1-A^{2n})\cos^{2n}\Omega t\left(\frac{A}{2}\right)^{2n+1}\sum_{k=1}^{n}\binom{2n}{n-k}\times\frac{(2k+1)^2}{k(k+1)(n+k+1)}\times$$
$$\times\left[\cos(2k+1)\Omega t-\cos\Omega t\right]-\varepsilon^2 nA^{2n}\cos^{2n-1}\Omega t\sin\Omega t\left(\frac{A}{2}\right)^{2n-1}\times$$
$$\times\sum_{k=1}^{n}\binom{2n}{n-k}\frac{2k+1}{k(k+1)(n+k+1)}\times\left[\sin(2k+1)\Omega t-(2k+1)\sin\Omega t\right]$$

By the same manipulation and by means of Eqs. 7.97, 7.98 and 7.100_1, the last equation can be written as

$$\ddot{u}_2+\Omega^2 u_2=\left[-\Omega_2 A+\frac{\varepsilon^2}{2}\left(\frac{A}{2}\right)^{2n+1}\sum_{k=1}^{n}\binom{2n}{n-k}\frac{(2k+1)^2}{k(k+1)(n+k+1)}-\right.$$
$$-\frac{\varepsilon^2}{2}\left(\frac{A}{2}\right)^{4n+1}\binom{2n+1}{n}\sum_{k=1}^{n}\binom{2n}{n-k}\frac{(2k+1)^2}{k(k+1)(n+k+1)}+$$
$$+\frac{\varepsilon^2}{2}\left(\frac{A}{2}\right)^{4n+1}\sum_{k=1}^{n}\binom{2n}{n-k}^2\frac{(2k+1)^2}{k(k+1)(n+k+1)}-$$
$$-\varepsilon^2\left(\frac{A}{2}\right)^{4n+1}\sum_{k=1}^{n}\binom{2n}{n-k}^2\frac{2k+1}{(k+1)(n+k+1)}+$$
$$\left.+\varepsilon^2\left(\frac{A}{2}\right)^{4k+1}\binom{2n}{n-1}\sum_{k=1}^{n}\binom{2n}{n-k}\frac{(2k+1)^2}{k(k+1)(n+k+1)}\right]\times$$
$$\times\cos\Omega t+H.O.T. \tag{7.102}$$

Eliminating secular terms in Eq. 7.102 requires that

$$\Omega_2=\frac{\varepsilon^2}{4}\frac{n!(n+1)!}{(2n)!}\sum_{k=1}^{n}\binom{2n}{n-k}\frac{(2k+1)^2}{k(k+1)(n+k+1)}-$$
$$-\frac{\varepsilon^2}{4}\left[\frac{n!(n+1)!}{(2n)!}\right]^2\binom{2n+1}{n}\sum_{k=1}^{n}\binom{2n}{n-k}\frac{(2k+1)^2}{k(k+1)(n+k+1)}+$$
$$+\frac{\varepsilon^2}{4}\left[\frac{n!(n+1)!}{(2n)!}\right]^2\sum_{k=1}^{n}\binom{2n}{n-k}\frac{(2k+1)^2}{k(k+1)(n+k+1)}-$$
$$-\frac{\varepsilon^2}{2}\left[\frac{n!(n+1)!}{(2n)!}\right]^2\sum_{k=1}^{n}\binom{2n}{n-k}\frac{2k+1}{(k+1)(n+k+1)}+$$
$$+\frac{\varepsilon^2}{2}\left[\frac{n!(n+1)!}{(2n)!}\right]^2\binom{2n}{n-1}\sum_{k=1}^{n}\binom{2n}{n-k}\frac{(2k+!)^2}{k(k+1)(n+k+1)} \tag{7.103}$$

where Eq. 7.100_2 has been taken into account.

The frequency Ω given by Eq. 7.31 is obtained from Eqs. 7.100_1 and 7.103:

$$\begin{aligned}\Omega^2 = 1 &- \frac{\varepsilon^2}{4}\frac{n!(n+1)!}{(2n)!}\sum_{k=1}^{n}\binom{2n}{n-k}\frac{(2k+1)^2}{k(k+1)(n+k+1)}+\\ &+\frac{\varepsilon^2}{4}\frac{n!(n+1)!(2n+1)}{(2n)!}\sum_{k=1}^{n}\binom{2n}{n-k}\frac{(2k+1)^2}{k(k+1)(n+k+1)}-\\ &-\frac{\varepsilon^2}{4}\left[\frac{n!(n+1)!}{(2n)!}\right]^2\sum_{k=1}^{n}\binom{2n}{n-k}^2\frac{(2k+1)^2}{k(k+1)(n+k+1)}+\\ &+\frac{\varepsilon^2}{2}\left[\frac{n!(n+1)!}{(2n)!}\right]^2\sum_{k=1}^{n}\binom{2n}{n-k}^2\frac{2k+1}{k(k+1)(n+k+1)}-\\ &-\frac{\varepsilon^2}{2}\left[\frac{n!(n+1)!}{(2n)!}\right]^2\binom{2n}{n-k}\sum_{k=1}^{n}\binom{2n}{n-k}\frac{(2k+1)^2}{k(k+1)(n+k+1)}\end{aligned} \tag{7.104}$$

The first-order approximate solution is obtained from Eqs. 7.95, 7.101 and 7.30 and we obtain

$$\begin{aligned}u(t) =&A\cos\Omega t - \frac{\varepsilon}{2\Omega}\left(\frac{A}{2}\right)^{2n+1}\sum_{k=1}^{n}\binom{2n}{n-k}\frac{2k+1}{k(k+1)(n+k+1)}\times\\ &\times\left[\sin(2k+1)\Omega t-(2k+1)\sin\Omega t\right]\end{aligned} \tag{7.105}$$

where A and Ω are given by Eqs. 7.100_2 and 7.104 respectively.

We remark that for $n = 1$ (traditional Van der Pol equation) we obtain from Eqs. 7.104 and 7.105 respectively

$$\Omega^2 = 1 - \frac{\varepsilon^2}{8}$$

and

$$u(t) = 2\cos\Omega t - \frac{\varepsilon}{\Omega}(\sin 3\Omega t - 3\sin\Omega t)$$

which are identical to Eqs. 7.91 and 7.92 respectively.

7.3 Basic Idea of Optimal Homotopy Perturbation Method and Some Applications

We consider Eq. 7.3 with the boundary conditions (7.2) and with the solution (7.8). Generally speaking, the nonlinear operator $N(u)$ depend on $t, u, \dot{u}, \ddot{u}$ and therefore one can write

$$N(u) = F(t, u, \dot{u}, \ddot{u}, ...) \tag{7.106}$$

Applying the Taylor series theorem for real values α, β, γ, we obtain

$$F(t, u+\alpha, \dot{u}+\beta, \ddot{u}+\gamma, ...) = F(t, u, \dot{u}, \ddot{u}) + \frac{\alpha}{1!}F_u(t, u, \dot{u}, \ddot{u}, ...) + \\ + \frac{\beta}{1!}F_{\dot{u}}(t, u, \dot{u}, \ddot{u}, ...) + \frac{\gamma}{1!}F_{\ddot{u}}(t, u, \dot{u}, \ddot{u}, ...) + ... \tag{7.107}$$

where $F_u = \partial F/\partial u$.

At this moment we consider only two alternatives to construct the homotopy given by Eq. 7.5. In the case of the *first alternative*, we introduce a number of unknown auxiliary functions $K_{ij}(t, C_k), i, j, k = 1, 2, \ldots$ that depend on the variable t and some constants $C_1, C_2, \ldots$ and a new homotopy which satisfies the following equation:

$$H(v,p) = L(v) - L(u_0) - f(r) + p[L(u_0) + K_{11}(t, C_k)F(t, v_0, \dot{v}_0, \ddot{v}_0, ...)] + \\ + p^2[K_{21}(t, C_k)v_1F_v(r, v_0, \dot{v}_0, \ddot{v}_0, ...) + K_{22}(t, C_k)\dot{v}_1F_{\dot{v}}(t, v_0, \dot{v}_0, \ddot{v}_0, ...) + \\ + K_{23}(t, C_k)\ddot{v}_1F_{\ddot{v}}(t, v_0, \dot{v}_0, \ddot{v}_0, ...)] + p^3[K_{31}(t, C_k)v_2F_v(t, v_0, \dot{v}_0, \ddot{v}_0, ...) + \\ + K_{32}(t, C_k)\dot{v}_2F_{\dot{v}}(t, v_0, \dot{v}_0, \ddot{v}_0, ...) + \\ + K_{33}(t, C_k)\ddot{v}_2F_{\ddot{v}}(t, v_0, \dot{v}_0, \ddot{v}_0, ...)...] + ... = 0 \tag{7.108}$$

where v is given by Eq. 7.7.

A whole set of equations are obtained by equating the coefficients of like powers of p for Eq. 7.1 and 7.3. More precisely, we have the following equations:

$$L(v_0) - L(u_0) - f(r) = 0, \quad B(v_0, \dot{v}_0) = 0 \tag{7.109}$$

$$L(v_1) + L(u_0) + K_{11}(t, C_k)F(t, v_0, \dot{v}_0, \ddot{v}_0, ...) = 0, \quad B(v_1, \dot{v}_1) = 0 \tag{7.110}$$

$$L(v_2) + K_{21}(t, C_k)v_1F_v(t, v_0, \dot{v}_0, \ddot{v}_0, ...) + K_{22}(t, C_k)\dot{v}_1F_{\dot{v}}(t, v_0, \dot{v}_0, \ddot{v}_0, ...) + \\ + K_{23}(t, C_k)\ddot{v}_1F_{\ddot{v}}(t, v_0, \dot{v}_0, \ddot{v}_0, ...) = 0 \; B(v_2, \dot{v}_2) = 0 \tag{7.111}$$

and so on.

In the case of the *second alternative*, the new homotopy can satisfy the equation

$$H(v,p) = L(v) - L(u_0) - f(r) + p[L(u_0) + K_{11}^*(t, C_k)F(t, v_0\dot{v}_0, \ddot{v}_0, ...) + \\ + K_{12}^*(t, C_k)F_v(t, v_0, \dot{v}_0, \ddot{v}_0, ...) + K_{13}^*(t, C_k)F_{\dot{v}}(t, v_0\dot{v}_0, \ddot{v}_0, ...) + \\ + K_{14}^*(t, C_k)F_{\ddot{v}}(t, v_0\dot{v}_0, \ddot{v}_0, ...)] = 0 \tag{7.112}$$

In this case, we can write only two equations:

$$L(v_0) - L(u_0) - f(t) = 0, \quad B(v_0, \dot{v}_0) = 0 \tag{7.113}$$

$$\begin{aligned} & L(v_1) + L(u_0) + K_{11}^*(t, C_k)F(t, v_0, \dot{v}_0, \ddot{v}_0, ...) + \\ & + K_{12}^*(t, C_k)F_v(t, v_0, \dot{v}_0, \ddot{v}_0, ...) + K_{13}^*(t, C_k)F_{\dot{v}}(t, v_0, \dot{v}_0, \ddot{v}_0, ...) + \\ & + K_{14}^*(t, C_k)F_{\ddot{v}}(t, v_0, \dot{v}_0, \ddot{v}_0, ...) = 0, \quad B(v_1, \dot{v}_1) = 0 \end{aligned} \tag{7.114}$$

The functions K_{ij} from Eq. 7.111 are not unique and they can be chosen so that the products $K_{ij}F_a$ and F_a be of the same form. The same remark applies to the functions K_{ij}^* from Eq. 7.112. In this way, in general, we expect that only two iterations are needed to achieve accurate solutions using the first alternative and only one iteration in the case of using the second alternative.

The convergence-control constants C_k, $k = 1, 2, \ldots q$, which appear in the expression of the functions $K_{ij}(t, C_k)$ or $K_{ij}^*(t, C_k)$can be optimally determined. This can be done via various methods, such as the least squares method, the weighted residual, the collocation method, the Galerkin method and so on (see Chap. 6). For example, these constants can be optimally determined by imposing the residual functional

$$J(C_1, C_2, ..., C_q) = \int_a^b [L(\bar{v}) + N(\bar{v}) - f(r)]^2 dr \tag{7.115}$$

be minimum, which leads to the system of equations:

$$\frac{\partial J}{\partial C_i} = 0, \quad i = 1, 2, ..., q \tag{7.116}$$

where a and b are two values from the domain of interest and $\bar{v}$is the m-th order approximate solution

$$\bar{v} = v_0 + v_1 + ... + v_m \tag{7.117}$$

where $m = 2$ for the first alternative and $m = 1$ for the second alternative.

The convergence-control constants C_k can also be optimally determined using other procedures. For example if $t_i \in (a, b)$, $i = 1, 2, ..., q$, then by substituting t_i into the residual R of the initial Eq. (7.1) obtained for the approximate solution (7.117), we obtain a system of nonlinear algebraic equations:

$$R(t_1, C_k) = R(t_2, C_k) = ... = R(t_q, C_k) = 0, \quad k = 1, 2, ..., q \tag{7.118}$$

The solutions of Eq. 7.1 subject to the conditions (7.2) can immediately be determined by using the optimal homotopy perturbation method once the convergence-control constants C_k are known.

In short, the idea of the presented procedure namely the optimal homotopy perturbation method (OHPM) is to construct a new homotopy such as Eqs. 7.108 or 7.112. It is to remark the presence of the functions $K_{ij}(t, C_k)$ or $K_{ij}^*(t, C_k)$, which include the convergence-control constants C_k that can be determined optimally so that the convergence of the approximate solutions can be easily controlled.

In what follows, we consider a few examples approached by means of these alternatives.

7.4 A Heat Transfer Problem

Consider the one-dimensional conduction of heat in a slab of thickness L that is made of a material with temperature-dependent thermal conductivity k [150]. If the temperatures of the two opposite faces of the slab are uniformly maintained at T_1 and T_2, $T_1 > T_2$, then the governing equation and boundary conditions are [107]:

$$\frac{d}{dx}\left(k\frac{dT}{dx}\right) = 0 \tag{7.119}$$

$$x = 0, T = T_1; \qquad x = L, T = T_2 \tag{7.120}$$

If we further assume that the thermal conductivity varies linearly with temperature, i.e.

$$k = k_2[1 + \beta(T - T_2)] \tag{7.121}$$

where k_2 is the thermal conductivity at temperature T_2 for some constant β, then by introducing the dimensionless quantities

$$\theta = \frac{T - T_2}{T_1 - T_2}, \ \eta = \frac{x}{L}, \ \varepsilon = \beta(T_1 - T_2) = \frac{k_1 - k_2}{k_2} \tag{7.122}$$

into Eqs. 7.119 and 7.120, we obtain

$$\theta''(\eta) + \varepsilon\theta(\eta)\theta''(\eta) + \varepsilon\theta'^2(\eta) = 0 \tag{7.123}$$

$$\theta(0) = 1, \quad \theta(1) = 0 \tag{7.124}$$

where prime denotes differentiation with respect to η.

For this problem we consider the first alternative, where in accordance with Eq. 7.108, the linear operator is chosen as

$$L(\theta) = \theta'' \tag{7.125}$$

and a non-linear operator is defined as ($f(r) = 0$):

$$N(\theta) = \varepsilon\theta\theta'' + \varepsilon\theta'^2 \tag{7.126}$$

The initial approximation $\theta_0(\eta)$ is obtained from Eq. 7.109 for $L(u_0) = 0$ as

$$\theta_0'' = 0, \quad \theta_0(0) = 1, \quad \theta_0(1) = 0 \tag{7.127}$$

The solution of Eq. 7.127 is

$$\theta_0(\eta) = 1 - \eta \tag{7.128}$$

and the Eq. 7.110 becomes:

$$\theta_1'' + K_{11}(\eta, C_i)(\varepsilon\theta_0\theta''0 + \varepsilon\theta_0'^2) = 0, \theta_1(0) = \theta_1(1) = 0 \tag{7.129}$$

We choose $K_{11}(\eta, C_i) = 1$and obtain, from Eq. 7.129

$$\theta_1(\eta) = \frac{1}{2}\varepsilon\eta - \frac{1}{2}\varepsilon\eta^2 \tag{7.130}$$

Remark. The choice of $K_{11}(\eta, C_i)$ is not unique. We can choose for example, $K_{11}(\eta, C_i) = C_1$ or $K_{11}(\eta, C_i) = C_1' + C_2'\eta$ or $K_{11}(\eta, C_i) = C_1'' + C_2''\eta + C_3''\eta^2$ with constants C_1, C_1', C_2', C_1'', C_2'' and C_3''.

From Eq. 7.111, the second-order problem is

$$\theta_2'' + K_{21}(\eta, C_i)(2\varepsilon^2\eta - \varepsilon^2) + K_{22}(\eta, C_j)(\varepsilon^2\eta - \varepsilon^2) = 0, \theta_2(0) = \theta_2(1) = 0 \tag{7.131}$$

so by choosing

$$K_{21}(\eta, C_i) = C_1 + C_2\eta + C_3\eta^2, \quad K_{22}(\eta, C_i) = C_4 + C_5\eta + C_6\eta^2 \tag{7.132}$$

we have from Eq. 7.131

$$\begin{aligned}\theta_2(\eta) = &\left(\frac{\varepsilon^2 C_1}{6} - \frac{\varepsilon^2 C_3}{60} + \frac{\varepsilon^2 C_4}{3} + \frac{\varepsilon^2 C_5}{12} + \frac{\varepsilon^2 C_6}{30}\right)\eta - \\ &- \frac{\varepsilon^2 C_1 + \varepsilon^2 C_4}{2}\eta^2 + \frac{2\varepsilon^2 C_1 - \varepsilon^2 C_2 + \varepsilon^2 C_4 - \varepsilon^2 C_5}{6}\eta^3 + \\ &+ \frac{2\varepsilon^2 C_2 - \varepsilon^2 C_3 + \varepsilon^2 C_5 - \varepsilon^2 C_6}{12}\eta^4 + \frac{2\varepsilon C_3 + \varepsilon^2 C_6}{20}\eta^5\end{aligned} \tag{7.133}$$

Again, the choices of K_{21} and K_{22} are not unique and these could have been chosen to be, say, $K_{22}(\eta, C_i) = C_3' + C_4'\eta + C_5'\eta^2 + C_6'\eta^3$ and $K_{21}(\eta, C_i) = C_1' + C_2'\eta$, or even $K_{21}(\eta, C_i) = C_1'' + C_2''\eta + C_3''\eta^2 + C_4''\eta^3$ and $K_{22}(\eta, C_i) = C''_5 + C''_6\eta + C''_7\eta^2 + C''_8\eta^3$.

On substituting Eqs. 7.128, 7.130 and 7.133 into Eq. 7.117, we obtain the second-order approximate solution of Eqs. 7.123 and 7.124 in the form:

$$\bar{\theta} = \theta_0 + \theta_1 + \theta_2 = \Big[\frac{1}{60}(10\varepsilon^2 C_1 - \varepsilon^2 C_3 + 20\varepsilon^2 C_4 + 5\varepsilon^2 C_5 +$$
$$+ 2\varepsilon^2 C_6) + \frac{\varepsilon}{2} - 1\Big]\eta + 1 - \frac{\varepsilon + \varepsilon^2 C_1 + \varepsilon^2 C_4}{2}\eta^2 + \frac{1}{6}(2\varepsilon^2 C_1 + \varepsilon^2 C_2 +$$
$$+ \varepsilon^2 C_4 - \varepsilon^2 C_5)\eta^3 + \frac{1}{12}(2\varepsilon^2 C_2 - \varepsilon^2 C_3 + \varepsilon^2 C_5 - \varepsilon^2 C_6)\eta^4 +$$
$$+ \frac{1}{20}(2\varepsilon^2 C_3 + \varepsilon^2 C_6)\eta^5 \quad (7.134)$$

The exact solution, obtained by integrating Eq. 7.123 twice under the conditions (7.124) is

$$\theta_{ex}(\eta) = \frac{1}{\varepsilon}\left(\sqrt{(\varepsilon^2 + 2\varepsilon)(1 - \eta) + 1} - 1\right) \quad (7.135)$$

The approximate analytical solution obtained using OHPM shall now be compared with the above exact solution for different values of ε to demonstrate the efficiency and accuracy of the OHPM.

Case (a) For $\varepsilon = 0.5$

Following the procedure described above, we have from Eq. 7.116:

$$C_1 = -0.235565, C_2 = -0.128899, C_3 = -0.723479,$$
$$C_4 = -0.854192, C_5 = -0.12291, C_6 = -0.334734 \quad (7.136)$$

This yields an approximate solution, from Eq. 7.134, of the form

$$\bar{\theta}(\eta) = 1 - 0.833333\eta - 0.113780375\eta^2 - 0.0447297\eta^3 +$$
$$+ 0.014115\eta^4 - 0.022271\eta^5 \quad (7.137)$$

The accuracy of the solution obtained by OHPM can be assessed graphically as shown in Fig. 7.5.

Case (b) For ε=, we have from Eq. 7.116

$$C_1 = 0.153352, \quad C_2 = 0.47144, \quad C_3 = -1.69402,$$
$$C_4 = -0.912121, \quad C_5 = C_6 = 0 \quad (7.138)$$

This yields an approximate solution, from Eq. 7.134, of the form

$$\bar{\theta}(\eta) = 1 - 0.750248\eta - 0.1206155\eta^2 - 0.17947616\eta^3 +$$
$$+ 0.219741666\eta^4 - 0.169402\eta^5 \quad (7.139)$$

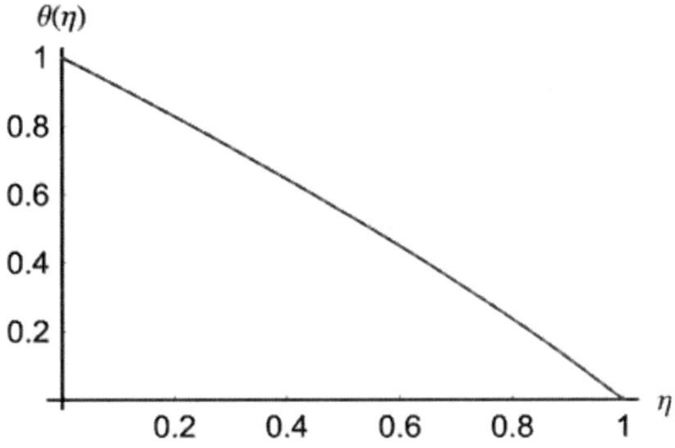

Fig. 7.5 Comparison between the present solution (7.137) and the exact solution (7.135) of Eq. 7.123 for ε = 0.5: ______ exact solution, _ _ _ _ approximate solution

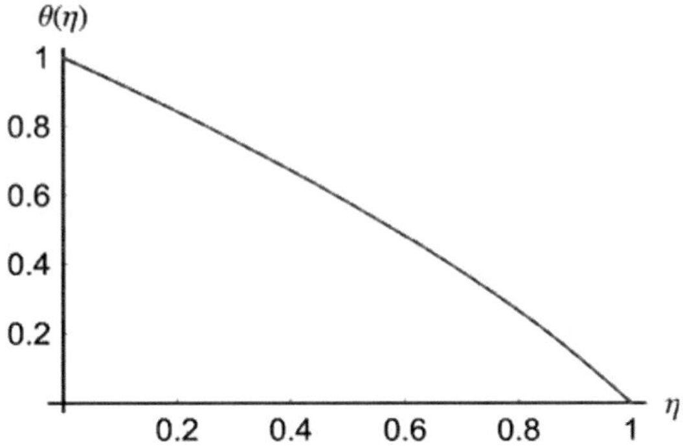

Fig. 7.6 Comparison between the present solution (7.139) and the exact solution (7.135) of Eq. 7.123 for ε = 1: ______ exact solution, _ _ _ _ approximate solution

Figure 7.6 shows a comparison between the approximate solution obtained by OHPM and the exact solution.

Case (c) For ε = 1.5, we obtain from Eq. (7.116),

$$\begin{aligned} C_1 &= -0.161965, \quad C_2 = 1.27736, \quad C_3 = -2.07241, \\ C_4 &= -0.425527, C_5 = -0.879691, \quad C_6 = 0.188495 \end{aligned} \tag{7.140}$$

This yields an approximate solution, from Eq. 7.134, of

$$\begin{aligned} \bar{\theta}(\eta) &= 1 - 0.581497937\eta - 0.0890715\eta^2 - 0.43017225\eta^3 \\ &\quad + 0.667302\eta^4 - 0.445086562\eta^5 \end{aligned} \tag{7.141}$$

Figure 7.7 shows the comparison between the approximate solution (7.141) and the exact solution (7.135).

The approximate solutions obtained above are almost identical to the exact ones, which show the high accuracy of the method presented.

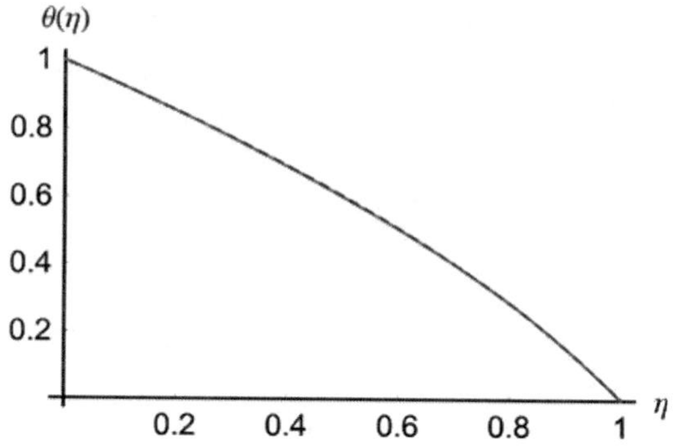

Fig. 7.7 Comparison between the present solution (7.141) and the exact solution (7.135) of Eq. 7.123 for $\varepsilon = 1.5$: ______ exact solution, _ _ _ _ approximate solution

7.5 Thin Film Flow of a Fourth Grade Fluid Down a Vertical Cylinder

Consider the nonlinear differential equation (see also Sect. 6.3):

$$\eta \frac{d^2 f}{d\eta^2} + \frac{df}{d\eta} + k\eta + 2b\left[\left(\frac{df}{d\eta}\right)^3 + 3\eta\left(\frac{df}{d\eta}\right)^2 \frac{d^2 f}{d\eta^2}\right] = 0 \tag{7.142}$$

with initial conditions

$$f(1) = 0, \quad f'(d) = 0 \tag{7.143}$$

where d is a parameter such that $d > 1$.

In accordance with Eq. 7.142, the linear operator is chosen as

$$Lf = \eta f'' + f' \tag{7.144}$$

and we define a non-linear operator as

$$Nf = 2b(f'^3 + 3\eta f'^2 f'') \tag{7.145}$$

where $f' = \frac{df}{d\eta}$.

The initial approximation $f_0'(\eta)$ is obtained from Eq. 7.109 where $L(u_0) = 0$:

$$\eta f_0'' + f_0' + k\eta = 0 \tag{7.146}$$

$$f_0'(d) = 0 \tag{7.147}$$

It is obtained:

$$f_0'(\eta) = \frac{k}{2}\left(\frac{d^2}{\eta} - \eta\right) \tag{7.148}$$

The first-order approximation $f'_1(\eta)$is obtained from Eq. 7.110

$$\eta f''_1 + f'_1 + K_{11}(\eta_1, C_k)F(f_0, f'_0, f''_0) = 0 \tag{7.149}$$

$$f'_1(d) = 0 \tag{7.150}$$

where

$$F(f_0, \dot{f}'_0, \dot{f}''_0) = 2b[f'^3_0 + 3\eta f'^2_0 f''_0] \tag{7.151}$$

We choose $K_{11}(\eta, C_k) = C_1$ (constant) because C_1F has the same form as F does. We remark that choosing of $K_{11}(\eta, C_k)$ is not unique. For example, we can choose $K_{11}(\eta, C_k) = C'_1 f'^2_0 + C'_2 f'_0$ or $K_{11}(\eta, C_k) = 1$ or $K_{11}(\eta, C_k) = C''_1 f'^3_0$ and so on.

From Eqs. 7.149, 7.150 and 7.151 we obtain:

$$f'_1(\eta) = -2bC_1 f'^3_0 \tag{7.152}$$

or using Eq. 7.148, Eq. 7.152 becomes:

$$f'_1(\eta) = -\frac{1}{4} bk^3 C_1 \left(\frac{d^2}{\eta} - \eta\right)^3 \tag{7.153}$$

The second-order approximation $f'_2(\eta)$ is obtained from Eq. 7.111

$$\begin{aligned} &\eta f''_2 + f'_2 + 2bK_{21}(\eta, C_i) f'_1 (3f'^2_0 + 6\eta f'_0 f''_0) + \\ &+ 2bK_{22}(\eta, C_i) f''_1 (3\eta f'^2_0) = 0 \end{aligned} \tag{7.154}$$

$$f'_2(d) = 0 \tag{7.155}$$

Substitutions of Eq. 7.152 into Eq. 7.154 yields

$$\begin{aligned} &(\eta f'_2)' - 12b^2 C_1 [K_{21}(\eta, C_i)(f'^5_0 + 2\eta f'^4_0 f''_0) + \\ &+ K_{22}(\eta, C_i)\eta f'^4_0 f''_0] = 0 \end{aligned} \tag{7.156}$$

In order to solve Eq. 7.156 we try to write its the second part in the form

$$\begin{aligned} &K_{21}(\eta, C_i)(f'^5_0 + 2\eta f'^4_0 f''_0) + K_{22}(\eta, C_i)\eta f'^4_0 f''_0 = \\ &= (C_2 \eta f'^i_0 + C_3 \eta f'^{i+1}_0 + C_4 \eta f'^{i+2}_0 + \dots + C_j \eta f'^{i+j-2}_0)' \end{aligned} \tag{7.157}$$

where i, j are positive integer numbers. From Eq. 7.157 it is obtained

$$\begin{aligned} K_{21} &= C_2 f'^{i-5}_0 + C_3 f'^{i-4}_0 + \dots + C_j f'^{i+j-7}_0 \\ K_{22} &= (i-2)C_2 f'^{i-5}_0 + (i-1)C_3 f'^{i-4}_0 + \dots + (i+j-4)C_j f'^{i+j-7}_0 \end{aligned} \tag{7.158}$$

Because K_{21} and K_{22} given by Eq. 7.158 are well-defined, from Eqs. 7.156 and 7.155 it is obtained

$$f_2' = 12b^2C_1(C_2f'^{\,i}_0 + C_3f'^{\,i+1}_0 + \ldots + C_jf'^{\,i+j-2}_0) \tag{7.159}$$

If we choose $i = 1$ and $j = 8$ into Eq. 7.159, it is obtained

$$\begin{aligned} f_2' = {} & 12b^2C_1C_2f_0' + 12b^2C_1C_3f'^2_0 + 12b^2C_1C_4f'^3_0 + 12b^2C_1C_5f'^4_0 + \\ & + 12b^2C_1C_6f'^5_0 + 12b^2C_1C_7f'^6_0 + 12b^2C_1C_8f'^7_0 \end{aligned} \tag{7.160}$$

We remark that the choice of the parameters $i = 1$ and $j = 8$ is not unique. Substituting Eq. 7.148 into Eq. 7.160 we obtain

$$\begin{aligned} f_2'(\eta) = {} & 6b^2kC_1C_2\left(\frac{d^2}{\eta} - \eta\right) + 3b^2k^2C_1C_3\left(\frac{d^2}{\eta} - \eta\right)^2 + \\ & + \frac{3}{2}b^2k^3C_1C_4\left(\frac{d^2}{\eta} - \eta\right)^3 + \frac{3}{4}b^2k^3C_1C_5\left(\frac{d^2}{\eta} - \eta\right)^4 + \\ & + \frac{3}{8}b^2k^5C_1C_6\left(\frac{d^2}{\eta} - \eta\right)^5 + \frac{3}{16}b^2k^6C_1C_7\left(\frac{d^2}{\eta} - \eta\right)^6 + \\ & + \frac{3}{32}b^2k^7C_1C_8\left(\frac{d^2}{\eta} - \eta\right)^7 \end{aligned} \tag{7.161}$$

The approximate differentiation solution of the second-order is given by

$$\bar{f}' = f_1' + f_1' + f_2' \tag{7.162}$$

or, using Eqs. 7.148, 7.153 and 7.161

$$\begin{aligned} \bar{f}'(\eta) = {} & \frac{k(1 + 12b^2C_1C_2)}{2}\left(\frac{d^2}{\eta} - \eta\right) + 3b^2k^2C_1C_3\left(\frac{d^2}{\eta} - \eta\right)^2 + \\ & + \frac{bk^3C_1(6bC_4 - 1)}{4}\left(\frac{d^2}{\eta} - \eta\right)^3 + \frac{3b^2k^4C_1C_5}{4}\left(\frac{d^2}{\eta} - \eta\right)^4 + \\ & + \frac{3b^2k^5C_1C_6}{8}\left(\frac{d^2}{\eta} - \eta\right)^5 + \frac{3b^2k^6C_1C_7}{16}\left(\frac{d^2}{\eta} - \eta\right)^6 + \\ & + \frac{3b^2k^7C_1C_8}{32}\left(\frac{d^2}{\eta} - \eta\right)^7 \end{aligned} \tag{7.163}$$

From Eqs. 7.163 and 7.143_1 we obtain the second-order approximate solution of Eq. 7.142 in the conditions given by Eq. 7.143

$$\begin{aligned}
\bar{f}(\eta) = & \frac{k(1+12b^2C_1C_2)}{2}\left(\frac{1-\eta^2}{2}+d^2\ln\eta\right)+3b^2k^2C_1C_3\left[\frac{\eta^3-1}{3}+\right.\\
& \left.+d^4\left(1-\frac{1}{\eta}\right)+2d^2(1-\eta)\right]+\frac{bk^3C_1(6bC_4-1)}{4}\left[\frac{d^6}{2}\left(1-\frac{1}{\eta^2}\right)-3d^4\ln\eta+\right.\\
& \left.+\frac{3d^2}{2}(\eta^2-1)+\frac{1-\eta^4}{4}\right]+\frac{3b^2k^4C_1C_5}{4}\left[\frac{d^8}{3}\left(1-\frac{1}{\eta^3}\right)+4d^6\left(\frac{1}{\eta}-1\right)+\right.\\
& \left.+6d^4(\eta-1)-\frac{4d^2}{3}(\eta^3-1)+\frac{1}{5}(\eta^5-1)\right]+\frac{3b^2k^5C_1C_6}{8}\left[\frac{d^{10}}{4}\left(1-\frac{1}{\eta^4}\right)+\right.\\
& \left.+\frac{5d^8}{2}\left(\frac{1}{\eta^2}-1\right)+10d^6\ln\eta+5d^4(1-\eta^2)+\frac{5d^2}{4}(\eta^4-1)+\frac{1}{6}(1-\eta^6)\right]+\\
& +\frac{3b^2k^6C_1C_7}{16}\left[\frac{d^{12}}{5}\left(1-\frac{1}{\eta^5}\right)+2d^{10}\left(\frac{1}{\eta^3}-1\right)+15d^8\left(1-\frac{1}{\eta}\right)+\right.\\
& \left.+20d^6(1-\eta)+5d^4(\eta^3-1)+\frac{6d^2}{5}(1-\eta^5)+\frac{1}{7}(\eta^7-1)\right]+\\
& +\frac{3b^2k^7C_1C_8}{32}\left[\frac{d^{14}}{6}\left(1-\frac{1}{\eta^6}\right)+\frac{7d^{12}}{4}\left(\frac{1}{\eta^4}-1\right)+\frac{21d^{10}}{2}\left(1-\frac{1}{\eta^2}\right)-\right.\\
& \left.-35d^8\ln\eta+\frac{35d^6}{2}(\eta^2-1)+\frac{21d^4}{4}(1-\eta^4)+\frac{7d^2}{6}(\eta^6-1)+\frac{1}{8}(1-\eta^8)\right]
\end{aligned} \tag{7.164}$$

The exact solution $f'(\eta)$ is obtained by integration of Eq. 7.142 with the conditions (7.143_2). We obtain:

$$f'(\eta)-\frac{k}{2}\left(\frac{d^2}{\eta}-\eta\right)+2bf'^3(\eta)=0 \tag{7.165}$$

$$f(1)=0 \tag{7.166}$$

With the notations:

$$\begin{aligned}
& D_1=C_1C_2;D_2=C_1C_3;D_3=C_1(6\beta C_4-1);D_4=C_1C_5;\\
& D_5=C_1C_6;D_6=C_1C_7;D_7=C_1C_8
\end{aligned} \tag{7.167}$$

the residual obtained substituting Eq. 7.163 into Eq. 7.144 becomes:

$$R(\eta,D_i)=\bar{f}'(\eta)-\frac{k}{2}\left(\frac{d^2}{\eta}-\eta\right)+2\beta\bar{f}'^3(\eta)=0 \tag{7.168}$$

If we locate the residual R given by Eq. 7.168 in 7 points: $\eta_i = 1 + \frac{id}{8}$, i = 1, 2, ..., 7, we obtain a system of seven algebraic equations:

$$R\left(1 + \frac{id}{8}, D_i\right) = 0 \tag{7.169}$$

whereas we obtain the unknown parameters $D_1, D_2, \ldots, D_7$, so that the approximate solution is well-defined.

We remark that the explicit analytic expression given by Eq. 7.164 contains the parameters $C_1, C_2, \ldots, C_8$ that can be expressed under the form given by Eq. 7.167, which give the convergence region and rate of approximation.

In order to prove the efficiency and accuracy of the optimal homotopy perturbation method we consider some cases for different value of the parameters k, b and d.

Case (a) Consider $k = 1, b = 1$

From the system of algebraic Eq. 7.169 we obtain:

$$\begin{aligned} &D_1 = -1.41 \cdot 10^{-28}; D_2 = -2.44 \cdot 10^{-27}; D_3 = -1; D_4 = 2.18 \cdot 10^{-26}; \\ &D_5 = 1; D_6 = 3.52 \cdot 10^{-25}; D_7 = -8 \end{aligned} \tag{7.170}$$

Therefore, the approximate solution of $f'(\eta)$ becomes

$$\begin{aligned} &\bar{f}'(\eta) = 0.5\left(\frac{d^2}{\eta} - \eta\right) - 0.25\left(\frac{d^2}{\eta} - \eta\right)^3 + 0.375\left(\frac{d^2}{\eta} - \eta\right)^5 - 0.75\left(\frac{d^2}{\eta} - \eta\right)^7 \\ &f(1) = 0 \end{aligned} \tag{7.171}$$

In the Tables 7.2 and 7.3 are presented some comparisons between the present solution $f'(\eta)$obtained from Eq. 7.171 and the exact solution $f'(\eta)$ of Eq. 7.142 given by Eq. 7.165 for $d = 1.02$ and $d = 1.04$, respectively, for different values of η.

It can be seen that the solution obtained by the present method is nearly identical with that given by the exact solution, demonstrating a very good accuracy.

Figures 7.8 and 7.9 present a comparison between the present solution (7.171) and the numerical solution of Eq. 7.142, for $d = 1.02$ and $d = 1.04$, respectively.

Case (b) Consider $k = 1, b = 1.5$

In this case we obtain from the system of algebraic Eq. 7.169:

$$\begin{aligned} &D_1 = 8.57 \cdot 10^{-18}; D_2 = 2.21 \cdot 10^{-16}; D_3 = -1; \\ &D_4 = 8.78 \cdot 10^{-15}; D_5 = 1; D_6 = 2.62 \cdot 10^{-14}; D_7 = -12 \end{aligned} \tag{7.172}$$

Table 7.2 Comparison between the present solution (7.171) and the exact solution (7.165) for $k = b = 1, d = 1.02$

η	$\bar{f}'$given by Eq. 7.171	f'exact given by Eq. 7.165
1	0.020183555411	0.020183555411
1.005	0.015105047488	0.015105048224
1.007	0.013079437568	0.013079438114
1.01	0.010047476329	0.010047476254
1.014	0.006017315728	0.006017314467
1.018	0.002001948589	0.002001948589

Table 7.3 Comparison between the present solution (7.171) and the exact solution (7.165) for k = b = 1, d = 1.04

η	$\bar{f}'$given by Eq. 7.171	f'exact given by Eq. 7.165
1	0.040665504003	0.040665504274
1.008	0.032439661905	0.032439661553
1.016	0.024254926148	0.024254940388
1.023	0.017131195967	0.017131196596
1.03	0.010046515649	0.010046510165
1.038	0.002001910736	0.002001910736

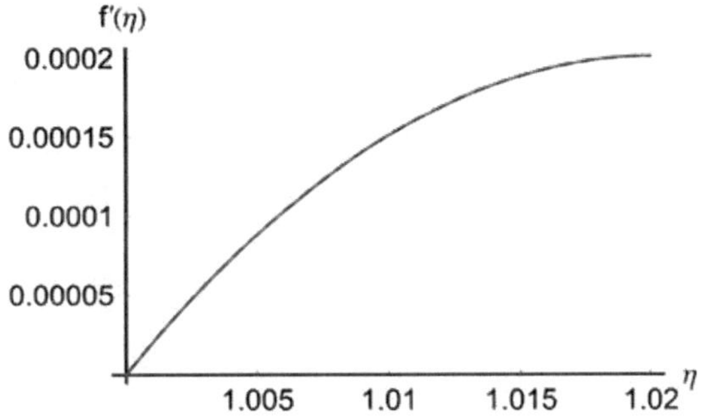

Fig. 7.8 Comparison between the present solution (7.171) and the numerical solution of Eq. 7.142 for $k = 1, b = 1, d = 1.02$

Therefore, the approximate solution of $f'(\eta)$becomes in this case

$$\bar{f}'(\eta) = 0.5\left(\frac{d^2}{\eta} - \eta\right) - 0.375003\left(\frac{d^2}{\eta} - \eta\right)^3 + $$
$$+ 0.84375\left(\frac{d^2}{\eta} - \eta\right)^5 - 2.53125\left(\frac{d^2}{\eta} - \eta\right)^7 \tag{7.173}$$
$$f(1) = 0$$

Tables 7.4 and 7.5 present a comparison between the present solution $f'(\eta)$given by Eq. 7.173, and the exact solution $f'(\eta)$ given by Eq. 7.165 for $d = 1.02$ and $d = 1.04$, respectively, for different values of η.

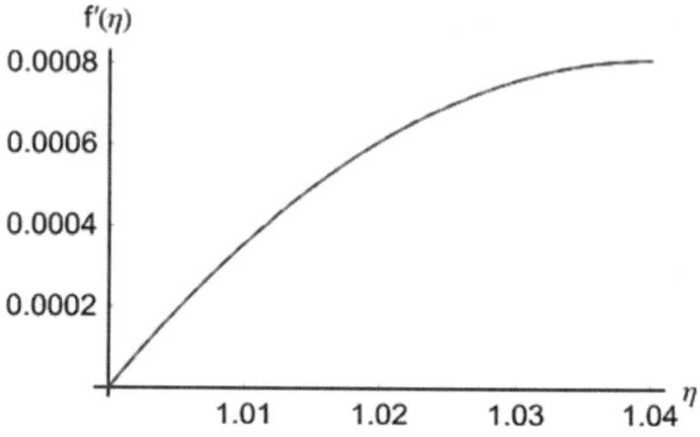

Fig. 7.9 Comparison between the present solution (7.171) and the numerical solution of Eq 7.142 for $k = 1, b = 1, d = 1.04$

Table 7.4 Comparison between the present solution (7.173) and the exact solution (7.165) for $k = 1, b = 1.5, d = 1.02$

η	$\bar{f}'$ given by Eq. 7.173	f' exact given by Eq. 7.165
1	0.020175362966	0.020175363141
1.005	0.015101608072	0.015101965160
1.007	0.013077203439	0.013077035778
1.01	0.010046462918	0.010046466732
1.014	0.006017097919	0.006017101651
1.018	0.002001940566	0.002001940795

Table 7.5 Comparison between the present solution (7.173) and the exact solution (7.165) for $k = 1, b = 1.5, d = 1.04$

η	$\bar{f}'$ given by Eq. 7.173	f' exact given by Eq. 7.165
1	0.040599239698	0.040599241011
1.008	0.032405844453	0.032406015043
1.016	0.024240732049	0.024240659597
1.023	0.017126181584	0.017126244676
1.03	0.010045502551	0.010045467225
1.038	0.002001902713	0.002001901376

It can be seen that the solution obtained by the present method is nearly identical with that given by the exact solution, demonstrating a very good accuracy.

It is easier to emphasize the accuracy of the obtained results if we graphically compare the analytical solutions obtained through OHPM with the numerical ones. Therefore, Figs. 7.10 and 7.11 present a comparison between the present solution (7.173) and the numerical solution of Eq. 7.165, for $d = 1.02$ and $d = 1.04$, respectively, in the case of $k = 1$ and $b = 1.5$.

7.6 Nonlinear Dynamics of an Electrical Machine Rotor-Bearing System

The rotating electrical machine considered in this section is modelled as a flexible rotor supported by two journal bearings with nonlinear suspension [151]. In order to simplify the mathematical modelling of such a system, one can assume that the rotor

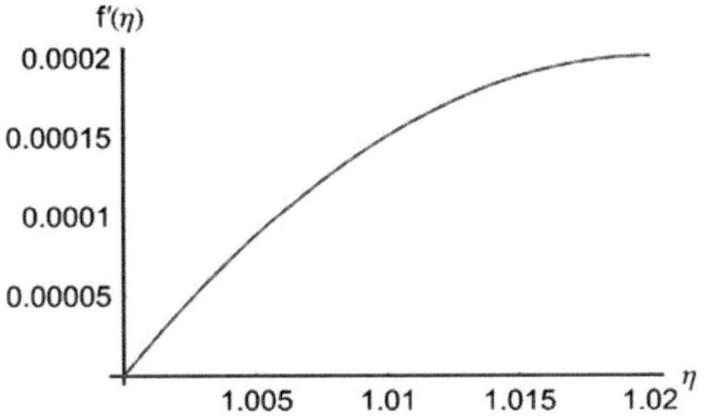

Fig. 7.10 Comparison between the present solution (7.173) and the numerical solution of Eq. 7.142 for $k = 1, b = 1.5, d = 1.02$

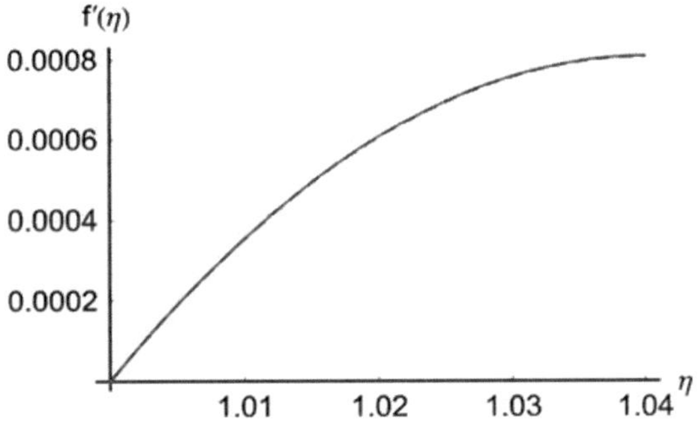

Fig. 7.11 Comparison between the present solution (7.173) and the numerical solution of Eq. 7.142 for $k = 1, b = 1.5, d = 1.04$

mass and the bearing mass are lumped at the midpoint, the rotor speed is constant, the axial and torsional vibrations and the mass of the shaft are negligible, as well as the damping. In these hypotheses the motion of the system is governed by the following non-linear system of four coupled non-linear differential equations [152]:

$$\begin{aligned}
&\ddot{u}_1 + \frac{1}{s_1^2}u_1 + \frac{\alpha}{s^2}u_1^3 - \frac{1}{2cs^2}(u_2 - u_1) = 0 \\
&\ddot{u}_2 + \frac{1}{s^2}(u_2 - u_1) = 0 \\
&\ddot{u}_3 + \frac{1}{s_1^2}u_3 + \frac{\alpha}{s^2}u_3^3 - \frac{1}{2cs^2}(u_4 - u_3) + \frac{f}{s^2} = 0 \\
&\ddot{u}_4 + \frac{1}{s^2}(u_4 - u_3) + \frac{f}{s^2} = 0
\end{aligned} \tag{7.174}$$

where u_1, u_2 and u_3, u_4 are the horizontal and vertical displacements of the bearing centre and rotor centre, respectively. The rest of the parameters are considered constants for a given working regime:

$$\begin{aligned}
&s_1^2 = c_{om}c_p s^2, \quad c_{om} = \frac{m_0}{m}, \quad c_p = \frac{k_p}{k_1}, \\
&s^2 = \frac{\omega_1}{\omega_n}, \quad \alpha = \frac{k_2 c^2}{k_p c_{om}}, \quad f = \frac{mg}{ck_p}
\end{aligned} \tag{7.175}$$

where ω_1 is the rotational speed of the shaft, k_p is the stiffness of the shaft, k_1 and k_2 are the stiffness of the linear and nonlinear springs which support the bearing housing, m and m_0 are the masses lumped at the rotor mid-point and at the bearing mid-point and c is the clearance.

Using the notations

$$B_1 = \frac{1}{s^2} + \frac{1}{2cs^2}, \quad B_2 = \frac{\alpha}{s^2}, \quad B_3 = -\frac{1}{2cs^2}, \quad B_4 = \frac{f}{s^2}, \quad \omega^2 = \frac{1}{s^2} \tag{7.176}$$

we can rewrite Eq. 7.174 in the form

$$\begin{aligned}
&\ddot{u}_1 + \Omega_1^2 u_1 + (B_1 - \Omega_1^2)u_1 + B_2 u_1^3 + B_3 u_2 = 0 \\
&\ddot{u}_2 + \Omega_2^2 u_2 + (\omega^2 - \Omega_2^2)u_2 - \omega^2 u_1 = 0 \\
&\ddot{u}_3 + \Omega_3^2 u_3 + (B_1 - \Omega_3^2)u_3 + B_2 u_3^3 + B_3 u_4 + B_4 = 0 \\
&\ddot{u}_4 + \Omega_4^2 u_4 + (\omega^2 - \Omega_4^2)u_4 - \omega^2 u_3 + B_4 = 0
\end{aligned} \tag{7.177}$$

with the initial conditions

$$u_i(0) = A_i, \quad \dot{u}_i(0) = 0, \quad i = 1, 2, 3, 4 \tag{7.178}$$

where Ω_i are the frequencies of u_i, respectively.

In this section we apply the second alternative of the OHPM, thus for the Eq. 7.177_1, the homotopy (7.112) becomes

$$\begin{aligned}
&\ddot{u}_1 + \Omega_1^2 u_1 + p\{K_{11}^*(t, C_i)[(B_1 - \Omega_1^2)u_{10} + B_1 u_{10}^3 + B_3 u_{20}] + \\
&\quad + K_{12}^*(t, C_j)(B_1 - \Omega_1^2 + 3B_2 u_{10}^2) + K_{13}^*(t, C_k)B_3\} = 0
\end{aligned} \tag{7.179}$$

where u_{10} is obtained from Eqs. 7.178 and 7.133, $u_1 = u_{10} + pu_{11}$ and

$$u_{10} = A_1 \cos \Omega_1 t \tag{7.180}$$

We choose

$$\begin{aligned}
&K_{11}^*(t, C_i) = C_1 + 2C_2 \cos 2\Omega_1 t + 2C_3 \cos 4\Omega_1 t \\
&K_{12}^*(t, C_i) = C_4 + 2C_5 \cos \Omega_1 t + 2C_6 \cos 3\Omega_2 t \\
&K_{13}^*(t, C_i) = C_7
\end{aligned} \tag{7.181}$$

where C_i, $i = 1, 2, \ldots, 7$ are unknown constants. In this case $K^*_{1i}F_x$ has the same form as F_x. We can also choose

$$\begin{aligned}
K^*_{11}(t, C_i) &= C_1 + 2C_2 \cos 2\Omega_1 t \\
K^*_{12}(t, C_i) &= C_3 + 2C_4 \cos \Omega_1 t \\
K^*_{13}(t, C_i) &= C_5 + C_6 \cos \Omega_1 t
\end{aligned} \tag{7.182}$$

and so on.

From Eqs. 7.180 and 7.181 we obtain a linear differential equation of the form:

$$\begin{aligned}
&\ddot{u}_{11} + \Omega_1^2 u_{11} + [(B_1 - \Omega_1^2)(A_1 C_1 + A_1 C_2 + 2C_5)+ \\
&\quad + \frac{3C_1 + 4C_2 + C_3}{4} A_1^3 B_2 + \frac{9C_5 + 3C_6}{2} A_1^2 B_2] \cos \Omega_1 t + M_3 \cos 3\Omega_1 t + \\
&\quad + M_5 \cos 5\Omega_1 t + M_7 \cos 7\Omega_1 t + (A_2 B_3 C_1 + B_3 C_7) \cos \Omega_2 t + \\
&\quad + A_2 B_3 C_2 [\cos(2\Omega_1 + \Omega_2)t + \cos(2\Omega_1 - \Omega_2)t] + A_2 B_3 C_3 [\cos(4\Omega_1 + \\
&\quad + \Omega_2)t + \cos(4\Omega_1 - \Omega_2)t] + (B_1 - \Omega_1^2 + \frac{3}{2} A_1^2 B_2) C_4 + \\
&\quad + \frac{3}{2} A_1^2 B_2 C_4 \cos 2\Omega_1 t = 0
\end{aligned} \tag{7.183}$$

where

$$\begin{aligned}
M_3 &= (B_1 - \Omega_1^2)(A_1 C_2 + A_1 C_3 + 2C_6) + \\
&\quad + \frac{1}{4} A_1^2 B_2 (A_1 C_1 + 3A_1 C_2 + 3A_1 C_3 + 6C_5 + 12C_6) \\
M_5 &= (B_1 - \Omega_1^2) A_1 C_3 + \frac{1}{4} B_2 A_1^2 (A_1 C_2 + 3A_1 C_3 + 6C_6) \\
M_7 &= \frac{1}{4} A_1^3 B_2 C_3
\end{aligned} \tag{7.184}$$

Avoiding the presence of secular terms in the Eq. 7.183, we obtain the frequency

$$\Omega_1^2 = B_1 + \frac{A_1^2 B_2 [(3C_1 + 4C_2 + C_3) A_1 + 18C_5 + 6C_6]}{4(A_1 C_1 + A_1 C_2 + 2C_5)} \tag{7.185}$$

From Eqs. 7.183, 7.179 and 7.180 one can obtain the first-order approximate solution for the first variable:

$$\bar{u}_1 = u_{10} + u_{11} = \left[A_1 - \frac{M_3}{8\Omega_1^2} - \frac{M_5}{24\Omega_1^2} - \frac{M_7}{48\Omega_1^2} + \frac{A_2B_3C_1 + B_3C_7}{\Omega_1^2 - \omega^2} - \right.$$
$$\left. - \frac{2(3\Omega_1^2 + \Omega_2^2)A_2B_3C_2}{9\Omega_1^4 - 10\Omega_1^2\Omega_2^2 + \Omega_2^4} - \frac{2(15\Omega_1^2 + \Omega_2^2)A_2B_3C_3}{225\Omega_1^4 - 34\Omega_1^2\Omega_2^2 + \Omega_2^4}\right] \cos\Omega_1 t +$$
$$+ \frac{M_3}{8\Omega_1^2}\cos 3\Omega_1 t + \frac{M_5}{24\Omega_1^2}\cos 5\Omega_1 t + \frac{M_7}{48\Omega_1^2}\cos 7\Omega_1 t -$$
$$- \frac{A_2B_3C_1 + B_3C_7}{\Omega_1^2 - \Omega_2^2}\cos\Omega_2 t + \frac{A_2B_3C_2}{3\Omega_1^2 + 4\Omega_1\Omega_2 + \Omega_2^2}\cos(2\Omega_1 + \Omega_2)t +$$
$$+ \frac{A_2B_3C_2}{3\Omega_1^2 - 4\Omega_1\Omega_2 + \Omega_2^2}\cos(2\Omega_1 - \Omega_2)t + \frac{A_2B_3C_3\cos(4\Omega_1 + \Omega_2)t}{15\Omega_1^2 + 8\Omega_1\Omega_2 + \Omega_2^2} +$$
$$+ \frac{A_2B_3C_3\cos(4\Omega_1 - \Omega_2)t}{15\Omega_1^2 - 8\Omega_1\Omega_2 + \Omega_2^2} - \frac{B_1 - \Omega_1^2 + \frac{3}{2}A_1^2B_2}{\Omega_1^2}(1 - \cos\Omega_1 t) +$$
$$+ \frac{A_1^2B_2C_4(\cos 2\Omega_1 t - \cos\Omega_1 t)}{2\Omega_1^2} \tag{7.186}$$

For Eq. 7.177_2, the homotopy (7.112) can be written in the form:

$$\ddot{u}_2 + \Omega_2^2 u_2 + p\{K^*_{21}(t, C_i)[(\omega^2 - \Omega_2^2)A_2\cos\Omega_2 t - \omega^2 A_1 \cos\Omega_1 t] +$$
$$+ K^*_{22}(t, C_j)(\omega^2 - \Omega_2^2) + K^*_{23}(t, C_k)(-\omega^2)\} = 0 \tag{7.187}$$

where we used the expression $u_{20}(t) = A_2\cos\Omega_2 t$ and $\bar{u}_2 = u_{20} + pu_{21}$.

If we choose

$$K^*_{21}(t, C_i) = C_8 + 2C_9\cos 2\Omega_1 t + 2C_{10}\cos 4\Omega_1 t$$
$$K^*_{22}(t, C_j) = C_{11} + C_{12}\cos\Omega_1 t \tag{7.188}$$
$$K^*_{23}(t, C_k) = C_{13} + C_{14}\cos\Omega_2 t + C_{15}\cos\omega t$$

with the constants C_j, $j = 8, 9, \ldots, 15$, then from Eqs. 7.187 and 7.188 we obtain

$$\ddot{u}_{21} + \Omega_2^2 u_{21} + [(\omega^2 - \Omega_2^2)A_2C_8 - \omega^2 C_{14}]\cos\Omega_2 t$$
$$+ (\omega^2 - \Omega_2^2)A_2C_9[\cos(2\Omega_1 + \Omega_2)t + \cos(2\Omega_1 - \Omega_2)t]$$
$$+ (\omega^2 - \Omega_2^2)A_2C_{10}[\cos(4\Omega_1 + \Omega_2)t + \cos(4\Omega_1 - \Omega_2)t] - \omega^2 A_1(C_8 +$$
$$+ C_9)\cos\Omega_1 t - \omega^2 A_1(C_9 + C_{10})\cos 3\Omega_1 t - \omega^2 A_1 C_{10}\cos 5\Omega_1 t +$$
$$+ (\omega^2 - \Omega_2^2)C_{11} - \omega^2 C_{13} + (\omega^2 - \Omega_2^2)C_{12}\cos 2\Omega_1 t - \omega^2 C_{15}\cos\omega t = 0 \tag{7.189}$$

Avoiding the presence of secular terms in Eq. 7.189, we obtain

$$\Omega_2^2 = \omega^2\left(1 - \frac{C_{14}}{A_2 C_8}\right) \tag{7.190}$$

The solution of Eq. 7.189 with the initial conditions (7.178) has the form

$$\begin{aligned}
u_{21}(t) = {} & \frac{\omega^2 C_{15}}{\Omega_2^2 - \omega^2}(\cos\omega t - \cos\Omega_2 t) + \\
& + \frac{\omega^2 A_1 (C_8 + C_9)}{\Omega_2^2 - \Omega_1^2}(\cos\Omega_1 t - \cos\Omega_2 t) + \frac{\omega^2 A_1 (C_9 + C_{10})}{\Omega_2^2 - 9\Omega_1^2} \times \\
& \times (\cos 3\Omega_1 t - \cos\Omega_2 t) + \frac{\omega^2 A_1 C_{10}}{\Omega_2^2 - 25\Omega_1^2}(\cos 5\Omega_1 t - \cos\Omega_2 t) + \\
& + \frac{A_2(\omega^2 - \Omega_2^2)C_9}{4\Omega_1(\Omega_1 + \Omega_2)}[\cos(2\Omega_1 + \Omega_2)t - \cos\Omega_2 t] + \\
& + \frac{A_2(\omega^2 - \Omega_2^2)C_9}{4\Omega_1(\Omega_1 - \Omega_2)}[\cos(2\Omega_1 - \Omega_2)t - \cos\Omega_2 t] + \\
& + \frac{(\omega^2 - \Omega_2^2)A_2 C_{10}}{8\Omega_1(2\Omega_1 + \Omega_2)}[\cos(4\Omega_1 + \Omega_2)t - \cos\Omega_2 t] + \\
& + \frac{(\omega^2 - \Omega_2^2)A_2 C_{10}}{8\Omega_1(2\Omega_1 - \Omega_2)}[\cos(4\Omega_1 - \Omega_2)t - \cos\Omega_2 t] + \\
& + \frac{\omega^2 C_{13} - (\omega^2 - \Omega_2^2)C_{11}}{\Omega_2^2}(1 - \cos\Omega_2 t) - \\
& - \frac{(\omega^2 - \Omega_2^2)C_{12}}{\Omega_2^2 - 4\Omega_1^2}(\cos 2\Omega_1 t - \cos\Omega_2 t)
\end{aligned} \tag{7.191}$$

In this way, the first-order approximate solution $\bar{u}_2 = u_{20} + u_{21}$ becomes

$$\begin{aligned}
\bar{u}_2 = {} & \left[A_2 - \frac{\omega^2 C_{15}}{\Omega_2^2 - \omega^2} - \frac{\omega^2 A_1 (C_8 + C_9)}{\Omega_2^2 - \Omega_1^2} - \frac{\omega^2 A_1 (C_9 + C_{10})}{\Omega_2^2 - 9\Omega_1^2} - \right. \\
& - \frac{\omega^2 A_1 C_{10}}{\Omega_2^2 - 25\Omega_1^2} - \frac{(\omega^2 - \Omega_2^2)A_2 C_9}{4\Omega_1(\Omega_1 + \Omega_2)} - \frac{(\omega^2 - \Omega_2^2)A_2 C_9}{4\Omega_1(\Omega_1 - \Omega_2)} - \\
& - \frac{(\omega^2 - \Omega_2^2)A_2 C_{10}}{8\Omega_1(2\Omega_1 + \Omega_2)} - \frac{(\omega^2 - \Omega_2^2)A_2 C_{10}}{8\Omega_1(2\Omega_1 - \Omega_2)} - \frac{\omega^2 C_{13} - (\omega^2 - \Omega_2^2)C_{11}}{\Omega_2^2} + \\
& \left. + \frac{(\omega^2 - \Omega_2^2)C_{12}}{\Omega_2^2 - 4\Omega_1^2}\right]\cos\Omega_2 t + \frac{\omega^2 C_{15}}{\Omega_2^2 - \omega^2}\cos\omega t + \\
& + \frac{\omega^2 A_1 (C_8 + C_9)}{\Omega_2^2 - \Omega_1^2}\cos\Omega_1 t + \frac{\omega^2 A_1 (C_9 + C_{10})}{\Omega_2^2 - 9\Omega_1^2}\cos 3\Omega_1 t +
\end{aligned}$$

$$+\frac{\omega^2 A_1 C_{10}}{\Omega_2^2-25\Omega_1^2}\cos 5\Omega_1 t+\frac{(\omega^2-\Omega_2^2)A_2C_9}{4\Omega_1(\Omega_1+\Omega_2)}\cos(2\Omega_1+\Omega_2)t+$$
$$+\frac{(\omega^2-\Omega_2^2)A_2C_9}{4\Omega_1(\Omega_1-\Omega_2)}\cos(2\Omega_1-\Omega_2)t+\frac{(\omega^2-\Omega_2^2)A_2C_{10}}{8\Omega_1(2\Omega_1+\Omega_2)}\cos(4\Omega_1+$$
$$+\Omega_2)t+\frac{(\omega^2-\Omega_2^2)A_2C_{10}}{8\Omega_1(2\Omega_1-\Omega_2)}\cos(4\Omega_1-\Omega_2)t+$$
$$+\frac{\omega^2 C_{13}-(\omega^2-\Omega_2^2)C_{11}}{\Omega_2^2}-\frac{(\omega^2-\Omega_2^2)C_{12}}{\Omega_2^2-4\Omega_1^2}\cos 2\Omega_1 t \tag{7.192}$$

Taking into account the initial conditions (7.178) and Eq. 7.113, we obtain $u_{30}=A_3\cos\Omega_3 t$. In this case Eq. 7.114 can be written for the variable u_3 in the form

$$\ddot{u}_3+\Omega_3^2u_3+p\{K_{31}^*(t,C_i)[(B_1-\Omega_3^2)x_{30}+B_2x_{30}^3+B_3x_{40}+B_4]+$$
$$+K_{32}^*(t,C_j)(B_1-\Omega_3^2+3B_2x_{30}^2)+K_{33}^*(t,C_k)B_3\}=0 \tag{7.193}$$

The auxiliary functions can be chosen in the form

$$K_{31}^*(t,C_i)=C_{16}+2C_{17}\cos 2\Omega_3 t+2C_{18}\cos 4\Omega_3 t$$
$$K_{32}^*(t,C_j)=C_{19}+2C_{20}\cos\Omega_3 t+2C_{21}\cos 3\Omega_3 t$$
$$K_{33}^*(t,C_k)=C_{22}\cos\Omega_4 t \tag{7.194}$$

The second term of the approximate solution $\bar{u}_3=u_{30}+u_{31}$is obtained from Eqs. 7.193 and 7.194:

$$\ddot{u}_{31}+\Omega_3^2u_{31}+[(B_1-\Omega_3^2)(A_3C_{16}+A_3C_{17}+2C_{20})+$$
$$+\frac{3C_{16}+4C_{17}+C_{18}}{4}A_3^2B_2+\frac{9C_{20}+3C_{21}}{2}A_3^2B_2]\cos\Omega_3 t+$$
$$+N_3\cos 3\Omega_3 t+N_5\cos 5\Omega_3 t+N_7\cos 7\Omega_3 t+(A_4B_3C_{16}+$$
$$+B_3C_{22})\cos\Omega_4 t+2B_4C_{18}\cos 4\Omega_3 t+B_4C_{16}+(B-\Omega_3^2+$$
$$+\frac{3}{2}B_2A_3^2)C_{19}+B_3A_4C_{17}[\cos(2\Omega_3+\Omega_4)t+\cos(2\Omega_3-\Omega_4)t]+$$
$$+A_4B_3C_{18}[\cos(4\Omega_3+\Omega_4)t+\cos(4\Omega_3-\Omega_4)t]+(\frac{3}{2}A_3^2B_2C_{19}+$$
$$+2B_4C_{17})\cos 2\Omega_3 t=0 \tag{7.195}$$

where

$$
\begin{aligned}
N_3 &= \frac{1}{4}A_3^2B_2C_{16} + A_3(C_{17} + C_{18})(B_1 - \Omega_3^2 + \frac{3}{4}A_3^2B_2) + \\
&+ \frac{3}{2}C_{19}A_3^2B_2 + 2C_{21}(B_1 - \Omega_3^2 + \frac{3}{2}A_3^2B_2) \\
N_5 &= \frac{1}{4}C_{17}A_3^3B_2 + A_3(B_1 - \Omega_3^2 + \frac{3}{4}A_3^2B_2)C_{18} + \frac{3}{2}A_3^2B_2C_{21} \\
N_7 &= \frac{1}{4}A_3^3B_2C_{18}
\end{aligned} \tag{7.196}
$$

Avoiding the presence of secular terms in Eq. 7.195, the frequency becomes:

$$
\Omega_3^2 = B_1 + \frac{A_3^2B_2[(3C_{16} + 4C_{17} + C_{18})A_3 + 18C_{20} + 6C_{21}]}{4(A_3C_{16} + A_3C_{17} + 2C_{20})} \tag{7.197}
$$

The first-order approximate solution $\bar{u}_3 = u_{30} + u_{31}$ is obtained from Eq. 7.195:

$$
\begin{aligned}
\bar{u}_3(t) = {} & [A_3 - \frac{N_3}{8\Omega_3^2} - \frac{N_5}{24\Omega_3^2} - \frac{N_7}{48\Omega_3^2} - \frac{C_{16}B_3A_4 + C_{21}B_3}{\Omega_4^2 - \Omega_3^2} - \frac{2B_4C_{17}}{3\Omega_3^2} - \\
& - \frac{2B_4C_{18}}{15\Omega_3^2} - \frac{B_4C_{16} + (B_1 - \Omega_3^2 + \frac{3}{2}A_3^2B_2)C_{19}}{\Omega_3^2} - \frac{A_4B_3C_{17}}{3\Omega_3^2 - 4\Omega_3\Omega_4 + \Omega_4^2} - \\
& - \frac{A_4B_3C_{18}}{15\Omega_3^2 + 8\Omega_3\Omega_4 + \Omega_4^2} - \frac{A_4B_3C_{18}}{15\Omega_3^2 - 8\Omega_3\Omega_4 + \Omega_4^2} - \frac{A_3^2B_2C_{19}}{2\Omega_3^2}] \cos\Omega_3 t + \\
& + \frac{N_3}{8\Omega_3^2}\cos 3\Omega_3 t + \frac{N_5}{24\Omega_3^2}\cos 5\Omega_3 t + \frac{N_7}{48\Omega_3^2}\cos 7\Omega_3 t + \\
& + \frac{A_4B_3C_{16} + B_3C_{22}}{\Omega_4^2 - \Omega_3^2}\cos\Omega_4 t + \frac{2B_4C_{17}}{3\Omega_3^2}\cos 2\Omega_3 t + \frac{2B_4C_{18}}{15\Omega_3^2}\cos 4\Omega_3 t + \\
& + \frac{B_4C_{16} + (B_1 - \Omega_3^2 + \frac{3}{2}A_3^2B_2)C_{19}}{\Omega_3^2} + \frac{B_3A_4C_{17}}{3\Omega_3^2 + 4\Omega_3\Omega_4 + \Omega_4^2}\cos(2\Omega_3 + \\
& + \Omega_4)t + \frac{A_4B_3C_{17}}{3\Omega_3^2 - 4\Omega_3\Omega_4 + \Omega_4^2}\cos(2\Omega_3 - \Omega_4)t + \\
& + \frac{A_4B_3C_{18}}{15\Omega_4^2 + 8\Omega_3\Omega_4 + \Omega_4^2}\cos(4\Omega_3 + \Omega_4)t + \\
& + \frac{A_4B_3C_{18}}{15\Omega_3^2 - 8\Omega_3\Omega_4 + \Omega_4^2}\cos(4\Omega_3 - \Omega_4)t + \frac{A_3^2B_2C_{19}}{2\Omega_3^2}\cos 2\Omega_3 t
\end{aligned} \tag{7.198}
$$

For Eq. 7.177$_4$ in the same manner like above, and taking into account that $\bar{u}_4 = u_{40} + u_{41}$, where $u_{40} = A_4 \cos \Omega_4 t$, we can write the homotopy

$$\begin{aligned}\ddot{u}_4 + \Omega_4^2 u_4 + p\{K_{41}^*(t, C_i)[(\omega^2 - \Omega_4^2)A_4 \cos \Omega_4 t - \omega^2 A_3 \cos \Omega_3 t + \\ + B_4] + K_{42}^*(t, C_j)(\omega^2 - \Omega_4^2) + K_{43}^*(t, C_k)(-\omega^2)\} = 0\end{aligned} \tag{7.199}$$

with

$$\begin{aligned}K_{41}^*(t, C_i) &= C_{23} + 2C_{24} \cos 2\Omega_3 t + 2C_{25} \cos 4\Omega_3 t \\ K_{42}^*(t, C_j) &= C_{26} + C_{27} \cos 2\Omega_3 t \\ K_{43}^*(t, C_k) &= C_{28} + C_{29} \cos \Omega_2 t + C_{30} \cos \omega t\end{aligned} \tag{7.200}$$

For the variable u_{41} we obtain the differential equation

$$\begin{aligned}&\ddot{u}_{41} + \Omega_4^2 u_{41} + [(\omega^2 - \Omega_4^2)A_4 C_{23} - \omega^2 C_{29}] \cos \Omega_2 t + (\omega^2 - \Omega_4^2)C_{26} - \\ &\quad - \omega^2 C_{28} + B_4 C_{23} + [(\omega^2 - \Omega_4^2)C_{27} + 2B_4 C_{24}] \cos 2\Omega_3 t + \\ &\quad + 2B_4 C_{25} \cos 4\Omega_3 t + (\omega^2 - \Omega_4^2)A_4 C_{24}[\cos(2\Omega_3 + \Omega_4)t + \cos(2\Omega_3 - \\ &\quad - \Omega_4)t] + (\omega^2 - \Omega_4^2)A_4 C_{25}[\cos(4\Omega_3 + \Omega_4)t + \cos(4\Omega_3 - \Omega_4)t] - \\ &\quad - \omega^2 A_3(C_{23} + C_{24}) \cos \Omega_3 t - \omega^2 A_3(C_{24} + \\ &\quad + C_{25}) \cos 3\Omega_1 t - \omega^2 A_3 C_{25} \cos 5\Omega_3 t - \omega^2 C_{30} \cos \omega t = 0\end{aligned} \tag{7.201}$$

In Eq. 7.201 the coefficient of $\cos \Omega_2 t$must be zero and thus:

$$\Omega_4^2 = \omega^2 \left(1 - \frac{C_{29}}{A_4 C_{23}}\right) \tag{7.202}$$

The first-order approximate solution $\bar{u}_4$ becomes

$$\begin{aligned}\bar{u}_4 = \Bigg[A_4 &- \frac{2B_4 C_{25}}{\Omega_4^2 - 16\Omega_3^2} - \frac{\omega^2 A_3(C_{23} + C_{24})}{\Omega_4^2 - \Omega_3^2} - \frac{\omega^2 A_3(C_{24} + C_{25})}{\Omega_4^2 - 9\Omega_3^2} - \\ &- \frac{\omega^2 A_3 C_{25}}{\Omega_4^2 - 25\Omega_3^2} - \frac{\omega^2 C_{30}}{\Omega_4^2 - \omega^2} + \frac{(\omega^2 - \Omega_4^2)C_{26} - \omega^2 C_{28} + B_4 C_{23}}{\Omega_4^2} + \\ &+ \frac{(\omega^2 - \Omega_4^2)C_{27} + 2B_4 C_{24}}{\Omega_4^2 - 4\Omega_3^2} + \frac{(\Omega_4^2 - \omega^2)A_4 C_{25}}{8\Omega_4(2\Omega_3 - \Omega_4)} + \frac{(\Omega_2^2 - \omega^2)A_4 C_{25}}{8\Omega_3(2\Omega_3 + \Omega_4)} +\end{aligned}$$

$$+\frac{(\Omega_4^2-\omega^2)A_4C_{24}}{4\Omega_3(\Omega_3+\Omega_4)}+\frac{(\Omega_4^2-\omega^2)A_4C_{24}}{4\Omega_3(\Omega_3-\Omega_4)}\Bigg]\cos\Omega_4 t+\frac{\omega^2C_{30}}{\Omega_4^2-\omega^2}\cos\omega t+$$
$$+\frac{\omega^2A_3(C_{23}+C_{24})}{\Omega_4^2-\Omega_3^2}\cos\Omega_3 t+\frac{\omega^2A_3(C_{24}+C_{25})}{\Omega_4^2-9\Omega_3^2}\cos 3\Omega_3 t+$$
$$+\frac{\omega^2A_3C_{25}}{\Omega_4^2-25\Omega_3^2}\cos 5\Omega_3 t+\frac{(\omega^2-\Omega_4^2)A_4C_{24}}{4\Omega_3(\Omega_3+\Omega_4)}\cos(2\Omega_3+\Omega_4)t+$$
$$+\frac{(\omega^2-\Omega_2^2)A_2C_{24}}{4\Omega_3(\Omega_3-\Omega_4)}\cos(2\Omega_3-\Omega_4)t+\frac{(\omega^2-\Omega_4^2)A_4C_{25}}{8\Omega_3(2\Omega_3+\Omega_4)}\cos(4\Omega_3+$$
$$+\Omega_4)t+\frac{(\omega^2-\Omega_4^2)A_4C_{25}}{8\Omega_3(2\Omega_3-\Omega_4)}\cos(4\Omega_3-\Omega_4)t+$$
$$+\frac{\omega^2C_{28}+(\Omega_4^2-\omega^2)C_{26}-2B_4C_{24}}{\Omega_4^2}-$$
$$-\frac{(\omega^2-\Omega_4^2)C_{27}+2B_4C_{24}}{\Omega_4^2-4\Omega_3^2}\cos 2\Omega_3 t+\frac{2B_4C_{25}}{\Omega_4^2+16\Omega_3^2}\cos 4\Omega_3 t \qquad (7.203)$$

In order to prove the efficiency and accuracy of the OHPM, we consider the specific case of a certain working regime:

$$B_1=1.99525; B_2=1.61287; B_3=-249406;$$
$$B_4=0.00318067; \omega=0.481239 \qquad (7.204)$$

In these conditions, following the procedure described above, we obtain the optimal values of the constants $C_1, C_2, \ldots, C_{30}$:

$$C_1=-2.41254; C_2=0.00684487; C_3=-0.0089363; C_4=0.0044225;$$
$$C_5=0.39103; C_6=0.125232; C_7=1.24942; C_8=3.03089040;$$
$$C_9=0.88468565; C_{10}=0.01875489; C_{11}=0.00000101;$$
$$C_{12}=-0.00326318; C_{13}=-0.00213231; C_{14}=1.43347182;$$
$$C_{15}=-2.0062441; C_{16}=-37.878; C_{17}=0.0857472; C_{18}=-0.0631161;$$
$$C_{19}=0.0515417; C_{20}=6.29335; C_{21}=0.14684; C_{22}=18.9817;$$
$$C_{23}=2.98545183; C_{24}=0.92218677; C_{25}=0.00443001;$$
$$C_{26}=0.00732412; C_{27}=-0.01226206; C_{28}=0.0292462;$$
$$C_{29}=1.41194106; C_{30}=-2.01619285 \qquad (7.205)$$

From Eqs. 7.185, 7.190, 7.197 and 7.202 we obtain the frequencies:

$$\Omega_1=0.49021; \Omega_2=0.111925; \Omega_3=0.490398; \Omega_4=0.111953 \qquad (7.206)$$

The first-order approximate solutions of the system (7.177) obtained by OHPM are:

$$\begin{aligned}\bar{u}_1(t) &= 0.0205389\cos\Omega_1 t + 0.324567\cos 3\Omega_1 t + 0.007228\cos 5\Omega_1 t- \\ &- 0.00001998\cos 7\Omega_1 t + 0.047243\cos\Omega_2 t - 0.0008957\cos(2\Omega_1 + \Omega_2)t- \\ &- 0.001666\cos(2\Omega_1 - \Omega_2)t + 0.0002747\cos(4\Omega_1 + \Omega_2)t+ \\ &+ 0.0003506\cos(4\Omega_1 - \Omega_2)t - 0.03942(1 - \cos\Omega_1 t) + 0.002374\cos 2\Omega_1 t\end{aligned} \tag{7.207}$$

$$\begin{aligned}\bar{u}_2(t) &= 2.120972632\cos\omega t - 1.59244446\cos\Omega_1 t - 0.038922185\cos 3\Omega_1 t- \\ &- 0.0002898\cos 5\Omega_1 t + 0.049944606\cos\Omega_2 t + 0.000215708\cos(2\Omega_1+ \\ &+ \Omega_2)t + 0.00073813\cos(2\Omega_1 - \Omega_2)t - 0.00001565\cos(4\Omega_1 + \Omega_2)t- \\ &- 0.000025477\cos(4\Omega_1 - \Omega_2)t - 0.03942 - 0.000753504\cos 2\Omega_1 t\end{aligned} \tag{7.208}$$

$$\begin{aligned}\bar{u}_3(t) &= 0.0608147\cos\Omega_3 t + 0.33216\cos 3\Omega_3 t + 0.00170912\cos 5\Omega_3 t- \\ &- 0.000141098\cos 7\Omega_3 t + 0.0466726\cos\Omega_4 t + 0.0284096\cos 2\Omega_3 t- \\ &- 0.000111301\cos 4\Omega_3 t - 0.0112131\cos(2\Omega_3 + \Omega_4)t- \\ &- 0.0207873\cos(2\Omega_3 - \Omega_4)t + 0.00193904\cos(4\Omega_3 + \Omega_4)t+ \\ &+ 0.00247456\cos(4\Omega_3 - \Omega_4)t - 0.0419269\end{aligned} \tag{7.209}$$

$$\begin{aligned}\bar{u}_4(t) &= 2.13155024\cos\omega t - 1.5879748\cos\Omega_3 t - 0.039890004\cos 3\Omega_3 t- \\ &- 0.0000684\cos 5\Omega_3 t + 0.0000028\cos 7\Omega_3 t + 0.0493427\cos\Omega_4 t- \\ &- 0.009008\cos 2\Omega_3 t + 0.0000071\cos 4\Omega_3 t - 0.0556602+ \\ &+ 0.0026979\cos(2\Omega_2 + \Omega_3)t + 0.0091995\cos(2\Omega_3 - \Omega_4)t- \\ &- 0.0001103\cos(4\Omega_3 + \Omega_4)t - 0.0001796\cos(4\Omega_3 - \Omega_4)t\end{aligned} \tag{7.210}$$

In order to assess the validity and accuracy of the obtained results, a comparison was performed between the approximate results and the results obtained through a fourth-order Runge–Kutta method. Figures 7.12–7.15 show this comparison for the independent variables $u_1, .., u_4$. It can be seen that the approximate results are nearly identical with the numerical solutions.

In this section a new analytical technique, called the optimal homotopy perturbation method is applied to nonlinear dynamic systems. The proposed procedure starts from the basics of He's homotopy perturbation method, but the construction of the new homotopy is different especially referring to the auxiliary functions $K_{ij}(t, C_k)$ and to the presence of some parameters C_1, $C_2 \ldots$, which ensure a rapid convergence of the solution when they are optimally determined.

A very good agreement was found between the analytical and numerical results. It is evident that overall results come very close to the exact solution even using

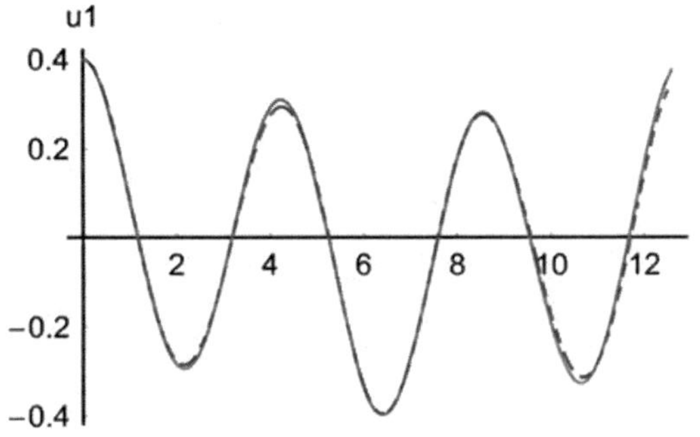

Fig. 7.12 Displacement u_1 obtained for Eq. 7.174: _____ numerical results; _ _ _ _ analytical results (7.207)

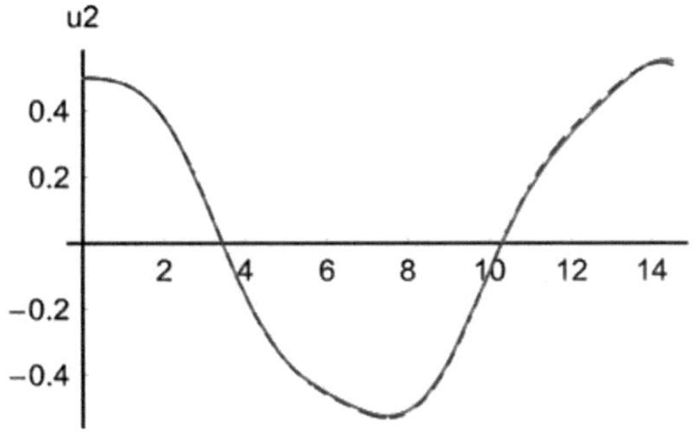

Fig. 7.13 Displacement u_2 obtained for Eq. 7.174: ________ numerical results; _ _ _ _ _ analytical results (7.208)

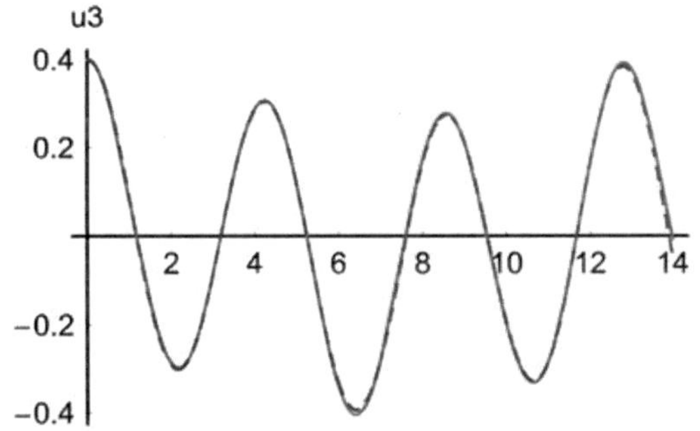

Fig. 7.14 Displacement u_3 obtained for Eq. 7.174: ________ numerical results; _ _ _ _ _ analytical results (7.209)

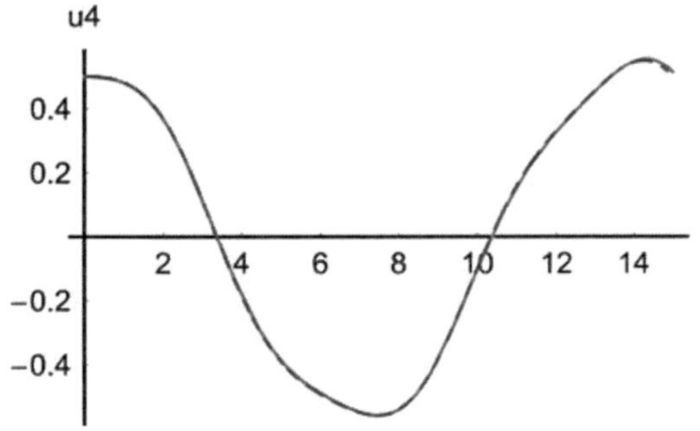

Fig. 7.15 Displacement u_4 obtained for Eq. 7.174: ________ numerical results; _ _ _ _ _ analytical results (7.210)

only two terms of the iteration formula. The proposed procedure is valid even if the nonlinear equations do not contain small or large parameters.

7.7 A Non-conservative Oscillatory System of a Rotating Electrical Machine

Rotating electrical machines are complex engineering systems which combine electrical and mechanical concepts. From mechanical point of view, these systems inherently involve the presence of a rotor, which often raises some dynamical problems, typical for rotor dynamics. The most common dynamic problem encountered by these systems is generated by unbalanced forces and sometimes even a small amount of unbalance can cause vibration that can reach undesirably high values, very detrimental for a properly work of the mechanical system. It is known that no absolutely perfect balancing can be realized and therefore there will always be a residual unbalance. Other dynamical problems could be also generated by bad or nonlinear bearings, mechanical looseness, misalignment and even some electrical problems.

From engineering point of view it is very important to make available some reliable tools intended to predict, analyze and correct these problems in order to obtain higher speed machines, to prevent unexpected failures or to assure longer periods between downtimes. Obviously, when an improvement of dynamic characteristics of these systems is needed, analytical developments and numerical simulations are preferable to experimental investigations, because of lower costs.

In what follows, we consider that the rotating electrical machine is modeled as a non-conservative oscillatory system subject to a parametric excitation caused by an axial thrust and a forcing excitation caused by an unbalanced force of the rotor. This system can be described by the second-order differential equation with variable coefficients:

$$m\ddot{u} + c\dot{u} + k(1 - \lambda\sin(\omega_2 t))u = f\sin(\omega_1 t) \tag{7.211}$$

subject to the initial conditions

$$u(0) = A, \quad \dot{u}(0) = 0 \tag{7.212}$$

where m is the lumped mass of the rotor, c is the linear damping coefficient, k is the shaft stiffness, ω_1 and ω_2 are the forcing and the parametric frequencies, respectively, and f and $k\lambda$ are the amplitudes of forcing and parametric excitations, respectively and A is a known parameter.

Equation 7.211 can be written in the dimensionless form

$$\ddot{u} + 2\mu\dot{u} + \omega^2 u - \alpha u\sin(\omega_2 t) - \beta\sin(\omega_1 t) = 0 \tag{7.213}$$

where

$$\mu = \frac{c}{2m}, \quad \omega^2 = \frac{k}{m}, \quad \alpha = \frac{k\lambda}{m}, \quad \beta = \frac{f}{m} \tag{7.214}$$

In accordance with Eq. 7.213, the linear operator is chosen as

$$L(u) = \ddot{u} + 2\mu\dot{u} + \omega^2 u \tag{7.215}$$

and the operator with variable coefficients as:

$$V(u) = -\alpha u \sin\omega_2 t - \beta \sin\omega_1 t \tag{7.216}$$

We apply the second alternative of the OHPM. The initial approximation $v_0(t)$is obtained from Eq. 7.113:

$$\ddot{v}_0 + 2\mu\dot{v}_0 + \omega^2 v_0 = 0 \tag{7.217}$$

with initial conditions

$$v_0(0) = A, \quad \dot{v}_0(0) = 0 \tag{7.218}$$

The solution of Eqs. 7.217 and 7.218 is

$$v_0(t) = \left(A\cos\Omega t + \frac{\mu A}{\Omega}\sin\Omega t\right)\exp(-\mu t) \tag{7.219}$$

where $\Omega = \sqrt{\omega^2 - \mu^2}$.

Equation 7.114 becomes

$$\begin{aligned} &\ddot{v}_1 + 2\mu\dot{v}_1 + \omega^2 v_1 + K_{11}(t, C_k)(-\alpha v_0 \sin\omega_2 t - \beta\sin\omega_1 t) + \\ &+ K_{12}(t, C_k)(-\alpha\sin\omega_2 t) = 0, \\ &v_1(0) = 0, \quad \dot{v}_1(0) = 0 \end{aligned} \tag{7.220}$$

We choose the functions $K_{11}(t, C_k)$ and $K_{12}(t, C_k)$ in the form

$$K_{11}(t, C_k) = C_1\exp(-\mu t) \tag{7.221}$$

$$K_{12}(t, C_k) = (C_2 + 2C_3\cos\Omega t + 2C_4\sin\Omega t)\exp(-\mu t) \tag{7.222}$$

Alternatively, we can also choose other forms of these functions, such as for example

$$K_{11}(t, C_k) = (C_1 + 2C_2\cos\Omega t + 2C_3\sin\Omega t)\exp(-\mu t) \tag{7.223}$$

$$K_{12}(t, C_k) = (C_4 + 2C_5 \cos \Omega t + 2C_6 \sin \Omega t) \exp(-\mu t) \tag{7.224}$$

or

$$K_{11}(t, C_k) = (C_1 + 2C_2 \cos \Omega t) \exp(-\mu t) \tag{7.225}$$

$$K_{12}(t,C_k) = (C_3 + 2C_4 \sin \Omega t + 2C_5 \cos \Omega t + +2C_6 \cos 3\Omega t + 2C_7 \sin 3\Omega t) \times \exp(-\mu t) \tag{7.226}$$

and so on.

On substituting Eqs. 7.219, 7.221 and 7.222 into Eq. 7.220, we obtain the following equation

$$\begin{aligned}
&\ddot{v}_1 + 2\mu \dot{v}_1 + \omega^2 v_1 + [-\beta C_1 \sin \omega_1 t - \alpha C_2 \sin \omega_2 t - \alpha C_3 \sin(\Omega + \omega_2)t + \\
&\quad + \alpha C_3 \sin(\Omega - \omega_2)t - \alpha C_4 \cos(\Omega - \omega_2)t + \alpha C_4 \cos(\Omega + \omega_2)t] \exp(-\mu t) + \\
&\quad + \frac{1}{2}\alpha C_1 [A \sin(\Omega - \omega_2)t - A \sin(\Omega + \omega_2)t + \frac{\mu A}{\Omega} \cos(\Omega + \omega_2)t - \\
&\quad - \frac{\mu A}{\Omega} \cos(\Omega - \omega_2)t] \exp(-2\mu t) = 0, \qquad v_1(0) = \dot{v}_1(0) = 0
\end{aligned} \tag{7.227}$$

The first-order approximation $v_1(t)$is obtained from Eq. 7.227 in the form

$$\begin{aligned}
v_1(t) = &\left[a \cos \Omega t + b \sin \Omega t + \frac{\beta C_1}{\Omega^2 - \omega_1^2} \sin \omega_1 t + \frac{\alpha C_2}{\Omega^2 - \omega_2^2} \sin \omega_2 t - \right. \\
&- \frac{\alpha C_3}{\omega_2(2\Omega + \omega_2)} \sin(\Omega + \omega_2)t - \frac{\alpha C_3}{\omega_2(2\Omega - \omega_2)} \sin(\Omega - \omega_2)t + \\
&\left. + \frac{\alpha C_4}{\omega_2(2\Omega - \omega_2)} \cos(\Omega - \omega_2)t + \frac{\alpha C_4}{\omega_2(2\Omega + \omega_2)} \cos(\Omega + \omega_2)t \right] \exp(-\mu t) + \\
&+ \frac{1}{2}\alpha C_1 \left\{ \frac{\frac{\mu A}{\Omega}[2\Omega(\Omega + \omega_2) - \omega^2 + (\Omega + \omega_2)^2] \cos(\Omega + \omega_2)t}{[\omega^2 - (\Omega + \omega_2)^2]^2 + 4\mu^2(\Omega + \omega_2)^2} + \right. \\
&+ \frac{A[\omega^2 - (\Omega + \omega_2)^2 + \frac{2\mu^2}{\Omega}(\Omega + \omega_2)] \sin(\Omega + \omega_2)t}{[\omega^2 - (\Omega + \omega_2)^2]^2 + 4\mu^2(\Omega + \omega_2)^2} + \\
&+ \frac{\frac{\mu A}{\Omega}[\omega^2 - 2\Omega(\Omega - \omega_2) - (\Omega - \omega_2)^2] \cos(\Omega - \omega_2)t}{[\omega^2 - (\Omega - \omega_2)^2]^2 + 4\mu^2(\Omega - \omega_2)^2} - \\
&\left. - \frac{A[\omega^2 - (\Omega - \omega_2)^2 + \frac{2\mu^2}{\Omega}(\Omega - \omega_2)] \sin(\Omega - \omega_2)t}{[\omega^2 - (\Omega - \omega_2)^2]^2 + 4\mu^2(\Omega - \omega_2)^2} \right\} \exp(-2\mu t)
\end{aligned} \tag{7.228}$$

where

$$a = -\frac{4\alpha\Omega C_4}{\omega_2(4\Omega^2 - \omega_2^2)} - \frac{\alpha\mu A C_1}{2\Omega}\left\{\frac{2\Omega(\Omega+\omega_2) - \omega^2 + (\Omega+\omega_2)^2}{\left[\omega^2 - (\Omega+\omega_2)^2\right]^2 + 4\mu^2(\Omega+\omega_2)^2} - \frac{2\Omega(\Omega-\omega_2) - \omega^2 + (\Omega-\omega_2)^2}{\left[\omega^2 - (\Omega-\omega_2)^2\right]^2 + 4\mu^2(\Omega-\omega_2)^2}\right\} \tag{7.229}$$

$$\begin{aligned} b = {} & \frac{\mu a}{\Omega} - \frac{\beta\omega_1 C_1}{\Omega(\Omega^2 - \omega_1^2)} - \frac{\alpha\omega_2 C_2}{\Omega(\Omega^2 - \omega_2^2)} + \\ & + \frac{2\alpha(2\Omega^2 - \omega_2^2)C_3}{\Omega\omega_2(4\Omega^2 - \omega_2^2)} + \frac{4\mu\alpha C_4}{\omega_2(4\Omega^2 - \omega_2^2)} + \\ & + \frac{\alpha A C_1}{2\Omega}\left\{\frac{(\Omega-\omega_2)\left[\omega^2 - (\Omega-\omega_2)^2\right] - 4\mu^2(\Omega-\omega_2) + \frac{2\mu^2\omega^2}{\Omega}}{\left[\omega^2 - (\Omega-\omega_2)^2\right]^2 + 4\mu^2(\Omega-\omega_2)^2} - \right. \\ & \left. - \frac{(\Omega+\omega_2)\left[\omega^2 - (\Omega+\omega_2)^2\right] - 4\mu^2(\Omega+\omega_2) + \frac{2\mu^2\omega^2}{\Omega}}{\left[\omega^2 - (\Omega+\omega_2)^2\right]^2 + 4\mu^2(\Omega+\omega_2)^2}\right\} \end{aligned} \tag{7.230}$$

The first-order approximate solution of Eq. 7.213 is obtained from Eqs. 7.117, 7.219 and 7.228:

$$\begin{aligned} \bar{u}(t) = {} & \left[(A+a)\cos\Omega t + \left(\frac{\mu A}{\Omega} + b\right)\sin\Omega t + \frac{\beta C_1}{\Omega^2 - \omega_1^2}\sin\omega_1 t + \right. \\ & + \frac{\alpha C_2}{\Omega^2 - \omega_2^2}\sin\omega_2 t - \frac{\alpha C_3}{\omega_2(2\Omega + \omega_2)}\sin(\Omega+\omega_2)t - \\ & - \frac{\alpha C_3}{\omega_2(2\Omega - \omega_2)}\sin(\Omega-\omega_2)t \frac{\alpha C_4}{\omega_2(2\Omega - \omega_2)}\cos(\Omega-\omega_2)t + \\ & \left. + \frac{\alpha C_4}{\omega_2(2\Omega + \omega_2)}\cos(\Omega+\omega_2)t\right]\exp(-\mu t) + \\ & + \frac{1}{2}\alpha C_1\left\{\frac{\frac{\mu A}{\Omega}\left[2\Omega(\Omega+\omega_2) - \omega^2 + (\Omega+\omega_2)^2\right]\cos(\Omega+\omega_2)t}{\left[\omega^2 - (\Omega+\omega_2)^2\right]^2 + 4\mu^2(\Omega+\omega_2)^2} + \right. \\ & + \frac{A\left[\omega^2 - (\Omega+\omega_2)^2 + \frac{2\mu^2}{\Omega}(\Omega+\omega_2)\right]\sin(\Omega+\omega_2)t}{\left[\omega^2 - (\Omega+\omega_2)^2\right]^2 + 4\mu^2(\Omega+\omega_2)^2} + \\ & + \frac{\frac{\mu A}{\Omega}\left[\omega^2 - 2\Omega(\Omega-\omega_2) - (\Omega-\omega_2)^2\right]\cos(\Omega-\omega_2)t}{\left[\omega^2 - (\Omega-\omega_2)^2\right]^2 + 4\mu^2(\Omega-\omega_2)^2} - \\ & \left. - \frac{A\left[\omega^2 - (\Omega-\omega_2)^2 + \frac{2\mu^2}{\Omega}(\Omega-\omega_2)\right]\sin(\Omega-\omega_2)t}{\left[\omega^2 - (\Omega-\omega_2)^2\right]^2 + 4\mu^2(\Omega-\omega_2)^2}\right\}\exp(-2\mu t) \end{aligned} \tag{7.231}$$

In order to prove the efficiency and accuracy of the OHPM, we have considered the special case of a certain working regime characterized by

$$\begin{aligned} &m = 2, \quad c = 0.8, \quad k = 100, \quad \lambda = 0.03, \\ &\omega_1 = 1, \quad \omega_2 = 3, \quad f = 2, \quad \Omega = 7.068238819 \end{aligned} \tag{7.232}$$

In these conditions, following the procedure described above, we obtain the optimal values of the convergence-control constants:

$$\begin{aligned} &C_1 = 1.76194, \quad C_2 = -0.249164, \quad C_3 = -0.335675, \\ &C_4 = 0.0316235 \end{aligned} \tag{7.233}$$

From Eqs. 7.231, 7.232 and 7.233 we can determine the first-order approximate solution of Eqs. 7.211 and 7.212 in the form

$$\begin{aligned} \bar{u}(t) = &[0.995731795\cos\Omega t + 0.063560129\sin\Omega t + 0.035987335\sin t - \\ &- 0.009124658\sin 3t + 0.009794165\sin(\Omega+3)t + 0.015070967\sin(\Omega-3)t + \\ &+ 0.001419816\cos(\Omega-3)t + 0.000922695\cos(\Omega+3)t]\exp(-0.2t) + \\ &+ [0.002727893\cos(\Omega+3)t - 0.025510674\sin(\Omega+3)t - \\ &- 0.0008022\cos(\Omega-3)t - 0.039467024\sin(\Omega-3)t]\exp(-0.4t) \end{aligned} \tag{7.234}$$

Figure 7.16 shows the comparison between the numerical solution obtained using a fourth-order Runge–Kutta method and the approximate solution given by Eq. 7.234.

One can observe that the displacement curve obtained for this non-conservative oscillatory system through OHPM is quasi-identical to that obtained via numerical simulation, which proves the validity of the proposed procedure for non-conservative oscillators.

In this section, our construction of the new homotopy is different from that employed in traditional HPM, especially referring to the auxiliary functions

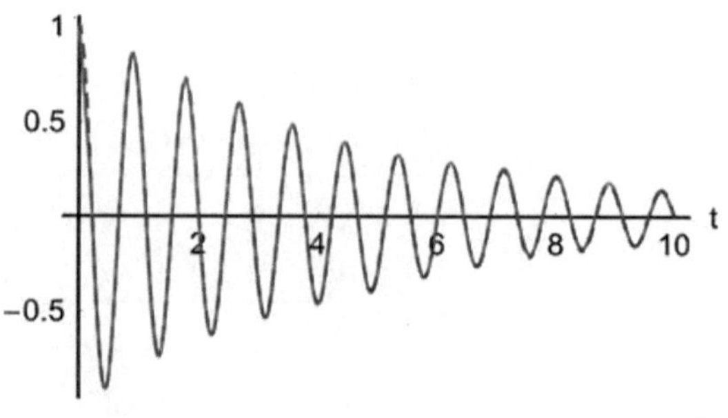

Fig. 7.16 Comparison between the present solution (7.234) and the numerical solution of Eqs. 7.211 and 7.212 for $m = 2$, $c = 0.8$, $k = 100$, $\lambda = 3$, $\omega_1 = 1$, $\omega_2 = 3, f = 2$

$K_{11}(t, C_k)$and $K_{12}(t, C_k)$and also to the involvement of some convergence-control constants C_1, C_2,... These constants ensure a rapid convergence of the solution when they are optimally determined.

Very good agreement was found between the analytical and numerical results. The proposed procedure is valid even if the nonlinear differential equations or the differential equations with variable coefficients do not contain small or large parameters.

The main feature of OHPM is that it provides a simple and rigorous way to control and adjust the convergence of a solution through several parameters C_{i} which are optimally determined. A main strength of the OHPM is its fast convergence since after only two iterations the solutions converge to the exact ones, which proves that this method is very efficient in practice.

Chapter 8
The Optimal Variational Iteration Method

8.1 The Variational Iteration Method and Applications

The variational iteration method was proposed by J. H. He in 1999 [153–155]. The method introduces a reliable and efficient process for a wide variety of scientific and engineering applications, linear and nonlinear, homogeneous or inhomogeneous, equations and systems of equations as well. The variational iteration method has no specific requirements, such as linearization, small variations, etc. for nonlinear operators. The power of the method gives it a wider applicability in handling a huge number of analytical and numerical applications.

To illustrate the basic concept and idea of the variational iteration method, we consider the following general nonlinear differential equation

$$Lu + Nu = g(t) \tag{8.1}$$

where L is a linear operator, N a nonlinear operator and $g(t)$ is inhomogeneous forcing term. According to the variational iteration method [153], we can construct a correction functional as follows

$$u_{n+1}(t) = u_n(t) + \int_0^t \lambda(s,t)[Lu_n(s) + N\tilde{u}_n(s) - g(s)]ds \tag{8.2}$$

where λ is a Lagrange multiplier, which can be identified optimally via the variational iteration method. The subscript n denotes the n-th approximation, $\tilde{u}_n$is considered as a restricted variation, i.e. $\delta\tilde{u}_n = 0$.

For the sake of simplicity, we consider only two cases of differential equations. In the first case, for the following example

$$\ddot{u} + \omega^2 u = f(t) \tag{8.3}$$

V. Marinca and N. Herisanu, *Nonlinear Dynamical Systems in Engineering*,
DOI 10.1007/978-3-642-22735-6_8,

its correction functional can be written down as follows

$$u_{n+1}(t) = u_n(t) + \int_0^t \lambda(s,t)[u_n''(s) + \omega^2 u_n(s) - f(s)]ds \tag{8.4}$$

Making the above correction functional stationary and noting that $\delta u(0) = 0$

$$\begin{aligned}\delta u_{n+1}(t) &= \delta u_n(t) + \delta \int_0^t \lambda(s,t)[u''_n(s) + \omega^2 u_n(s) - f(s)]ds = \\ &= \delta u_n(t) + \lambda(s,t)\delta u'_n(s)|_{s=t} - \\ &- \frac{\partial \lambda(s,t)}{\partial s}\delta u_n(s)|_{s=t} + \int_0^t [\frac{\partial^2 \lambda(s,t)}{\partial s^2} + \omega^2 \lambda(s,t)]ds = 0\end{aligned}$$

yields the following stationary conditions

$$\begin{aligned}\delta u_n :&\quad \frac{\partial^2 \lambda(s,t)}{\partial s^2} + \omega^2 \lambda(s,t) = 0 \\ \delta u'_n :&\quad \lambda(s,t)|_{s=t} = 0 \\ \delta u_n :&\quad 1 - \left.\frac{\partial \lambda(s,t)}{\partial s}\right|_{s=t} = 0\end{aligned} \tag{8.5}$$

The Lagrange multiplier, therefore, can be readily identified

$$\lambda(s,t) = \frac{1}{\omega} \sin \omega(s-t) \tag{8.6}$$

As a result, we obtain the following iteration formula

$$u_{n+1}(t) = u_n(t) + \frac{1}{\omega}\int_0^t \sin \omega(s-t)[u''_n(s) + \omega^2 u_n(s) - f(s)]ds \tag{8.7}$$

We start with an initial approximation $u_0(t)$given by Eq. 8.3 and using the above iteration formula. After identifying the Lagrange multiplier λ in Eq. 8.2, one can show [154] that we can construct the iteration formula

$$u_{n+1}(t) = u_0(t) + \int_0^t \lambda(s,t)Nu_n(s)ds \tag{8.8}$$

or

$$u_{n+2}(t) = u_0(t) + \int_0^t \lambda(s,t)Nu_{n+1}(s)ds \tag{8.9}$$

Now, subtracting Eq. 8.9 from Eq. 8.8, we obtain the following variational iteration

$$u_{n+2}(t) = u_{n+1}(t) + \int_0^t \lambda(s,t)[Nu_{n+1}(s) - Nu_n(s)]ds \quad (8.10)$$

J.H. He [155] named Eq. 8.2 Variational Iteration Algorithm – I, Eq. 8.8 is named Variational Iteration Algorithm – II and Eq. 8.10 is named Variational Iteration Algorithm – III.

On the other hand, by taking into consideration the identity

$$\int_0^t \sin\omega(s-t)[u_n''(s) + \omega^2 u_n(s)]ds = -\omega u_n(t) + \omega u_n(0)\cos\omega t + \dot{u}_n(0)\sin\omega t$$

it follows that

$$u_{n+1}(t) = u_n(0)\cos\omega t + \frac{1}{\omega}\dot{u}_n(0)\sin\omega t + \frac{1}{\omega}\int_0^t \sin\omega(s-t)f(s)ds \quad (8.11)$$

In the second case, we consider the following nonlinear system of differential equations

$$\dot{u}_i(t) = f_i(u_j) + q_i(t), \quad i,j = 1,2,...,n \quad (8.12)$$

where $q_i(t), \quad i = 1,2,...,n$are inhomogeneous forcing terms.

The correction functional for the nonlinear system (8.12) can be expressed as follows

$$u_1^{(m+1)}(t) = u_1^{(m)}(t) + \int_0^t \lambda_1(s,t)[u_1'(m)(s) -$$
$$- f_1(\tilde{u}_1^{(m)}(s), \tilde{u}_2^{(m)}(s), \ldots, \tilde{u}_n^{(m)}(s)) - q_1(s)]ds$$
$$u_2^{(m+1)}(t) = u_2^{(m)}(t) + \int_0^t \lambda_2(s,t)[u_2'(m)(s) -$$
$$- f_2(\tilde{u}_1^{(m)}(s), \tilde{u}_2^{(m)}(s), \ldots, \tilde{u}_n^{(m)}(s)) - q_2(s)]ds$$
$$\cdots\cdots\cdots\cdots\cdots\cdots$$
$$u_n^{(m+1)}(t) = u_n^{(m)}(t) + \int_0^t \lambda_n(s,t)[u_n'(m)(s) -$$
$$- f_n(\tilde{u}_1^{(m)}(s), \tilde{u}_2^{(m)}(s), \ldots, \tilde{u}_n^{(m)}(s)) - q_n(s)]ds \quad (8.13)$$

where $\lambda_i(s,t)$, $i = 1,2,...,n$ are Lagrange multipliers, $\tilde{u}_1, \tilde{u}_2, ..., \tilde{u}_n$ denote the restricted variations.

Making the above functional stationary, we obtain the following conditions

$$\left.\frac{\partial \lambda_1(s,t)}{\partial s}\right|_{s=t} = 0, \quad \left.\frac{\partial \lambda_2(s,t)}{\partial s}\right|_{s=t} = 0, ..., \left.\frac{\partial \lambda_n(s,t)}{\partial s}\right|_{s=t} = 0$$
$$1 + \lambda_1(s,t)|_{s=t} = 0, \quad 1 + \lambda_2(s,t)|_{s=t} = 0, \quad , 1 + \lambda_n(s,t)|_{s=t} = 0 \tag{8.14}$$

The Lagrange multipliers can be easily identified as

$$\lambda_1(s,t) = \lambda_2(s,t) = ... = \lambda_n(s,t) = -1 \tag{8.15}$$

Substituting Eq. 8.15 into functional (8.13), we obtain the following iteration formulas:

$$u_1^{(m+1)}(t) = u_1^{(m)}(t) - \int_0^t [u_1'(m)(s) - f_1(u_1^{(m)}(s), u_2^{(m)}(s), ..., u_n^{(m)}(s)) - q_1(s)]ds$$

$$u_2^{(m+1)}(t) = u_2^{(m)}(t) - \int_0^t [u_2'(m)(s) - f_2(u_1^{(m)}(s), u_2^{(m)}(s), ..., u_n^{(m)}(s)) - q_2(s)]ds$$

....

$$u_n^{(m+1)}(t) = u_n^{(m)}(t) - \int_0^t [u_n'(m)(s) - f_n(u_1^{(m)}(s), u_2^{(m)}(s), ..., u_n^{(m)}(s)) - q_n(s)]ds$$
$$\tag{8.16}$$

It can easily show that Eq. 8.16 are equivalent to equations

$$u_1^{(m+1)}(t) = u_1^{(0)}(t) + \int_0^t f_1(u_1^{(m)}(s), u_2^{(m)}(s), ..., u_n^{(m)}(s)) + q_1(s)]ds$$

$$u_2^{(m+1)}(t) = u_2^{(0)}(t) + \int_0^t f_2(u_1^{(m)}(s), u_2^{(m)}(s), ..., u_n^{(m)}(s)) + q_2(s)]ds$$

....

$$u_n^{(m+1)}(t) = u_n^{(0)}(t) + \int_0^t f_n(u_1^{(m)}(s), u_2^{(m)}(s), ..., u_n^{(m)}(s)) + q_n(s)]ds \tag{8.17}$$

where $u_1^{(0)}(t), u_2^{(0)}(t), ..., u_n^{(0)}(t)$are initial approximations with or without unknown parameters which are determined from initial conditions. The number of unknown parameter is the same with the number of initial or boundary conditions.

In [154, 155] are given the variational iteration algorithms for some frequently used differential equations. We present only the Variational Iteration Algorithm – I as follows:

$$\dot{u} + f(t, u, \dot{u}) = 0, \quad u_{n+1}(t) = u_n(t) - \int_0^t [u'_n(s) + f(s, u_n(s), u'_n(s))]ds \tag{8.18}$$

$$\dot{u} + \alpha u + f(t, u, \dot{u}) = 0$$

$$u_{n+1}(t) = u_n(t) - \int_0^t e^{\alpha(s-t)}[u'_n(s) + \alpha u_n(s) + f(s, u_n(s), u'_n(s))]ds \tag{8.19}$$

$$\ddot{u} + f(t, u, \dot{u}, \ddot{u}) = 0$$

$$u_{n+1}(t) = u_n(t) + \int_0^t (s-t)[u''_n(s) + f(s, u_n(s), u'_n(s), u''(s))]ds \tag{8.20}$$

$$\ddot{u} - \alpha^2 u + f(t, u, \dot{u}, \ddot{u}) = 0$$

$$u_{n+1}(t) = u_n(t) + \int_0^t \frac{1}{\alpha} sh\alpha(s-t)[u''_n(s) - \alpha^2 u_n(s) + f(s, u(s), u'(s), u''(s))]ds \tag{8.21}$$

$$\dddot{u} + f(t, u, \dot{u}, \ddot{u}, u) = 0$$

$$u_{n+1}(t) = u_n(t) - \frac{1}{2}\int_0^t (s-t)^2[u'''_n(s) + f(s, u(s), u'(s), u''(s), u'''(s))]ds \tag{8.22}$$

$$u^{(k)}(t) + f(t, u, \dot{u}, ..., u^{(k)}) = 0$$

$$u_{n+1}(t) = u_n(t) + \frac{(-1)^k}{(k-1)!}\int_0^t (s-t)^{k-1}[u_n^{(k)}(s) + f(s, u(s), u'(s), ..., u^{(k)}(s))]ds \tag{8.23}$$

The variational iteration method has successfully been applied to many situations. For example, Soliman [156] solved KdV-Burger's and Lax's seventh-order KdV equations, Abulwafa et al. [157] solved nonlinear coagulation problem with mass loss, Slota [158] solved direct and inverse one-phase Stefan problem, Momani et al. [159] solved the Helmholtz equation, Bildick et al. [160] solved different types of nonlinear partial differential equations, Drăgănescu et al. [161] solved nonlinear relaxation phenomena in polycrystalline solids, and so on.

In the following, we consider some applications of the variational iteration method.

8.1.1 Nonlinear Oscillator with Quadratic and Cubic Nonlinearities

We consider the free oscillation of a nonlinear oscillator with quadratic and cubic nonlinearities [162]:

$$\ddot{x} + \omega^2 x + ax^2 + bx^3 = 0, x(0) = A, \dot{x}(0) = 0 \tag{8.24}$$

where a and b are constants.

By introducing the dependent variable change

$$x = u - \frac{a}{3b} \tag{8.25}$$

the quadratic non-linearity in Eq. 8.24 is eliminated to obtain:

$$\ddot{u} + \left(\omega^2 - \frac{a^2}{3b}\right)u + bu^3 - k = 0, \quad k = \frac{\omega^2 a}{3b} - \frac{2a^3}{27b^2} \tag{8.26}$$

Equation 8.26 can be written in the form

$$\ddot{u} + \Omega^2 u - F(u) = 0 \tag{8.27}$$

where

$$F(u) = \lambda u - bu^3 + k, \quad \lambda = \Omega^2 - \omega^2 + \frac{a^2}{3b} \tag{8.28}$$

The iteration formula (8.11) for Eq. 8.27 becomes for $n = 0$:

$$u_1(t) = B\cos\Omega t + \frac{1}{\Omega}\int_0^1 \sin\Omega(t-\tau)F(u_0(\tau))d\tau \tag{8.29}$$

with the input function as (from $\dot{u}_0(0) = 0$)

$$u_0(t) = B\cos\Omega t \tag{8.30}$$

where B is obtained from (8.24_2) and (8.25):

$$B = A + \frac{a}{3b} \tag{8.31}$$

Expanding $F(u_0)$ in a Fourier series, we have:

$$F(u_0) = B\left(\lambda - \frac{3}{4}bB^2\right)\cos\Omega t - \frac{1}{4}bB^3\cos 3\Omega t + k \tag{8.32}$$

In order to ensure that no secular terms appear in the next iteration, resonance must be avoided. To do so, the coefficient of $\cos\Omega t$ into Eq. 8.32 must be zero, i.e.:

$$\lambda_1 = \frac{3}{4}bB^2 \tag{8.33}$$

such that Ω^2 becomes from (8.28_2):

$$\Omega^2 = \omega^2 + \frac{3}{4}bB^2 - \frac{a^2}{3b} \tag{8.34}$$

From Eqs. 8.29, 8.30 and 8.32 we obtain:

$$u_1(t) = \left(B - \frac{bB^3}{32\Omega^2} - \frac{k}{\Omega^2}\right)\cos\Omega t + \frac{k}{\Omega^2} + \frac{bB^3}{32\Omega^2}\cos 3\Omega t \tag{8.35}$$

and therefore:

$$F(u_1) = \left(B - \frac{bB^3}{32\Omega^2} - \frac{k}{\Omega^2}\right)\left[\lambda - \frac{3}{4}b\left(B - \frac{bB^3}{32\Omega^2} - \frac{k}{\Omega^2}\right)^2 - \frac{3bk^2}{\Omega^4}\right.$$
$$\left. - \frac{3b^2B^3}{128\Omega^2}\left(B - \frac{bB^3}{32\Omega^2} - \frac{k}{\Omega^2}\right) - \frac{3b^3B^6}{2048\Omega^4}\right]\cos\Omega t + k + \frac{\lambda k}{\Omega^2} - \frac{bk^3}{\Omega^6} -$$

$$-\frac{3kb}{2\Omega^2}\left(B-\frac{bB^3}{32\Omega^2}-\frac{k}{\Omega^2}\right)^2-\frac{3kb^3B^6}{2048\Omega^6}-\left[\frac{3kb}{2\Omega^2}\left(B-\frac{bB^3}{32\Omega^2}-\frac{k}{\Omega^2}\right)^2+\right.$$
$$\left.+\frac{3kb^2B^3}{32\Omega^4}\left(B-\frac{bB^3}{32\Omega^2}-\frac{k}{\Omega^2}\right)\right]\cos 2\Omega t+\left[\frac{\lambda bB^3}{32\Omega^2}-\frac{b}{4}\left(B-\frac{bB^3}{32\Omega^2}-\frac{k}{\Omega^2}\right)^3\right.$$
$$\left.-\frac{3b^4B^9}{131072\Omega^6}-\frac{3b^2B^3}{64\Omega^2}\left(B-\frac{bB^3}{32\Omega^2}-\frac{k}{\Omega^2}\right)^2-\frac{3k^2b^2A^3}{96\Omega^8}\right]\cos 3\Omega t-$$
$$-\frac{3kb^2B^3}{32\Omega^4}\left(B-\frac{bB^3}{32\Omega^2}-\frac{k}{\Omega^2}\right)\cos 4\Omega t-\left[\frac{3b^2B^3}{128\Omega^2}\left(B-\frac{bB^3}{32\Omega^2}-\frac{k}{\Omega^2}\right)^2+\right.$$
$$\left.+\frac{3b^3B^6}{4096\Omega^4}\left(B-\frac{bB^3}{32\Omega^2}-\frac{k}{\Omega^2}\right)\right]\cos 5\Omega t-\frac{3kb^3B^6}{2048\Omega^6}\cos 6\Omega t-$$
$$-\frac{3b^3B^6}{4096\Omega^4}\left(B-\frac{bB^3}{32\Omega^2}-\frac{k}{\Omega^2}\right)\cos 7\Omega t-\frac{b^4B^9}{131072\Omega^6}\cos 9\Omega t \tag{8.36}$$

Avoiding the presence of a secular term in Eq. 8.11 for $n = 1$, needs

$$\lambda_2=\frac{3b}{4}\left(B-\frac{bB^3}{32\Omega^2}-\frac{k}{\Omega^2}\right)^2+\frac{3bk^2}{\Omega^4}+\frac{3b^2B^3}{128\Omega^2}\left(B-\frac{bB^3}{32\Omega^2}-\frac{k}{\Omega^2}\right)+\frac{3b^3B^6}{2048\Omega^4} \tag{8.37}$$

Substituting Eq. 8.37 into Eq. 8.28_2, we obtain the equation for Ω:

$$\Omega^2=\omega^2-\frac{a^2}{3b}+\frac{3bB^2}{4}-\frac{3bB(64k+bB^2)}{128\Omega^2}+\frac{3b(2560k^2+16kbB^3+b^2B^6)}{2048\Omega^4} \tag{8.38}$$

In order to illustrate the remarkable accuracy of this method, we compare the approximate results given by Eq. 8.38 with numerical integration results for the following numerical examples:

(a) Consider the equation

$$\ddot{x}+x+4x^2+50x^3=0,\ x(0)=2,\ \dot{x}(0)=0 \tag{8.39}$$

In this case $a = 4$, $b = 50$, $\omega = 1$, $A = 2$ and from Eqs. 8.26_2 and 8.31 we obtain $k = 0{,}02477037$, $B = 2{,}02666667$, and from Eq. 8.38 we have

$$\Omega = 12.19942839 \tag{8.40}$$

Therefore, from Eq. 8.36 we obtain

$$F(u_1) = 0.002608936 - 0.05090743\cos 2\Omega t - 102.0046467\cos 3\Omega t - \\ - 0.004219655\cos 4\Omega t - 12.81158791\cos 5\Omega t - 0.000095343\cos 6\Omega t \\ - 0.553925714\cos 7\Omega t - 0.008343994\cos 9\Omega t \quad (8.41)$$

Setting $n = 2$ in Eq. 8.11 we obtain the second-order approximate solution:

$$u_2(t) = 2.02666667\cos\Omega t + 0.00001753(1 - \cos\Omega t) - \\ - 0.000114019(\cos\Omega t - \cos 2\Omega t) - 0.085674386(\cos\Omega t - \cos 3\Omega t) \\ - 0.003586846(\cos\Omega t - \cos 5\Omega t) - 0.00007754(\cos\Omega t - \cos 7\Omega t) \quad (8.42)$$

From Eqs. 8.42 and 8.25 we obtain the following result:

$$x(t) = -0.026649136 + 1.937195679\cos\Omega t + 0.000114019\cos 2\Omega t + \\ +0.085694386\cos 3\Omega t + 0.003586846\cos 5\Omega t + 0.00007754\cos 7\Omega t \quad (8.43)$$

where Ω is given by Eq. 8.40.

Figure 8.1 shows the comparison between the present solution and the numerical integration results obtained by a fourth-order Runge–Kutta method.

It can be seen from Fig. 8.1 that the solution obtained by the present method is nearly identical with that given by the numerical method.

(b) As the second case, we consider the equation

$$\ddot{x} + x + 6x^2 + 8x^3 = 0,\ x(0) = 2,\ \dot{x}(0) = 0, (a = 6, b = 8, \omega = 1, A = 2) \quad (8.44)$$

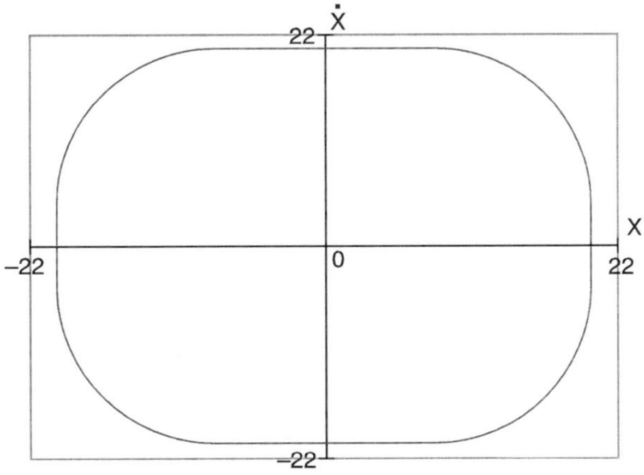

Fig. 8.1 Phase plane for Eq. 8.39: ______ numerical simulation; _ _ _ _ _ present method given by Eq. 8.43

It is important that $k = 0$. It follows that $B = 2.25$ and for the input function:

$$u_0 = B\cos\Omega t \tag{8.45}$$

we have:

$$F(u_0) = \lambda u_0 - bu_0^3 = B\left(\lambda - \frac{3bB^2}{4}\right)\cos\Omega t - \frac{bB^3}{4}\cos 3\Omega t \tag{8.46}$$

In order to avoid secular term, we set:

$$\lambda_1 = \frac{3}{4}bB^2, \quad \Omega^2 = -\frac{1}{2} + \frac{3bB^2}{4} \tag{8.47}$$

The first-order approximate solution is given by the relation:

$$u_1(t) = B\left(1 - \frac{bB^2}{32\Omega^2}\right)\cos\Omega t + \frac{bB^3}{32\Omega^2}\cos 3\Omega t \tag{8.48}$$

and therefore

$$\begin{aligned} F(u_1) = {} & B\left(1 - \frac{bB^2}{32\Omega^2}\right)\left[\lambda - \frac{3bB^2}{4}\left(1 - \frac{bB^2}{32\Omega^2}\right)^2 + \frac{3b^2B^4}{128\Omega^2}\left(1 - \frac{bB^2}{32\Omega^2}\right) - \right. \\ & \left. - \frac{3b^3B^6}{2048\Omega^4}\right]\cos\Omega t + \left[-\frac{\lambda bB^3}{32\Omega^2} - \frac{bB^3}{4}\left(1 - \frac{bB^2}{32\Omega^2}\right)^3 - \frac{3b^2B^5}{64\Omega^2}\left(1 - \frac{bB^2}{32\Omega^2}\right)^2 \right. \\ & \left. - \frac{3b^4B^9}{131072\Omega^6}\right]\cos 3\Omega t + \frac{3b^2B^5}{128\Omega^2}\left(1 - \frac{bB^2}{32\Omega^2}\right)\cos 5\Omega t - \\ & - \frac{3b^3B^7}{4096\Omega^4}\left(1 - \frac{bB^2}{32\Omega^2}\right)\cos 7\Omega t - \frac{b^4B^9}{131072\Omega^6}\cos 9\Omega t \end{aligned} \tag{8.49}$$

Avoiding no secular term requires that

$$\lambda_2 = \frac{3}{4}bB^2\left(1 - \frac{bB^2}{32\Omega^2}\right) + \frac{3b^3B^6}{2048\Omega^4} \tag{8.50}$$

From Eqs. 8.28_2 and 8.50 we obtain the equation for Ω:

$$\Omega^2 + \frac{1}{2} - \frac{3}{4}bB^2 + \frac{3b^2B^4}{128\Omega^2} - \frac{3b^3B^6}{2048\Omega 4} = 0 \tag{8.51}$$

Substituting Eq. 8.50 into 8.49, we have:

$$\begin{aligned} F(u_1) = &-\frac{bB^3}{4}\left(1+\frac{3bB^2}{16\Omega^2}-\frac{3b^2B^4}{256\Omega^4}+\frac{7b^3B^6}{1638\Omega^6}\right)\cos 3\Omega t+ \\ &+\frac{3b^2B^5}{128\Omega^2}\left(1-\frac{b^2B^5}{128\Omega^2}\right)\cos 5\Omega t- \\ &-\frac{3b^3B^7}{4096\Omega^4}\left(1-\frac{bB^2}{32\Omega^2}\right)\cos 7\Omega t-\frac{b^4B^9}{131072\Omega^6}\cos 9\Omega t \end{aligned} \tag{8.52}$$

From Eq. 8.51 we obtain

$$\Omega = 5.462517735 \tag{8.53}$$

and from Eqs. 8.52 and 8.29 for $n = 2$, we obtain

$$\begin{aligned} u_2(t) = &\, 2.124701943\cos\Omega t + 0.123522681\cos 3\Omega t+ \\ &+ 0.001815473\cos 5\Omega t - 0.00040097\cos 7\Omega t \end{aligned} \tag{8.54}$$

From Eq. 8.25 we obtain the second-order approximation;

$$\begin{aligned} x_2(t) = &-0.25 + 2.124701943\cos\Omega t + 0.123522681\cos 3\Omega t+ \\ &+0.001815473\cos 5\Omega t - 0.00040097\cos 7\Omega t \end{aligned} \tag{8.55}$$

where Ω is given by Eq. 8.53.

As in the previous case, the results obtained are remarkably good when compared to the results obtained numerically using a fourth-order Runge–Kutta method (Fig. 8.2).

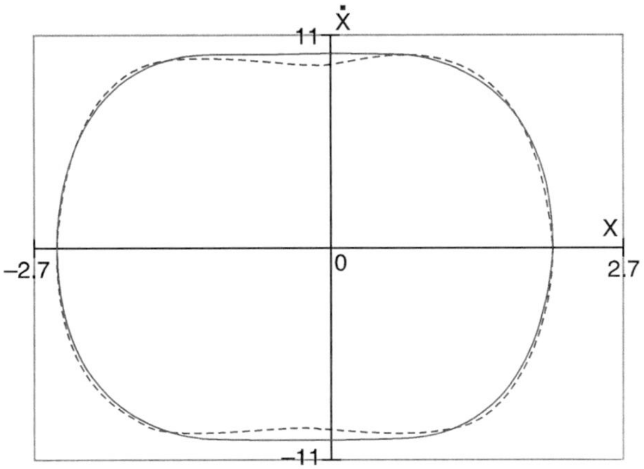

Fig. 8.2 Phase plane for Eq. 8.44: ______ numerical simulation; _ _ _ _ _ present method given by Eq. 8.55

8.1.2 A Family of Nonlinear Differential Equations

As the second example, let us consider a family of nonlinear differential equations

$$\ddot{u} + \alpha u + \gamma u^{2n+1} = 0, \alpha \geq 0, \gamma > 0, n = 1, 2, 3, \ldots \tag{8.56}$$

with the initial conditions

$$u(0) = A, \quad \dot{u}(0) = 0 \tag{8.57}$$

The corresponding exact period T is

$$T_{ex} = 4 \int_0^{\frac{\pi}{2}} \frac{d\theta}{\sqrt{\alpha + \frac{\gamma}{n+1} A^{2n} \left(1 + \sin^2\theta + \sin^4\theta + \ldots + \sin^{2n}\theta\right)}} \tag{8.58}$$

The Eq. 8.56 can be written as:

$$\ddot{u} + \Omega^2 u = F(u) \tag{8.59}$$

where

$$F(u) = (\Omega^2 - \alpha)u - \gamma u^{2n+1} \tag{8.60}$$

With the initial conditions (8.57), the input function is

$$u_0(t) = A \cos \Omega t \tag{8.61}$$

Expanding $F(u_0)$ in a Fourier series, we have

$$F(u_0) = A \cos \Omega t \left(\Omega^2 - \alpha - \gamma A^{2n} \cos^{2n} \Omega t\right) \tag{8.62}$$

By using the identity

$$\cos^{2n+1} \Omega t = \frac{1}{4^n} \sum_{k=0}^{n} \binom{2n+1}{n-k} \cos(2k+1)\Omega t \tag{8.63}$$

Equation 8.62 becomes:

$$F(u_0) = A\left[\Omega^2 - \alpha - \frac{\gamma A^{2n}}{4^n} \binom{2n+1}{n}\right] \cos \Omega t - \frac{\gamma A^{2n+1}}{4^n} \sum_{k=1}^{n} \binom{2n+1}{n-k} \cos(2k+1)\Omega t \tag{8.64}$$

Avoiding the secular term requires that

$$\Omega_1(n) = \sqrt{\alpha + \frac{\gamma A^{2n}}{4^n}\binom{2n+1}{n}} \tag{8.65}$$

The first-order approximate solution (8.11) of Eq. 8.49 is

$$u_1(t) = A\cos\Omega_1 t - \frac{\gamma A^{2n+1}}{4^{n+1}}\sum_{k=1}^{n}\frac{1}{k(k+1)}\binom{2n+1}{n-k}[\cos\Omega_1 t - \cos(2k+1)\Omega_1 t] \tag{8.66}$$

where Ω_1 is given by Eq. 8.65.

For integer n, the approximate period in the first-order approximation is

$$T_{ap}(n) = \frac{2\pi}{\Omega_1(n)} = 2\pi\left[\alpha + \frac{\gamma A^{2n}}{4^n}\binom{2n+1}{n}\right]^{-\frac{1}{2}} \tag{8.67}$$

Formula (8.67) is valid for any possible amplitude and gives the maximum errors as the dimensionless amplitude γA^{2n} tends to infinity. Note that even for $n = 9$, the maximum error given by Eq. 8.67 is less than 4% for an amplitude $A \in [0, \infty]$

$$\lim_{\gamma A^{18}\to\infty}\frac{T_{ap}(9)}{T_{ex}} \approx \frac{16\pi\sqrt{\frac{5}{46189}}}{0,5406369} \approx 0,967341 \tag{8.68}$$

8.1.3 The Duffing Equation

The use of this procedure may be illustrated by the following Duffing equation

$$\ddot{u} + u + au^3 = 0 \tag{8.69}$$

where a is a real parameter. The initial conditions are:

$$u(0) = A, \quad \dot{u}(0) = 0 \tag{8.70}$$

So, we can rewrite Eq. 8.69 in the form:

$$\ddot{u} + \Omega^2 u = F(u) \tag{8.71}$$

where

$$F(u) = a(\lambda u - u^3), \quad a(\lambda) = \Omega^2 - 1 \tag{8.72}$$

and λ is a parameter.

The left hand side of Eq. 8.71 in the conditions (8.70) has the solution (the input start function):

$$u_0 = A\cos\Omega t \tag{8.73}$$

Expanding $F(u_0)$ in a Fourier series, we have:

$$F(x_0) = aA\left(\lambda - \frac{3A^2}{4}\right)\cos\Omega t - \frac{aA^3}{4}\cos 3\Omega t \tag{8.74}$$

No secular terms in Eq. 8.74 iteration requires that

$$\lambda_1 = \frac{3}{4}A^2 \tag{8.75}$$

and therefore from Eq. 8.72_1 we obtain

$$\Omega^2 = 1 + \frac{3}{4}aA^2 \tag{8.76}$$

From Eq. 8.11 for $n = 0$ we have the following first-order approximate solution:

$$u_1(t) = A\left(1 - \frac{aA^2}{32\Omega^2}\right)\cos\Omega t + \frac{aA^3}{32\Omega^2}\cos 3\Omega t \tag{8.77}$$

where the frequency Ω is listed in Eq. 8.76, and therefore:

$$\begin{aligned} F(u_1) = aA\Bigg[\lambda - \frac{\lambda aA^2}{32\Omega^2} - \frac{3}{4}A^2\left(1 - \frac{aA^2}{32\Omega^2}\right)^3 - \frac{3}{4}\left(1 - \frac{aA^2}{32\Omega^2}\right)^2\frac{aA^4}{32\Omega^2} - \frac{3a^2A^6}{2048\Omega^4}\times \\ \times\left(1 - \frac{aA^2}{32\Omega^2}\right)\Bigg]\cos\Omega t - aA\left[\frac{1}{4}\left(1 - \frac{aA^2}{32\Omega^2}\right) + \frac{3aA^4}{64\Omega^2}\left(1 - \frac{aA^2}{32\Omega^2}\right)^2 + \\ + \frac{3a^3A^8}{131072\Omega^6} - \frac{\lambda aA^2}{3\Omega^2}\right]\cos 3\Omega t - a\left[\frac{3aA^3}{128\Omega^2}\left(A - \frac{aA^3}{32\Omega^2}\right)^2 + \frac{3a^2A^6}{4096\Omega^4}\times \\ \times\left(A - \frac{aA^3}{32\Omega^2}\right)\right]\cos 5\Omega t - \frac{3a^3A^6}{4096\Omega^4}\left(A - \frac{aA^3}{32\Omega^2}\right)\cos 7\Omega t - \frac{a^4A^9}{131072\Omega^6}\cos 9\Omega t \end{aligned}$$

Avoiding the presence of a secular term in the next iteration, needs:

$$\lambda_2 = \frac{3}{4}A^2 - \frac{3aA^4}{128\Omega^2} + \frac{3a^2A^6}{2048\Omega^4} \tag{8.79}$$

Substituting Eq. 8.79 into Eq. 8.72, we obtain the equation in Ω:

$$\Omega^2 = 1 + \frac{3}{4}aA^2 - \frac{3a^2A^4}{128\Omega^2} + \frac{3a^2A^6}{2048\Omega^4} \tag{8.80}$$

The second-order approximate solution is given by ($n = 1$ into Eq. 8.11):

$$\begin{aligned} u_2(t) = &\left(A + \frac{aM}{8\Omega^2} + \frac{aN}{24\Omega^2} + \frac{aP}{48\Omega^2} + \frac{aQ}{80\Omega^2}\right)\cos\Omega t - \frac{aM}{8\Omega^2}\cos 3\Omega t - \\ &- \frac{aN}{24\Omega^2}\cos 5\Omega t - \frac{aP}{48\Omega^2}\cos 7\Omega t - \frac{aQ}{80\Omega^2}\cos 9\Omega t \end{aligned} \tag{8.81}$$

where

$$\begin{aligned} M &= -\frac{A}{4} + \frac{aA^3}{128\Omega^2} - \frac{3aA^5}{128\Omega^2} + \frac{9a^2A^7}{4096\Omega^4} - \frac{3a^3A^9}{131072\Omega^6} \\ N &= -\frac{3aA^4}{128\Omega^2}\left(A - \frac{aA^3}{32\Omega^2}\right) \\ P &= -\frac{3a^2A^6}{4096\Omega^4}\left(A - \frac{aA^3}{32\Omega^2}\right) \\ Q &= -\frac{a^3A^9}{131072\Omega^6} \end{aligned} \tag{8.82}$$

The accuracy of these results is illustrated by Table 8.1 which provides a comparison between the exact frequency of Eq. 8.69 and the approximate frequency computed by Eq. 8.80. For any value of a, the maximal relative error of the second-order approximate frequency with respect to the exact solution is less than 0.028%.

On the other hand, the second-order approximate solution given by Eq. 8.81 becomes for $a = 1.25$ and $A = 2$:

$$\begin{aligned} u_2(t) = &\ 1.989351555\cos\Omega t + 0.008334403\cos 3\Omega t + 0.002206664\cos 5\Omega t \\ &+ 0.000053689\cos 7\Omega t + 0.00000026\cos 9\Omega t \end{aligned} \tag{8.83}$$

where $\Omega = 2{,}15031$.

Table 8.1 Comparison of approximate frequency with exact frequency for Duffing equation

aA^2	Ω_{ex}	Ω_{ap} (8.80)
5	2,15042	2,15031
1,000	26,8107	26,8026
10,000	84,7245	84,7015

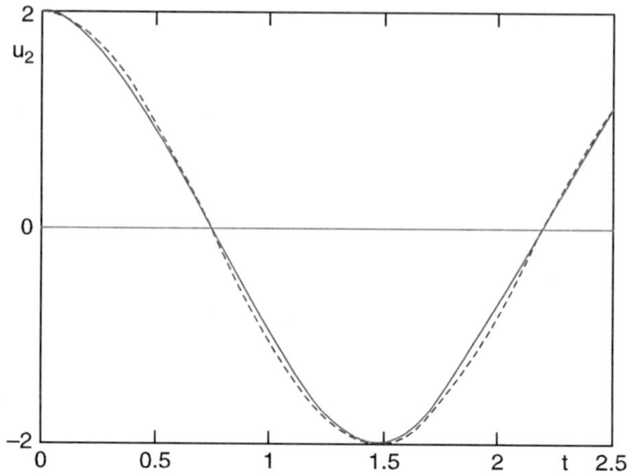

Fig. 8.3 Comparison between numerical solution of Eq. 8.69 and approximate solution (8.83) for $a = 1.25$, $A = 2$: ________ numerical results; _ _ _ _ _ _ approximate results

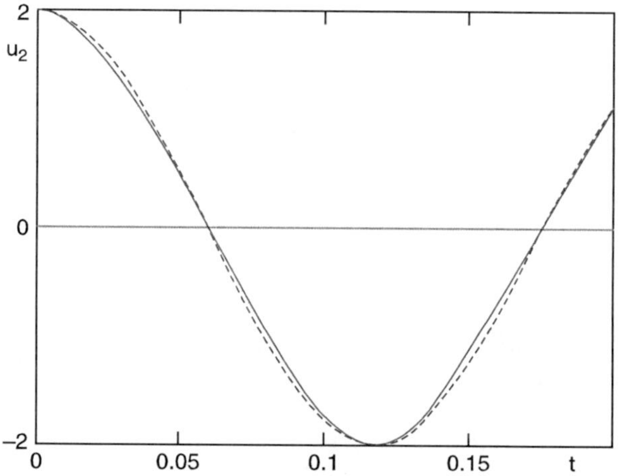

Fig. 8.4 Comparison between numerical solution of Eq. 8.69 and approximate solution (8.84) for $a = 250$, $A = 2$: ________ numerical results; _ _ _ _ _ _ approximate results

For $a = 250$ and $A = 2$ we obtain

$$\begin{aligned} u_2(t) = {} & 1.977507645 \cos \Omega t + 0.018844215 \cos \Omega t + 0.003619986 \cos 5\Omega t \\ & + 0.000027438 \cos 7\Omega t + 0.000164633 \cos 9\Omega t \end{aligned} \tag{8.84}$$

where $\Omega = 26.8026$.

Figures 8.3 and 8.4 show the comparison between the present solution and numerical integration results obtained using a fourth-order Runge–Kutta method. It can be seen that the solution obtained by the present method shows a remarkable accuracy.

8.2 Mathematical Description of the Optimal Variational Iteration Method

An alternative approach for solving nonlinear differential equations is the Optimal Variational Iteration Method (OVIM), which is a combination of the classical variational iteration method with a rigorous computational algorithm for minimizing the residual functional, which provides a simple procedure to control the convergence of the results. The proposed procedure yields a rapid convergence with respect to the exact solution and has some distinct advantages over usual approximate methods in that the approximate solutions obtained are valid not only for small values of the parameters but also for large ones. Our procedure is very effective and accurate for nonlinear problems with approximations converging rapidly to accurate solutions after only one iteration.

For the nonlinear differential Eq. 8.1 has been constructed the correction functional (8.2) or (8.8) with λ known via the variational iteration method. In Eq. 8.8, the initial approximation $u_0(t)$contains several unknown parameters C_i, $i = 1,2,\ldots$, m which are determined from m initial/boundary conditions. In our procedure, the initial approximation $u_0(t)$contains $m + n$ unknown parameters C_{i+j}, $i = 1,2,\ldots,m$, $j = 1,2,\ldots,n$ which will be determined optimally like in Chaps. 6 and 7 for example: from the stationary conditions of the residual functional

$$J(C_{m+1}, C_{m+2}, \ldots, C_{m+n}) = \int_a^b [Lu(s) + Nu(s) - g(s)]^2 ds \tag{8.85}$$

These conditions are well-known:

$$\frac{\partial J}{\partial C_{m+1}} = \frac{\partial J}{\partial C_{m+2}} = \ldots = \frac{\partial J}{\partial C_{m+n}} = 0 \tag{8.86}$$

Note that the unknown parameters C_{m+j} can be identified via various methodologies such as the collocation method, the least squares method, the Galerkin method etc.

In what follows, the optimal variational iteration method is illustrated by some applications.

8.3 Duffing-Harmonic Oscillator

Conservative nonlinear oscillatory systems can often be modelled by potentials having a rational form for the potential energy [163], which lead to the equation of motion

$$\ddot{x} + \frac{ax^3}{b + cx^2} = 0 \tag{8.87}$$

for which the parameters a, b, c are non-negative. Defining

$$u = \sqrt{\frac{b}{c}}x, \quad \eta = \sqrt{\frac{c}{a}}t \tag{8.88}$$

equation 8.87 is reduced to the following non-dimensional equation

$$\ddot{u} + \frac{u^3}{1 + u^2} = 0 \tag{8.89}$$

Equation 8.89 is an example of conservative nonlinear oscillatory system having a rational form of the restoring force. For small u the equation is that of a Duffing-type nonlinear oscillator, while for large u, the equation approximates that of a linear harmonic oscillator, hence Eq. 8.89 is called the Duffing-harmonic oscillator. The restoring force in Eq. 8.89 is the same for both negative and positive amplitudes. At the same time, the above system has no possible small parameters. Therefore, the classical perturbation methods do not apply to such a problem. Eq. 8.89 can be rewritten as

$$u'' + u^2u'' + u^3 = 0 \tag{8.90}$$

If ω is the frequency of Duffing-harmonic oscillator, Eq. 8.90 can be written as

$$u''(\tau) + u(\tau) + f(\tau, u, u', u'') = 0, \quad f(\tau, u, u', u'') = u^2(\tau)u''(\tau) + \frac{1}{\omega^2}u^3(\tau) - u(\tau) \tag{8.91}$$

where

$$\tau = \omega\eta, \quad u' = \frac{du}{d\tau} \tag{8.92}$$

We choose the initial solution in the form:

$$u_0(\tau) = C_1 \cos\tau + C_2 \cos 3\tau + C_3 \cos 5\tau \tag{8.93}$$

where C_1, C_2 and C_3 are unknown parameters that will be determined later. We have:

$$\begin{aligned} f(\tau, u_0, u'_0, u''_0) = \Bigg[& -\frac{3}{4}C_1^3 - \frac{19}{2}C_1C_2^2 - \frac{51}{2}C_1C_3^2 - \frac{11}{4}C_1^2C_2 - \frac{35}{2}C_1C_2C_3 - \\ & -\frac{43}{4}C_2^2C_3 - C_1 + \frac{1}{\omega^2}\left(\frac{3}{4}C_1^3 + \frac{3}{4}C_1^2C_2 + \frac{3}{2}C_1C_2^2 + \frac{3}{2}C_1C_3^2 + \right. \\ & \left.\frac{3}{4}C_2^2C_3 + \frac{3}{2}C_1C_2C_3\right)\Bigg] \cos\tau + A_1\cos 3\tau + A_2\cos 5\tau + A_3\cos 7\tau + \\ & + A_4\cos 9\tau + A_5\cos 11\tau + A_6\cos 13\tau + A_7\cos 15\tau \end{aligned} \tag{8.94}$$

where

$$\begin{aligned}
A_1 &= -\frac{1}{4}C_1^3 - \frac{11}{2}C_1^2C_2 - \frac{27}{4}C_1^2C_3 - \frac{35}{2}C_1C_2C_3 - \frac{27}{4}C_2^3 - \frac{59}{2}C_2C_3^2 - C_2 + \\
&\quad + \frac{1}{\omega^2}\left(\frac{1}{4}C_1^3 + \frac{3}{4}C_2^3 + \frac{3}{2}C_1^2C_2 + \frac{3}{4}C_1^2C_3 + \frac{3}{2}C_2C_3^2 + \frac{3}{2}C_1C_2C_3\right) \\
A_2 &= -\frac{19}{4}C_1C_2^2 - \frac{11}{4}C_1^2C_2 - \frac{27}{2}C_1^2C_3 - \frac{43}{2}C_2^2C_3 - \frac{75}{4}C_3^3 - C_3 + \\
&\quad + \frac{1}{\omega^2}\left(\frac{3}{4}C_3^3 + \frac{3}{4}C_1^2C_2 + \frac{3}{4}C_1C_2^2 + \frac{3}{2}C_1^2C_3 + \frac{3}{2}C_2^2C_3\right) \\
A_3 &= -\frac{19}{4}C_1C_2^2 - \frac{27}{4}C_1^2C_3 - \frac{35}{2}C_1C_2C_3 - \frac{59}{4}C_2C_3^2 + \\
&\quad + \frac{1}{\omega^2}\left(\frac{3}{4}C_1C_2^2 + \frac{3}{4}C_1^2C_3 + \frac{3}{4}C_2C_3^2 + \frac{3}{2}C_1C_2C_3\right) \\
A_4 &= -\frac{51}{4}C_1C_3^2 - \frac{35}{2}C_1C_2C_3 - \frac{9}{4}C_2^3 + \frac{1}{\omega^2}\left(\frac{1}{4}C_2^3 + \frac{3}{4}C_1C_3^2 + \frac{3}{2}C_1C_2C_3\right) \\
A_5 &= -\frac{51}{4}C_1C_3^2 - \frac{43}{4}C_2^2C_3 + \frac{1}{\omega^2}\left(\frac{3}{4}C_1C_3^2 + \frac{3}{4}C_2^2C_3\right) \\
A_6 &= -\frac{59}{4}C_2C_3^2 + \frac{3}{4\omega^2}C_2C_3^2, \quad A_7 = -\frac{25}{4}C_3^3 + \frac{1}{4\omega^2}C_3^3
\end{aligned} \tag{8.95}$$

In order to ensure that no secular terms appear in the next iteration, resonance must be avoided. So, the coefficient of $\cos\tau$ into Eq. 8.94 has to be zero; i.e.:

$$\omega^2 = \frac{3(C_1^3 + C_1^2C_2 + 2C_1C_2^2 + 2C_1C_3^2 + C_2^2C_3 + 2C_1C_2C_3)}{3C_1^3 + 4C_1 + 38C_1C_2^2 + 102C_1C_3^2 + 11C_1^2C_2 + 70C_1C_2C_3 + 43C_2^2C_3} \tag{8.96}$$

For $n = 0$ in Eq. 8.7 we obtain the first-order approximate solution:

$$\begin{aligned}
u_1(\tau) &= \left[C_1 - \frac{A_1}{8} - \frac{A_2}{24} - \frac{A_3}{48} - \frac{A_4}{80} - \frac{A_5}{120} - \frac{A_6}{168} - \frac{A_7}{224}\right]\cos\tau + \left[C_2 + \frac{A_1}{8}\right]\cos 3\tau \\
&\quad + \left[C_3 + \frac{A_2}{24}\right]\cos 5\tau + \frac{A_3}{48}\cos 7\tau + \frac{A_4}{80}\cos 9\tau + \frac{A_5}{120}\cos 11\tau + \frac{A_6}{168}\cos 13\tau \\
&\quad + \frac{A_7}{224}\cos 15\tau
\end{aligned} \tag{8.97}$$

The parameters C_1, C_2, C_3 can be determined from initial conditions

$$u(0) = A, \quad u'(0) = 0 \tag{8.98}$$

such that

$$C_1 + C_2 + C_3 = A \tag{8.99}$$

and the other two necessary conditions will be derived optimally by minimizing the residual functional

$$J = \int_0^{2\pi} \left[u''_1(\tau) + u_1^2 u''_1(\tau) + \frac{1}{\omega^2} u_1^3(\tau) \right]^2 d\tau \tag{8.100}$$

The condition that the functional J be minimized is given by the system:

$$C_3 = A - C_1 - C_2, \quad \frac{\partial J}{\partial C_1} = \frac{\partial J}{\partial C_2} = 0 \tag{8.101}$$

The conditions (8.101) provide a system of algebraic equation whose solution represents the optimal values of the constants C_i. In this way the approximate solution is well-determined.

Following this procedure, we need only the first-order approximate solution for solving Duffing-harmonic oscillator and its frequency (given by Eqs. 8.97 and 8.96, respectively).

We present some examples of obtaining analytical solutions and frequencies in order to show the efficiency of the method described in the previous section for solving Eq. 8.89 or Eq. 8.90.

(a) As a first example, we consider $A = 0.1$. From Eqs. 8.101 we obtain:

$$C_1 = 0.095973775, \quad C_2 = 0.0038674917, \quad C_3 = 0.0001587327$$

From Eqs. 8.97 and 8.96 we obtain therefore:

$$\begin{aligned} u_1(t) &= 0.0915236051 \cos \omega t + 0.0081483792 \cos 3\omega t + 0.000321646 \cos 5\omega t \\ &\quad + 0.0000062186 \cos 7\omega t + 0.000000168 \cos 9\omega t + 3.5 \cdot 10^{-9} \cos 11\omega t \\ &\quad + 5.2 \cdot 10^{-11} \cos 13\omega t + 5.1 \cdot 10^{-13} \cos 15\omega t \\ \omega &= 0.0845707 \end{aligned} \tag{8.102}$$

In Fig. 8.5 we compare the present solution and the numerical integration results obtained by a fourth-order Runge–Kutta method.

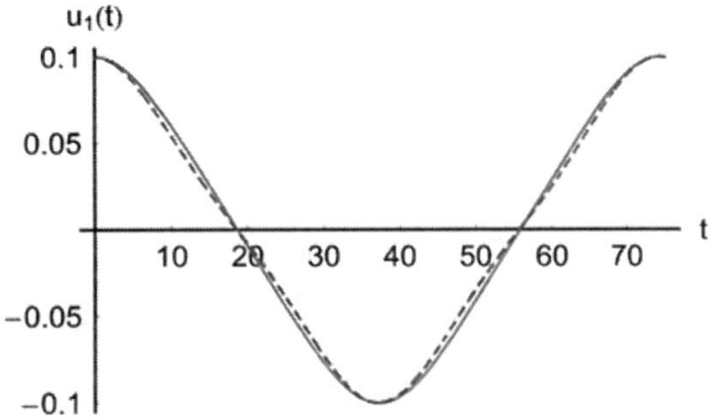

Fig. 8.5 The behaviour of the present solution (8.102) and the numerical integration results of Eq. 8.90 for $A = 0.1$: _______ numerical solution; _ _ _ _ _ _ approximate solution

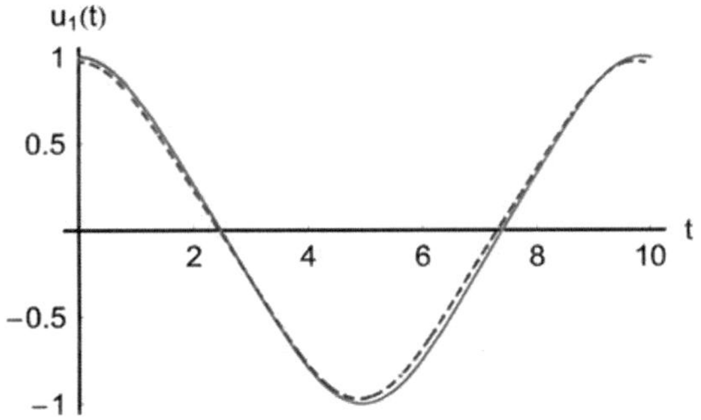

Fig. 8.6 The behavior of the present solution (8.103) and the numerical integration results of Eq. 8.90 for $A = 1$: _______ numerical solution; _ _ _ _ _ _ approximate solution

(b) In this example we solve Eq. 8.90 with $A = 1$. The parameters C_i are in this case

$$C_1 = 0.9728243593, \quad C_2 = -0.0001966401, \quad C_3 = -0.0008579578$$

with the solution given by Eq. 8.97

$$\begin{aligned} u_1(t) &= 0.93126560626 \cos \omega t + 0.0409011098 \cos 3\omega t + \\ &+ 0.0004804024 \cos 5\omega t + 0.0000835741 \cos 7\omega t \\ &- 0.0000001264 \cos 9\omega t - 6.5 \cdot 10^{-8} \cos 11\omega t \\ &+ 1.11 \cdot 10^{-11} \cos 13\omega t + 1.6 \cdot 10^{-11} \cos 15\omega t \\ \omega &= 0.63895601 \end{aligned} \tag{8.103}$$

In Fig. 8.6 the present solution is compared with the numerical integration results.

(c) In this case $A = 5$ and we obtain:

$$C_1 = 4.9479890585, \quad C_2 = -0.0002277091, \quad C_3 = -0.0000711966$$

and the solution is:

$$\begin{aligned} u_1(t) &= 4.786381392 \cos \omega t + 0.2095410409 \cos 3\omega t + +0.0012524974 \cos 5\omega t \\ &\quad + 0.00021635625 \cos 7\omega t - 1.971 \cdot 10^{-8} \cos 9\omega t - 2.498 \cdot 10^{-9} \cos 11\omega t \\ &\quad + 9.59 \cdot 10^{-14} \cos 15\omega t \\ \omega &= 0.97095656 \end{aligned} \tag{8.104}$$

In Fig. 8.7 the present solution is compared with the numerical integration results.

(d) For $A = 10$ we obtain:

$$C_1 = 10.0128057685, \quad C_2 = -0.0012160379, \quad C_3 = -0.0001158973$$

and the solutions becomes:

$$\begin{aligned} u_1(t) &= 9.51883831227 \cos \omega t + 0.475390212 \cos 3\omega t + 0.015796936 \cos 5\omega t \\ &\quad + 0.0014483729 \cos 7\omega t - 3.02 \cdot 10^{-7} \cos 9\omega t - 1.34 \cdot 10^{-8} \cos 11\omega t \\ &\quad + 1.36 \cdot 10^{-12} \cos 13\omega t + 4.16 \cdot 10^{-14} \cos 15\omega t \\ \omega &= 0.99109273 \end{aligned} \tag{8.105}$$

In Fig. 8.8 the present solution is compared with the numerical integration results.

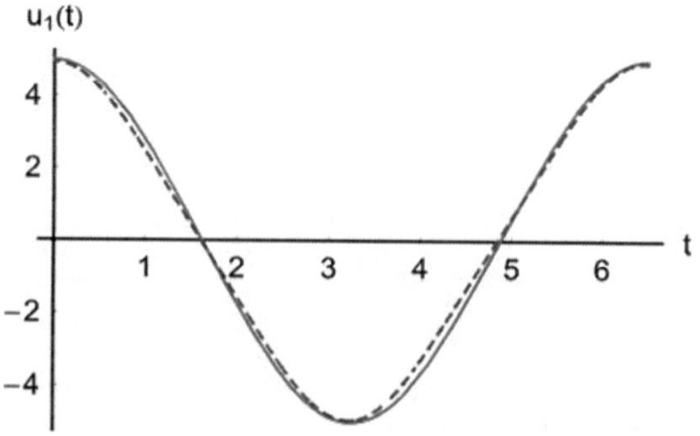

Fig. 8.7 The behavior of the present solution (8.104) and the numerical integration results of Eq. 8.90 for $A = 5$: ________ numerical solution; ------ approximate solution

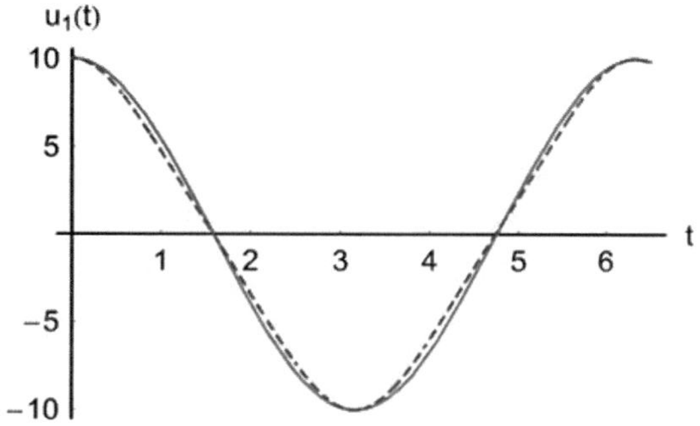

Fig. 8.8 The behavior of the present solution (8.104) and the numerical integration results of Eq. 8.90 for $A = 10$: _______ numerical solution; _ _ _ _ _ _ approximate solution

Table 8.2 Comparison of approximate frequencies with the exact frequencies

A	ω_{ex}	$\frac{\omega}{\omega_{ex}}$
0.1	0.08439	1.00214124
1	0.63678	1.00341721
5	0.96698	1.00911236
10	0.99092	1.00017432
50	0.99961	1.00013105
100	0.99990	1.00003500

For the Duffing-harmonic oscillator, the exact angular frequency is

$$\omega_{ex}(A) = \frac{\pi}{2}\left(\int_0^{\frac{\pi}{2}} A\cos\tau\left[A^2\cos^2\tau + \ln(1 - A^2\cos^2\tau(1 + A^2)^{-1}\right]^{-\frac{1}{2}} d\tau\right) \tag{8.106}$$

Table 8.2 shows the ratios of the approximate angular frequencies obtained through the proposed procedure to the exact angular frequency ω_{ex} in Eq. 8.106.

The obtained analytical expression for the angular frequency yields very good approximations for both small and large values of amplitude A. Graphical comparison of the approximate and numerical solutions show that the proposed method provides excellent approximations to periodic solutions for both small and large amplitude.

8.4 Oscillations of a Uniform Cantilever Beam Carrying an Intermediate Lumped Mass and Rotary Inertia

Consider the nonlinear oscillator (see Sect. 6.7)

$$\ddot{u} + u + \alpha u^2\ddot{u} + \alpha u\dot{u}^2 + \beta u^3 = 0 \tag{8.107}$$

subject to the initial conditions

$$u(0) = A, \quad \dot{u}(0) = 0 \tag{8.108}$$

We rewrite Eq. 8.107 as

$$\ddot{u} + \Omega^2 u + (1 - \Omega^2)u + \alpha(u\dot{u}^2 + u^2\ddot{u}) + \beta u^3 = 0 \tag{8.109}$$

with the initial conditions (8.108), where Ω is the frequency of the system. As an initial guess for $u_0(t)$ we choose

$$u_0(t) = C_1 \cos \Omega t + C_2 \cos 3\Omega t + C_3 \cos 5\Omega t \tag{8.110}$$

where C_1, C_2 and C_3 are unknown constants which partially can be determined from Eq. 8.108:

$$C_1 + C_2 + C_3 = A \tag{8.111}$$

In Eq. 8.109 it can be written

$$Nu := f(t, u, \dot{u}, \ddot{u}) = (1 - \Omega^2)u + \alpha(u\dot{u}^2 + u^2\ddot{u}) + \beta u^3 \tag{8.112}$$

From Eq. 8.8 for $n = 0$ it is obtained

$$u_1(t) = u_0(t) + \frac{1}{\Omega}\int_0^t \sin \Omega(\tau - t) f(\tau, u_0(\tau), u'_0(\tau), u''_0(\tau))d\tau \tag{8.113}$$

where $f(\tau, u_0(\tau), u_0'(\tau), u_0''(\tau))$is obtained substituting Eq. 8.110 into Eq. 8.112 and therefore

$$\begin{aligned} f(\tau, u_0(\tau), u'_0(\tau), u''_0(\tau)) = {} & [(1 - \Omega^2)C_1 + \frac{1}{4}\alpha\Omega^2(-2C_1^3 - 20C_1C_2^2 - \\ & - 6C_1^2C_2 - 52C_1C_3^2 - 36C_1C_2C_3 - 22C_2^2C_3) \\ & + \frac{1}{4}\beta(3C_1^3 + 3C_1^2C_2 + 6C_1C_2^2 + 6C_1C_3^2 + 3C_2^2C_3 \\ & + 6C_1C_2C_3)] \cos \Omega t + [(1 - \Omega^2)C_2 - \frac{1}{2}\alpha\Omega^2(C_1^3 + 9C_1^2C_3 \\ & + 22C_1C_2C_3 + 8C_1^2C_2 + 9C_2^3 + 34C_2C_3^2) + \\ & + \frac{1}{4}\beta(C_1^3 + 3C_2^3 + 6C_1^2C_2 + 3C_1^2C_3 + 6C_2C_3^2 \\ & + 6C_1C_2C_3)] \cos 3\Omega t + [(1 - \Omega^2)C_3 \end{aligned}$$

$$
\begin{aligned}
&-\frac{1}{2}\alpha\Omega^2(11C_1C_2^2+9C_1^2C_2+26C_1^2C_3+34C_2^2C_3+50C_3^3)\\
&+\frac{3}{4}\beta(C_3^3+C_1^2C_2+C_1C_2^2+2C_1^2C_3+2C_2^2C_3)]\cos 5\Omega t+\\
&+[-\frac{1}{2}\alpha\Omega^2(17C_1C_2^2+19C_1^2C_3+42C_1C_2C_3+27C_2C_3^2)\\
&+\frac{3}{4}\beta(C_1C_2^2+C_1^2C_3+C_2C_3^2+2C_1C_2C_3)]\cos 7\Omega t\\
&+[-\frac{1}{2}\alpha\Omega^2(33C_1C_3^2+58C_1C_2C_3+9C_2^3)\\
&+\frac{1}{4}\beta(C_2^3+3C_1C_3^2+6C_1C_2C_3)]\cos 9\Omega t\\
&+[-\frac{1}{2}\alpha\Omega^2(43C_1C_3^2+41C_2^2C_3)+\frac{3}{4}\beta(C_1C_3^2+C_2^2C_3)]\times\\
&\times\cos 11\Omega t+[-\frac{57}{2}\alpha\Omega^2C_2C_3^2+\frac{3}{4}\beta C_2C_3^2]\cos 13\Omega t\\
&+[-\frac{25}{2}\alpha\Omega^2C_3^3+\frac{1}{4}\beta C_3^3]\cos 15\Omega t
\end{aligned}
\tag{8.114}
$$

Avoiding the presence of secular terms in the right-hand side of Eq. 8.113, we obtain from Eq. 8.114 the frequency of the system

$$
\Omega^2=\frac{4C_1+3\beta(C_1^3+C_1^2C_2+2C_1C_2^2+2C_1C_3^2+C_2^2C_3+2C_1C_2C_3)}{4C_1+2\alpha(C_1^3+10C_1C_2^2+3C_1^2C_2+26C_1C_3^2+18C_1C_2C_3+11C_2^2C_3)}
\tag{8.115}
$$

Substituting Eq. 8.114 into Eq. 8.113 we obtain

$$
\begin{aligned}
u_1(t)=&\ (C_1-\frac{a}{8\Omega^2}-\frac{b}{24\Omega^2}-\frac{c}{48\Omega^2}-\frac{d}{80\Omega^2}-\frac{e}{120\Omega^2}-\frac{f}{168\Omega^2}-\\
&-\frac{g}{224\Omega^2})\cos\Omega t+\left(C_2+\frac{a}{8\Omega^2}\right)\cos 3\Omega t+\left(C_3+\frac{b}{24\Omega^2}\right)\cos 5\Omega t+\\
&+\frac{c}{48\Omega^2}\cos 7\Omega t+\frac{d}{80\Omega^2}\cos 9\Omega t+\frac{e}{120\Omega^2}\cos 11\Omega t+\\
&+\frac{f}{168\Omega^2}\cos 13\Omega t+\frac{g}{224\Omega^2}\cos 15\Omega t
\end{aligned}
\tag{8.116}
$$

where

$$a = (1-\Omega^2)C_2 - \frac{1}{2}\alpha\Omega^2(C_1^3 + 9C_1^2C_3 + 22C_1C_2C_3 + 8C_1^2C_2 + 9C_2^3 + \\ + 34C_2C_3^2) + \frac{1}{4}\beta(C_1^3 + 3C_2^3 + 6C_1^2C_2 + 3C_1^2C_3 + 6C_2C_3^2 + 6C_1C_2C_3)$$

$$b = (1-\Omega^2)C_3 - \frac{1}{2}\alpha\Omega^2(11C_1C_2^2 + 9C_1^2C_2 + 26C_1^2C_3 + 34C_2^2C_3 + \\ + 50C_3^3) + \frac{3}{4}\beta(C_3^3 + C_1^2C_2 + C_1C_2^2 + 2C_1^2C_3 + 2C_2^2C_3)$$

$$c = -\frac{1}{2}\alpha\Omega^2(17C_1C_2^2 + 19C_1^2C_3 + 42C_1C_2C_3 + 27C_2C_3^2) + \\ + \frac{3}{4}\beta(C_1C_2^2 + C_1^2C_3 + C_2C_3^2 + 2C_1C_2C_3) \tag{8.117}$$

$$d = -\frac{1}{2}\alpha\Omega^2(33C_1C_3^2 + 58C_1C_2C_3 + 9C_2^3) + \frac{1}{4}\beta(C_2^3 + 3C_1C_3^2 + 6C_1C_2C_3)$$

$$e = -\frac{1}{2}\alpha\Omega^2(43C_1C_3^2 + 41C_2^2C_3) + \frac{3}{4}\beta(C_1C_3^2 + C_2^2C_3)$$

$$f = -\frac{57}{2}\alpha\Omega^2C_2C_3^2 + \frac{3}{4}\beta C_2C_3^2$$

$$g = -\frac{25}{2}\alpha\Omega^2C_3^3 + \frac{1}{4}\beta C_3^3$$

with Ω given by Eq. 8.115.

Substituting Eq. 8.116 into Eq. 8.107, it results the residual

$$R(t) = \ddot{u}_1 + u_1 + \alpha(u_1^2\ddot{u}_1 + u_1\dot{u}_1^2) + \beta u_1^3 \tag{8.118}$$

The residual functional is given by

$$J = \int_0^T R^2(t)dt, \quad T = \frac{2\pi}{\Omega} \tag{8.119}$$

The parameters C_1, C_2 and C_3 are derived from Eq. 8.111 and from the condition that the residual functional J be minimum (conditioned minimum):

$$\frac{\partial J}{\partial C_1} = \frac{\partial J}{\partial C_2}, \quad \frac{\partial J}{\partial C_2} = \frac{\partial J}{\partial C_3} \tag{8.120}$$

In this way, the solution (8.116) in the first approximation is well-determined.

We illustrate the accuracy of our procedure comparing the approximate solutions previously obtained with the numerical integration results obtained using a fourth-order Runge–Kutta method.

(a) When $\alpha = 0.5$, $\beta = 0.5$, $A = 2$, it is obtained
$C_1 = 2.1542$, $C_2 = -0.115724$, $C_3 = -0.038479$

$$\begin{aligned} u_1(t) = {} & 2.113788817 \cos 1.14348t - 0.1710453 \cos 3.43044t + \\ & + 0.0471591 \cos 5.7174t + 0.012261 \cos 8.00436 - \\ & - 0.00194563 \cos 10.29132t - 0.000235328 \cos 12.57828t + \\ & + 0.0000142413 \cos 14.86524t \end{aligned} \tag{8.121}$$

(b) When $\alpha = 1$, $\beta = 1$, $A = 1$, it is obtained
$C_1 = 0.99041$, $C_2 = -0.00652485$, $C_3 = -0.016108$

$$\begin{aligned} u_1(t) = {} & 0.994931286 \cos 1.09036t - 0.01076994 \cos 3.27108t + \\ & + 0.01583921 \cos 5.45181t - 0.000050524 \cos 7.63252t - \\ & - 0.000000185 \cos 9.81324t \end{aligned} \tag{8.122}$$

It is easy to verify the accuracy of the obtained results if we graphically compare these analytical solutions with the numerical ones. Figures 8.9 and 8.10 show the

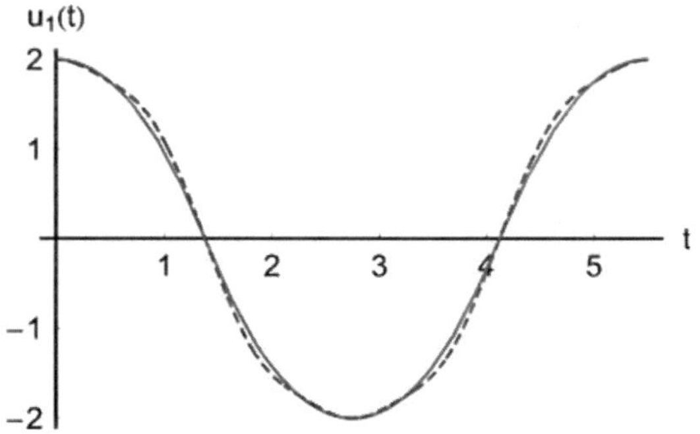

Fig. 8.9 Comparison between the results obtained for Eq. 8.107 in case (a), $\alpha = \beta = 0.5$, $A = 2$: ________ numerical integration results; _ _ _ _ _ _ approximate results given by Eq. 8.121

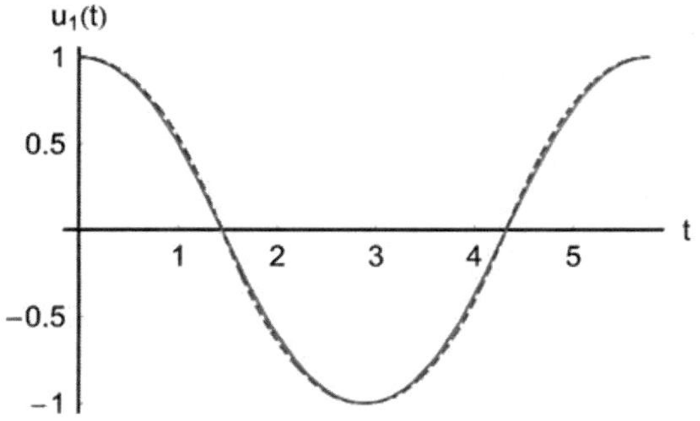

Fig. 8.10 Comparison between the results obtained for Eq. 8.107 in case (b), $\alpha = \beta = 1$, $A = 1$: ______ numerical integration results; _ _ _ _ _ _ approximate results given by Eq. 8.122

comparison between the present solutions and the numerical integration results obtained using a fourth-order Runge–Kutta method.

From Figs. 8.9 and 8.10 one can see that the errors between the approximate solutions and numerical integration results are very good.

8.5 Thin Film Flow of a Fourth-Grade Fluid Down a Vertical Cylinder

Consider the nonlinear differential equation (see Sects. 6.3 and 7.5)

$$\eta\frac{d^2f}{d\eta^2}+\frac{df}{d\eta}+k\eta+2b\left[\left(\frac{df}{d\eta}\right)^3+3\eta\left(\frac{df}{d\eta}\right)^2\frac{d^2f}{d\eta^2}\right]=0 \qquad (8.123)$$

with the initial conditions

$$f(1)=0,\quad f'(d)=0,\quad d>1 \qquad (8.124)$$

The correction functional for Eq. 8.123 can be written down as follows:

$$f_{n+1}(\eta)=f_n(\eta)+\int_0^\eta \lambda(s,\eta)\left\{s\frac{d^2f}{ds^2}+\frac{df}{ds}+ks+2b\left[\left(\frac{df}{ds}\right)^3+3s\left(\frac{df}{ds}\right)^2\frac{d^2f}{ds^2}\right]\right\}ds \qquad (8.125)$$

But Eq. 8.123 has variable coefficients and thus with notations

$$F(s,\eta,f,f',f'')=\lambda(s,\eta)\left\{sf''(s)+f'(s)+ks+2b\left[f'^3(s)+3\mu f'^2(s)f''(s)\right]\right\} \qquad (8.126)$$

the variation of the functional

$$v=\int_0^\eta F(s,\eta,f,f',f'')ds$$

symbolized by δvbecomes [165]:

$$\delta v=\delta\int_0^\eta F(s,\eta,f,f',f'')ds=\left(F_{f'}-\frac{d}{ds}F_{f''}\right)\delta f|_0^\eta+ \\ +F_{f''}\delta f'|_0^\eta+\int_0^\eta\left(F_f-\frac{d}{ds}F_{f'}+\frac{d^2}{ds^2}F_{f''}\right)\delta f ds \qquad (8.127)$$

Making the above functional stationary, noticing that $\delta\tilde{f} = 0$, yield the following stationary conditions:

$$\begin{aligned}
\delta f_n :\quad & F_f(s,\eta,f_n,f_n',f_n'') - \frac{d}{ds}F_{f'}(s,\eta,f_n,f_n',f_n'') \\
& + \frac{d^2}{ds^2}F_{f''}(s,\eta,f_n,f_n',f_n'') = 0 \\
\delta f'_n :\quad & F_{f''}(s,\eta,f_n,f'_n,f_n'')\big|_0^\eta = 0 \\
\delta f_n :\quad & 1 + \left(F_{f'} - \frac{d}{ds}F_{f'}\right)\bigg|_0^\eta = 0
\end{aligned} \tag{8.128}$$

The Lagrange multiplier therefore can be readily identified from Eq. 8.128. For Eqs. 8.123 and 8.124 the correction functional becomes

$$\begin{aligned}
f_{n+1}(\eta) = f_n(\eta) + \int_0^\eta \lambda(s,\eta)\{ & s f_n''(s) + f_n'(s) + ks \\
& + 2b[\tilde{f}'^3_n(s) + 3s\tilde{f}'^2_n(s)\tilde{f}''_n(s)]\}ds
\end{aligned} \tag{8.129}$$

The stationary conditions (8.128) in the case of Eq. 8.123 are

$$\frac{\partial\lambda(s,n)}{\partial s} + s\frac{\partial^2\lambda(s,\eta)}{\partial s^2} = 0 \tag{8.130}$$

$$\left[1 - \lambda(s,\eta) - s\frac{\partial\lambda(s,\eta)}{\partial s}\right]_{s=\eta} = 0 \tag{8.131}$$

$$\lambda(s,\eta)\big|_{s=\eta} = 0 \tag{8.132}$$

As a result, we obtain the following iteration formula [166]

$$\begin{aligned}
f_{n+1}(\eta) = f_n(\eta) + \int_d^\eta \left(1 - \frac{\eta}{s}\right)\{ & s f''_n(s) + f'_n(s) + \\
& + ks + 2b\left[{f'_n}^3(s) + 3s{f'_n}^2(s)f''_n(s)\right]\Big\}ds
\end{aligned} \tag{8.133}$$

Integrating part by part, from Eq. 8.133 we obtain the following two identities

$$\int_d^\eta \left(1 - \frac{\eta}{s}\right)[s f''_n(s) + f'_n(s)]ds = -\eta\int_d^\eta \frac{1}{s}f'_n(s)ds \tag{8.134}$$

$$\int_d^\eta \left(1 - \frac{\eta}{s}\right)\left[s{f'_n}^3(s) + 3s{f'_n}^2(s)f''_n(s)\right]ds = -\eta\int_d^\eta \frac{1}{s}{f'_n}^3(s)ds \tag{8.135}$$

From Eqs. 8.134, 8.135 and 8.133 we obtain

$$f_{n+1}(\eta) = f_n(\eta) - \eta \int_d^{\eta} \frac{1}{s}\left[f_n'(s) + 2bf_n'^{\,3}(s)\right] - \frac{1}{2}k(d-\eta)^2 \tag{8.136}$$

By differentiating into Eq. 8.136, we obtain the following iteration formula

$$f_{n+1}'(\eta) = k(d-\eta) - 2\beta f_n'^3(\eta) - \int_d^{\eta} \frac{1}{s}\left[f_n'(s) + 2\beta f_n'^{\,3}(s)\right] ds \tag{8.137}$$

For $n = 0$ into Eq. 8.137 we have

$$f_1'(\eta) = k(d-\eta) - 2bf_0'(\eta) - \int_d^{\eta} \frac{1}{s}\left[f_0'(s) + 2bf_0'^{\,3}(s)\right] ds \tag{8.138}$$

where $f_0'(\eta)$is the initial approximation which must satisfy the boundary conditions (8.124):

$$f_0(1) = 0, f_0'(d) = 0 \tag{8.139}$$

The initial approximation $f_0'(\eta)$ is chosen such as it depends on several unknown constants. For example, if we take into account only the linear part of Eq. 8.123, we can write

$$nf'' + f' + k\eta = 0, \quad f'(d) = 0 \tag{8.140}$$

Therefore, we obtain from Eq. 8.140

$$f_0' = \frac{k}{2}\left(\frac{d^2}{\eta} - \eta\right) \tag{8.141}$$

The initial approximation f_0'which appears in the right side of Eq. 8.138 is chosen in the form

$$f_0'(\eta) = C_1\left(\frac{d^2}{\eta} - \eta\right) + C_2\left(\frac{d^2}{\eta} - \eta\right)^2 + C_3\left(\frac{d^2}{\eta} - \eta\right)^3 \tag{8.142}$$

Alternatively, this initial approximation can be chosen in the form:

$$f_0'(\eta) = C_1\left(\frac{d^2}{\eta} - \eta\right) + C_2\left(\frac{d^2}{\eta} - \eta\right)^3 + C_3\left(\frac{d^2}{\eta} - \eta\right)^5 + C_4\left(\frac{d^2}{\eta} - \eta\right)^7 \tag{8.142$_1$}$$

or

$$f_0'(\eta) = C_1\left(\frac{d^2}{\eta} - \eta\right) + C_2\left(\frac{d^2}{\eta} - \eta\right)^2 + C_3\left(\frac{d^2}{\eta} - \eta\right)^4 \tag{8.142$_2$}$$

and so on. At this moment, the constants C_1, C_2 and C_3 which appear in Eq. 8.142 are unknown.

Having in view Eq. 8.138, from Eq. 8.142 we obtain:

$$
\begin{aligned}
f_0'(s) + 2bf_0'^3(s) = {} & C_1\left(\frac{d^2}{s}-s\right) + C_2\left(\frac{d^2}{s}-s\right)^2 (C_3+2bC_1^3)\left(\frac{d^2}{s}-s\right)^3 + \\
& +6bC_1^2C_2\left(\frac{d^2}{s}-s\right)^4 + 2b(C_1C_2^2+C_1^2C_3)\left(\frac{d^2}{s}-s\right) + 2bC_2^3\left(\frac{d^2}{s}-s\right)^6 + \\
& +6b(C_1C_3^2+C_2^2C_3)\left(\frac{d^2}{s}-s\right)^7 + 6bC_2C_3^2\left(\frac{d^2}{s}-s\right)^8 + 2bC_3^3\left(\frac{d^2}{s}-s\right)^9
\end{aligned}
\tag{8.143}
$$

and therefore

$$
\begin{aligned}
\int_d^\eta \frac{1}{s}[f_0'(s) + 2bf_0'^3(s)]ds = {} & C_1\left(2d-\eta-\frac{d^2}{\eta}\right) + C_2\big[d^2(1+2\ln d) - 2d^2\eta - \\
& -\frac{\eta^2}{2} - \frac{d^4}{2\eta^2}\Big] + (C_3+2bC_1^3)\left(-\frac{d^6}{3\eta^3} + \frac{3d^4}{\eta} + 3d^2\eta - \frac{\eta^3}{3} - \frac{16d^2}{3}\right) + \\
& + 6bC_1^2C_2\left(-\frac{d^8}{4\eta^2} + \frac{2d^6}{\eta^2} + 6d^4\ln\eta - 2d^2\eta^2 + \frac{\eta^4}{4} - 6d^4\right) + 2b(C_1C_2^2 + \\
& + C_1^2C_3)\left(-\frac{d^{10}}{5\eta^5} + \frac{5d^8}{3\eta^3} - \frac{10d^6}{\eta} - 10d^4\eta + \frac{5d^2}{3}\eta^3 - \frac{\eta^5}{5} + \frac{256d^5}{15}\right) + \\
& + 2bC_2^3\left(-\frac{d^{12}}{6\eta^6} + \frac{3d^{10}}{2\eta^4} - \frac{15d^8}{2\eta^2} - 20d^6\ln\eta + \frac{15}{2}d^4\eta^2 - \frac{3}{2}d^2\eta^4 + \right.\\
& \left. + \frac{\eta^4}{6} + 20d^6\right) + 2b(C_1C_3^2 + C_2^2C_3)\left(-\frac{d^{14}}{7\eta^7} + \frac{7d^{12}}{5\eta^5} - \frac{7d^{10}}{\eta^3} + \frac{35d^8}{\eta} + \right.\\
& \left. + 35d^6\eta - 21d^4\eta^3 + 7d^2\eta^5 - \frac{\eta^7}{7} - \frac{1754d^7}{35}\right) + 6bC_2C_3^2\left(-\frac{d^{16}}{8\eta^8} + \right.\\
& + \frac{4d^{14}}{3\eta^6} - \frac{14d^{12}}{3\eta^4} + \frac{28d^{10}}{\eta^2} + 70d^8\ln\eta - 28d^6\eta^2 + 7d^4\eta^4 - \frac{4}{3}d^2\eta^6 + \\
& \left. + \frac{\eta^8}{8} - \frac{7}{3}d^8 - 70d^8\ln d\right) + 2bC_3^3\left(-\frac{d^{18}}{9\eta^4} + \frac{9d^{16}}{7\eta^7} - \frac{36d^{14}}{5\eta^5} + \frac{28d^{12}}{\eta^3} - \right.\\
& \left. - \frac{126d^{10}}{\eta} - 126d^8\eta + 28d^6\eta^3 - \frac{36}{5}d^4\eta^5 + \frac{9}{7}d^2\eta^7 - \frac{\eta^9}{9} + \frac{64536d^9}{315}\right)
\end{aligned}
\tag{8.144}
$$

Substituting Eq. 8.144 into Eq. 8.138 we obtain

$$f_1(\eta) = C_1\frac{d^2}{\eta} + (C_1 - k)\eta + (k - 2C_1)d + C_2\left[\frac{d^4}{2\eta^2} + \frac{\eta^2}{2} + 2d\ln\eta - \right.$$
$$\left. - d^2(1 + 2\ln d)\right] - 2bC_1^3\left(\frac{d^2}{\eta} - \eta\right)^3 + 6bC_1^3C_2\left(\frac{d^2}{\eta} - \eta\right)^4 -$$
$$- 2b(C_1C_2^2 + C_1^2C_3)\left(\frac{d^2}{\eta} - \eta\right)^5 - 2bC_2^3\left(\frac{d^2}{\eta} - \eta\right)^6 - 6b(C_1C_3^2 +$$
$$+ C_2^2C_3\left(\frac{d^2}{\eta} - \eta\right)^7 + (C_3 + 2\beta C_1^3)\left(\frac{d^6}{3\eta^3} - \frac{3d^4}{\eta} - 3d^2\eta + \frac{\eta^3}{3} + \frac{16d^2}{3}\right) +$$
$$+ 6bC_1^2C_2\left(\frac{d^8}{4\eta^2} - \frac{2d^6}{\eta^2} - 6d^4\ln\eta + 2d^2\eta^2 - \frac{\eta^4}{4} + 6d^4\right) + 2b(C_1C_2^2 +$$
$$+ C_1^2C_3)\left(\frac{d^{10}}{5\eta^5} - \frac{5d^8}{3\eta^3} + \frac{10d^6}{\eta} + 10d^4\eta - \frac{5d^2}{3}\eta^3 + \frac{\eta^5}{5} - \frac{256d^5}{15}\right) +$$
$$+ 2bC_2^3\left(\frac{d^{12}}{6\eta^6} - \frac{3d^{10}}{2\eta^4} + \frac{15d^8}{2\eta^2} + 20d^6\ln\eta - \frac{15}{2}d^4\eta^2 + \frac{3}{2}d^2\eta^4 - \right.$$
$$\left. - \frac{\eta^4}{6} - 20d^6\right) + 2b(C_1C_3^2 + C_2^2C_3)\left(\frac{d^{14}}{7\eta^7} - \frac{7d^{12}}{5\eta^5} + \frac{7d^{10}}{\eta^3} - \frac{35d^8}{\eta} - \right.$$
$$\left. - 35d^6\eta + 21d^4\eta^3 - 7d^2\eta^5 + \frac{\eta^7}{7} + \frac{1754d^7}{35}\right) + \ldots \tag{8.145}$$

The residual functional J given by Eq. 8.85 is

$$J = \int_1^d \left\{\eta f''_1(\eta) + f'_1(\eta) + k\eta + 2b[f'^3_1(\eta) + 3\eta f'^2_1(\eta)f''_1(\eta)]\right\}^2 d\eta \tag{8.146}$$

The constants C_1, C_2 and C_3 are determined from Eq. 8.86 and thus from the conditions ($m = 0, n = 3$):

$$\frac{\partial J}{\partial C_1} = \frac{\partial J}{\partial C_2} = \frac{\partial J}{\partial C_3} = 0 \tag{8.147}$$

The explicit analytic expression given by Eq. 8.145 contains the parameters C_1, C_2 and C_3 which give the convergence region and rate of approximation for the OVIM. In order to prove the efficiency of the OVIM we consider different cases for some values of the parameters k, β and d [166].

(a) The case $k = 1, \beta = 1$
Form the system (8.147) we obtain

$$C_1 = 0.5, C_2 = 0.00001127, C_3 = 0.74789281$$

Table 8.3 Comparison between the present solution (8.145) and the exact solution for $k = \beta = 1$, $d = 1.02$

η	$\bar{f}'$given by Eq. 8.145	f'exact
1	0.020183555	0.020183555
1.005	0.015105047	0.015105047
1.007	0.013079437	0.013079437
1.01	0.010047475	0.010047475
1.0105	0.009542918	0.009542918
1.0108	0.009240289	0.009240289

Table 8.4 Comparison between the present solution (8.145) and the exact solution for $k = \beta = 1$, $d = 1.04$

η	$\bar{f}'$given by (8.145)	f'exact
1	0.040665504	0.040665504
1.008	0.032439661	0.032439661
1.016	0.024254926	0.024254926
1.023	0.017131195	0.017131195
1.03	0.010046515	0.010046515
1.038	0.00200191	0.00200191

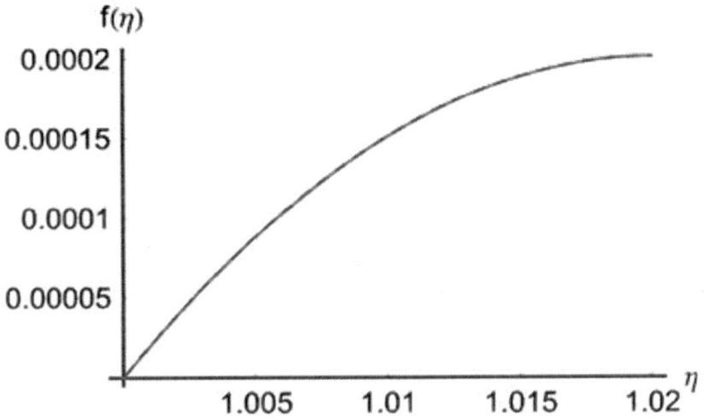

Fig. 8.11 Comparison between the present solution (8.145) of Eq. 8.123 and the numerical solution for $k = 1$, $\beta = 1, d = 1.02$, ________ numerical integration results; _ _ _ _ _ _ approximate results

In Tables 8.3 and 8.4 are presented some comparisons between the present solution obtained from formula (8.145) and the exact solution of Eq. 8.123 for $d = 1.02$ and $d = 1.04$ respectively. It can be seen that the solution obtained by the present method is identical with that given by the exact solution, demonstrating a very good accuracy.

Figures 8.11 and 8.12 present a comparison between the present solution given by Eq. 8.145 and the exact solution of Eq. 8.123 for $d = 1.02$ and $d = 1.04$, respectively.

(b) The case $k = 1, \beta = 1.5$

In this case we obtain

$$C_1 = 0.5, C_2 = 0.000012321, C_3 = 0.74791023$$

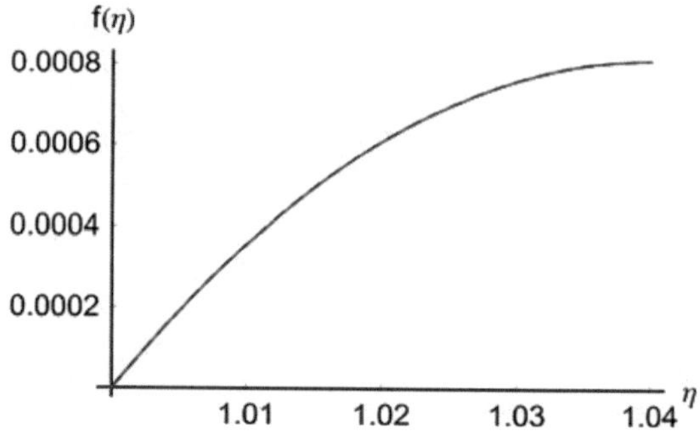

Fig. 8.12 Comparison between the present solution (8.145) of Eq. 8.123 and the numerical solution for k = 1, β = 1, d = 1.04, ________ numerical integration results; _ _ _ _ _ _ approximate results

Table 8.5 Comparison between the present solution (8.145) and the exact solution for $k = 1$, $\beta = 1.5$, $d = 1.02$

η	$\bar{f}'$given by Eq. 8.145	f'exact
1	0.020175363	0.020175363
1.005	0.015101608	0.015101607
1.007	0.013077203	0.013077203
1.01	0.010046462	0.010046462
1.0105	0.009542049	0.009542049
1.0108	0.009239501	0.009239501

Table 8.6 Comparison between the present solution (8.145) and the exact solution for $k = 1$, $\beta = 1.5$, $d = 1.04$

η	$\bar{f}'$given by Eq. 8.145	f'exact
1	0.040599239	0.040599241
1.008	0.032405844	0.032405844
1.016	0.024240732	0.024240731
1.023	0.017126181	0.017126181
1.03	0.010045502	0.010045502
1.038	0.002001902	0.002001901

In Tables 8.5 and 8.6 are presented some comparisons between the present solution (8.145) and the exact solution for $d = 1.02$ and $d = 1.04$, respectively.

It is easy to verify the accuracy of the obtained solutions if we graphically compare these analytical solutions with the exact ones (Figs. 8.13 and 8.14).

The results obtained through the proposed method reveal very good agreement with the exact solution. It is to observe that we need only one iteration to obtain a remarkable good accuracy.

8.6 Dynamic Analysis of a Rotating Electric Machine

Rotating electric machines are elastic systems exhibiting nonlinear vibrations in their working regime. From the engineering point of view it is important to accurately predict the behaviour of such a system. This prediction is the key to

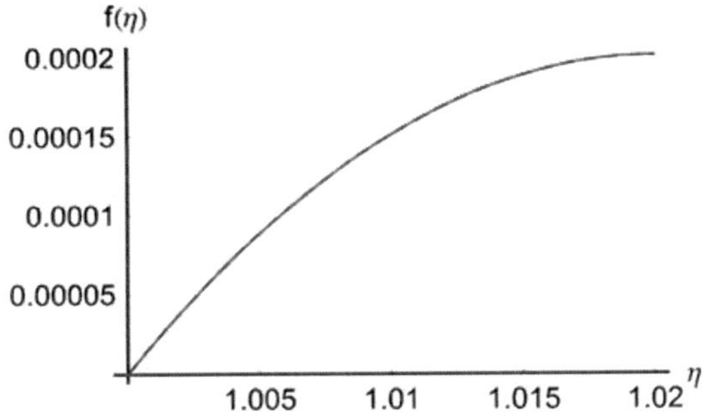

Fig. 8.13 Comparison between the present solution (8.145) of the Eq. 8.123 and the numerical solution for $k = 1$, $\beta = 1.5$, $d = 1.02$, ________ numerical integration results; _ _ _ _ _ _ approximate results

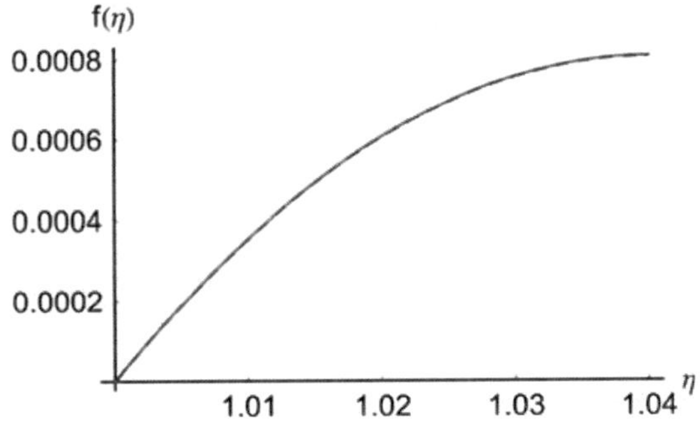

Fig. 8.14 Comparison between the present solution (8.145) of the Eq. 8.123 and the numerical solution for $k = 1$, $\beta = 1.5$, $d = 1.04$, ________ numerical integration results; _ _ _ _ _ _ approximate results

the design of high-performance electric machines with higher speeds or longer periods between downtimes. The main causes that determine the occurrences of undesirable vibration in these dynamic systems are the nonlinear bearings, which support the rotating machine, the unbalanced forces, the mechanical looseness or misalignments. All these factors can affect the integrity of the system, therefore being highly detrimental [167].

In this section, a rotating electrical machine supported by nonlinear bearings is investigated. The main sources of undesirable nonlinear vibration for this dynamic system are the nonlinear bearings, which support the rotating machine, the unbalanced forces and misalignments. All these factors can affect the integrity of the system, being therefore highly detrimental. The rotating machine under study is modelled as a 2-DOF system, assuming that the rotor mass is lumped at the midpoint, the rotor speed is constant, the axial and torsional vibrations and the mass of the shaft are negligible, as well as the damping. In these simplifying hypotheses, the equations of motion describing the dynamic behaviour of the system are:

$$\begin{aligned} &\ddot{x}_1 + \left(\frac{1}{s_1^2} + \frac{1}{2cs^2}\right)x_1 + \frac{\alpha}{s^2}x_1^3 = 0 \\ &\ddot{y}_1 + \left(\frac{1}{s_1^2} + \frac{1}{2cs^2}\right)y_1 + \frac{\alpha}{s^2}y_1^3 + \frac{f}{s^2} = 0 \end{aligned} \tag{8.148}$$

where x_1 is the horizontal and y_1 the vertical displacement and s_1, s, c, α and f are known parameters.

Using the notations

$$B_1 = \frac{1}{s_1^2} + \frac{1}{2cs^2}, \quad B_2 = \frac{\alpha}{s^2}, \quad B_3 = \frac{f}{s^2} \tag{8.149}$$

we can rewrite Eq. 8.148 as a system of differential equations

$$\begin{aligned} &\ddot{x}_1 + \Omega_1^2 x_1 + (B_1 - \Omega_1^2)x_1 + B_2 x_1^3 = 0 \\ &\ddot{y}_1 + \Omega_2^2 y_1 + (B_1 - \Omega_2^2)y_1 + B_2 y_1^3 + B_3 = 0 \end{aligned} \tag{8.150}$$

where Ω_1 and Ω_2 are the frequencies of x_1 and y_1 respectively. The following variational formulas can be obtained

$$\begin{aligned} x_{1,n}(t) &= x_{1,0}(t) + \frac{1}{\Omega_1}\int_0^t \sin\Omega_1(s-t) f_1(x_{1,0})ds \\ y_{1,n}(t) &= y_{1,0}(t) + \frac{1}{\Omega_2}\int_0^t \sin\Omega_2(s-t) f_2(y_{1,0})ds \end{aligned} \tag{8.151}$$

where

$$\begin{aligned} f_1(x_{10}) &= (B_1 - \Omega_1^2)x_{10} + B_2 x_{10}^3 \\ f_2(y_{10}) &= (B_1 - \Omega_2^2)y_{10} + B_2 y_{10}^3 + B_3 \end{aligned} \tag{8.152}$$

We start with the initial approximations

$$\begin{aligned} x_{10}(t) &= C_1\cos\Omega_1 t + C_2\cos 3\Omega_1 t + C_3\cos 5\Omega_1 t \\ y_{10}(t) &= C_4\cos\Omega_2 t + C_5\cos\Omega_2 t + C_6\cos 5\Omega_2 t \end{aligned} \tag{8.153}$$

where $C_1, C_2, \ldots, C_6$ are unknown constants at this moment.

Substituting Eq. 8.153 into Eq. 8.152 we obtain the following results:

$$\begin{aligned} f_1(x_{10}) = {} & [C_1(B_1 - \Omega_1^2) + \frac{3}{4}B_2(C_1^3 + C_1^2C_2 + 2C_1C_2^2 + 2C_1C_3^2 + \\ & + C_2^2C_3 + 2C_1C_2C_3)]\cos\Omega_1 t + M_3\cos 3\Omega_1 t + M_5\cos 5\Omega_1 t + \\ & + M_7\cos 7\Omega_1 t + M_9\cos 9\Omega_1 t + M_{11}\cos 11\Omega_1 t + \\ & + M_{13}\cos 13\Omega_1 t + M_{15}\cos 15\Omega_1 t \end{aligned} \tag{8.154}$$

$$\begin{aligned} f_2(y_{10}) = {} & [C_4(B_1 - \Omega_2^2) + \frac{3}{4}B_2(C_4^3 + C_4^2C_5 + 2C_4C_5^2 + 2C_4C_6^2 + \\ & + C_5^2C_6 + 2C_4C_5C_6)]\cos\Omega_2 t + N_3\cos 3\Omega_2 t + N_5\cos 5\Omega_2 t + \\ & + N_7\cos 7\Omega_2 t + N_9\cos 9\Omega_2 t + N_{11}\cos 11\Omega_2 t + \\ & + N_{13}\cos 13\Omega_2 t + N_{15}\cos 15\Omega_2 t + B_3 \end{aligned} \tag{8.155}$$

where

$$M_3 = (B_1 - \Omega_1^2)C_2 + \frac{1}{4}B_2(C_1^3 + 3C_2^3 + 6C_1^2C_2 + 3C_1^2C_3 + 6C_2C_3^2 + 6C_1C_2C_3)$$

$$M_5 = (B_1 - \Omega_1^2)C_3 + \frac{3}{4}B_2(C_3^3 + C_1^2C_2 + C_1C_2^2 + 2C_1^2C_3 + 2C_2^2C_3)$$

$$M_7 = \frac{3}{4}B_2(C_1C_2^2 + C_1^2C_3 + C_2C_3^2 + 2C_1C_2C_3)$$

$$M_9 = \frac{3}{4}B_2(C_1C_3^2 + C_2^3 + 2C_1C_2C_3)$$

$$M_{11} = \frac{3}{4}B_2(C_1C_3^2 + C_2^2C_3), \quad M_{13} = \frac{3}{4}B_2C_2C_3^2, \quad M_{15} = \frac{1}{4}B_2C_3^3$$

$$N_3 = (B_1 - \Omega_2^2)C_5 + \frac{1}{4}B_2(C_4^3 + 3C_5^3 + 6C_4^2C_5 + 3C_4^2C_5 + 6C_5C_6^2 + 6C_4C_5C_6)$$

$$N_5 = (B_1 - \Omega_2^2)C_6 + \frac{3}{4}B_2(C_6^3 + C_4^2C_5 + C_4C_5^2 + 2C_4^2C_6 + 2C_5^2C_6)$$

$$N_7 = \frac{3}{4}B_2(C_4C_5^2 + C_4^2C_6 + C_5C_6^2 + 2C_4C_5C_6)$$

$$N_9 = \frac{3}{4}B_2(C_4C_6^2 + C_5^2 + 2C_4C_5C_6), \; N_{11} = \frac{3}{4}B_2(C_4C_6^2 + C_5^2C_6)$$

$$N_{13} = \frac{3}{4}B_2C_5C_6^2, \quad N_{15} = \frac{1}{4}B_2C_6^3 \tag{8.156}$$

Avoiding the presence of secular terms in the right-hand side of Eq. 8.151, we obtain the frequencies

$$\Omega_1^2 = B_1 + \frac{3B_2}{4C_1}(C_1^3 + C_1^2C_2 + 2C_1C_2^2 + 2C_1C_3^2 + 2C_2^2C_3 + 2C_1C_2C_3) \tag{8.157}$$

$$\Omega_2^2 = B_1 + \frac{3B_2}{4C_4}(C_4^3 + C_4^2C_5 + 2C_4C_5^2 + 2C_4C_6^2 + 2C_5^2C_6 + 2C_4C_5C_6) \tag{8.158}$$

Finally, the first iteration ($n = 1$ into Eq. 8.151) results in

$$\begin{aligned}x_{1,1}(t) = {} & C_1 \cos \Omega_1 t + C_2 \cos 3\Omega_1 t + C_3 \cos 5\Omega_1 t + \frac{M_3}{8\Omega_1^2}(\cos 3\Omega_1 t - \\ & - \cos \Omega_1 t) + \frac{M_5}{24\Omega_1^2}(\cos 5\Omega_1 t - \cos \Omega_1 t) + \frac{M_7}{48\Omega_1^2}(\cos 7\Omega_1 t - \cos \Omega_1 t) + \\ & + \frac{M_9}{80\Omega_1^2}(\cos 9\Omega_1 t - \cos \Omega_1 t) + \frac{M_{11}}{120\Omega_1^2}(\cos 11\Omega_1 t - \cos \Omega_1 t) + \\ & + \frac{M_{13}}{168\Omega_1^2}(\cos 13\Omega_1 t - \cos \Omega_1 t) + \frac{M_{15}}{224\Omega_1^2}(\cos 15\Omega_1 t - \cos \Omega_1 t)\end{aligned} \tag{8.159}$$

$$\begin{aligned}y_{1,1}(t) = {} & C_4 \cos \Omega_2 t + C_5 \cos 3\Omega_2 t + C_6 \cos 5\Omega_2 t + \frac{N_3}{8\Omega_2^2}(\cos 3\Omega_2 t - \\ & - \cos \Omega_2 t) + \frac{N_5}{24\Omega_2^2}(\cos 5\Omega_2 t - \cos \Omega_2 t) + \frac{N_7}{48\Omega_2^2}(\cos 7\Omega_2 t - \cos \Omega_2 t) + \\ & + \frac{N_9}{80\Omega_2^2}(\cos 9\Omega_2 t - \cos \Omega_2 t) + \frac{N_{11}}{120\Omega_2^2}(\cos 11\Omega_2 t - \cos \Omega_2 t) + \\ & + \frac{N_{13}}{168\Omega_2^2}(\cos 13\Omega_2 t - \cos \Omega_2 t) + \frac{N_{15}}{224\Omega_2^2}(\cos 15\Omega_2 t - \cos \Omega_2 t) + \\ & + \frac{B_3}{\Omega_2^2}(\cos \Omega_2 t - 1)\end{aligned} \tag{8.160}$$

Applying $x_{1,1}(0) = A_1, \dot{x}_{1,1}(0) = 0, y_{1,1}(0) = A_2, \dot{y}_{1,1}(0) = 0$ as initial conditions on the approximate solutions (8.159) and respectively (8.160), we obtain the following system of algebraic equations:

$$x_{1,1}(0) = C_1 + C_2 + C_3 = A_1; \quad y_{1,1}(0) = C_4 + C_5 + C_6 = A_2 \tag{8.161}$$

From stationary conditions and Eq. 8.161 we can obtain the following system of equations with the unknown C_1, C_2,. . .,C_6

$$\begin{aligned}& C_3 = A_1 - C_1 - C_2, \quad C_6 = A_2 - C_4 - C_5 \\ & \frac{\partial J}{\partial C_1} = \frac{\partial J}{\partial C_2} = 0; \quad \frac{\partial J}{\partial C_4} = \frac{\partial J}{\partial C_5} = 0\end{aligned} \tag{8.162}$$

where

$$\begin{aligned}J(C_1, C_2, ..., C_6) = {} & \int_0^{T_1} [x''_{1,1} + \Omega_1^2 x_{1,1} + f_1(x_{1,1})]^2 ds + \\ & + \int_0^{T_2} [y''_{1,1} + \Omega_2^2 y_{1,1} + f_2(y_{1,1})]^2 ds\end{aligned} \tag{8.163}$$

In this way, the approximate solutions (8.159) and (8.160) are well-determined. In order to assess the advantages and the accuracy of OVIM for solving this nonlinear problem, some numerical applications were performed for a certain working regime characterized by $s_1 = 0.756829$, $s = 2.07797$, $\alpha = 6.96429$, $c = 0.5, f = 0.013734$. In these conditions, we obtained for the constants $C_1 \ldots C_6$ the values:

$$C_1 = 0.299582;\ C_2 = 0.000188573;\ C_3 = 0.00022966;\ C_4 = 0.40038;$$
$$C_5 = 0.000585701;\ C_6 = -0.000965953$$

which must be replaced in Eqs. 8.159 and 8.160 to obtain the analytic solutions

$$\begin{aligned} x_1 = {} & 0.299582 \cos[1.35982\ t] + 0.000188573 \cos[4.07946\ t] + \\ & + 0.000646829\ (-\cos[1.35982\ t] + \cos[4.07946\ t]) + \\ & + 0.00022966 \cos[6.79911\ t] \end{aligned} \tag{8.164}$$

$$\begin{aligned} y_1 = {} & 0.40038 \cos[1.3872\ t] + 0.000585701 \cos[4.16159\ t] + \\ & + 0.00147327\ (\cos[4.16159\ t] - \cos[1.3872\ t]) - \\ & - 0.000965953 \cos[6.93599\ t] - 0.000165287616 \end{aligned} \tag{8.165}$$

In order to assess the validity and accuracy of the obtained results, a comparison between the approximate results and the results obtained through a fourth-order Runge–Kutta method was developed. Figures 8.15 and 8.16 show this comparison, where one can observe that analytical and numerical results are nearly identical.

8.7 Oscillators with Fractional-Power Nonlinearities

Among nonlinear oscillators, an interesting category is represented by the oscillators with fractional-power nonlinearities which were studied using different procedures [168].

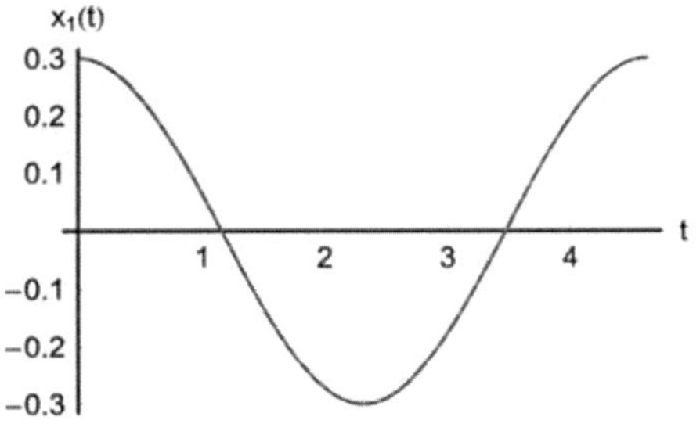

Fig. 8.15 Displacement x_1(t) given by Eq. 8.164: _ _ _ _ approximate results; ______ numerical integration results for Eq. 8.150

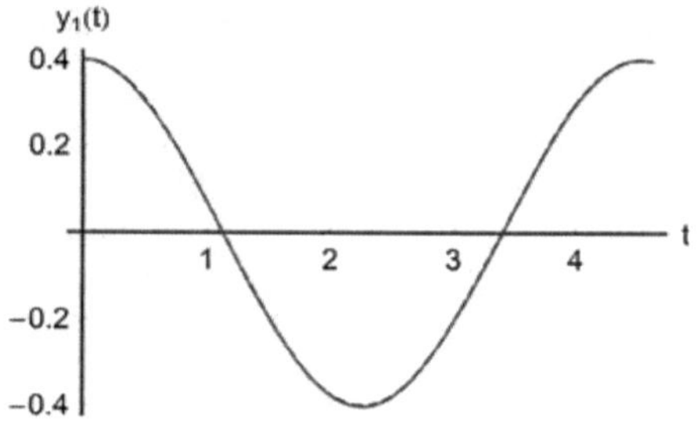

Fig. 8.16 Displacement $y_1(t)$ given by Eq. 8.165: _ _ _ _ approximate results; _______ numerical integration results for Eq. 8.150

The purpose of this section is to construct accurate approximations to periodic solutions and frequencies of nonlinear oscillators with fractional-power restoring force by using the OVIM. In order to develop an application of the proposed method, we consider the nonlinear oscillator with fractional-power restoring force:

$$\ddot{u} + \omega_0^2 u + \alpha^2 u|u|^{n-1} - f\cos\Omega t = 0 \tag{8.166}$$

where $\omega_O, \alpha, f, \Omega$ are constants and $n > 0$ rational.

Initial conditions are given by

$$u(0) = a, \quad \dot{u}(0) = 0 \tag{8.167}$$

There exists no small parameter in the equation and therefore the traditional perturbation methods cannot be applied directly.

It can be shown that

$$u|u|^{n-1} = |u|^n signu \tag{8.168}$$

so that Eq. 8.166 can be rewritten as

$$\ddot{u} + \omega^2 u + (\omega_0^2 - \omega^2)u + \alpha^2|u|^n signu - f\cos\Omega t = 0 \tag{8.169}$$

where ω is the frequency of the system.

The Eq. 8.169 describes a system oscillating with the unknown frequency ω and the period T, we switch to a scalar time$\tau = \frac{2\pi t}{T} = \omega t$. Under the transformation

$$\tau = \omega t \tag{8.170}$$

we can rewrite Eq. 8.169 in the form

$$u''(\tau) + u(\tau) + h(\tau, u(\tau)) = 0 \tag{8.171}$$

where prime denotes differentiation with respect to τ and

$$h(\tau, u(\tau)) = \left(\frac{\omega_0^2}{\omega^2} - 1\right)u + \frac{\alpha^2}{\omega^2}|u|^n signu - \frac{f}{\omega^2}\cos\frac{\Omega}{\omega}\tau \tag{8.172}$$

As an initial approximation for $u_0(\tau)$we choose:

$$u_0(\tau) = C_1 \cos\tau + C_2 \cos 3\tau + C_3 \cos 5\tau + C_4 \cos 7\tau \tag{8.173}$$

where C_1, C_2, C_3 and C_4 are unknown constants which partially can be determined from Eq. 8.167:

$$C_1 + C_2 + C_3 + C_4 = a \tag{8.174}$$

It is known that if g is an analytic function, then

$$g(y + p) = g(y) + \frac{p}{1!}g'(y) + \frac{p^2}{2!}g''(y) + ..., \quad k! = 1 \cdot 2 \cdot 3 \cdot .... \cdot k \tag{8.175}$$

for any real y and p, where prime denotes differentiation with respect to y.

In our case:

$$\begin{aligned} g(u) &= |u|^n signu, \quad g'(u) = n|u|^{n-1}, \\ y &= C_1 \cos\tau, \quad p = C_2 \cos 3\tau + C_3 \cos 5\tau + C_4 \cos 7\tau \end{aligned} \tag{8.176}$$

From Eq. 8.8 for $n = 0$ it is obtained:

$$u_1(\tau) = u_0(\tau) + \int_0^\tau \sin(s - \tau) h(s, u_0(s)) ds \tag{8.177}$$

where $h(s, u_0(s))$ is obtained substituting Eq. 8.173 into Eq. 8.172. On the other hand, using only the first two terms into Eq. 8.175, we can approximate g in the form:

$$g(u_0(s)) = g(C_1 \cos s) + (C_2 \cos 3s + C_3 \cos 5s + C_4 \cos 7s) g'(C_1 \cos s) \tag{8.178}$$

For $g(C_1 \cos s)$, we obtain the following Fourier series expansions:

$$g(C_1 \cos s) = C_1^n (a_{1n} \cos s + a_{3n} \cos 3s + a_{5n} \cos 5s + a_{7n} \cos 7s + ...) \tag{8.179}$$

where

$$a_{2j+1,n} = \frac{4}{\pi} \int_0^{\frac{\pi}{2}} (\cos s)^n \cos(2j+1) s ds, \quad j = 0, 1, 2, ...n \in \Re \tag{8.180}$$

The last term in Eq. 8.178 can be written in the form

$$
\begin{aligned}
&(C_2 \cos 3s + C_3 \cos 5s + C_4 \cos 7s) g'(C_1 \cos s) = \\
&\quad = nC_1^{-1}(C_1 \cos s)^n [C_2(2\cos 2s - 1) + \\
&\quad + C_3(2\cos 4s - 2\cos 2s + 1) + C_4(2\cos 6s - 2\cos 4s + 2\cos 2s - 1)]
\end{aligned} \tag{8.181}
$$

From Eqs. 8.179 and 8.181 we obtain

$$
\begin{aligned}
&(C_2 \cos 3s + C_3 \cos 5s + C_4 \cos 7s) g'(C_1 \cos s) = \\
&\quad = nC_1^{n-1}\{(a_{3n}C_2 + a_{5n}C_3 + a_{7n}C_4)\cos s + \\
&\quad + [a_{1n}C_2 + a_{3n}(C_3 - C_2) + a_{5n}(C_2 - C_3 + C_4) + a_{7n}(C_3 - C_4)]\cos 3s + \\
&\quad + [a_{1n}C_3 + a_{3n}(C_2 - C_3 + C_4) + a_{5n}(C_3 - C_2 - C_4) + a_{7n}(C_2 - \\
&\quad - C_3 + C_4)]\cos 5s + [a_{1n}C_4 + a_{3n}(C_3 - C_4) + a_{5n}(C_2 - C_3 + C_4) + \\
&\quad + a_{7n}(C_3 - C_2 - C_4)]\cos 7s + [a_{3n}C_4 + a_{5n}(C_3 - C_4) + \\
&\quad + a_{7n}(C_2 - C_3 + C_4)]\cos 9s + [a_{5n}C_4 + a_{7n}(C_3 - C_4)]\cos 11s + \\
&\quad + a_{7n}C_4 \cos 13s\}
\end{aligned} \tag{8.182}
$$

Substituting Eqs. 8.182 and 8.179 into Eq. 8.178 and then substituting Eq. 8.178 in Eq. 8.172, we obtain

$$
\begin{aligned}
h(s, u_0(s)) = \Bigg[\left(\frac{\omega_0^2}{\omega^2} - 1\right)C_1 + \frac{\alpha^2}{\omega^2}(a_{1n}C_1^n + na_{3n}C_1^{n-1}C_2 + \\
+ na_{5n}C_1^{n-1}C_3 + na_{7n}C_1^{n-1}C_4)\Bigg]\cos s + H.O.T.
\end{aligned} \tag{8.183}
$$

Avoiding the presence of secular terms in the right-hand side of Eq. 8.177, we obtain from Eq. 8.183 the frequency ω as

$$
\omega^2 = \omega_0^2 + \alpha^2 C_1^{n-2}(a_{1n}C_1 + na_{3n}C_2 + na_{5n}C_3 + na_{7n}C_4) \tag{8.184}
$$

With this requirement, Eq. 8.183 becomes

$$
\begin{aligned}
h(s, u_0(s)) = A\cos 3s + B\cos 5s + C\cos 7s + D\cos 9s + \\
+ E\cos 11s + F\cos 13s - \frac{f}{\omega^2}\cos\frac{\Omega}{\omega}s
\end{aligned} \tag{8.185}
$$

where:

$$
\begin{aligned}
A = \left(\frac{\omega_0^2}{\omega^2} - 1\right)C_2 + \frac{\alpha^2 C_1^{n-1}}{\omega^2}[a_{3n}C_1 + n(a_{1n} - a_{3n} + a_{5n})C_2 + \\
+ n(a_{3n} - a_{5n} + a_{7n})C_3 + n(a_{5n} - a_{7n})C_4]
\end{aligned}
$$

$$B = \left(\frac{\omega_0^2}{\omega^2} - 1\right)C_3 + \frac{\alpha^2 C_1^{n-1}}{\omega^2}[a_{5n}C_1 + n(a_{3n} - a_{5n} + a_{7n})C_2 + $$
$$+n(a_{1n} - a_{3n} + a_{5n} - a_{7n})C_3 + n(a_{3n} - a_{5n} + a_{7n})C_4]$$

$$C = \left(\frac{\omega_0^2}{\omega^2} - 1\right)C_4 + \frac{\alpha^2 C_1^{n-1}}{\omega^2}[a_{7n}C_1 + n(a_{5n} - a_{7n})C_2 + $$
$$+ n(a_{3n} - a_{5n} + a_{7n})C_3 + n(a_{1n} - a_{3n} + a_{5n} - a_{7n})C_4]$$

$$D = \frac{\alpha^2}{\omega^2}nC_1^{n-1}[a_{7n}C_2 + (a_{5n} - a_{7n})C_3 + (a_{3n} - a_{5n} + a_{7n})C_4]$$

$$E = \frac{\alpha^2}{\omega^2}nC_1^{n-1}[a_{7n}C_3 + (a_{5n} - a_{7n})C_4], \quad F = a_{7n}nC_1^{n-1}C_4 \tag{8.186}$$

Substituting Eqs. 8.185 and 8.170 into Eq. 8.177 yields:

$$\begin{aligned} u_1(t) &= C_1\cos\omega t + C_2\cos 3\omega t + C_3\cos 5\omega t + C_4\cos 7\omega t + \\ &+ \frac{A}{8}(\cos 3\omega t - \cos\omega t) + \frac{B}{24}(\cos 5\omega t - \cos\omega t) + \\ &+ \frac{C}{48}(\cos 7\omega t - \cos\omega t) + \frac{D}{80}(\cos 9\omega t - \cos\omega t) + \\ &+ \frac{E}{120}(\cos 11\omega t - \cos\omega t) + \frac{F}{168}(\cos 13\omega t - \cos\omega t) + G(t) \end{aligned} \tag{8.187}$$

where ω is given by Eq. 8.184 and

$$G(t) = \begin{cases} \dfrac{f\Omega}{\omega(\omega^2 - \Omega^2)}[1 - \cos(\omega - \Omega)t], & \text{if } \omega \neq \Omega \\ \dfrac{2f}{\omega^2\varepsilon(2+\varepsilon)}\sin\dfrac{1}{2}\varepsilon\omega t\sin(\omega + \dfrac{1}{2}\varepsilon\omega)\text{t}, & \text{if } \Omega = \omega(1+\varepsilon),\ 0<\varepsilon<<1 \end{cases} \tag{8.188}$$

At this moment the first approximation given by Eq. 8.187 depends on the parameters C_1, C_2, C_3 and C_4. If $R(t, C_1, C_2, C_3, C_4)$ is the residual obtained substituting the first approximation (8.187) into Eq. 8.166:

$$R(t, C_1, C_2, C_3, C_4) = \ddot{u}_1 + \omega_0 u_1 + \alpha^2 u_1|u_1|^{n-1} - f\cos\Omega t \tag{8.189}$$

then the constants C_1, C_2, C_3 and C_4 can be determined from Eq. 8.174 and by means of the collocation method:

$$R(t_1, C_1, C_2, C_3, C_4) = R(t_2, C_1, C_2, C_3, C_4) = R(t_3, C_1, C_2, C_3, C_4) = 0 \tag{8.190}$$

where t_i, $i = 1,2,3$ are three arbitrary values, such that

$$0 < t_i < \frac{2\pi}{\omega}, \quad i = 1, 2, 3 \tag{8.191}$$

We illustrate the accuracy of the OVIM by comparing the previously obtained approximate solutions (8.187) with the numerical integration results obtained by a fourth-order Runge–Kutta method.

I. In the first case we consider $n = \frac{1}{3}$ and from Eq. 8.180 it is obtained

$$a_{1,\frac{1}{3}} = 1.159595266; a_{3,\frac{1}{3}} = -0.231919053;$$
$$a_{5,\frac{1}{3}} = 0.115959526; a_{7,\frac{1}{3}} = -0.073792426$$

(a) For $a = 5$, $\alpha = 1$, $\omega_0 = 0, f = 0$, from Eqs. 8.174 and 8.190 we obtain

$$C_1 = 5.083874107; \; C_2 = 0.005746695; \; C_3 = -0.057474928;$$
$$C_4 = -0.032145873$$

The first-order approximate solution (8.187) becomes:

$$u_1(t) = 5.194375003 \cos \omega t - 0.12115924 \cos 3\omega t -$$
$$- 0.034843933 \cos 5\omega t - 0.038371881 \cos 7\omega t + 0.000007998 \cos 9\omega t \tag{8.192}$$

where $\omega = 0.6269145$.

(b) For $a = 5$, $\alpha = 1$, $\omega_0 = 1, f = 0$, the following results are obtained:

$$C_1 = 5.018214118; C_2 = 0.01968471; C_3 = -0.015158401; C_4 = -0.022740428$$

$$u_1(t) = 5.05176786 \cos \omega t - 0.016254795 \cos 3\omega t -$$
$$- 0.011945177 \cos 5\omega t - 0.023567886 \cos 7\omega t + 0.000006078 \cos 9\omega t \tag{8.193}$$

where $\omega = 1.1817241$.

(c) For $a = 5$, $\alpha = 1$, $\omega_0 = 1, f = 0.1$, $\Omega = 1$, it is obtained

$$C_1 = 5.020224969; C_2 = 0.018260618; C_3 = 0.007441779; C_4 = -0.045927367$$

$$u_1(t) = 5.053982615 \cos \omega t - 0.01779699 \cos 3\omega t +$$
$$+ 0.010444517 \cos 5\omega t - 0.046645334 \cos 7\omega t + 0.000020455 \cos 9\omega t +$$
$$+ 0.223011649(1 - \cos 0.17537309t) \tag{8.194}$$

where $\omega = 1.17537309$.

II. In the second case we consider $n = 2$. From Eq. 8.180 it is obtained:

$$a_{1,2} = 0.848826363; a_{3,2} = 0.169765272; a_{5,2} = -0.02425218; a_{7,2} = 0.00808406$$

(a) For $a = 5$, $\alpha = 1$, $\omega_0 = 0, f = 0$, we have

$$C_1 = 4.929081167; C_2 = 0.000890717; C_3 = 0.030884506; C_4 = 0.039143609$$

$$\begin{aligned} u_1(t) &= 4.79659747 \cos \omega t + 0.12566055 \cos 3\omega t + \\ &+ 0.032727712 \cos 5\omega t + 0.044810445 \cos 7\omega t + 0.000203823 \cos 9\omega t \end{aligned} \tag{8.195}$$

where $\omega = 2.045312$.

(b) For $a = 5$, $\alpha = 1$, $\omega_0 = 1, f = 0$, we obtain

$$C_1 = 4.952586959; C_2 = -0.015000551; C_3 = 0.034838155; C_4 = 0.027575436$$

$$\begin{aligned} u_1(t) = 4.8543132899 \cos \omega t + 0.085782596 \cos 3\omega t + \\ +0.030887963 \cos 5\omega t + 0.028913121 \cos 7\omega t + 0.000103034 \cos 9\omega t \end{aligned} \tag{8.196}$$

where $\omega = 2.2807431$.

(c) For $a = 5$, $\alpha = 1$, $\omega_0 = 1, f = 0.1$, $\Omega = 1$, we have

$$C_1 = 4.936194326; C_2 = -0.025541995; C_3 = 0.059682125; C_4 = 0.029665543$$

$$\begin{aligned} u_1(t) &= 4.837168737 \cos \omega t + 0.075535083 \cos 3\omega t + \\ &+ 0.056053186 \cos 5\omega t + 0.031232993 \cos 7\omega t + 0.000091955 \cos 9\omega t + \\ &+ 0.010503799(1 - \cos 1.2764321t) \end{aligned} \tag{8.197}$$

where $\omega = 2.2764321$.

III. In the third case, we consider$n = \frac{5}{3}$. From Eq. 8.187 it is obtained:

$$\begin{aligned} a_{1,\frac{5}{3}} &= 0.891467659; a_{3,\frac{5}{3}} = 0.127352522; \\ a_{5,\frac{5}{3}} &= -0.025470504; a_{7,\frac{5}{3}} = 0.009796347 \end{aligned}$$

(a) For $a = 2$, $\alpha = 1$, $\omega_0 = 0, f = 0$, we have:

$$C_1 = 1.971343187; C_2 = 0.020411906; C_3 = 0.001490562; C_4 = 0.006754343$$

$$u_1(t) = 1.958991\cos\omega t + 0.0334889\cos 3\omega t + 0.000654408\cos 5\omega t + \\ + 0.0068656\cos 7\omega t + 0.000001134\cos 9\omega t \tag{8.198}$$

where $\omega = 1.1854322$.

(b) For $a = 2$, $\alpha = 1$, $\omega_0 = 1, f = 0$, it is obtained

$$C_1 = 1.985630001; C_2 = 0.00505791; C_3 = 0.0000205581; C_4 = 0.00929309$$

$$u_1(t) = 1.977958874\cos\omega t + 0.013179562\cos 3\omega t - \\ - 0.000490736\cos 5\omega t + 0.009346505\cos 7\omega t + 0.000008539\cos 9\omega t \tag{8.199}$$

where $\omega = 1.553241$.

(c) For $a = 2$, $\alpha = 1$, $\omega_0 = 1, f = 0.1$, $\Omega = 1$, we obtain:

$$C_1 = 1.975849538; C_2 = 0.037915719; C_3 = 0.0.013202094; \\ C_4 = -0.026967352$$

$$u_1(t) = 1.904864454\cos\omega t + 0.01624387\cos 3\omega t + \\ + 0.047954663\cos 5\omega t + 0.031098074\cos 7\omega t + 0.000019391\cos 9\omega t + \\ + 0.047282731(1 - \cos 0.540437t) \tag{8.200}$$

where $\omega = 1.540437$.

Figures 8.17–8.25 show the comparison between the present solutions and the numerical integration results obtained by a fourth-order Runge–Kutta method in the above cases.

It is clear that the solutions obtained by OVIM are nearly identical with the solutions given by the numerical method.

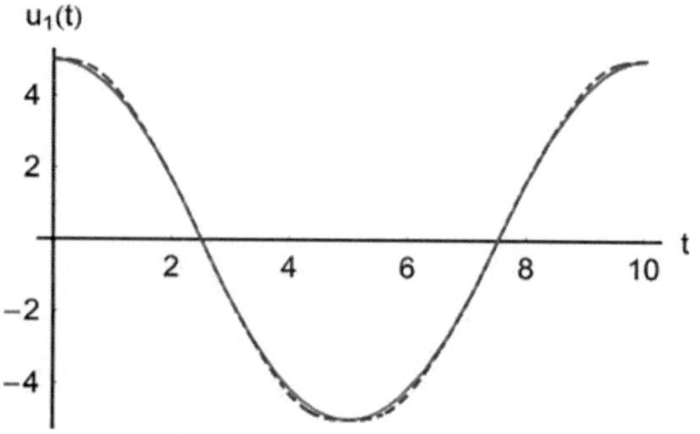

Fig. 8.17 Comparison between the approximate solution and numerical results obtained for Eq. 8.166 in case $n = 1/3$, $a = 5$, $\omega_0 = 0$, $f = 0$, $\alpha = 1$: ______ numerical solution;_ _ _ _ approximate solution (8.192)

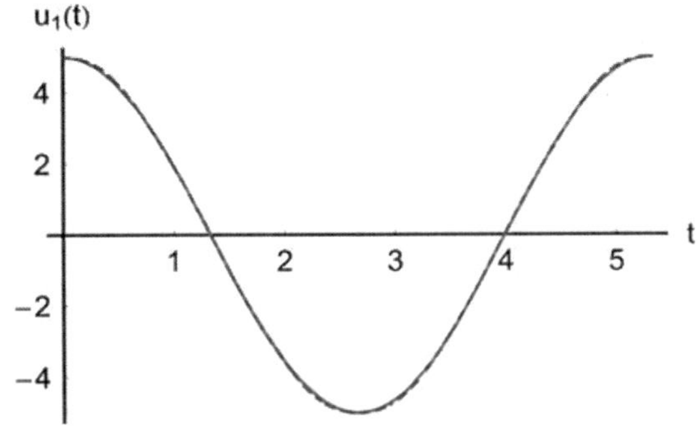

Fig. 8.18 Comparison between the approximate solution and numerical results obtained for Eq. 8.166 in case $n = 1/3$, $a = 5$, $\omega_0 = 1$, $f = 0$, $\alpha = 1$: ______ numerical solution; _ _ _ _ approximate solution (8.193)

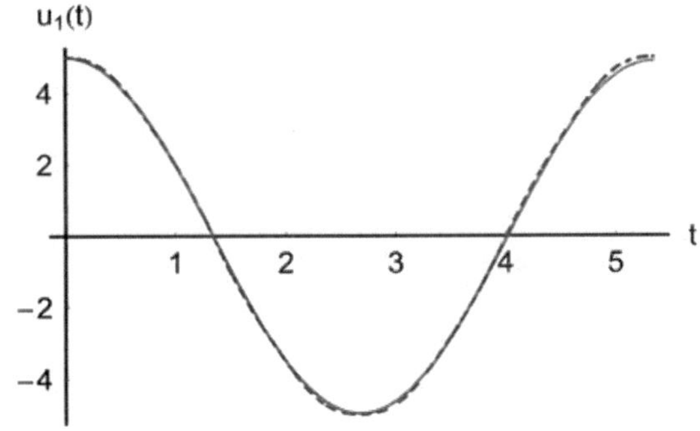

Fig. 8.19 Comparison between the approximate solution and numerical results obtained for Eq. 8.166 in case $n = 1/3$, $a = 5$, $\omega_0 = 1$, $f = 0.1$, $\alpha = 1$, $\Omega = 1$: ______ numerical solution; _ _ _ _ approximate solution (8.194)

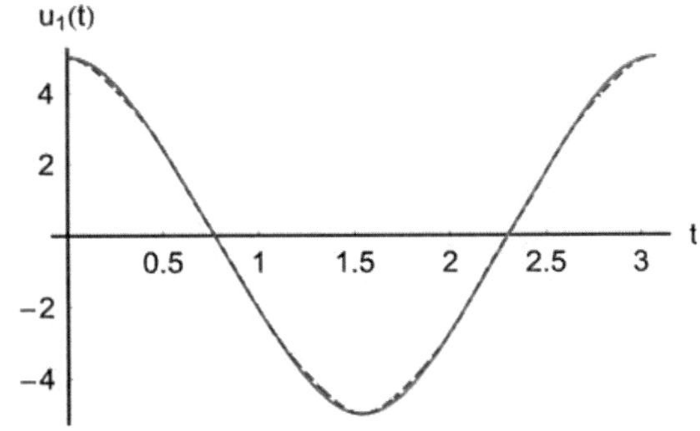

Fig. 8.20 Comparison between the approximate solution and numerical results obtained for Eq. 8.166 in case $n = 2$, $a = 5$, $\omega_0 = 0, f = 0$, $\alpha = 1$: ______ numerical solution; _ _ _ _ approximate solution (8.195)

8.8 A Boundary Layer Equation in Unbounded Domain

In this section we apply the OVIM to solve a boundary layer equation in unbounded domain [169–172]:

$$f'''(x) + (n-1)f(x)f''(x) - 2n[f'(x)]^2 = 0, \quad n>0 \tag{8.201}$$

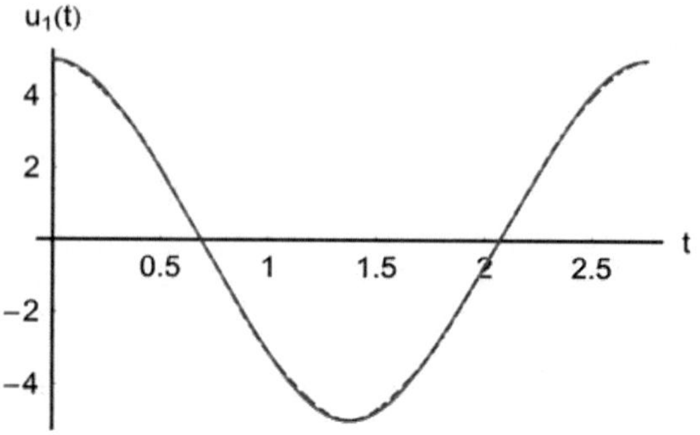

Fig. 8.21 Comparison between the approximate solution and numerical results obtained for Eq. 8.166 in case $n = 2, a = 5, \omega_0 = 1, f = 0, \alpha = 1$: ______ numerical solution; _ _ _ _ approximate solution (8.196)

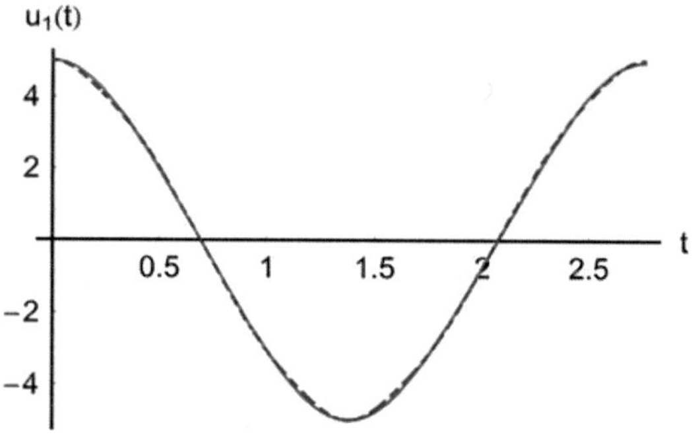

Fig. 8.22 Comparison between the approximate solution and numerical results obtained for Eq. 8.166 in case $n = 2, a = 5, \omega_0 = 1, f = 0.1, \alpha = 1, \Omega = 1$: ______ numerical solution; _ _ _ _ approximate solution (8.197)

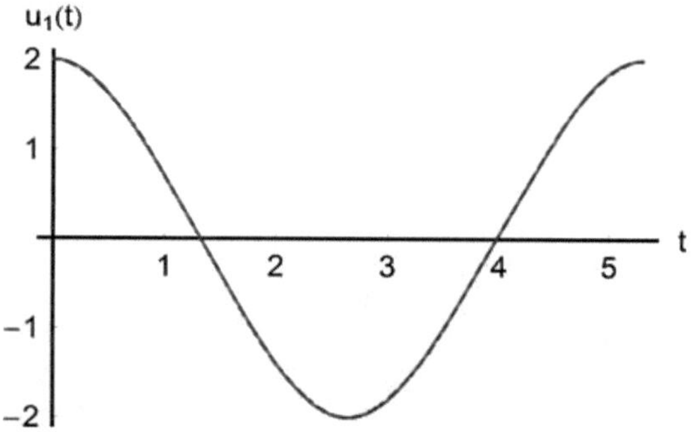

Fig. 8.23 Comparison between the approximate solution and numerical results obtained for Eq. 8.166 in case $n = 5/3, a = 2, \omega_0 = 0, f = 0, \alpha = 1$: ______ numerical solution; _ _ _ _ approximate solution (8.198)

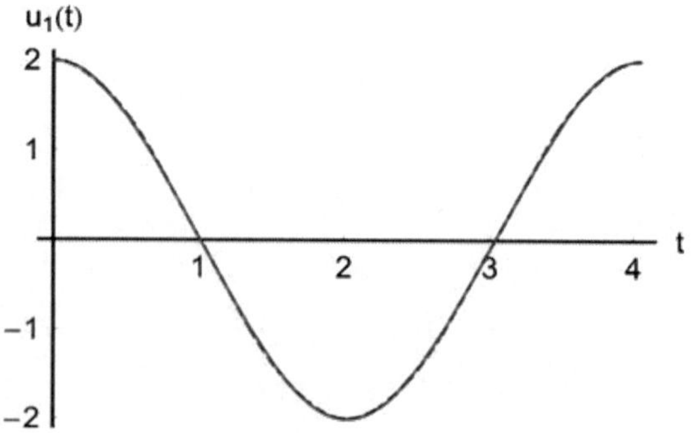

Fig. 8.24 Comparison between the approximate solution and numerical results obtained for Eq. 8.166 in case $n = 5/3, a = 2, \omega_0 = 1, f = 0, \alpha = 1$: ______ numerical solution; _ _ _ _ approximate solution (8.199)

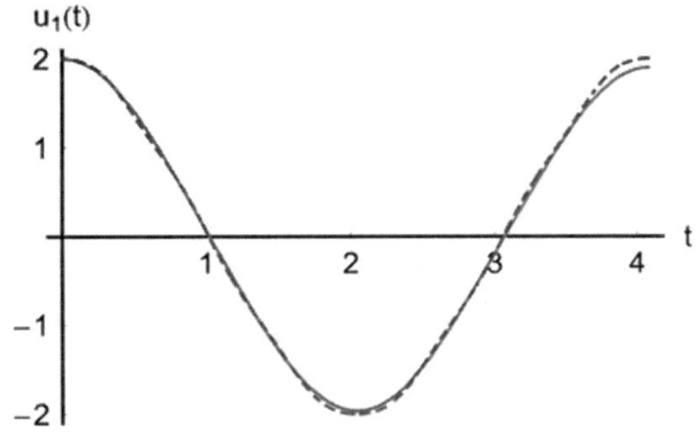

Fig. 8.25 Comparison between the approximate solution and numerical results obtained for Eq. 8.166 in case n = 5/3, $a = 2$, $\omega_0 = 1$, $f = 0.1$, $\alpha = 1$, $\Omega = 1$: ______ numerical solution; _ _ _ _ approximate solution (8.200)

with boundary conditions:

$$f(0) = 0,\ f'(0) = 1,\ f'(\infty) = 0 \tag{8.202}$$

This equation appears in boundary layers in fluid mechanics, where$f''(0)<0$. An analytic treatment combined with numerical computations will be approached to find the numerical values of $f''(0)$for several values of x.

It is interesting to point out that Kuiken [169] investigated this problem for three cases of n, namely for $0 < n < 1$, $n = 1$ and for $n > 1$. Kuiken gave such an asymptotic expression

$$f(x) = (x - x_0)^{\alpha} \sum_{i=0}^{N} C_0^{1-i} \bar{C}_i (x - x_0)^{-i(1+\alpha)},\ \alpha = \frac{1-n}{1+n} \tag{8.203}$$

where the coefficients $\bar{C}_i$ are given by recursive formulas and the coefficients C_0, x_0 are determined by an iterative numerical approach. Thus, rigorously speaking, Kuiken's solution is a semi-numerical one. Besides, the above expression is valid only for $x> > x_0 + 1$. It was shown in [172] that the algebraic behaviour at the outer edge of a boundary layer in fluid mechanics can sometimes be allowed in regions of non-vanishing extent.

To solve this problem by OVIM, we make the transformation $\frac{df}{dx} = g$, $\frac{dg}{dx} = h$ and thus we can rewrite Eq. 8.201 as a system of differential equations

$$\begin{cases} \dfrac{df}{dx} = g(x) \\ \dfrac{dg}{dx} = h(x) \\ \dfrac{dh}{dx} = 2ng^2(x) + (1-n)f(x)h(x) \end{cases} \tag{8.204}$$

The following Variational Iteration Algorithm – II can be obtained from the Eqs. 8.204 and 8.17:

$$f_{n+1}(x) = f_0(x) + \int_0^x g_n(s)ds \tag{8.205}$$

$$g_{n+1}(x) = g_0(x) + \int_0^x h_n(s)ds \tag{8.206}$$

$$h_{n+1}(x) = h_0(x) + \int_0^x \left[2ng_n^2(s) + (1-n)f_n(s)h_n(s)\right]ds \tag{8.207}$$

From the initial conditions (8.202) it is natural to express $f(x)$ from (8.201) by the set of base functions

$$\left\{x^m e^{-\lambda nx} \middle| m \geq 0, n \geq 0, \lambda > 0\right\} \tag{8.208}$$

in the form

$$f(x) = \sum_{m,n \geq 0} a_{m,n} x^m e^{-\lambda nx} \tag{8.209}$$

where $a_{\mathrm{m,n}}$ and $\lambda$are unknown coefficients.

We start with the initial approximations:

$$\begin{aligned} f_0(x) &= C_1 e^{-\lambda x} + C_2 x e^{-2\lambda x}, \\ g_0(x) &= C_3 e^{-2\lambda x} + C_4 x e^{-3\lambda x}, \\ h_0(x) &= C_5 e^{-3\lambda x} + C_6 x e^{-\lambda x} \end{aligned} \tag{8.210}$$

where C_1, C_2,. . ..C_6 and $\lambda > 0$ are unknown coefficients at this moment.

Substituting Eq. 8.210 into 8.205, 8.206 and 8.207, we obtain the following results ($n = 0$):

$$f_1(x) = \frac{C_4}{9\lambda^2} + \frac{C_3}{2\lambda} + C_1 e^{-\lambda x} + \left(C_2 x - \frac{C_3}{2\lambda}\right)e^{-2\lambda x} - \frac{3\lambda C_4 x + C_4}{9\lambda^2} e^{-3\lambda x} \tag{8.211}$$

$$g_1(x) = \frac{C_5}{3\lambda} + \frac{C_6}{\lambda^2} - \frac{\lambda C_6 x + C_6}{\lambda^2} e^{-\lambda x} + C_3 e^{-2\lambda x} + \left(C_4 x - \frac{C_5}{3\lambda}\right) e^{-3\lambda x} \tag{8.212}$$

$$\begin{aligned} h_1(x) = {} & 2n\left(\frac{C_3^2}{4\lambda} + \frac{2C_3C_4}{25\lambda^2} + \frac{C_1^2}{108\lambda^3}\right) + (1-n)\left(\frac{C_1C_6}{4\lambda^2} + \frac{2C_2C_6}{27\lambda^3} + \frac{C_1C_5}{4\lambda} + \right.\\ & \left. + \frac{C_2C_5}{25\lambda^2}\right) + C_6 x e^{-\lambda x} + \frac{C_1C_6(n-1)(2\lambda x+1)}{4\lambda^2} e^{-2\lambda x} + \\ & + \left[C_5 + \frac{C_2C_6(n-1)(9\lambda^2x^2+6\lambda x+2)}{27\lambda^3}\right] e^{-3\lambda x} + \left[\frac{C_1C_5(n-1)}{4\lambda} - \frac{nC_3^2}{2\lambda}\right] e^{-4\lambda x} + \\ & + \frac{C_2C_5(n-1) - 4C_3C_4 n}{5\lambda^2}(5\lambda+1)e^{-5\lambda x} - \frac{nC_4^2}{108\lambda^3}(36\lambda^2x^2 + 12\lambda x + 1)e^{-6\lambda x} \end{aligned} \tag{8.213}$$

For the second iteration, $(n = 1)$ we obtain the following results:

$$\begin{aligned} f_2(x) = {} & \left(\frac{C_5}{3\lambda} + \frac{C_6}{\lambda^2}\right)x + \frac{C_3}{2\lambda} + \frac{C_4}{9\lambda^2} - \frac{C_5}{9\lambda^2} - \frac{2C_6}{\lambda^3} + \left(\frac{C_6}{\lambda^2}x + C_1 + \right.\\ & \left. + \frac{2C_6}{\lambda^3}\right)e^{-\lambda x} + \left(C_2 x - \frac{C_3}{2\lambda}\right)e^{-2\lambda x} + \left(-\frac{C_4}{3\lambda}x + \frac{C_5 - C_4}{9\lambda^2}\right)e^{-3\lambda x} \end{aligned} \tag{8.214}$$

$$\begin{aligned} g_2(x) = {} & \left[2n\left(\frac{C_3^2}{4\lambda} + \frac{2C_3C_4}{25\lambda^2} + \frac{C_4^2}{108\lambda^3}\right) + (1-n)\left(\frac{C_1C_6}{4\lambda^2} + \frac{2C_2C_6}{27\lambda^3} + \right.\right.\\ & \left.\left. + \frac{C_1C_5}{4\lambda} + \frac{C_2C_5}{25\lambda}\right)\right]x + (n-1)\left(\frac{C_1C_6}{4\lambda^3} + \frac{2C_2C_6}{27\lambda^4} + \frac{C_1C_5}{16\lambda^2} + \frac{2C_2C_5}{125\lambda^3}\right) - \\ & - n\left(\frac{C_3^2}{8\lambda^2} + \frac{8C_3C_4}{125\lambda^3} + \frac{C_4^2}{108\lambda^4}\right) + \frac{C_5}{3\lambda} + \frac{C_6}{\lambda^2} - \left(\frac{C_6}{\lambda}x + \frac{C_6}{\lambda^2}\right)e^{-\lambda x} + \\ & + \left[\frac{C_1C_6(1-n)}{4\lambda^2}x + C_3 + \frac{C_1C_6(1-n)}{4\lambda^3}\right]e^{-2\lambda x} + \\ & + \left[\frac{C_2C_6(1-n)}{9\lambda^2}x^2 + \left(C_4 + \frac{4C_2C_6(1-n)}{27\lambda^3}\right)x + \frac{2C_2C_6(1-n)}{27\lambda^4} - \frac{C_5}{3\lambda}\right]e^{-3\lambda x} + \\ & + \frac{2nC_3^2 + C_1C_5(1-n)}{16\lambda^2}e^{-4\lambda x} + \frac{4C_3C_4 n + C_2C_5(1-n)}{25\lambda^2}\left(x + \frac{2}{5\lambda}\right)e^{-5\lambda x} + \\ & + \left(\frac{nC_4^2}{18\lambda^2}x^2 + \frac{nC_4^2}{27\lambda^3}x + \frac{nC_4^2}{108\lambda^4}\right)e^{-6\lambda x} \end{aligned} \tag{8.215}$$

Finally, the third iteration result in ($n = 2$):

$$\begin{aligned}
f_3(x) = &\left[n\left(\frac{C_3^2}{4\lambda}+\frac{2C_3C_4}{25\lambda^2}+\frac{C_4^2}{108\lambda^3}\right)+\frac{1-n}{2}\left(\frac{C_1C_6}{4\lambda^2}+\frac{2C_2C_6}{27\lambda^3}+\frac{C_1C_5}{4\lambda}+\right.\right.\\
&\left.\left.+\frac{C_2C_5}{25\lambda^2}\right)\right]x^2+\left[\frac{C_5}{3\lambda}+\frac{C_6}{\lambda^2}+(n-1)\left(\frac{C_1C_6}{4\lambda^3}+\frac{2C_2C_6}{27\lambda^4}+\frac{C_1C_5}{16\lambda^2}+\frac{2C_2C_5}{25\lambda^3}\right)-\right.\\
&\left.-n\left(\frac{C_3^2}{8\lambda^2}+\frac{8C_3C_4}{125\lambda^3}+\frac{C_4^2}{108\lambda^4}\right)\right]x+n\left(\frac{C_3^2}{32\lambda^3}+\frac{12C_3C_4}{625\lambda^4}+\frac{C_4^2}{324\lambda^5}\right)+\\
&+(1-n)\left(\frac{3C_1C_6}{16\lambda^4}+\frac{4C_2C_6}{81\lambda^3}+\frac{C_1C_5}{64\lambda^3}+\frac{3C_2C_5}{625\lambda^4}\right)+\frac{C_3}{2\lambda}-\frac{2C_6}{\lambda^3}-\frac{C_5}{9\lambda^2}+\frac{C_4}{9\lambda^2}+\\
&+\left(\frac{C_6}{\lambda^2}x+C_1+\frac{2C_6}{\lambda^3}\right)e^{-\lambda x}+\left[\left(C_2+\frac{C_1C_6(n-1)}{8\lambda^3}\right)x+\frac{3C_1C_6(n-1)}{16\lambda^4}-\right.\\
&\left.-\frac{C_3}{2\lambda}\right]e^{-2\lambda x}+\left[\frac{C_2C_6(n-1)}{27\lambda^3}x^2+\left(\frac{2C_2C_6(n-1)}{27\lambda^4}-\frac{C_4}{3\lambda}\right)x+\right.\\
&\left.+\frac{4C_2C_6(n-1)}{81\lambda^5}+\frac{C_5}{9\lambda^2}-\frac{C_4}{9\lambda^2}\right]e^{-3\lambda x}+\frac{C_1C_5(n-1)-2nC_3^2}{64\lambda^3}e^{-4\lambda x}+\\
&+\frac{C_2C_5(n-1)-4C_3C_4n}{125\lambda^3}\left(x+\frac{3}{5\lambda}\right)e^{-5\lambda x}-\frac{nC_4^2}{108\lambda^3}\left(x^2+\frac{x}{\lambda}+\frac{1}{3\lambda^2}\right)e^{-6\lambda x}
\end{aligned} \tag{8.216}$$

Applying the boundary conditions (8.202) on the approximate solution (8.216), we obtain the following algebraic system with unknowns C_1, C_2,...C_6 and λ:

$$C_1 = 0 \tag{8.217}$$

$$-\lambda C_1 + C_2 + C_3 = 1 \tag{8.218}$$

$$\begin{aligned}
&2n\left(\frac{C_3^2}{4\lambda}+\frac{2C_3C_4}{25\lambda^2}+\frac{C_4^2}{108\lambda^3}\right)+\\
&+(1-n)\left(\frac{C_1C_6}{4\lambda^2}+\frac{2C_2C_6}{27\lambda^3}+\frac{C_1C_5}{4\lambda}+\frac{C_2C_5}{25\lambda^2}\right)=0
\end{aligned} \tag{8.219}$$

$$\begin{aligned}
&n\left(\frac{C_3^2}{8\lambda^2}+\frac{8C_3C_4}{125\lambda^3}+\frac{C_4^2}{108\lambda^4}\right)-\frac{C_4}{3\lambda}-\frac{C_6}{\lambda^2}+\\
&+(1-n)\left(\frac{C_1C_6}{4\lambda^3}+\frac{2C_2C_6}{27\lambda^3}+\frac{C_1C_5}{16\lambda^2}+\frac{2C_2C_5}{125\lambda^3}\right)=0
\end{aligned} \tag{8.220}$$

From Eqs. 8.217, 8.218, 8.219 and 8.220 we can obtain C_2, C_5 and C_6 as functions of λ, C_3 and C_4.

From the residual functional:

$$J=\int_0^\infty\left[f_3'''(s)+(n-1)f_3(s)f_3''(s)-2nf_3'^2(s)\right]^2ds \tag{8.221}$$

Table 8.7 The coefficients $C_1, C_2, \ldots C_6$ and λ given by OVIM for several values of n

n	C_1	C_2	C_3	C_4	C_5	C_6	λ
1/3	0	0.924589	0.0754107	0.0027517	−0.003321	−5.53e-15	0.145537
4	0	1.0217901520	−0.021790	0.1	0	0	0.6473200694
10	0	1.007896023	−0.007896	0.0999958	0.0000852	−0.000383	1.0286048
100	0	0.9258295575	0.0741704	0.0662164	0.0126234	−0.225463	3.35582155
1,000	0	0.9862411165	0.0137588	0.0873533	0.0014285	−0.084459	10.256605438
5,000	0	0.9956307348	0.0043692	0.0949307	0.0003390	−0.043459	26.101014908

Table 8.8 Comparison of the results obtained for $f''(0)$through our procedure with results obtained using different methods, for several values of n

n	HPM [170]	Padé [171]	Homotopy-Padé [172]	HAM [172]	OVIM (present results)
1/3	-	−0.5614491934	−0.56144919	−0.56145	−0.56078
4	−2.5568	−2.483954	–	–	−2.5174906
10	−4.0476	−4.026385	–	–	−4.030582
100	−12.8501	−12.84334	–	–	−12.846640
1,000	−40.6556	−40.65538	–	–	−40.655400
5,000	−90.9127	−104.8420	–	–	−104.08070

we can obtain the following three equations:

$$\frac{\partial J}{\partial \lambda} = \frac{\partial J}{\partial C_3} = \frac{\partial J}{\partial C_4} = 0 \tag{8.222}$$

In this way, approximate solution (8.216) is obtained with the parameters C_1, $C_2, \ldots C_6$ and λ given by the system of Eqs. 8.217, 8.218, 8.219, 8.220 and 8.222.

In order to assess the advantages and accuracy of OVIM for solving nonlinear problems, some numerical applications were performed for several distinct values of n, as shown in Table 8.7.

In Table 8.8 we compare the results obtained for $f''(0)$through the proposed procedure with the results available in the literature, obtained by Homotopy Perturbation Method (HPM) [170], Padé approximants [171], Homotopy-Padé approach [172] and Homotopy Analysis Method (HAM) [172].

Different from other classical perturbation techniques, the proposed procedure is independent upon small parameters and provides us with a convenient way to control and adjust the convergence region, when necessary. OVIM can be employed to efficiently approximate a nonlinear problem by choosing different sets of base functions.

Chapter 9
Optimal Parametric Iteration Method

9.1 Short Considerations

The most common and most widely studied methods for determining analytical approximate solutions of a nonlinear dynamical system are iteration methods. These methods, in principle, do not require the presence of a parameter ε which should be small. Several researchers have studied different nonlinear problems by means of iteration procedures [173–179].

Firstly, consider a nonlinear conservative oscillator described as

$$\ddot{u} + f(u) = 0 \quad u(0) = A, \quad \dot{u}(0) = 0 \tag{9.1}$$

where $f(u)$ is an odd function and its derivative near $u = 0$ is positive. Equation 9.1 can be written as

$$\ddot{u} + \omega^2 u = \omega^2 u - f(u) := g(u) \tag{9.2}$$

where ω is a priori unknown frequency of the periodic solution $u(t)$ being sought. The original Mickens procedure is given as [173]

$$\ddot{u}_k + \omega^2 u_k = g(\omega, u_{k-1}), \quad k = 1, 2, ... \tag{9.3}$$

where the input of starting function is

$$u_0(t) = A \cos \omega t \tag{9.4}$$

This iteration scheme was used to solve many nonlinear oscillating equations [173–178]. Lim et al. [175] proposed a modified iteration scheme

$$\ddot{u}_{k+1} + \omega^2 u_{k+1} = g(\omega, u_{k-1}) + g_u(\omega, u_{k-1})(u_k - u_{k-1}), \quad k = 0, 1, 2, ... \tag{9.5}$$

V. Marinca and N. Herisanu, *Nonlinear Dynamical Systems in Engineering*, DOI 10.1007/978-3-642-22735-6_9, © Springer-Verlag Berlin Heidelberg 2011 with the inputs of starting functions as

$$u_{-1}(t) = u_0(t) = A\cos\omega t \tag{9.6}$$

where $g_u = \frac{\partial g}{\partial u}$. In principle, the approximations can be obtained to any desired order. However, more and more complicated nonlinear algebraic equations in ω have to be solved. Chen et al. [177] proposed a new iteration scheme, considering ω as ω_k:

$$\ddot{u}_k + \omega_{k-1}^2 u_k = g(\omega_{k-1}, u_{k-1}), \quad k = 1, 2, ... \tag{9.7}$$

where the right-hand side of Eq. 9.7 can be expanded in the Fourier series:

$$g(\omega_{k-1}, u_{k-1}) = \sum_{i=1}^{\varphi(k)} a_{k-1,i}(\omega_{k-1})\cos(i\omega_{k-1}t) \tag{9.8}$$

where the coefficients $a_{k-1,i}$ are functions of ω_{k-1} and $\varphi(k)$ is a positive integer. The (k-1) th-order approximation ω_{k-1} is obtained by eliminating the so-called secular terms, i.e. letting

$$a_{k-1,1}(\omega_{k-1}) = 0, \quad k = 1, 2, ... \tag{9.9}$$

Equation 9.9 is always a linear algebraic equation in ω_{k-1}^2.

In a different manner, J.H. He [179] proposed a new iteration scheme considering the following nonlinear oscillator:

$$\ddot{u} + u + \varepsilon f(u, \dot{u}) = 0, \quad u(0) = A, \quad \dot{u}(0) = 0 \tag{9.10}$$

We rewrite Eq. 9.10 in the following form

$$\ddot{u} + u + \varepsilon u g(u, \dot{u}) = 0 \tag{9.11}$$

where $g(u, \dot{u}) = \frac{f(u, \dot{u})}{u}$.

J.H. He has constructed an iteration formula for the above equation

$$\ddot{u}_{k+1} + u_{k+1} + \varepsilon u_{k+1} g(u_k, \dot{u}_k) = 0 \tag{9.12}$$

For nonlinear oscillation, Eq. 9.12 is of Mathieu type. This technique is called iteration perturbation method.

At the begging of this section, we consider two variants of iteration schemes.

9.1.1 A Combination of Mickens and He Iteration Methods

Consider the following, in general nonlinear oscillation:

$$\ddot{u} + \omega^2 u = f(u, \dot{u}, \ddot{u}), \quad u(0) = A, \quad \dot{u}(0) = 0 \tag{9.13}$$

We rewrite Eq. 9.13 in the following form:

$$\ddot{u} + \Omega^2 u = u\left(\Omega^2 - \omega^2 + \frac{f(u, \dot{u}, \ddot{u})}{u}\right) := ug(u, \dot{u}, \ddot{u}) \tag{9.14}$$

where Ω is a priori unknown frequency of the periodic solution $u(t)$ being sought. The proposed iteration scheme is [180]:

$$\begin{aligned}\ddot{u}_{n+1} + \Omega^2 u_{n+1} = u_{n-1}[&g(u_{n-1}, \dot{u}_{n-1}, \ddot{u}_{n-1}) + \\ &+ g_u(u_{n-1}, \dot{u}_{n-1}, \ddot{u}_{n-1})(u_n - u_{n-1}) + \\ &+ g_{\dot{u}}(u_{n-1}, \dot{u}_{n-1}, \ddot{u}_{n-1})(\dot{u}_n - \dot{u}_{n-1}) + \\ &+ g_{\ddot{u}}(u_{n-1}, \dot{u}_{n-1}, \ddot{u}_{n-1})(\ddot{u}_n - \ddot{u}_{n-1})], \ \mathrm{n} = 0, 1, 2, 3, \ldots\end{aligned} \tag{9.15}$$

where the inputs of starting functions are [175]:

$$u_{-1}(t) = u_0(t) = A\cos\Omega t \tag{9.16}$$

It is further required that for each n, the solution to Eq. 9.15, is to satisfy the initial conditions

$$u_n(0) = A, \quad \dot{u}_n(0) = 0, \quad n = 1, 2, 3, \ldots \tag{9.17}$$

Note that for given $u_{n-1}(t)$ and $u_n(t)$, Eq. 9.15 is a second order, inhomogeneous differential equation for $u_{n+1}(t)$. Right side of Eq. 9.15 can be expanded into the following Fourier series:

$$\begin{aligned}&u_{n-1}[g(u_{n-1}, \dot{u}_{n-1}, \ddot{u}_{n-1}) + g_u(u_{n-1}, \dot{u}_{n-1}, \ddot{u}_{n-1})(u_n - u_{n-1}) + \\ &\quad + g_{\dot{u}}(u_{n-1}, \dot{u}_{n-1}, \ddot{u}_{n-1})(\dot{u}_n - \dot{u}_{n-1}) + g_{\ddot{u}}(u_{n-1}, \dot{u}_{n-1}, \ddot{u}_{n-1})(\ddot{u}_n - \ddot{u}_{n-1})] = \\ &= a_1(A, \Omega, \omega)\cos\Omega t + \sum_{n=2}^{N} a_n(A, \Omega, \omega)\cos n\Omega t + \\ &\quad + b_1(A, \Omega, \omega)\sin\Omega t + \sum_{n=2}^{N} b_n(A,\Omega, \omega)\sin n\Omega t\end{aligned} \tag{9.18}$$

where the coefficients $a_n(A, \Omega, \omega)$ and $b_n(A, \Omega, \omega)$ are known, and the integer N depends upon the function $g(u, \dot{u}, \ddot{u})$ on the right hand side of Eq. 9.14. In Eq. 9.18, the requirement of no secular term needs

$$a_1(A, \Omega, \omega) = 0, \quad b_1(A, \Omega, \omega) = 0 \tag{9.19}$$

The solution of the Eq. 9.15 with the initial conditions (9.17) is taken to be:

$$\begin{aligned} u_{n+1}(t) =& A\cos\Omega t - \sum_{n=2}^{N} \frac{a_n(A, \Omega, \omega)}{(n^2-1)\Omega^2}(\cos n\Omega t - \cos\Omega t) - \\ & - \sum_{n=2}^{N} \frac{b_n(A, \Omega, \omega)}{(n^2-1)\Omega^2}(\sin n\Omega t - n\sin\Omega t) \end{aligned} \tag{9.20}$$

Equation 9.19 allows the determination of the frequency Ω as a function of A and ω. This procedure can be performed to any desired iteration step n. As we will show in the following examples, an excellent approximate analytical representation to the exact solution, valid for small as well as large values of oscillation amplitude is obtained.

(a) *The motion of a simple pendulum*

When damping is neglected, the differential equation governing the free oscillation of the mathematical pendulum is given by

$$ml\ddot{\theta} + mg\sin\theta = 0 \tag{9.21}$$

or

$$\ddot{\theta} + a\sin\theta = 0 \tag{9.22}$$

Here m is the mass, l the length of the pendulum, g the gravitational acceleration and $a = g/l$. The angle θ designates the deviation from the vertical equilibrium position.

We rewrite Eq. 9.22 in the form

$$\ddot{\theta} + \Omega^2\theta = \theta\left(\Omega^2 - a\frac{\sin\theta}{\theta}\right) \tag{9.23}$$

where Ω is an unknown frequency of the periodic solution. Here the initial conditions are $\theta(0) = A$, $\dot{\theta}(0) = 0$, the inputs of starting function are $\theta_{-1}(t) = \theta_0(t) = A\cos\Omega t$ and $g(t, \theta, \dot{\theta}, \ddot{\theta}) = \Omega^2 - a\frac{\sin\theta}{\theta}$.

The first iteration is given by the equation:

$$\ddot{\theta}_1 + \Omega^2\theta_1 = \Omega^2 A\cos\Omega t - a\sin(A\cos\Omega t) \tag{9.24}$$

The term $\sin(A\cos\Omega t)$ can be expanded in the power series:

$$\sin(A\cos\Omega t) = A\cos\Omega t - \frac{A^3\cos^3\Omega t}{3!} + \frac{A^5\cos^5\Omega t}{5!} - \frac{A^7\cos^7\Omega t}{7!} + \frac{A^9\cos^9\Omega t}{9!} + \ldots \tag{9.25}$$

We rewrite powers of $\cos\Omega t$ in (9.25) in terms of the cosine of multiples of Ωt with the aid of the known identity

$$\cos^{2n+1}\Omega t = \frac{1}{4^n}\sum_{k=0}^{n}\binom{2n+1}{n-k}\cos(2k+1)\Omega t \tag{9.26}$$

By using Eq. 9.26, Eq. 9.25 may be expressed in the form

$$\begin{aligned}\sin(A\cos\Omega t) &= A\cos\Omega t - \frac{A^3}{24}(\cos 3\Omega t + 3\cos\Omega t) + \\ &+ \frac{A^5}{1920}(\cos 5\Omega t + 5\cos 3\Omega t + 10\cos\Omega t) - \\ &- \frac{A^7}{322560}(\cos 7\Omega t + 7\cos 5\Omega t + 21\cos 3\Omega t + 35\cos\Omega t) + \\ &+ \frac{A^9}{92897280}(\cos 9\Omega t + 9\cos 7\Omega t + 36\cos 5\Omega t + 84\cos 3\Omega t + 126\cos\Omega t)\end{aligned} \tag{9.27}$$

Substituting Eq. 9.27 into Eq. 9.24, this can be rewritten as:

$$\begin{aligned}\ddot{\theta}_1 + \Omega^2\theta_1 &= \left[A\Omega^2 - a\left(A - \frac{A^3}{8} + \frac{A^5}{192} - \frac{A^7}{9216} + \frac{A^9}{737280} + \ldots\right)\right]\cos\Omega t - \\ &- \frac{A^3}{24}\cos 3\Omega t + \frac{A^5}{1920}(\cos 5\Omega t + 5\cos 3\Omega t) - \frac{A^7}{322560}(\cos 7\Omega t + 7\cos 5\Omega t + \\ &+ 21\cos 3\Omega t) + \frac{A^9}{92897280}(\cos 9\Omega t + 9\cos 7\Omega t + 36\cos 5\Omega t + 84\cos 3\Omega t)\end{aligned} \tag{9.28}$$

No secular terms in θ_1 requires that

$$\Omega_1^2 = a\left(1 - \frac{A^2}{8} + \frac{A^4}{192} - \frac{A^6}{9216} + \frac{A^8}{737280} + \ldots\right) \tag{9.29}$$

and solving Eq. 9.28 with the initial conditions $\theta_1(0) = A$, $\dot{\theta}_1(0) = 0$, we obtain

$$\begin{aligned}\theta_1(t) &= A\cos\Omega_1 t + \frac{A^3}{192\Omega_1^2}(\cos 3\Omega_1 t - \cos\Omega_1 t) - \\ &- \frac{A^5}{46080\Omega_1^2}(\cos 5\Omega_1 t - \cos\Omega_1 t) - \frac{A^5}{3072\Omega_1^2}(\cos 3\Omega_1 t - \cos\Omega_1 t) + \\ &+ \frac{A^7}{15482880\Omega_1^2}(\cos 7\Omega_1 t - \cos\Omega_1 t) + \frac{A^7}{1105920\Omega_1^2}(\cos 5\Omega_1 t - \cos\Omega_1 t) + \\ &+ \frac{A^7}{122880\Omega_1^2}(\cos 3\Omega_1 t - \cos\Omega_1 t) - \frac{A^9}{7431782400\Omega_1^2}(\cos 9\Omega_1 t - \cos\Omega_1 t) - \\ &- \frac{A^9}{495452160\Omega_1^2}(\cos 7\Omega_1 t - \cos\Omega_1 t) - \frac{A^9}{61931520\Omega_1^2}(\cos 5\Omega_1 t - \cos\Omega_1 t) - \\ &- \frac{A^9}{8847360\Omega_1^2}(\cos 3\Omega_1 t - \cos\Omega_1 t)\end{aligned} \tag{9.30}$$

or

$$\begin{aligned}\theta_1(t) = &- \frac{A^9}{7431782400\Omega_1^2}\cos 9\Omega_1 t + \left(\frac{A^7}{15482880\Omega_1^2} - \right. \\ &\left. - \frac{A^9}{495452160\Omega_1^2}\right)\cos 7\Omega_1 t + \left(-\frac{A^5}{46080\Omega_1^2} + \frac{A^7}{1105920\Omega_1^2} - \right. \\ &\left. - \frac{A^9}{61931520\Omega_1^2}\right)\cos 5\Omega_1 t + \left(\frac{A^3}{192\Omega_1^2} - \frac{A^5}{3072\Omega_1^2} + \frac{A^7}{122880\Omega_1^2} - \right. \\ &\left. - \frac{A^9}{8847360\Omega_1^2}\right)\cos 3\Omega_1 t + \left(A - \frac{A^3}{192\Omega_1^2} + \frac{A^5}{2880\Omega_1^2} - \frac{141A^7}{15482880\Omega_1^2} + \right. \\ &\left. + \frac{61A^9}{464486400\Omega_1^2}\right)\cos\Omega_1 t\end{aligned} \tag{9.31}$$

The approximate period can be expressed from Eq. 9.29: $T_{app} = 2\pi/\Omega_1$ and we obtain:

$$T_{app} = \frac{2\pi}{\sqrt{a}}\left(1 + \frac{A^2}{16} + \frac{5A^4}{1536} + \frac{13A^6}{73728} + \frac{239A^8}{23592960} + \ldots\right) \tag{9.32}$$

while the exact period reads:

$$T_{ex} = \frac{4}{\sqrt{a}} \int_0^{\frac{\pi}{2}} \frac{d\phi}{\sqrt{1 - k^2 \sin^2\varphi}}; \ \mathrm{k} = \sin\frac{\mathrm{A}}{2} \tag{9.33}$$

A comparison with the exact period is shown in Table 9.1.

Figure 9.1 shows the comparison between the present solution obtained from Eqs. 9.29 and 9.31 and the numerical integration results obtained by using a fourth-order Runge-Kutta method

(b) *Sliding mass*

We consider the motion of a ring of mass m sliding freely on the wire described by the parabola $y = qu^2$ which rotates with a constant angular velocity λ about the y-axis as shown in Fig. 9.2 [22].

The equation describing the motion of the ring is (see also Sect.5.12):

$$\ddot{u} + \omega^2 u = -4qu(u\ddot{u} + \dot{u}^2) \tag{9.34}$$

where $\omega^2 = 2gq - \lambda^2$ and the initial conditions are $u(0) = A$, $\dot{u}(0) = 0$.

In case q <<1, the traditional perturbation methods can be applied. In our study, q does not need to be small, i.e. it follows $0 < q < +\infty$. For $n = 0$ into Eq. 9.15,

Table 9.1 Comparison between the approximate period T_{app} given by Eq. 9.32 and the exact solution T_{ex} given by Eq. 9.33

A	$\frac{\pi}{10}$	$\frac{\pi}{9}$	$\frac{\pi}{8}$	$\frac{\pi}{7}$	$\frac{\pi}{4}$	$\frac{\pi}{3}$	$\frac{\pi}{2}$
$\frac{T_{app}}{T_{ex}}$	0.99999681	0.99999512	0.99999211	0.99976763	0.99961425	0.99951134	0.99830437

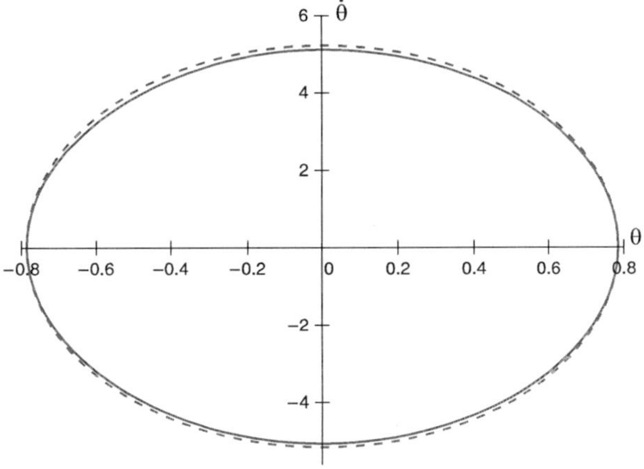

Fig. 9.1 Phase plane for Eq. 9.22: $a = 100$, $A = \pi/4$: ------ present method, Eq. 9.31 ; ________ numerical solution

Fig. 9.2 Particle on a rotating parabola

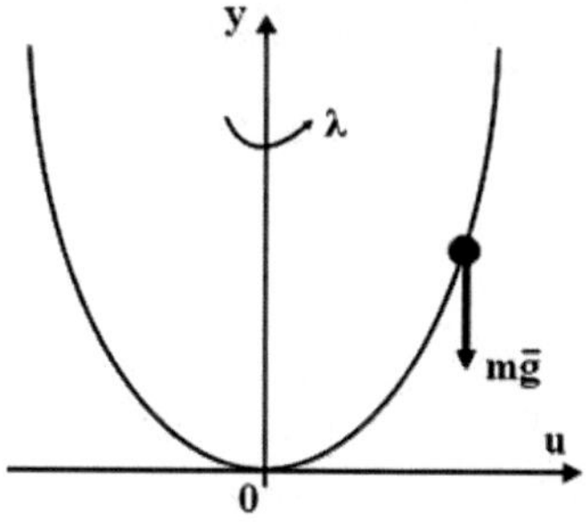

with initial functions $u_{-1}(t) = u_0(t) = A\cos\Omega t$ and $g(u,\dot{u},\ddot{u}) = \Omega^2 - \omega^2 - 4q \times$ $\times(u\ddot{u} + \dot{u}^2)$ we obtain first iteration as:

$$\ddot{u}_1 + \Omega^2 u_1 = A\cos\Omega t[\Omega^2 - \omega^2 + 4q(A^2\Omega^2\cos^2\Omega t - A^2\Omega^2\sin^2\Omega t)]$$

which can be rewritten in a more convenient form as

$$\ddot{u}_1 + \Omega^2 u_1 = A\cos\Omega t(\Omega^2 - \omega^2 + 2qA^2\Omega^2) + 2qA^3\Omega^2\cos 3\Omega t)] \tag{9.35}$$

In order to ensure that no secular terms appear in the next iteration, resonance must be avoided. To do so, the coefficient of $\cos\Omega t$ in Eq. 9.35 requires being zero i.e.

$$\Omega_1^2 = \frac{\omega^2}{1 + 2qA^2} \tag{9.36}$$

So, from Eq. 9.35, with initial conditions (9.17), we have the following first-order approximate solution:

$$u_1(t) = A\cos\Omega t + \frac{1}{4}qA^3(\cos\Omega t - \cos 3\Omega t) \tag{9.37}$$

For $n = 1$ into Eq. 9.15, with the initial functions (9.16) and u_1 given by (9.37) we obtain the following differential equation for u_2:

$$\begin{aligned}\ddot{u}_2 + \Omega^2 u_2 =& A\cos\Omega t(\Omega^2 - \omega^2 + 2qA^2\Omega^2) + \\ &+ (2qA^3\Omega^2 - 4q^2A^5\Omega^2)\cos 3\Omega t - 4q^2A^5\Omega^2\cos 5\Omega t\end{aligned} \tag{9.38}$$

Avoiding the presence of a secular term needs

$$\Omega_2^2 = \frac{\omega^2}{1 + 2qA^2} \tag{9.39}$$

Solving Eq. 9.38 with the initial conditions (9.17), we obtain

$$u_2(t) = A\cos\Omega t + \frac{1}{4}qA^3(\cos\Omega t - \cos 3\Omega t) + \\ + \frac{1}{6}q^2A^5(\cos 5\Omega t + 3\cos 3\Omega t - 4\cos\Omega t) \tag{9.40}$$

It is interesting to note that Ω_1 and Ω_2 are the same. The approximate period can be expressed as

$$T_{app} = \frac{2\pi}{\omega}\sqrt{1 + 2qA^2} \tag{9.41}$$

while the exact period reads

$$T_{ex} = \frac{4}{\omega}\int_0^{\frac{\pi}{2}} \sqrt{1 + 4qA^2\cos^2 t dt} \tag{9.42}$$

As mentioned before, this method gives good approximations in case $q \ll 1$. But our results are valid even in case of q tending to infinity. In the case$q \to \infty$, we have

$$\lim_{q\to\infty}\frac{T_{app}}{T_{ex}} = \frac{2\sqrt{2}}{\pi} = 0,90031 \tag{9.43}$$

Therefore, for any values of q, it can be easily seen that the maximal relative error of the period (9.41) is less than 10%.

(c) *Buckling of a column*

Now, we consider the structure shown in Fig. 9.3 [22].

The mass m moves in the horizontal direction only. Neglecting the weight of all but the mass, show that governing equation for the motion of m is

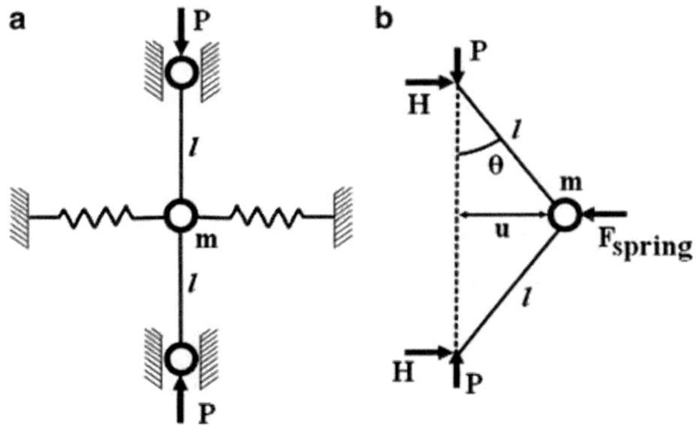

Fig. 9.3 Model for the buckling of a column

$$m\ddot{u} + \left(k_1 - \frac{2p}{1}\right)u + \left(k_3 - \frac{p}{1^3}\right)u^3 + \ldots = 0$$

where the spring force is given by:

$$F_{spring} = k_1 u + k_3 u^3 + \ldots$$

The governing equation of the motion of m can be put in the general form:

$$\ddot{u} + \alpha_1 u + \alpha_3 u^3 + \ldots = 0 \tag{9.44}$$

The oscillator with cubic elastic restoring force occupies an important place in the theory of nonlinear systems, since it is the simplest oscillator displaying specific nonlinear properties. On the other hand, it provides a first approximation for the behaviour of a much larger class of oscillators.

For simplicity, we consider the well-known Duffing equation

$$\ddot{u} + u + \varepsilon u^3 = 0, \quad u(0) = A, \quad \dot{u}(o) = 0 \tag{9.45}$$

In this case, the first iteration can be written in the form ($u_{-1}(t) = u_0(t) = A\cos\Omega t$):

$$\ddot{u}_1 + \Omega^2 u_1 = A\cos\Omega t(\Omega^2 - 1 - \frac{3}{4}\varepsilon A^2) - \frac{1}{4}\varepsilon A^3 \cos 3\Omega t \tag{9.46}$$

No secular terms in u_1 requires that

$$\Omega_1 = \sqrt{1 + \frac{3}{4}\varepsilon A^2} \tag{9.47}$$

From Eq. 9.46, we have the following first-order approximate solution:

$$u_1(t) = A\cos\Omega t + \frac{\varepsilon A^3}{32\Omega^2}(\cos 3\Omega t - \cos\Omega t) \tag{9.48}$$

The second iteration is obtained by substituting this result into Eq. 9.15 for n = 1:

$$\ddot{u}_2 + \Omega^2 u_2 = A\cos\Omega t(\Omega^2 - 1 - \frac{3}{4}\varepsilon A^2 + \frac{\varepsilon^2 A^4}{32\Omega^2}) - \\ - \frac{\varepsilon A^3}{4}\cos 3\Omega t - \frac{\varepsilon^2 A^5}{64\Omega^2}(\cos 5\Omega t + \cos 3\Omega t) \tag{9.49}$$

Avoiding the presence of a secular term needs:

$$\Omega_2^2 = \frac{3\varepsilon A^2 + 4 + \sqrt{7\varepsilon^2 A^4 + 24\varepsilon A^2 + 16}}{8} \tag{9.50}$$

Solving Eq. 9.49 with the initial conditions (9.17), we obtain

$$u_2(t) = A\cos\Omega t + \frac{\varepsilon A^3}{32\Omega^2}(\cos 3\Omega t - \cos\Omega t) + \\ + \frac{\varepsilon^2 A^5}{1536\Omega^4}(\cos 5\Omega t + 3\cos 3\Omega t - 4\cos\Omega t) \tag{9.51}$$

Substituting Eqs. 9.48 and 9.51 into Eq. 9.15, we have the third iteration ($n = 2$):

$$\ddot{u}_3 + \Omega^2 u_3 = \left[A\left(1 - \frac{\varepsilon A^2}{32\Omega^2}\right)\cos\Omega t + \frac{\varepsilon A^3}{32\Omega^2}\cos 3\Omega t\right]\{\Omega^2 - 1 - \\ \varepsilon\left[A\left(1 - \frac{\varepsilon A^2}{32\Omega^2}\right)\cos\Omega t + \frac{\varepsilon A^3}{32\Omega^2}\cos 3\Omega t\right]^2 - 2\varepsilon\left[A\left(1 - \frac{\varepsilon A^2}{32\Omega^2}\right)\times \\ \times\cos\Omega t + \frac{\varepsilon A^3}{32\Omega^2}\cos 3\Omega t\right]\left[\frac{\varepsilon^2 A^5}{1536\Omega^2}(\cos 5\Omega t + 3\cos 3\Omega t - 4\cos\Omega t\right]\right\} \tag{9.52}$$

In order to eliminate the secular term arising in the third iteration, the coefficient of $\cos\Omega t$ in this equation requires to be zero i.e.

$$\Omega_3^2 - 1 - \frac{25}{32}\varepsilon A^2 + \frac{\varepsilon A^2}{32\Omega_3^2} + \frac{3\varepsilon^2 A^4}{64\Omega_3^2} + \frac{3\varepsilon^3 A^6}{4096\Omega_3^4} - \frac{13\varepsilon^4 A^8}{65536\Omega_3^6} + \frac{5\varepsilon^5 A^{10}}{1048576\Omega_3^8} = 0 \tag{9.53}$$

The exact frequency of the periodic motion of the Duffing equation is given by:

$$\Omega_{ex} = \frac{\pi}{2}\sqrt{1 + \varepsilon A^2}\left(\int_0^{\frac{\pi}{2}} \frac{d\theta}{\sqrt{1 - m\sin^2\theta}}\right)^{-1}, m = \frac{\varepsilon A^2}{2(1 + \varepsilon A^2)} \tag{9.54}$$

For comparison between the exact frequency Ω_{ex} obtained by integrating Eq. 9.54 and the approximate frequencies Ω_1, Ω_2, Ω_3 computed by Eqs. 9.47, 9.50 and 9.53, respectively, we have

$$\lim_{\varepsilon A^2\to\infty}\frac{\Omega_1}{\Omega_{ex}}=\frac{\sqrt{3}}{\pi}\int_0^{\frac{\pi}{2}}\frac{d\theta}{\sqrt{1-\frac{1}{2}\sin^2\theta}}=1.02223 \tag{9.55a}$$

$$\lim_{\varepsilon A^2\to\infty}\frac{\Omega_2}{\Omega_{ex}}=\frac{\sqrt{3+\sqrt{7}}}{2\sqrt{2}\pi}\int_0^{\frac{\pi}{2}}\frac{d\theta}{\sqrt{1-\frac{1}{2}\sin^2\theta}}=0.9935672 \tag{9.55b}$$

$$\lim_{\varepsilon A^2\to\infty}\frac{\Omega_3}{\Omega_{ex}}=\frac{2\sqrt{0,719601}}{\pi}\int_0^{\frac{\pi}{2}}\frac{d\theta}{\sqrt{1-\frac{1}{2}\sin^2\theta}}=0.999134 \tag{9.55c}$$

On the other hand, for large value of εA^2 in Eq. 9.53 we obtain $\Omega_3^2 \approx 1+0.719601\varepsilon A^2$.

Therefore the approximate analytical solutions to the Duffing equation with respect to the exact solution for the first, the second and the third iteration never exceed 2.2, 0.64 and 0.08%, respectively.

Our result (9.51) is compared in Fig. 9.4 with the numerical integration results obtained by using a fourth-order Runge-Kutta method.

(d) As last example, we consider *a particle having mass m moving under the influence of the central force field* of magnitude k/r^{2n+3}. The equation of the orbit in the polar coordinates (r,θ) is

$$\frac{d^2u}{d\theta^2}+u=-cu^{2n+1}$$

where k and c are constants and $u=1/r$. In this case, let us consider a family of nonlinear differential equation (see Sect. 7.1.2):

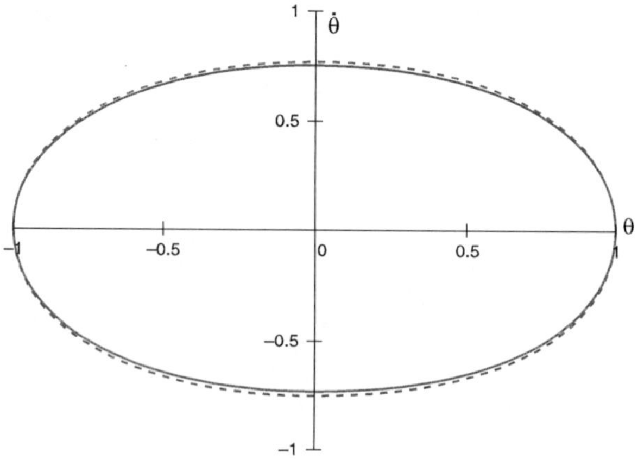

Fig. 9.4 Phase plane for Eq. 9.45 : $\varepsilon = 1/2$, $A = 1$: - - - - - - present method, Eq. 9.51; _____ numerical solution

$$u'' + \alpha u + \gamma u^{2n+1} = 0, \alpha>0, \gamma>0, n = 1,2,3,\ldots,$$
$$u(0) = A; u'(0) = 0 \tag{9.56}$$

The corresponding exact period T is

$$T_{ex} = 4\int_0^{\frac{\pi}{2}} \frac{d\theta}{\sqrt{\alpha + \frac{\gamma}{n+1}A^{2n}(1+\sin^2\theta + \sin^4\theta + \ldots + \sin^{2n}\theta)}} \tag{9.57}$$

In this case we have: $g(u) = \Omega^2 - \alpha - \gamma u^{2n}, u_{-1}(t) = u_0(t) = A\cos\Omega t$. The first iteration becomes:

$$\ddot{u}_1 + \Omega^2 u_1 = A\cos\Omega t(\Omega^2 - 1 - \gamma A^{2n}\cos^{2n}\Omega t) \tag{9.58}$$

By using (9.26), equation may be expressed in the form:

$$\ddot{u}_1 + \Omega^2 u_1 = A\cos\Omega t\left[\Omega^2 - \alpha - \frac{\gamma A^{2n}}{4^n}\binom{2n+1}{n}\right] -$$
$$-\frac{1}{4^n}\gamma A^{2n+1}\sum_{k=1}^{n}\binom{2n+1}{n-k}\cos(2k+1)\Omega t \tag{9.59}$$

Avoiding the secular term needs that:

$$\Omega_1(n) = \sqrt{\alpha + \frac{\gamma A^{2n}}{4^n}\binom{2n+1}{n}} \tag{9.60}$$

The solution of Eq. 9.59 reads:

$$u_1(t) = A\cos\Omega t + \frac{\gamma A^{2n+1}}{4^{n+1}}\sum_{k=1}^{n}\frac{1}{k(k+1)}\binom{2n+1}{n-k}[\cos(2k+1)\Omega t - \cos\Omega t] \tag{9.61}$$

For $n = 1$ in Eqs. 9.56, 9.60, 9.61 we recover Eqs. 9.45, 9.47 and 9.48 respectively ($\alpha = 1, \gamma = \varepsilon$). For $n = 2$ in Eqs. 9.56 and 9.60 we obtain

$$\Omega_1(2) = \sqrt{1 + \frac{5}{8}\varepsilon A^4} \tag{9.62}$$

The same result is obtained in [181] using the variational iteration method. Its approximate period can be calculated as follows

$$T_{1app} = \frac{2\pi}{\sqrt{1 + 5\varepsilon A^4/8}} \tag{9.63}$$

It should be specially pointed out that the iterative perturbation formula (9.63) is valid not only for a small parameter, but also for a very large parameter: even in the case $\varepsilon A^4 \to \infty$, we have

$$\lim_{\varepsilon A^4 \to \infty} \frac{T_{1\text{app}}}{\text{T}_{\text{ex}}} =$$

$$= \lim_{\varepsilon A^4 \to \infty} \frac{\pi}{2\sqrt{1 + \frac{5}{8}\varepsilon A^4}} \left(\int_0^{\frac{\pi}{2}} \frac{d\theta}{\sqrt{1 + \frac{\varepsilon A^4}{3}(1 + \sin^2\theta + \sin^4\theta)}} \right)^{-1} \approx$$

$$\approx \sqrt{\frac{2}{15}} \cdot \pi \cdot 0,87099 = 0,99916 \tag{9.64}$$

Therefore, for any value of ε it can be easily proved that the relative error of the period (9.63) is less than 0.08%. This rather extraordinary virtue of this procedure has first been exploited in this Section. And this might be the reason that the obtained approximate solution is valid for all $\varepsilon > 0$.

For integer n, the approximate period in the first-order approximate is

$$T_{app} = \frac{2\pi}{\Omega_1(n)} = 2\pi \left[\alpha + \frac{\gamma A^{2n}}{4^n} \binom{2n+1}{n} \right]^{-\frac{1}{2}} \tag{9.65}$$

Formula (9.65) is valid for any possible amplitudes and gives the maximum errors as the dimensionless amplitude γA^{2n} tends to infinity. Note that, even for $n = 9$, the maximum error given by (9.65) is less than 4% for amplitude $A \in [0, \infty]$:

$$\lim_{\gamma A^{18} \to \infty} \frac{T_{app}}{T_{ex}} \approx \frac{16\pi\sqrt{\frac{5}{46189}}}{0.5406369} \approx 0.967341 \tag{9.66}$$

9.1.2 *An Iteration Procedure with Application to Van der Pol Oscillator*

We consider the following nonlinear equation

$$\ddot{u} + \omega^2 u = \varepsilon f(\Omega t, u, \dot{u}) \tag{9.67}$$

where ω and Ω are positive constants; in general f is assumed to be non-linear function of both u and $\dot{u}$ which may be expanded in a Fourier series, and $\dot{u} = du/dt$ [182].

Assume that $\omega \approx \Omega$ and let us denote (σ is detuning parameter):

$$\Omega^2 - \omega^2 = \varepsilon\sigma \ (0<\varepsilon<<1) \tag{9.68}$$

The Eq. 9.67 may then be written:

$$\ddot{u} + \Omega^2 u = \varepsilon[\sigma u + f(\Omega t, u, \dot{u})] \tag{9.69}$$

For $\varepsilon = 0$, Eq. 9.69 has the solution $u = A\cos(\Omega t + \varphi)$, where A and φ are constants. For $\varepsilon \neq 0$ we try the input of starting function as:

$$u_0 = A_0 \cos(\Omega t + \varphi_0) \tag{9.70}$$

We propose the following iteration formula for Eq. 9.69:

$$\begin{aligned} u_n(t) = {} & A_n \cos(\Omega t + \varphi_n) + \\ & + \frac{\varepsilon}{\Omega} \int_0^t [\sigma u_{n-1}(\tau) + f(\Omega t, u_{n-1}(\tau), ux_{n-1}(\tau))] d\tau, \text{ n} = 1, 2, \ldots \end{aligned} \tag{9.71}$$

where A_{n} and φ_{n} are constants and u_0 is given by Eq. 9.70.

Expanding $\sigma u_{n-1} + f$ in a Fourier series, we have:

$$\begin{aligned} \sigma u_{n-1}(\tau) + f(\Omega t, u_{n-1}(\tau), u'_{n-1}(\tau)) = {} & \sum_{p=0}^{P} a_p^{n-1}(A_{n-1}, \varphi_{n-1}, \Omega, \sigma) \cos p\Omega\tau + \\ & + \sum_{r=0}^{R} b_r^{n-1}(A_{n-1}, \varphi_{n-1}, \Omega, \sigma) \sin r\Omega\tau \end{aligned} \tag{9.72}$$

and therefore, the approximation of n-th order (9.71) becomes:

$$\begin{aligned} u_n(t) = {} & A_n \cos(\Omega t + \varphi_n) + \frac{\varepsilon}{\Omega} \left[\frac{1}{2} a_1^{n-1} (t \sin \Omega t) + \right. \\ & + \frac{1}{2} b_1^{n-1} \left(t \cos \Omega t + \frac{1}{2\Omega} \sin \Omega t \right) + a_0^{n-1} + \\ & \left. + \sum_{p=2}^{P} \frac{a_p^{n-1}(\cos \Omega t - \cos p\Omega t)}{(p^2 - 1)\Omega} + \sum_{r=2}^{R} \frac{b_r^{n-1}(r \sin \Omega t - \sin r\Omega t)}{(r^2 - 1)\Omega} \right] \end{aligned} \tag{9.73}$$

The solution (9.73) is chosen such that it contains no secular terms, which requires that coefficients a_1^{n-1} and b_1^{n-1} into (9.73) disappear, i.e.:

$$a_1^{n-1}(A_{n-1}, \varphi_{n-1}, \Omega, \sigma) = 0;$$
$$\mathrm{b}_1^{n-1}(A_{n-1}, \varphi_{n-1}, \Omega, \sigma) = 0 \qquad (9.74)$$

For real systems, the expansion of the function $\sigma u_{n-1} + f$ usually contains only a small number of harmonics.

If we consider the nonlinear oscillator governed by the Van der Pol equation

$$\ddot{u} + u = \varepsilon(1 - u^2)\dot{u} \qquad (9.75)$$

in Eq. 9.71 we have $\omega = 1$ and assuming $\Omega \approx 1$, Eq. 9.69 becomes:

$$\ddot{u} + \Omega^2 u = \varepsilon F(u, \dot{u}); \quad F(u, \dot{u}) = \sigma u + \dot{u} - \dot{u}u^2 \qquad (9.76)$$

Assuming that the input of starting function is (9.70), we obtain

$$F(u_0, \dot{u}_0) = \sigma A_0 \cos(\Omega t + \varphi_0) + \Omega A_0 \left(\frac{A_0^2}{4} - 1\right) \sin(\Omega t + \varphi_0) +$$
$$+ \frac{1}{4}\Omega A_0^3 \sin(3\Omega t + \varphi_0) \qquad (9.77)$$

From the conditions (9.74) we have:

$$a_1^0 = \sigma A_0 = 0; \quad b_1^0 = \Omega A_0 \left(\frac{A_0^2}{4} - 1\right) = 0 \qquad (9.78)$$

and therefore

$$\sigma = 0, \quad \mathrm{A}_0 = 2 \qquad (9.79)$$

The solution $u_0(t)$ given by (9.70) becomes:

$$u_0(t) = 2\cos(\Omega t + \varphi_0) \qquad (9.80)$$

An additional condition to determine the constant φ_0 of the homogenous solution (9.80) is simplified to $\dot{u}_0(0) = 0$, such that $\varphi_0 = 0$. The solution (9.80) and $F(u_0, \dot{u}_0)$ become respectively:

$$u_0(t) = 2\cos\Omega t, \quad F(u_0, \dot{u}_0) = 2\Omega \sin 3\Omega t$$

The solution (9.73) for $n = 1$ is written in the form:

$$u_1(t) = A_1\cos(\Omega t + \varphi_1) + \frac{3\varepsilon}{4}\sin\Omega t - \frac{\varepsilon}{4\Omega}\sin 3\Omega t$$

or

$$u_1(t) = A_1'\cos(\Omega t + \varphi_1') - \frac{\varepsilon}{4\Omega}\sin 3\Omega t \tag{9.81}$$

where A_1' and φ_1' are constants and

$$A_1'\cos(\Omega t + \varphi_1') = A_1\cos(\Omega t + \varphi_1) - \frac{\varepsilon}{4\Omega}\sin 3\Omega t$$

The homogenous solution given by (9.81) is simplified to $\dot{u}_{1h}(0) = 0$, such that $\varphi_1' = 0$ and Eq. 9.81 becomes

$$u_1(t) = A_1'\cos\Omega t - \frac{\varepsilon}{4\Omega}\sin 3\Omega t \tag{9.82}$$

and therefore

$$\begin{aligned} F(u_1, \dot{u}_1) = &\left(\sigma A'_1 + \frac{\varepsilon A_1'2}{16}\right)\cos\Omega t + \Omega A'_1\left(\frac{A_1'2}{4} - 1 + \frac{\varepsilon^2}{32\Omega^2}\right)\sin\Omega t + \\ &+ \left(\frac{3}{8}\varepsilon A_1'2 - \frac{3}{4}\varepsilon + \frac{3\varepsilon^3}{256\Omega^2}\right)\cos 3\Omega t + \left(\frac{1}{4}\Omega A_1'3 - \frac{\varepsilon\sigma}{4\Omega}\right)\sin 3\Omega t + \\ &+ \frac{5}{16}\varepsilon A_1'2\cos 5\Omega t - \frac{5\varepsilon^2 A'_1}{64\Omega}\sin 5\Omega t - \frac{7\varepsilon^2 A'_1}{64\Omega}\sin 7\Omega t - \frac{3\varepsilon^3}{256\Omega^2}\cos 9\Omega t \end{aligned} \tag{9.83}$$

Eliminating the secular terms in Eq. 9.83 needs

$$A_1' = 2 - \frac{\varepsilon^2}{32\Omega^2} - \frac{\varepsilon^4}{4096\Omega^4}, \quad \sigma = -\frac{1}{8}\varepsilon + \frac{\varepsilon^3}{512\Omega^2} + \frac{\varepsilon^5}{65536\Omega^4} \tag{9.84}$$

such that Eqs. 9.82 and 9.83 become respectively

$$u_1(t) = \left(2 - \frac{\varepsilon^2}{32\Omega^2} - \frac{\varepsilon^4}{4096\Omega^4}\right)\cos\Omega t - \frac{\varepsilon}{4\Omega}\sin 3\Omega t$$

$$F(u_1,\dot{u}_1)=\left(\frac{3}{4}\varepsilon-\frac{9\varepsilon^3}{256\Omega^2}\right)\cos 3\Omega t+$$
$$+\left(2\Omega-\frac{\varepsilon^2}{16\Omega}-\frac{\varepsilon^4}{4096\Omega^3}\right)\sin 3\Omega t+\left(\frac{5}{4}\varepsilon-\frac{5\varepsilon^3}{128\Omega^2}\right)\cos 5\Omega t-$$
$$-\left(\frac{5\varepsilon^2}{32\Omega}-\frac{5\varepsilon^4}{2048\Omega^3}\right)\sin 5\Omega t-\left(\frac{7\varepsilon^2}{32\Omega}-\frac{7\varepsilon^4}{2048\Omega^3}\right)\sin 7\Omega t-\frac{3\varepsilon^3}{256\Omega^2}\cos 9\Omega t$$

For $n=2$ into Eq. 9.73, we obtain:

$$u_2(t)=A_2\cos(\Omega t+\varphi_2)+\frac{\varepsilon}{\Omega}\left[\left(\frac{3}{4}\varepsilon-\frac{9\varepsilon^3}{256\Omega^2}\right)\frac{\cos\Omega t-\cos 3\Omega t}{8\Omega}+\right.$$
$$+\left(2\Omega-\frac{\varepsilon^2}{16\Omega}-\frac{\varepsilon^4}{4096\Omega^3}\right)\frac{3\sin\Omega t-\sin 3\Omega t}{8\Omega}+$$
$$+\left(\frac{5}{4}\varepsilon-\frac{5\varepsilon^3}{128\Omega^2}\right)\frac{\cos\Omega t-\cos 5\Omega t}{24\Omega}-\left(\frac{5\varepsilon^2}{32\Omega}-\frac{5\varepsilon^4}{2048\Omega^3}\right)\frac{5\sin\Omega t-\sin 5\Omega t}{24\Omega}-$$
$$\left.-\left(\frac{7\varepsilon^2}{32\Omega}-\frac{7\varepsilon^4}{2048\Omega^3}\right)\frac{7\sin\Omega t-\sin 7\Omega t}{48\Omega}-\frac{3\varepsilon^3}{256\Omega^2}\frac{\cos\Omega t-\cos 9\Omega t}{80\Omega}\right] \tag{9.85}$$

We insist that $\dot{u}_{2h}=0$ and it follows that:

$$u_2(t)=A'_2\cos\Omega t-\left(\frac{3\varepsilon^2}{32\Omega^2}-\frac{9\varepsilon^4}{2048\Omega^4}\right)\cos 3\Omega t-$$
$$-\left(\frac{\varepsilon}{4\Omega}-\frac{\varepsilon^3}{128\Omega^3}-\frac{\varepsilon^5}{32768\Omega^5}\right)\sin 3\Omega t-\left(\frac{5\varepsilon^2}{96}-\frac{5\varepsilon^4}{3072\Omega^4}\right)\cos 5\Omega t+$$
$$+\left(\frac{5\varepsilon^3}{768\Omega^3}-\frac{5\varepsilon^5}{49152\Omega^5}\right)\sin 5\Omega t+\left(\frac{7\varepsilon^3}{1536\Omega^3}-\frac{7\varepsilon^5}{98304\Omega^5}\right)\sin 7\Omega t+$$
$$+\frac{3\varepsilon^4}{20480\Omega^4}\cos 9\Omega t \tag{9.86}$$

and substituting $u_2(t)$ in (9.76$_2$), yields the result

$$
\begin{aligned}
F(u_2,\dot{u}_2) = \Bigg[& \sigma A'_2 + A'^2_2\left(\frac{\varepsilon}{16} - \frac{\varepsilon^3}{512\Omega^2} - \frac{\varepsilon^5}{131072\Omega^4}\right) + \\
& + A'_2\left(-\frac{5\varepsilon^3}{384\Omega^2} + \frac{5\varepsilon^5}{294912\Omega^4}\right) + \frac{19\varepsilon^5}{32768\Omega^4}\Bigg] \cos\Omega t + \\
& + \Bigg[\Omega\left(\frac{A'_2 3}{4} - A'_2\right) + A'^2_2\left(-\frac{3\varepsilon^2}{128\Omega} + \frac{9\varepsilon^4}{8192\Omega^3}\right) + \\
& + A'_2\left(\frac{\varepsilon^2}{32\Omega} + \frac{115\varepsilon^4}{18432\Omega^3}\right) + \frac{5\varepsilon^4}{6144\Omega^3}\Bigg] \sin\Omega t + H.O..T.
\end{aligned}
\tag{9.87}
$$

Avoiding the secular terms in Eq. 9.87 requires that:

$$
\begin{aligned}
& \sigma A'_2 + A'^2_2\left(\frac{\varepsilon}{16} - \frac{\varepsilon^3}{512\Omega^2} - \frac{\varepsilon^5}{131072\Omega^4}\right) + \\
& + A'_2\left(-\frac{5\varepsilon^3}{384\Omega^2} + \frac{5\varepsilon^5}{294912\Omega^4}\right) + \frac{19\varepsilon^5}{32768\Omega^4} = 0 \\
& \Omega\left(\frac{A'^3_2}{4} - A'_2\right) + A'_2\left(-\frac{3\varepsilon^2}{128\Omega} + \frac{9\varepsilon^4}{8192\Omega^3}\right) + \\
& + A'_2\left(\frac{\varepsilon^2}{32\Omega} + \frac{115\varepsilon^4}{18432\Omega^3}\right) + \frac{5\varepsilon^4}{6144\Omega^3} = 0
\end{aligned}
\tag{9.88}
$$

From Eq. 9.88 if we keep terms only up to $0(\varepsilon^6)$, we have:

$$
\Omega = 1 - \frac{\varepsilon^2}{16} + \frac{37\varepsilon^4}{6144} + 0(\varepsilon^6) \tag{9.89}
$$

$$
A'_2 = 2 + \frac{\varepsilon^2}{64} + \frac{6265\varepsilon^4}{73728} + 0(\varepsilon^6) \tag{9.90}
$$

which lead to:

$$
\begin{aligned}
u_2(t) = & \left(2 + \frac{\varepsilon^2}{64} + \frac{6265\varepsilon^4}{73728}\right)\cos\Omega t - \left(\frac{3\varepsilon^2}{32} + \frac{15\varepsilon^4}{2048}\right)\cos 3\Omega t - \left(\frac{\varepsilon}{4} - \frac{\varepsilon^3}{128} - \right. \\
& \left. - \frac{223\varepsilon^5}{98304}\right)\sin 3\Omega t - \left(\frac{5\varepsilon^2}{96} + \frac{5\varepsilon^4}{1024}\right)\cos 5\Omega t + \left(\frac{5\varepsilon^3}{768} + \right. \\
& \left. + \frac{5\varepsilon^5}{24576}\right)\sin 5\Omega t + \left(\frac{7\varepsilon^3}{1536} + \frac{7\varepsilon^5}{49152}\right)\sin 7\Omega t + \frac{3\varepsilon^4}{20480}\cos 9\Omega t
\end{aligned}
\tag{9.91}
$$

It is worthy pointing out that Mickens [183] and Chen et al. [184] obtained the following result

$$\Omega = 1 - \frac{\varepsilon^2}{16} + \frac{17\varepsilon^4}{3072} + 0(\varepsilon^6) \quad (9.92)$$

by using the classical Lindstedt-Poincaré method under the same initial conditions, respectively by using a perturbation method in two variables.

Figures 9.5–9.8 illustrate that the displacements and velocities are very accurate, when comparing to numerical results.

Figures 9.9–9.10 show the comparison in terms of the phase plane obtained for $\varepsilon = 0.6$ and $\varepsilon = 0.8$, between the present solution given by Eq. 9.91 and numerical integration results obtained using a fourth-order Runge-Kutta method.

It can be seen that the displacement-time curve and velocity-time curve of the present solution are very close to the numerical solution.

Therefore one can conclude that adopting the present procedure for the solution of the Van der Pol equation, satisfactory results are obtained even for moderate values of the parameter ε.

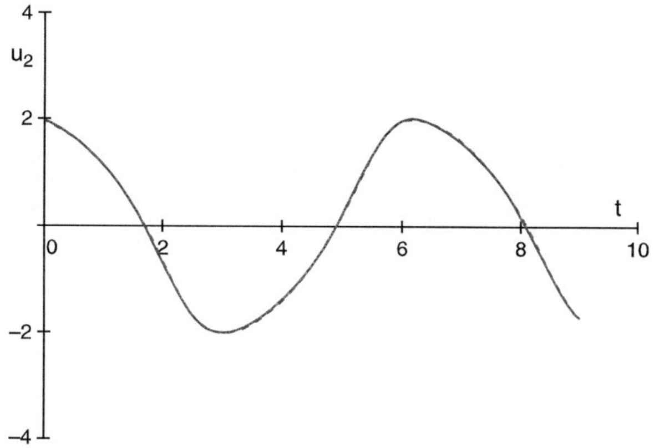

Fig. 9.5 Solutions of the Van der Pol Eq. 9.75 for $\varepsilon = 0.6$: ______ numerical solution, - - - - - proposed method (9.91)

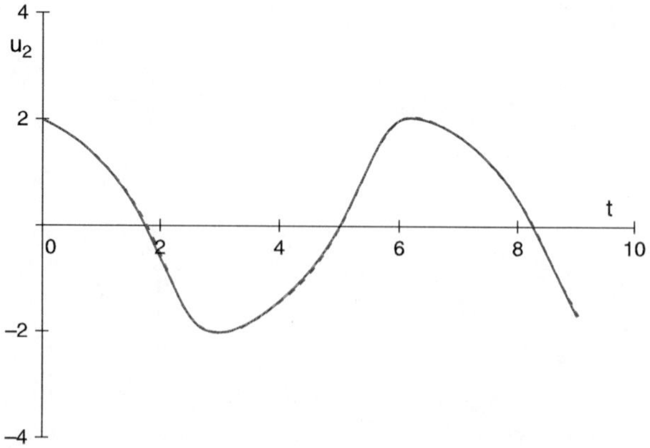

Fig. 9.6 Solutions of the Van der Pol Eq. 9.75 for $\varepsilon = 0.8$: ______ numerical solution; - - - - - proposed method (9.91)

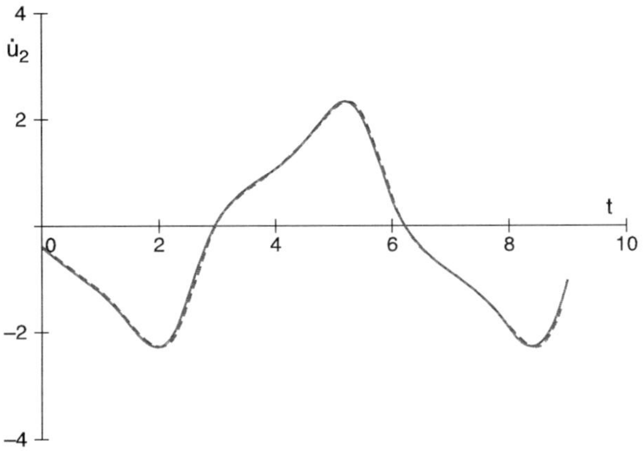

Fig. 9.7 Velocity-time curves obtained from Eq. 9.75 for $\varepsilon = 0.6$: ______ numerical solution; - - - - - proposed method (9.91)

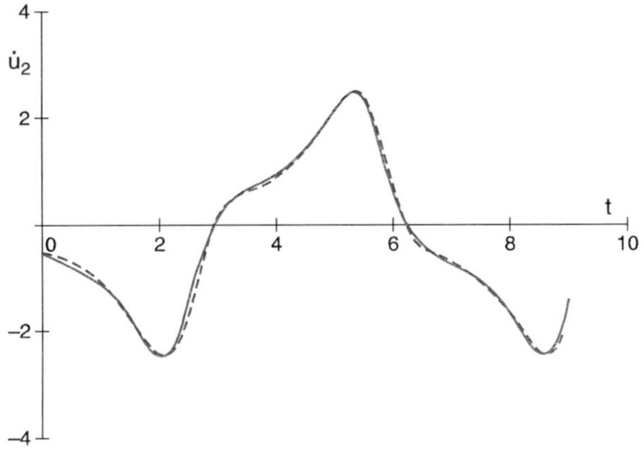

Fig. 9.8 Velocity-time curves obtained from Eq. 9.75 for $\varepsilon = 0.8$: ______ numerical solution; - - - - - proposed method (9.91)

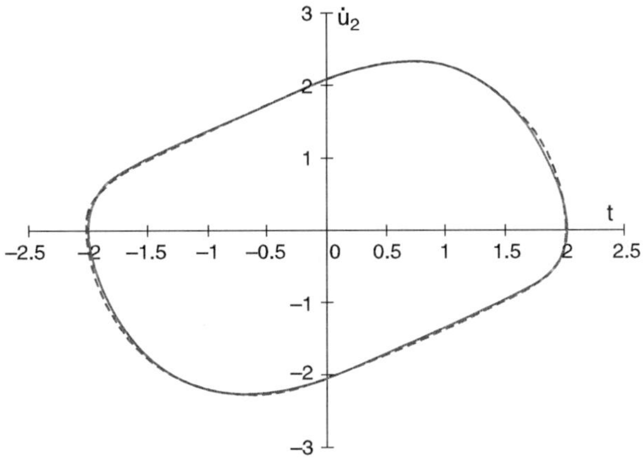

Fig. 9.9 Phase plane for Van der Pol Eq. 9.75, $\varepsilon = 0.6$: ______ numerical solution; - - - - - proposed method (9.91)

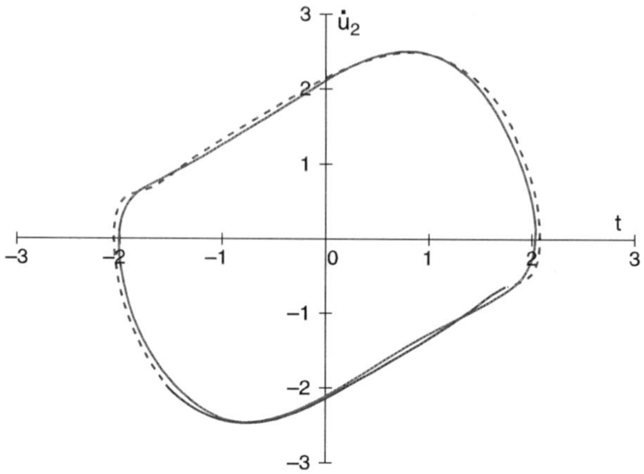

Fig. 9.10 Phase plane for Van der Pol Eq. 9.75, $\varepsilon = 0.8$: ______ numerical solution (9.91); - - - - - proposed method

9.2 Basic Idea of Optimal Parametric Iteration Method

We consider the nonlinear differential equation

$$L(u) + N(t, u, \dot{u}, \ddot{u}, ...) - g(r) = 0 \tag{9.93}$$

with L a linear operator and N a nonlinear operator and g a known function.

Initial conditions are

$$B(u, \dot{u}) = 0 \tag{9.94}$$

If α, β, γ are real values, then applying the well-known Taylor formula for an analytic function F, we obtain

$$\begin{aligned} &F(t, u+\alpha, \dot{u}+\beta, \ddot{u}+\gamma, ...) = F(t, u, \dot{u}, \ddot{u}, ...) + \frac{\alpha}{1!}F_u(t, u, \dot{u}, \ddot{u}, ...)+ \\ &\frac{\beta}{1!}F_{\dot{u}}(t, u, \dot{u}, \ddot{u}, ...) + \frac{\gamma}{1!}F_{\ddot{u}}(t, u, \dot{u}, \ddot{u}, ...) + ... \end{aligned} \tag{9.95}$$

where $F_u = \partial F/\partial u$

Instead of solving the nonlinear differential Eq. 9.93, one can solve another equation, making recourse to Eq. 9.95 and to the following scheme

$$\begin{aligned} &L(u_{n+1}(t)) + N(t, u_n, \dot{u}_n, \ddot{u}_n, ...) + \alpha_n(t, C_i)N_u(t, u_n, \dot{u}_n, \ddot{u}_n, ...)+ \\ &+ \beta_n(t, C_j)N_{\dot{u}}(t, u_n, \dot{u}_n, \ddot{u}_n, ...) + \gamma_n(t, C_k)N_{\ddot{u}}(t, u_n, \dot{u}_n, \ddot{u}_n, ...) + ... - g(r) = 0, \\ &B(u_{n+1}, \dot{u}_{n+1}) = 0 \end{aligned} \tag{9.96}$$

where α_n, β_n, and γ_n are auxiliary functions and the initial approximation $u_0(t)$can be determined at least in the following two alternatives:

In the first alternative $u_0(t)$ is the solution of the equation

$$L(u_0(t)) - g(r) = 0 \quad B(u_0, \dot{u}_0) = 0 \tag{9.97}$$

In the second alternative we can choose the initial approximation in the general form

$$u_0(t) = \sum_{i=1}^{m} C_i f_i \tag{9.98}$$

where C_i are unknown constants, the function f_i depends on the form of linear operator L and also on the coefficients of given Eq. 9.93 and m is an integer number. For both alternatives, the auxiliary functions $\alpha(t, C_i)$, $\beta(t, C_j)$, $\gamma(t, C_k)$... can be chosen such that the products $\alpha_n N_u$, $\beta_n N_{\dot{u}}$, γ_n, $N_{\ddot{u}}$,... and N_u, $N_{\dot{u}}$, $N_{\ddot{u}}$, ... be of the same form, respectively. The auxiliary functions f_i, α_n, β_n, γ_n,... are not unique. In this way, we obtain an approximation of the solution of Eq. 9.93 given by a truncated series (9.96). To improve the order of convergence of the sequence $u_n(t)$ as given in Eq. 9.96 we propose that the constants C_1, C_2,... which appear in Eq. 9.98 and in the auxiliary functions α_n, β_n, γ_n,... (which are unknown at this moment) be determined optimally (see Chaps. 6–8) using the least square method, the Galerkin method, the Ritz method, the collocation method etc. In this way, the solution of Eq. 9.96 is well-determined after only one iteration.

The basic ideas of the proposed procedure are the construction of a new iteration scheme (9.96) and the involvement of the convergence-control constants C_i, i = 1,2,..., through the auxiliary functions and Eq. 9.98, which lead to an excellent agreement between the approximate and exact solutions. The presence of a finite number of initially unknown parameters C_i, which are optimally determined later provides a rigorous way to control the convergence of the solution.

In what follows, we apply the optimal parametric iteration method (OPIM) to some nonlinear problems using the above described alternatives.

9.3 Thin Film Flow of a Fourth Grade Fluid Down a Vertical Cylinder

Consider the nonlinear differential equation with variable coefficients [166] (See also Sects. 6.3, 7.5, 8.5)

$$\eta \frac{d^2 f}{d\eta^2} + \frac{df}{d\eta} + k\eta + 2b\left[\left(\frac{df}{d\eta}\right)^3 + 3\eta\left(\frac{df}{d\eta}\right)^2 \frac{d^2 f}{d\eta^2}\right] = 0 \tag{9.99}$$

with the initial conditions

$$f(1) = 0, f'(d) = 0 \tag{9.100}$$

In this case a linear operator, known function and a nonlinear operator are respectively

$$\begin{aligned}
&Lf(\eta) = \eta\frac{d^2f}{d\eta^2} + \frac{df}{d\eta} + k\eta \\
&f(r) = -k\eta \\
&N\left(\frac{df}{d\eta}, \frac{d^2f}{d\eta^2}\right) = 2b\left[\left(\frac{df}{d\eta}\right)^3 + 3\eta\left(\frac{df}{d\eta}\right)^2\frac{d^2f}{d\eta^2}\right]
\end{aligned} \tag{9.101}$$

where k and b are known constants.

Using the first alternative, the initial approximation $f_0(\eta)$ is given by Eq. 9.97 and thus

$$L(f_0(\eta)) = 0, \quad f_0'(d) = 0 \tag{9.102}$$

where $f' = df/d\eta$. In our case, Eq. 9.102 can be written as

$$\eta f''_0 + f'_0 + k\eta = 0, \quad f'(d) = 0$$

It is obtained

$$f_0'(\eta) = \frac{k}{2}\left(\frac{d^2}{\eta} - \eta\right) \tag{9.103}$$

The first-order approximate solution $f_1(\eta)$is obtained from Eq. 9.96 for $n = 0$:

$$\begin{aligned}
&\eta f''_1 + f'_1 + k\eta + 2b[f_0'^3 + 3\eta f_0'^2 f''_0 + \\
&+ \alpha(\eta, C_i)(3f_0'^2 + 6\eta f'_0 f''_0) + \beta(\eta, C_j)(3\eta f_0'^2)] = 0 \\
&f'_1(d) = 0
\end{aligned} \tag{9.104}$$

The Eq. 9.104 can be written in the following form

$$\begin{aligned}
&(\eta f'_1)' + (\frac{k}{2}\eta^2)' + 2b(\eta f_0'^3)' + \\
&\quad + 6b[\alpha(\eta, C_i)(\eta f_0'^2)' + \beta(\eta, C_j)(\eta f_0'^2)] = 0
\end{aligned} \tag{9.105}$$

The Eq. 9.105 can be easily solved if the functions $\alpha(\eta, C_i)$and $\beta(\eta, C_i)$are chosen such as

$$\alpha(\eta, C_i)(\eta f_0'^2)' + \beta(\eta, C_j)(\eta f_0'^2) = \\ = \frac{1}{3}(C_1 \eta f_0'^i + C_2 \eta f_0'^j + C_3 \eta f_0'^k)' \tag{9.106}$$

where C_1, C_2 and C_3 are unknown constants and i, j, k are integers.
From Eq. 9.106 it is obtained:

$$\alpha(\eta, C_i) = \frac{1}{6}(C_1 f_0'^{i-2} + C_2 f_0'^{j-2} + C_3 f_0'^{k-2}) \tag{9.107}$$

$$\beta(\eta, C_i) = \frac{1}{6\eta}(C_1 f_0'^{i-2} + C_2 f_0'^{j-2} + C_3 f_0'^{k-2}) \tag{9.108}$$

From Eqs. 9.104$_2$, 9.105, 9.107 and 9.108 we obtain

$$f_1'(\eta) = \frac{k}{2}\left(\frac{d^2}{\eta} - \eta\right) - 2b[f_0'^3(\eta) + C_1 f_0'^i + C_2 f_0'^j + C_3 f_0'^k] \tag{9.109}$$

For $i = 4, j = 5, k = 6$ and using Eq. 9.103, from Eq. 9.109 we obtain

$$f_1'(\eta) = \frac{k}{2}\left(\frac{d^2}{\eta} - \eta\right) - \frac{1}{4}bk^3\left(\frac{d^2}{\eta} - \eta\right)^3 - \frac{1}{8}bk^4 C_1\left(\frac{d^2}{\eta} - \eta\right)^4 - \\ - \frac{1}{16}bk^5 C_2\left(\frac{d^2}{\eta} - \eta\right)^5 - \frac{1}{32}bk^6 C_3\left(\frac{d^2}{\eta} - \eta\right)^6 \tag{9.110}$$

The residual functional J given by Eq. 6.115 becomes:

$$J = \int_1^d \{\eta f''_1(\eta) + f'_1(\eta) + k\eta + 2b[f_1'^3(\eta) + 3\eta f_1'^2(\eta) f''_1(\eta)]\}^2 d\eta \tag{9.111}$$

From Eq. 6.116 for $\beta = k = 1$ we obtain

$$C_1 = -0.00135812; \quad C_2 = -5.98314207; \quad C_3 = 0.02175281$$

Therefore, the explicit analytic expression given by Eq. 9.110 of the first-order approximate solution becomes:

$$f_1'(\eta) = 0.5\left(\frac{d^2}{\eta} - \eta\right) - 0.25\left(\frac{d^2}{\eta} - \eta\right)^3 + 0.000169765\left(\frac{d^2}{\eta} - \eta\right)^4 + \\ + 0.373946379\left(\frac{d^2}{\eta} - \eta\right)^5 - 0.000679775\left(\frac{d^2}{\eta} - \eta\right)^6 \tag{9.112}$$

Table 9.2 Comparison between the present solution (9.112) and the exact solution for $d = 1.02$

η	$\bar{f}'$ given by Eq. 9.112	f' exact
1	0.020183555	0.020183555
1.005	0.015105047	0.015105047
1.007	0.013079437	0.013079437
1.01	0.010047475	0.010047475
1.0105	0.009542918	0.009542918
1.0108	0.009240289	0.009240289

Table 9.3 Comparison between the present solution (9.112) and the exact solution for $d = 1.04$

H	$\bar{f}'$ given by (9.112)	f' exact
1	0.040665504	0.040665504
1.008	0.032439661	0.032439661
1.016	0.024254926	0.024254926
1.023	0.017131195	0.017131195
1.03	0.010046515	0.010046515
1.038	0.00200191	0.00200191

Tables 9.2 and 9.3 present a comparison between the present solution obtained from formula (9.112) and the exact solution of Eq. 9.99.

It can be seen that the solution obtained through the present method is identical with that given by exact solution, demonstrating a very good accuracy.

9.4 Thermal Radiation on MHD Flow over a Stretching Porous Sheet

The study of magnetohydrodynamic flow (MHD) and heat transfer of an electrically conducted fluid finds useful applications in many engineering problems such as nuclear power plants, gas turbines and various propulsion devices for aircraft, missiles, satellites and space vehicles. It serves as the basis for understanding some of the important phenomena occurring in heat exchanger devices. Magnetohydrodynamics can be regarded as a combination of fluid mechanics and electromagnetism. The effects of thermal radiation on MHD flow and heat transfer problems have become more important industrially. At high operating temperature, thermal radiation effects could be quite significant.

In this section we determine the approximate solution of thermal radiation effects on MHD steady asymmetric flow of an electrically conducting fluid past a stretching porous sheet.

Consider a steady two-dimensional incompressible flow caused by a moving sheet, which is placed in a quiescent, electrically conducting fluid. A magnetic field of uniform strength is applied perpendicular to the stretching sheet. The magnetic Reynolds number is taken to be small enough so that the induced magnetic field can be neglected. The equation governing the dimensionless temperature $\theta(\eta)$ is given by [185]:

$$\theta''(\eta) + \beta f(\eta)\theta'(\eta) - \beta S f'(\eta)\theta(\eta) = 0 \tag{9.113}$$

where η is the similarity variable, $\beta = \frac{3RP_r}{4+3R}$, R is the radiation parameter, P_r is the modified Prandl number, S is the wall temperature parameter and $f(\eta)$ is a dimensionless stream given by the solution of nonlinear differential equation:

$$f'''(\eta) + f(\eta)f''(\eta) - f'^2(\eta) - Mf'(\eta) = 0 \tag{9.114}$$

where M is the magnetic parameter and prime denotes derivative with respect toη. The boundary conditions for Eq. 9.114 are:

$$f(0) = \lambda, \quad f'(0) = 1, \quad f'(\infty) = 0 \tag{9.115}$$

where λ is the injection/suction parameter.

In order to satisfy the above boundary conditions, the solution of Eq. 9.114 for arbitrary M and λ is sought in the form

$$f(\eta) = \lambda + \frac{1}{\lambda}(1 - e^{-\gamma\eta}) \tag{9.116}$$

where

$$\gamma = \frac{1}{2}(\lambda + \sqrt{\lambda^2 + 4M + 4}) \tag{9.117}$$

Substituting Eq. 9.116 into Eq. 9.113 we obtain:

$$\theta''(\eta) + (a - be^{-\gamma\eta})\theta'(\eta) - ce^{-\gamma\eta}\theta(\eta) = 0 \tag{9.118}$$

where

$$a = \beta\lambda + \frac{\beta}{\gamma}, \quad b = \frac{\beta}{\lambda}, \quad c = \beta S \tag{9.119}$$

The boundary conditions are given by:

$$\theta(0) = 1, \quad \theta(\infty) = 0 \tag{9.120}$$

The exact solution of Eq. 9.118 with boundary conditions (9.120) is graphically known (in terms of Kummer's confluent hyper-geometric function) [186].

For Eq. 9.118, which is a differential equation with variable coefficients, with the conditions (9.120) we propose an approximate analytical solution by OPIM using the first alternative.

For $n = 0$ into Eq. 9.96 and with $\theta_0 = 0$from Eq. 9.97 we obtain the first-order problem:

$$\theta_1''(\eta) = ce^{-\gamma\eta}\alpha(\eta, C_i) + (be^{-\gamma\eta} - a)\beta(\eta, C_j) \tag{9.121}$$

$$\theta_1(0) = 1, \theta_1(\infty) = 0 \tag{9.122}$$

If we choose

$$\begin{aligned} \alpha(\eta, C_i) &= (C_1\eta + C_2)e^{-k\eta} + C_5 \\ \beta(\eta, C_j) &= (C_3\eta + C_4)e^{-k\eta} \end{aligned} \tag{9.123}$$

where C_1, C_2, C_3, C_4, C_5 and $k > 0$ are unknown constants at this moment, then Eq. 9.121 can be written as

$$\theta_1''(\eta) = ce^{-\gamma\eta}(C_1\eta + C_2)e^{-k\eta} + (be^{-\gamma\eta} - a)(C_3\eta + C_4)e^{-k\eta} + cC_5e^{-\gamma\eta} \tag{9.124}$$

Also, we can consider other expressions for α and β, because these functions which appear in Eq. 9.121 are not unique. For example, one can choose

$$\begin{aligned} \alpha(\eta, C_i) &= C_1^*e^{-k\eta}; \\ \beta(\eta, C_j) &= (C_2^*\eta^2 + C_3^*\eta + C_4^*)e^{-k\eta} \end{aligned} \tag{9.125}$$

or

$$\begin{aligned} \alpha(\eta, C_i) &= (C_1'\eta^2 + C_2'\eta)e^{-k\eta}; \\ \beta(\eta, C_j) &= C_3'e^{-k\eta} \end{aligned} \tag{9.126}$$

and so on.

Now, Eq. 9.124 can be written in the following form

$$\begin{aligned} \theta''_1(\eta) = {} & [(cC_1 + bC_3)\eta + cC_2 + bC_4]e^{-(\gamma+k)\eta} + \\ & + (-aC_3\eta - aC_4)e^{-k\eta} + cC_5e^{-\gamma\eta} \end{aligned} \tag{9.127}$$

From Eq. 9.127, with the boundary conditions (9.122), we obtain the first-order approximation

$$\theta_1(\eta) = \left[\frac{cC_1 + bC_3}{(\gamma + k)^2}\eta + \frac{2aC_3}{k^3} + \frac{aC_4}{k^2} - \frac{c}{\gamma^2}C_5 + 1\right]e^{-(\gamma+k)\eta} + \\ + \left(-\frac{aC_3}{k^2}\eta - \frac{2aC_3}{k^3} - \frac{aC_4}{k^2}\right)e^{-k\eta} + \frac{c}{\alpha^2}C_5 e^{-\gamma\eta} \tag{9.128}$$

In accordance with Eq. 7.115, the residual functional J becomes

$$J = \int_0^\infty [\theta''_1(\eta) + (a - be^{-\lambda\eta})\theta'_1(\eta) - ce^{-\gamma\eta}\theta(\eta)]^2 d\eta \tag{9.129}$$

The unknown constants k, C_1, C_3, C_4 and C_5 are given by the conditions (6.116)

$$\frac{\partial J}{\partial k} = \frac{\partial J}{\partial C_1} = \frac{\partial J}{\partial C_3} = \frac{\partial J}{\partial C_4} = \frac{\partial J}{\partial C_5} = 0 \tag{9.130}$$

The exact solution of Eq. 9.118 satisfying Eq. 9.119 in terms of Kummer's confluent hyper-geometric function ${}_1F_1(\cdot,\cdot,\cdot)$ is [186]

$$\theta(\eta) = e^{-\alpha\gamma\eta}\frac{{}_1F_1(\delta - s, \delta + 1, -\frac{\beta}{\gamma^2}e^{-\gamma\eta})}{{}_1F_1(\delta - s, \delta + 1, -\frac{\beta}{\gamma^2})} \tag{9.131}$$

where $\delta = \frac{\beta(1+\gamma\lambda)}{\gamma^2}$.

Consider $P_r = 0.71$, $\lambda = 0.1$, $s = 1$, $R = 1$ and $M = 1$. In this case, for the above specified parameters it is obtained: $\gamma = 1.465097$, $\beta = 0.3042857$, $a = 0.238118$, $b = 0.207689$, $c = 0.304286$.

From Eq. 9.130 it results

$C_1 = 0.175774$; $C_3 = 0.00146048$; $C_4 = -0.230031$; $C_5 = -0.71168$, $k = 0.2381184$

Therefore, the approximate solution (9.128) is well determined:

$$\theta_1(\eta) = (0.0185419\eta + 2.01233875)e^{-1.7032154\eta} + \\ + (-0.0014604\eta - 0.91145205)e^{-0.2381184\eta} - 0.100886713e^{-1.465097\eta} \tag{9.132}$$

In Fig. 9.11 we compared the exact solution (9.131) and the approximate solution (9.132) in the considered particular case, and a very good agreement can be observed. In Fig. 9.12 the residual obtained from the initial equation using the approximate solution (9.132) is given in the considered case. As it can be seen, a very good error is achieved.

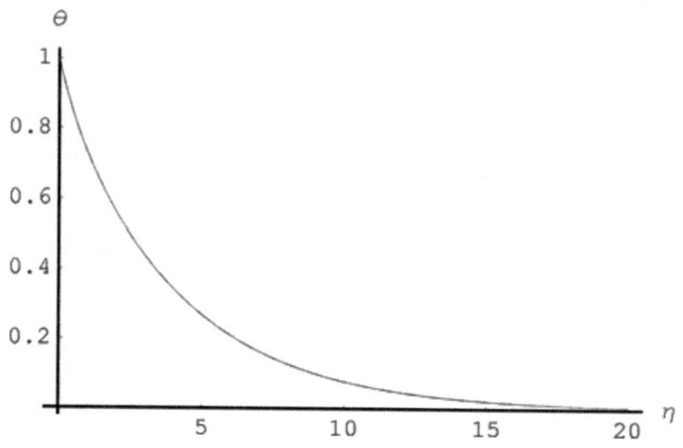

Fig. 9.11 Dimensionless temperature for $R = 1$, $P_r = 0.71$, $\lambda = 0.1$, $s = 1$, $R = 1$ and $M = 1$ ____ exact solution (9.131); _ _ _ approximate solution (9.132) of the Eq. 9.113

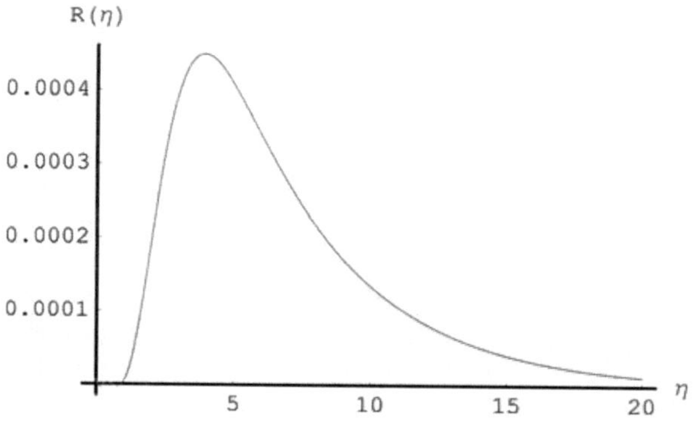

Fig. 9.12 The residual of the initial equation (9.113) obtained using (9.132) in the above case

9.5 The Oscillator with Cubic and Harmonic Restoring Force

Consider the nonlinear differential oscillator governed by the equation

$$\ddot{u} + u + au^3 + b\sin u = 0 \tag{9.133}$$

subject to the initial conditions

$$u(0) = A, \quad \dot{u}(0) = 0 \tag{9.134}$$

If Ω is the frequency of the system described by (9.133) and introducing a new independent variable

$$\tau = \Omega t \tag{9.135}$$

then Eq. 9.133 becomes

$$u'' + u + N(u) = 0 \qquad (9.136)$$

where $' = d/d\tau$ and

$$L(u) = u'' + u, \quad N(u) = \left(\frac{1}{\Omega^2} - 1\right)u + \frac{a}{\Omega^2}u^3 + \frac{b}{\Omega^2}\sin u, \quad f(r) = 0 \qquad (9.137)$$

We use the second alternative. We consider the initial approximation and the auxiliary function in the form [187, 188]

$$\begin{aligned} &u_0(\tau) = C_1 \cos\tau \\ &\alpha(\tau, C_2, C_3, C_4) = 2C_2\cos 3\tau + 2C_3\cos 5\tau + 2C_4\cos 7\tau \\ &\beta(\tau) = \gamma(\tau) = \ldots = 0 \end{aligned} \qquad (9.138)$$

where C_1, C_2, C_3 and C_4 are unknown constants at this moment.

For $n = 0$ into Eq. 9.96 we obtain only one iteration given by

$$u_1'' + u_1 + N(u_0) + \alpha(\tau, C_2, C_3, C_4)N_u(u_0) = 0 \qquad (9.139)$$

where

$$N(u_0) = \left(\frac{1}{\Omega^2} - 1\right)C_1\cos\tau + \frac{a}{\Omega^2}C_1^3\cos^3\tau + \frac{b}{\Omega^2}\sin(C_1\cos\tau) \qquad (9.140)$$

$$N_u(u_0) = \frac{1}{\Omega^2} - 1 + \frac{3aC_1^2}{\Omega^2}\cos^2\tau + \frac{b}{\Omega^2}\cos(C_1\cos\tau) \qquad (9.141)$$

In Eq. 9.140 the terms $\cos^3\tau$ and $\sin(C_1\cos\tau)$can be expanded in the power series

$$\cos^3\tau = \frac{\cos 3\tau + 3\cos\tau}{4} \qquad (9.142)$$

$$\sin(C_1\cos\tau) = C_1\cos\tau - \frac{C_1^3\cos^3\tau}{3!} + \frac{C_1^5\cos^5\tau}{5!} - \frac{C_1^7\cos^7\tau}{7!} + \frac{C_1^9\cos^9\tau}{9!} + \ldots \qquad (9.143)$$

By using Eqs. 9.142 and 9.143 by simple manipulations, Eq. 9.140 can be expressed in the form

$$N(u_0) = \left[\left(\frac{1}{\Omega^2} - 1\right)C_1 + \frac{3aC_1^3}{4\Omega^2} + \frac{bC_1}{\Omega^2}\left(1 - \frac{C_1^2}{8} + \frac{C_1^4}{192} - \frac{C_1^6}{9216} + \right.\right.$$
$$\left.\left. + \frac{C_1^8}{737280}\right)\right] \cos\tau + \left[\frac{aC_1^3}{4\Omega^2} - \frac{bC_1^3}{\Omega^2}\left(\frac{1}{24} - \frac{C_1^2}{384} + \frac{C_1^4}{15360} - \right.\right.$$
$$\left.\left. - \frac{C_1^6}{1105920}\right)\right] \cos 3\tau + \left[\frac{bC_1^5}{\Omega^2}\left(\frac{1}{1920} - \frac{C_1^2}{46080} + \right.\right.$$
$$\left.\left. + \frac{C_1^4}{2580480}\right)\right] \cos 5\tau + \left[-\frac{bC_1^7}{\Omega^2}\left(\frac{1}{322560} - \frac{C_1^2}{10321920}\right)\right] \cos 7\tau +$$
$$+ \left[\frac{bC_1^9}{92897280}\right] \cos 9\tau + .. \tag{9.144}$$

The last term into Eq. 9.141 can be written as

$$\cos(C_1 \cos\tau) = 1 - \frac{C_1^2 \cos^2\tau}{2!} + \frac{C_1^4 \cos^4\tau}{4!} - \frac{C_1^6 \cos^6\tau}{6!} + \frac{C_1^8 \cos^8\tau}{8!} + ...$$

or

$$\cos(C_1 \cos\tau) = 1 - \frac{C_1^2}{4} + \frac{C_1^4}{64} - \frac{C_1^6}{2304} + \frac{C_1^8}{147456} + \left(-\frac{C_1^2}{4} + \frac{C_1^4}{48} - \frac{C_1^6}{1536} + \right.$$
$$\left. + \frac{C_1^8}{92160} + ...\right) \cos 2\tau + \left(\frac{C_1^4}{192} - \frac{C_1^6}{3840} + \frac{C_1^8}{184320}\right) \cos 4\tau + \left(-\frac{C_1^6}{23040} + \right.$$
$$\left. + \frac{C_1^8}{645120} + ...\right) \cos 6\tau + \left(\frac{C_1^8}{5160960} + ...\right) \cos 8\tau + ... \tag{9.145}$$

Therefore Eq. 9.141 becomes

$$N_u(u_0) = \frac{1}{\Omega^2} - 1 + \frac{3a^2C_1^3}{2\Omega^2} + \frac{b}{\Omega^2}\left(1 - \frac{C_1^2}{4} + \frac{C_1^4}{64} - \frac{C_1^6}{2304} + \right.$$
$$\left. + \frac{C_1^8}{147456} + ...\right) + \left[\frac{3a^2C_1^3}{2\Omega^2} + \frac{bC_1^2}{4\Omega^2}\left(-1 + \frac{C_1^2}{12} - \frac{C_1^4}{384} + \right.\right.$$
$$\left.\left. + \frac{C_1^6}{23040} + ...\right)\right] \cos 2\tau + \frac{bC_1^4}{192\Omega^2}\left(1 - \frac{C_1^2}{20} + \frac{C_1^4}{960} + ...\right)\Bigg] \cos 4\tau +$$
$$+ \left[-\frac{bC_1^6}{23040\Omega^2}\left(1 - \frac{C_1^2}{28} + ...\right)\right] \cos 6\tau + \left[\frac{bC_1^8}{5160960} + ...\right] \cos 8\tau \tag{9.146}$$

From (9.144) we remark that $N(u_0)$ is a combination of the functions $\cos\tau$, $\cos 3\tau, \cos 5\tau, \ldots$ and therefore we choose $\alpha(\tau, C_i)$ such as the term $\alpha(\tau, C_i)N_u(u_0)$

from (9.96) to be of the same form like $N(u_0)$ (see Eq. 9.138$_2$). In this way, Eq. 9.139 can be written in the form:

$$\begin{aligned} u''_1 + u_1 + \Bigg\{ \Bigg[\left(\frac{1}{\Omega^2} - 1 \right) C_1 + \frac{3aC_1^3}{4\Omega^2} + \frac{bC_1}{\Omega^2} \left(1 - \frac{C_1}{8} + \frac{C_1^4}{192} - \frac{C_1^6}{9216} + \right. \\ \left. + \frac{C_1^8}{737280} + \ldots \right) + C_2 \left[- \frac{bC_1^2}{4\Omega^2} \left(1 - \frac{5C_1^2}{48} + \frac{7C_1^4}{1920} - \right. \right. \\ \left. \left. - \frac{C_1^6}{15360} + .. \right) \right] + C_3 \left[\frac{bC_1^4}{192\Omega^2} \left(1 - \frac{7C_1^2}{120} + \frac{3C_1^4}{2240} + \ldots \right) + \right. \\ \left. + C_4 \left[- \frac{bC_1^6}{23040\Omega^2} \left(1 - \frac{9C_1^2}{224} + \ldots \right) \right] \right] \Bigg\} \cos\tau + H.O.T. = 0 \end{aligned} \tag{9.147}$$

The solution of (9.147) is chosen that it contains no secular terms, which requires that coefficient of $\cos\tau$ disappear, i.e.

$$\begin{aligned} \Omega^2 = 1 + \frac{3aC_1^2}{4} + b \left(1 - \frac{C_1^2}{8} + \frac{C_1^4}{192} - \frac{C_1^6}{9216} + \frac{C_1^8}{737280} + \ldots \right) + \\ + \frac{C_2}{C_1} \left[- \frac{bC_1^2}{4} \left(1 - \frac{5C_1^2}{48} + \frac{7C_1^4}{1920} - \frac{C_1^6}{15360} + \ldots \right) \right] + \\ + \frac{C_3}{C_1} \left[\frac{bC_1^4}{192} \left(1 - \frac{7C_1^2}{120} + \frac{3C_1^4}{2240} + \ldots \right) \right] + \\ + \frac{C_4}{C_1} \left[- \frac{bC_1^6}{23040} \left(1 - \frac{9C_1^2}{224} + \ldots \right) \right] \end{aligned} \tag{9.148}$$

The Eq. 9.147 becomes

$$\begin{aligned} u''_1 + u_1 + [\alpha_3 + (2\beta_0 + \beta_6)C_2 + (\beta_2 + \beta_8)C_3 + \beta_4 C_4] \cos 3\tau + \\ + [\alpha_5 + (\beta_2 + \beta_8)C_2 + 2\beta_0 C_3 + \beta_2 C_4] \cos 5\tau + [\alpha_7 + \beta_4 C_2 + \beta_0 C_4] \cos 7\tau + \\ + [\alpha_9 + \beta_6 C_2 + \beta_4 C_3 + \beta_2 C_4] \cos 9\tau = 0 \end{aligned} \tag{9.149}$$

where

$$\alpha_3 = \frac{aC_1^3}{4\Omega^2} - \frac{bC_1^3}{\Omega^2} \left(\frac{1}{24} - \frac{C_1^2}{384} + \frac{C_1^4}{15360} - \frac{C_1^6}{1105920} + \ldots. \right)$$

$$\alpha_5 = \frac{bC_1^5}{1920\Omega^2} \left(1 - \frac{C_1^2}{24} + \frac{C_1^4}{1344} + \ldots \right)$$

$$\alpha_7 = -\frac{bC_1^7}{322560\Omega^2}\left(1 - \frac{C_1^2}{32} + ...\right)$$

$$\alpha_9 = \frac{bC_1^9}{92897280\Omega^2}$$

$$\beta_0 = \frac{1}{\Omega^2} - 1 + \frac{b}{\Omega^2}\left(1 - \frac{C_1^2}{4} + \frac{C_1^4}{64} - \frac{C_1^6}{2304} + \frac{C_1^8}{147456} + ...\right) + \frac{3a^2C_1^3}{2\Omega^2}$$

$$\beta_2 = \frac{3a^2C_1^3}{2\Omega^2} - \frac{bC_1^2}{4\Omega^2}\left(1 - \frac{C_1^2}{12} + \frac{C_1^4}{384} - \frac{C_1^6}{23040} + ...\right)$$

$$\beta_4 = \frac{bC_1^4}{192\Omega^2}\left(1 - \frac{C_1^2}{20} + \frac{C_1^4}{960} + ..\right)$$

$$\beta_6 = -\frac{bC_1^6}{2304\Omega^2}\left(1 - \frac{C_1^2}{28} + ...\right)$$

$$\beta_8 = \frac{bC_1^8}{5160960\Omega^2} + ... \tag{9.150}$$

From Eq. 9.149, with initial conditions (9.134) we obtain the first-order approximate solution

$$\begin{aligned} u_1(t) = {} & A\cos\Omega t + \frac{1}{8}[\alpha_3 + (2\beta_0 + \beta_6)C_2 + (\beta_2 + \beta_8)C_3 + \\ & + \beta_4 C_4](\cos 3\Omega t - \cos\Omega t) + \frac{1}{24}[\alpha_5 + (\beta_2 + \beta_8)C_2 + 2\beta_0 C_3 + \\ & + \beta_2 C_4](\cos 5\Omega t - \cos\Omega t) + \frac{1}{48}[\alpha_7 + \beta_4 C_2 + \beta_2 C_3 + 2\beta_0 C_4](\cos 7\Omega t - \\ & - \cos\Omega t) + \frac{1}{80}[\alpha_9 + \beta_6 C_2 + \beta_4 C_3 + \beta_2 C_4](\cos 9\Omega t - \cos\Omega t) \end{aligned} \tag{9.151}$$

where Ω is given by Eq. 9.148

The parameters C_1, C_2, C_3 and C_4 can be determined optimally using several procedures. Thus, such optimal value for the parameters C_i can be found imposing that the residual functional given by

$$J = \int_0^{\frac{2\pi}{\Omega}} [u''_1(t) + u_1(t) + au_1^3(t) + b\sin u_1(t)]^2 dt \tag{9.152}$$

be minimum, i.e.

$$\frac{\partial J}{\partial C_1} = \frac{\partial J}{\partial C_2} = \frac{\partial J}{\partial C_3} = \frac{\partial J}{\partial C_4} = 0 \tag{9.153}$$

or by means of a collocation-type method involving the residual R of the initial equation:

$$R(t_i, C_i) = 0, \quad i = 1,2,3,4 \quad t_i \in \left(0, \frac{2\pi}{\Omega}\right) \tag{9.154}$$

In this way, solving the above system of algebraic equations, the solution (9.151) in the first approximation is well-determined.

We will illustrate the applicability, accuracy and effectiveness of the proposed approach by comparing the analytical approximate periodic solution with numerical integration results obtained using a fourth-order Runge-Kutta method.

The comparison was made in terms of displacements, phase plane and residuals of original equation. These comparisons are presented in the Figs. 9.13–9.24 for several cases, where the constants C_i were determined from Eq. 9.154. An error analysis is graphically performed plotting $Er(t) = u_{RK}(t) - u_{app}(t)$, where u_{RK} is the numerical solution of Eq. 9.133 obtained using a Runge-Kutta algorithm and u_{app} is the approximate solution given by (9.151)

Case (a) For $a = 1, b = 1, A = 1$, following the procedure described above we obtain the optimal values of the unknown parameters:

$$C_1 = 0.99614, C_2 = 0.0165173, \quad C_3 = 0.00555482, C_4 = -0.00603017 \tag{9.155}$$

and therefore $\Omega = 1.61911$.

Taking into account these optimal parameters, in Fig. 9.13 is presented a comparison between the approximate solution and the solution obtained through numerical simulations. Moreover, Fig. 9.14 presents a comparison between the approximate solution (9.151) and the numerical results in terms of phase plane.

In order to provide a comprehensive evidence of the accuracy of the results, the error between the numerical and approximate solution has been computed. A graphical representation of this error for the case (a) is presented in Fig. 9.15.

Case (b) For $a = 1, b = 1, A = 2$, following the same procedure we obtain:

$$C_1 = 1.9768, C_2 = 0.0066599, C_3 = 0.00831937, C_4 = -0.00835391 \tag{9.156}$$

and therefore $\Omega = 2.12459$. Comparisons between the approximate and numerical results for case (b) are presented in Figs. 9.16–9.18.

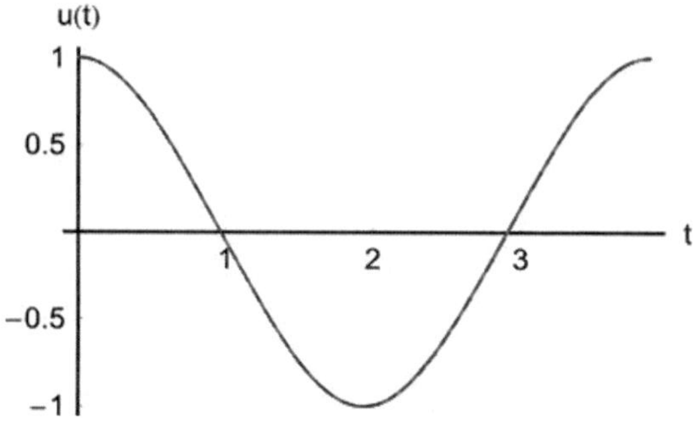

Fig. 9.13 Comparison between the approximate solution (9.151) and numerical solution of Eq. 9.133 in case (a): $a = b = A = 1$: _____numerical solution; _________ approximate solution

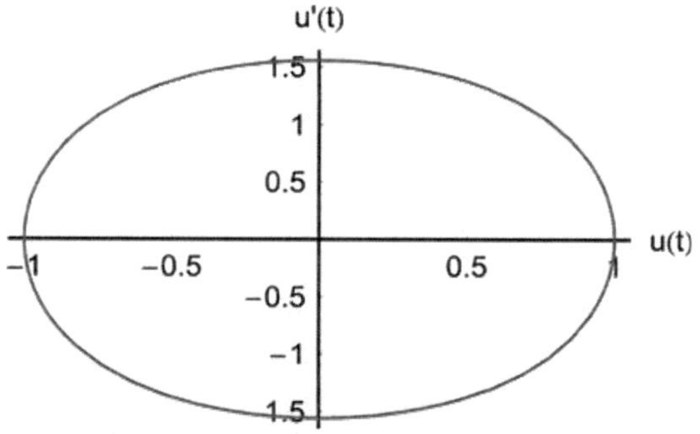

Fig. 9.14 Comparison between the approximate solution (9.151) and numerical results of Eq. 9.133 in terms of phase plane in case (a) a = b = A = 1: _____numerical solution; _________approximate solution

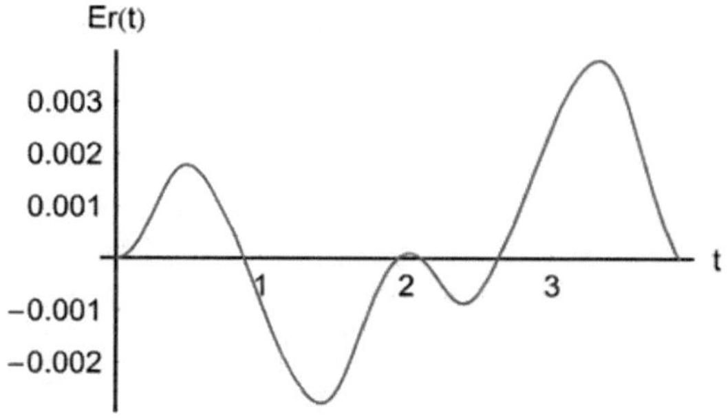

Fig. 9.15 The error between the numerical and approximate solution (9.151) in case (a) $a = b = A = 1$

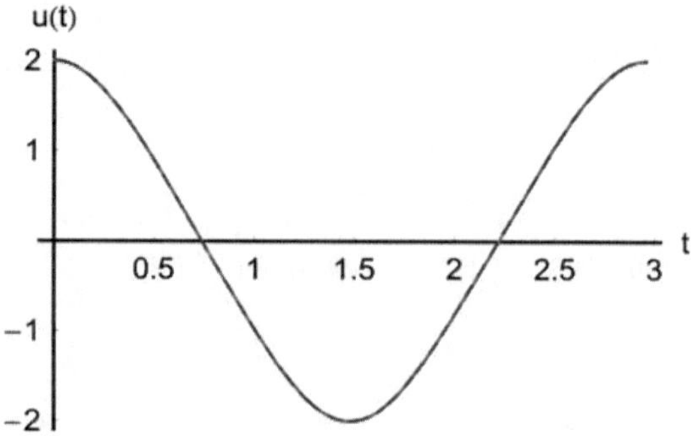

Fig. 9.16 Comparison between the approximate solution (9.151) and numerical solution of Eq. 9.133 in case (b) $a = b = 1, A = 2$: _____numerical solution; ________approximate solution

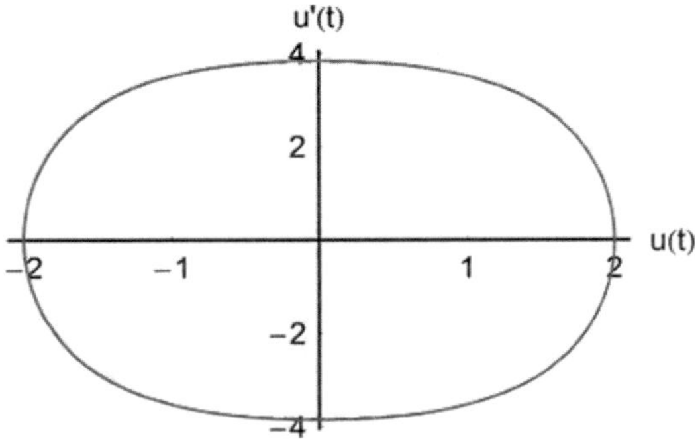

Fig. 9.17 Comparison between the approximate solution (9.151) and numerical results of Eq. 9.133 in terms of phase plane in case (b) $a = b = 1, A = 2$: _____numerical solution; _________approximate solution

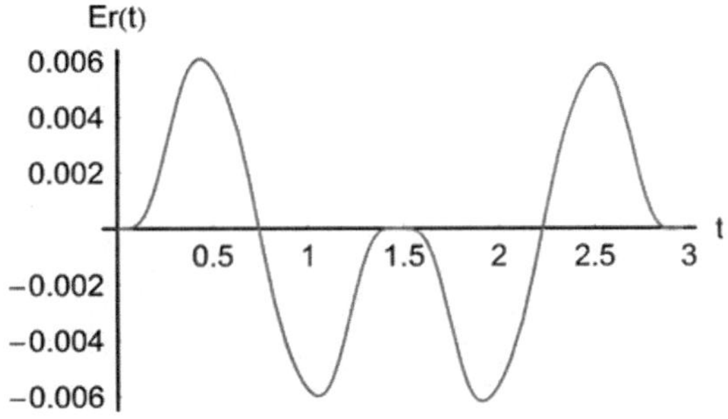

Fig. 9.18 The error between the numerical and approximate analytical solution (9.151) in case (b) $a = b = 1, A = 2$

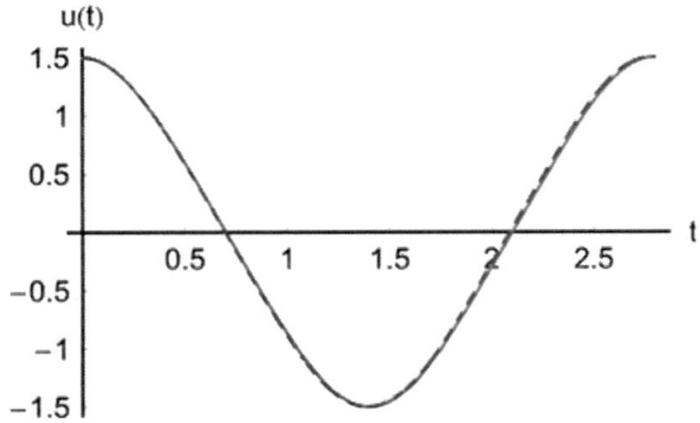

Fig. 9.19 Comparison between the approximate solution (9.151) and numerical solution of Eq. 9.133 in case (c) $a = 2$, $b = 1, A = 1.5$: _____numerical solution; _______approximate solution

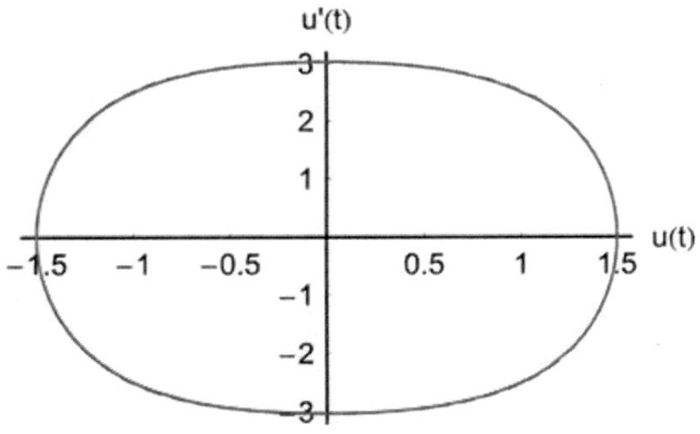

Fig. 9.20 Comparison between the approximate solution (9.151) and numerical results of Eq. 9.133 in terms of phase plane in case (c) $a = 2, b = 1$, $A = 1.5$: _____numerical solution; _________ approximate solution

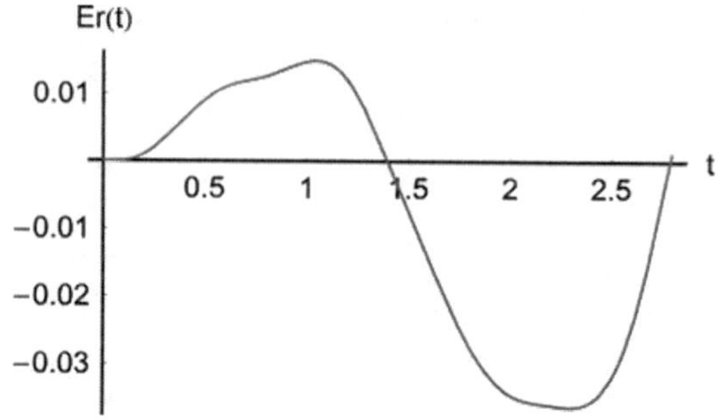

Fig. 9.21 The error between the numerical and approximate analytical solution (9.151) in case (c) $a = 2, b = 1, A = 1.5$

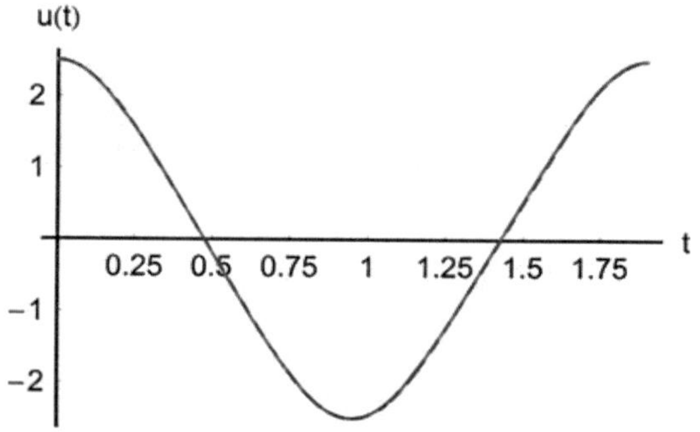

Fig. 9.22 Comparison between the approximate solution (9.151) and numerical solution of Eq. 9.133 in case (d) $a = b = 2, A = 2.5$: _____numerical solution; ______approximate solution

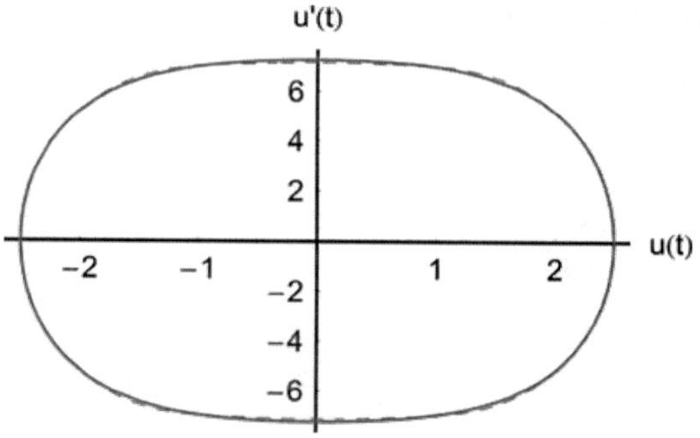

Fig. 9.23 Comparison between the approximate solution (9.151) and numerical results of Eq. 9.133 in terms of phase plane in case (d) $a = b = 2, A = 2.5$: _____numerical solution; _________approximate solution

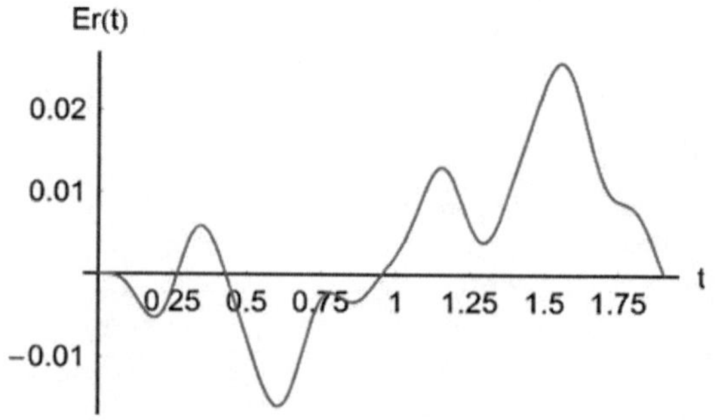

Fig. 9.24 The error between the numerical and approximate analytical solution (9.151) in case (d) $a = b = 2, A = 2.5$

Case (c) For a = 2, b = 1, A = 1.5, we obtain:

$$C_1 = 1.49429, C_2 = 0.000614432; C_3 = 0.00157958, C_4 = -0.00128292 \quad (9.157)$$

and therefore $\Omega = 2.25718$. Comparisons between the approximate results and numerical simulations results for case (c) are presented in Figs. 9.19–9.21.

Case (d) For a = 2, b = 2, A = 2.5, we obtain:

$$C_1 = 2.45728, \quad C_2 = 0.0140962, C_3 = 0.0178149, \quad C_4 = 0.0108985 \quad (9.158)$$

and therefore $\Omega = 3.29742$. Comparisons between the approximate results and numerical simulations results for case (d) are presented in Figs. 9.22–9.24.

It can be seen from Figs. 9.13–9.24 that the results obtained using OPIM are nearly identical with those obtained through numerical simulations for different initial amplitudes.

9.6 Oscillations of a Uniform Cantilever Beam Carrying an Intermediate Lumped Mass and Rotary Inertia

For the nonlinear oscillator described by the equation (see Sects. 6.7 and 8.4)

$$\ddot{u} + u + au^2\ddot{u} + au\dot{u}^2 + bu^3 = 0 \quad (9.159)$$

subject to the initial conditions

$$u(0) = A, \quad \dot{u}(0) = 0 \quad (9.160)$$

we introduce a new independent variable

$$\tau = \Omega t \quad (9.161)$$

Equations 9.159 and 9.160 become

$$u'' + u + \left(\frac{1}{\Omega^2} - 1\right)u + au^2u'' + auu'^2 + \frac{b}{\Omega^2}u^3 = 0 \quad (9.162)$$

$$u(0) = A, \quad u'(0) = 0 \quad (9.163)$$

where the prime denotes derivative with respect to the new variable τ

Having in view the second alternative of OPIM, we consider the following functions to construct the iteration (9.96)

$$\begin{aligned} u_0(\tau) &= C_1 \cos\tau \\ \alpha(t, C_i) &= C_2 \cos 3\tau \\ \beta(t, C_j) &= C_3 \cos 5\tau \\ \gamma(t, C_k) &= C_4 \cos 7\tau \end{aligned} \tag{9.164}$$

Equation 9.96 can be expressed for $n = 0$ in the form

$$\begin{aligned} &u''_1 + u_1 + N(C_1 \cos\tau) + (C_2 \cos 3\tau) N_u(C_1 \cos\tau) + \\ &\quad + (C_3 \cos 5\tau) N_{\dot u}(C_1 \cos\tau) + (C_4 \cos 7\tau) N_{\ddot u}(C_1 \cos\tau) = 0 \end{aligned} \tag{9.165}$$

where

$$N(u, \dot u, \ddot u) = au^2 \ddot u + au\dot u^2 + bu^3 \tag{9.166}$$

By simple manipulations, Eq. 9.165 can be written as

$$\begin{aligned} &u''_1 + u_1 - \left[\left(1 - \frac{1}{\Omega^2}\right)C_1 + \frac{1}{2}aC_1^3 + \frac{3b}{4\Omega^2}\left(C_1^2 C_2 - C_1^3\right) + \frac{3}{4}aC_1^2 C_2\right]\cos\tau + \\ &- \left[\frac{1}{2}a(C_1^3 - C_1^2 C_3) - \frac{b}{4\Omega^2}C_1^3 + \left(1 - \frac{1}{\Omega^2} + \frac{1}{2}aC_1^2 - \frac{3b}{2\Omega^2}C_1^2\right)C_2\right]\cos 3\tau + \\ &- \left[\frac{1}{4}a(3C_1^2 C_2 - C_1^2 C_4) + \frac{3b}{4\Omega^2}C_1^2 C_2\right]\cos 5\tau - \frac{1}{2}a(C_1^2 C_3 - C_1^2 C_4)\cos 7\tau - \\ &+ \frac{1}{4}aC_1^2 C_4 \cos 9\tau = 0 \end{aligned} \tag{9.167}$$

The solution of Eq. 9.167 is chosen so that it contains no secular terms, which requires that the coefficient of $\cos\tau$ disappears from this equation, i.e.

$$\Omega^2 = \frac{4 + 3b(C_1^2 - C_1 C_2)}{4 + a(2C_1^2 + 3C_1 C_2)} \tag{9.168}$$

From Eq. 9.167 we obtain the first-order approximate solution

$$\begin{aligned} u_1(\tau) &= \alpha\cos\tau + \beta\sin\tau + \left[-\frac{1}{16}a(C_1^3 - C_1^2C_3) + \frac{b}{32\Omega^2}C_1^3 - \right. \\ &\left. -\frac{1}{8}\left(1 - \frac{1}{\Omega^2} + \frac{1}{2}aC_1^2 - \frac{3b}{2\Omega^2}C_1^2\right)C_2\right]\cos 3\tau + \\ &+ \left[-\frac{1}{96}a(3C_1^2C_2 - C_1^2C_4) - \frac{b}{32\Omega^2}C_1^2C_2\right]\cos 5\tau - \\ &-\frac{1}{96}a(C_1^2C_3 - C_1^2C_4)\cos 7\tau + \frac{1}{320}aC_1^2C_4\cos 9\tau \end{aligned} \tag{9.169}$$

where α and β are constants.

From the initial conditions (9.163) and from Eq. 9.161, the first-order approximate solution (9.169) becomes

$$\begin{aligned} u_1(t) &= (A - \gamma_1 - \gamma_2 - \gamma_3 - \gamma_4)\cos\Omega t + \gamma_1\cos 3\Omega t + \\ &+ \gamma_2\cos 5\Omega t + \gamma_3\cos 7\Omega t + \gamma_4\cos 7\Omega t \end{aligned} \tag{9.170}$$

where Ω is given by Eq. 9.168 and the parameters γ_i, $i = 1,2,3,4$ are:

$$\begin{aligned} \gamma_1 &= -\frac{1}{16}a(C_1^3 - C_1^2C_3) + \frac{b}{32\Omega^2}C_1^3 - \\ &\quad -\frac{1}{8}\left(1 - \frac{1}{\Omega^2} + \frac{1}{2}aC_1^2 - \frac{3b}{2\Omega^2}C_1^2\right)C_2 \\ \gamma_2 &= -\frac{1}{96}a(3C_1^2C_2 - C_1^2C_4) - \frac{b}{32\Omega^2}C_1^2C_2 \\ \gamma_3 &= -\frac{1}{96}a(C_1^2C_3 - C_1^2C_4) \\ \gamma_4 &= \frac{1}{320}aC_1^2C_4 \end{aligned} \tag{9.171}$$

The parameters C_1, C_2, C_3 and C_4 can be determined optimally, i.e. the residual functional J given by

$$J = \int_0^T \left[u''_1 + u_1 - \left(1 - \frac{1}{\Omega^2}\right)u_1 + au_1^2u''_1 + au_1u_1'2 + \frac{b}{\Omega^2}u_1^3\right]^2 dt; T = \frac{2\pi}{\Omega} \tag{9.172}$$

be minimized:

$$\frac{\partial J}{\partial C_1} = \frac{\partial J}{\partial C_2} = \frac{\partial J}{\partial C_3} = \frac{\partial J}{\partial C_4} = 0 \tag{9.173}$$

From Eqs. 9.173 and 9.168, one can determine the parameters C_1, C_2, C_3, C_4 and the frequency Ω. In this way, the solution (9.170) in the first approximation is well-determined.

This procedure is tested for solving Eqs. 9.159 and 9.160 in different conditions. We consider different values of a, b and A, and we compare our results with known results from the literature [51] and also with the numerical integration results obtained using a fourth-order Runge-Kutta method.

Case (a) When $a = 1, b = 1, A = 5$ it is obtained:

$$C_1 = 3.1445891;\ C_2 = -0.348612116;\ C_3 = 1.537425988;\ C_4 = 1.208556244$$

$$\begin{aligned} u_1(t) &= 5.282603026\cos\Omega t - 0.578012411\cos 3\Omega t + \\ &+ 0.291938471\cos 5\Omega t - 0.033875093\cos 7\Omega t + 0.037346051\cos 9\Omega t \end{aligned} \tag{9.174}$$

where $\Omega = 1.343011012$.

The approximate solution obtained in [51] is

$$u_w(t) = 5.479790632\cos\omega t - 0.479790632\cos 3\omega t \tag{9.175}$$

where $\omega = 1.389279655$.

Case (b) When $a = 2, b = 2, A = 5$ it is obtained:

$$\begin{aligned} C_1 &= 0.2493232;\ C_2 = -0.840897196; \\ C_3 &= -85.34186347;\ C_4 = 10.02682709 \end{aligned}$$

$$\begin{aligned} u_1(t) &= 5.473754321\cos\Omega t - 0.619143788\cos 3\Omega t + \\ &+ 0.017987348\cos 5\Omega t + 0.123506543\cos 7\Omega t + 0.003895551\cos 9\Omega t \end{aligned} \tag{9.176}$$

where $\Omega = 1.372149301$.

The approximate solution obtained in [51] in this case is

$$u_w(t) = 5.532121689\cos\omega t - 0.532121689\cos 3\omega t \tag{9.177}$$

where $\omega = 1.335870935$.

Case (c) When $a = 1, b = 2, A = 5$ it is obtained:

$$\begin{aligned} C_1 &= 2.89302478;\ C_2 = -0.390415451; \\ C_3 &= 1.152442491;\ C_4 = 0.997740614 \end{aligned}$$

$$\begin{aligned} u_1(t) &= 5.329628381\cos\Omega t - 0.589431231\cos 3\Omega t + \\ &+ 0.247194381\cos 5\Omega t - 0.013487413\cos 7\Omega t + 0.026095882\cos 9\Omega t \end{aligned} \tag{9.178}$$

where $\Omega = 1.874934651$.

The approximate solution obtained in [51] is

$$u_{w}(t) = 5.518356604 \cos \omega t - 0.5183566 \cos 3\omega t \qquad (9.179)$$

where $\omega = 1.846429459$.

It is easy to assess the accuracy of the obtained results if we graphically compare these solutions with the numerical simulation results, or with other results known in the literature. Figures 9.25–9.27 show the comparison between the present solutions, the numerical integration results obtained using a fourth-order Runge-Kutta method and the results obtained in [51].

It can be seen from Figs. 9.25–9.27 that the solutions obtained through the proposed procedure are nearly identical with the numerical solutions obtained using a fourth-order Runge-Kutta method. Moreover, the analytical solutions obtained through our procedure prove to be more accurate than other known results obtained by combining the linearization of the governing equation with the method of harmonic balance [51].

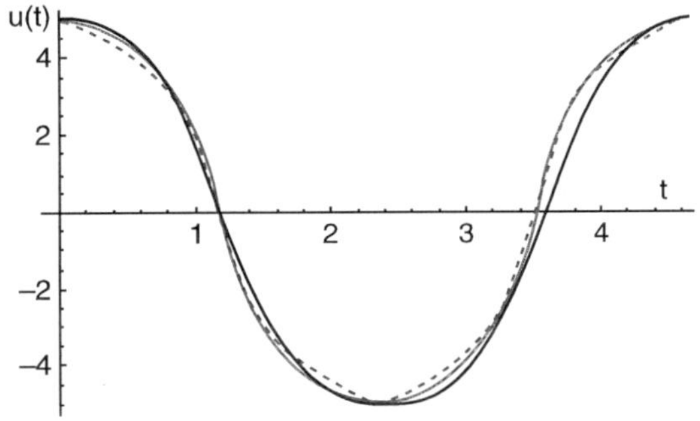

Fig. 9.25 Comparison between the results obtained for Eq. 9.159 in the case (a), $a = b = 1, A = 5$, ____ numerical simulation; _ _ _ _ present method (9.174); _____ $u_w(t)$ given in [51]

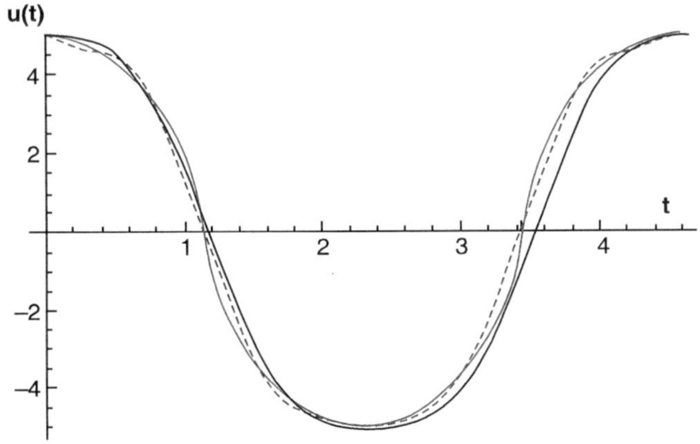

Fig. 9.26 Comparison between the results obtained for Eq. 9.159 in the case (b), $a = b = 2$, $A = 5$, ____ numerical simulation, _ _ _ _ present method (9.176), _____ u_w given in [51]

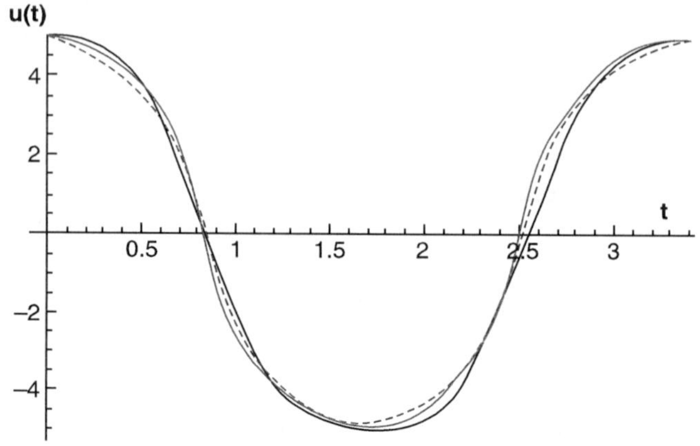

Fig. 9.27 Comparison between the results obtained for Eq. 9.159 in the case (c), $a = 1$, $b = 2$, $A = 5$, ____ numerical simulation; _ _ _ _ present method (9.178); _____ u_w given in [51]

9.7 A Modified Van der Pol Oscillator

The self-excited systems have a long history in the field of mechanics [22]. A self-excited oscillator is a system which has some external energy sources. Recently, self-excited systems have been proposed as fundamental tools for control and reduction of friction [189]. A modified Van der Pol oscillator has been proposed to describe a self-excited body sliding on a periodic potential [188]. This autonomous modified Van der Pol oscillator is described by the following equation [189]:

$$\ddot{u} + u + \varepsilon(u^2 - 1)\dot{u} + \Delta \sin u = 0 \tag{9.180}$$

with initial conditions

$$u(0) = A, \quad \dot{u}(0) = 0 \tag{9.181}$$

where ε and Δ are positive parameters.

Equation 9.180 can be written in the following form

$$\ddot{u} + \Omega^2 u + N(u) = 0 \tag{9.182}$$

where Ω is the frequency of the system (9.180) and

$$N(u, \dot{u}) = (1 - \Omega^2)u + \Delta \sin u + \varepsilon(u^2 - 1)\dot{u} \tag{9.183}$$

In this case we consider the second alternative with

$$\begin{aligned} u_0(t) &= C_1 \cos \Omega t \\ \alpha(t) &= C_2 \cos \Omega t + C_3 \cos 3\Omega t \\ \beta(t) &= C_4 \cos 3\Omega t + C_5 \cos 5\Omega t \end{aligned} \tag{9.184}$$

where C_1, C_2, C_3, C_4 and C_5 are unknown constants.

For $n = 0$ into Eq. 9.96 we obtain:

$$\begin{aligned} &\ddot{u}_1 + \Omega^2 u_1 + N(C_1 \cos \Omega t, -\Omega C_1 \sin \Omega t) + \\ &\quad + (C_2 \cos \Omega t + C_3 \cos 3\Omega t) N_u (C_1 \cos \Omega t, -\Omega C_1 \sin \Omega t) + \\ &\quad + (C_4 \cos 3\Omega t + C_5 \cos 5\Omega t) N_{\dot{u}} (C_1 \cos \Omega t, -\Omega C_1 \sin \Omega t) = 0 \end{aligned} \tag{9.185}$$

From Eqs. 9.183 and 9.184 we obtain:

$$\begin{aligned} N(C_1 \cos \Omega t, -\Omega C_1 \sin \Omega t) &= C_1 \left(1 - \Omega^2\right) \cos \Omega t + \Delta \sin(C_1 \cos \Omega t) + \\ &\quad + \varepsilon \left(C_1^2 \cos^2 \Omega t - 1\right) C_1 \Omega \sin \Omega t. \end{aligned} \tag{9.186}$$

The term $\sin(C_1 \cos \Omega t)$ can be expanded as:

$$\begin{aligned} \sin(C_1 \cos \Omega t) = C_1 \cos \Omega t - \frac{C_1^3 \cos^3 \Omega t}{3!} + \frac{C_1^5 \cos^5 \Omega t}{5!} - \\ - \frac{C_1^7 \cos^7 \Omega t}{7!} + \frac{C_1^9 \cos^9 \Omega t}{9!} \end{aligned} \tag{9.187}$$

We rewrite the powers of $\cos \Omega t$ in Eq. 9.187 in terms of multiples of Ωt:

$$\cos^{2n+1} \Omega t = \frac{1}{4^n} \sum_{k=0}^{n} \binom{2n+1}{n-k} \cos(2k+1) \Omega t \tag{9.188}$$

By using Eqs. 9.187 and 9.188, Eq. 9.186 can be expressed as:

$$\begin{aligned} N(C_1 \cos \Omega t, -\Omega C_1 \sin \Omega t) &= \left[(1 - \Omega^2) C_1 + \Delta C_1 - \frac{\Delta C_1^3}{8} + \frac{\Delta C_1^5}{192} - \frac{\Delta C_1^7}{9216} + \right. \\ &+ \left. \frac{\Delta C_1^9}{737280} + \ldots \right] \cos \Omega t - \left[\frac{\Delta C_1^3}{24} - \frac{\Delta C_1^5}{384} + \frac{\Delta C_1^7}{15360} - \frac{\Delta C_1^9}{1105920} + \ldots \right] \cos 3\Omega t + \\ &+ \left[\frac{\Delta C_1^5}{1920} - \frac{\Delta C_1^7}{46080} + \frac{\Delta C_1^9}{2580480} - \ldots \right] \cos 5\Omega t - \\ &- \left[\frac{\Delta C_1^7}{322560} - \frac{\Delta C_1^9}{10321920} + \ldots \right] \cos 7\Omega t + \left[\frac{\Delta C_1^9}{92897280} + \ldots \right] \cos 9\Omega t + \\ &+ \frac{\varepsilon C_1 \left(4 - C_1^2\right) \Omega}{4} \sin \Omega t - \frac{\varepsilon C_1^3 \Omega}{4} \sin 3\Omega t + \ldots \end{aligned} \tag{9.189}$$

and N_{u} and $N_{\dot{u}}$ are respectively:

$$N_u(C_1\cos\Omega t, -\Omega C_1\sin\Omega t) = 1-\Omega^2+\Delta-\frac{\Delta C_1^2}{4}+\frac{\Delta C_1^4}{64}-\frac{\Delta C_1^6}{2304}+ \\ +\frac{\Delta C_1^8}{147456}+\ldots-\left[\frac{\Delta C_1^2}{4}-\frac{\Delta C_1^4}{48}+\frac{\Delta C_1^6}{1536}-\frac{\Delta C_1^8}{9210}+\ldots\right]\cos 2\Omega t+\left[\frac{\Delta C_1^4}{192}- \\ -\frac{\Delta C_1^6}{3840}+\frac{\Delta C_1^8}{184320}-\ldots\right]\cos 4\Omega t-\left[\frac{\Delta C_1^6}{23040}-\frac{\Delta C_1^8}{645120}+\ldots\right]\cos 6\Omega t+ \\ +\left[\frac{\Delta C_1^8}{5160960}+\ldots\right]\cos 8\Omega t-\varepsilon C_1^2\Omega\sin 2\Omega t+\ldots \tag{9.190}$$

$$N_{\dot{u}}(C_1\cos\Omega t, -\Omega C_1\sin\Omega t) = 1-\frac{C_1^2}{2}-\frac{C_1^2}{2}\cos 2\Omega \tag{9.191}$$

Substitution of Eqs. 9.189, 9.190 and 9.191 into Eq. 9.185 results in:

$$\ddot{u}_1+\Omega^2 u_1-\Bigg[(\Omega^2-1)C_1-\Delta C_1+\frac{\Delta C_1^3}{8}-\frac{\Delta C_1^5}{192}+\frac{\Delta C_1^7}{9216}-\frac{\Delta C_1^9}{737280}+\ldots \\ +C_2\left(\Omega^2-1-\Delta+\frac{3\Delta C_1^2}{8}-\frac{5\Delta C_1^4}{192}+\frac{7\Delta C_1^6}{9216}-\frac{\Delta C_1^8}{81920}+\ldots\right)+ \\ +C_3\left(\frac{\Delta C_1^2}{8}-\frac{5\Delta C_1^4}{384}+\frac{7\Delta C_1^6}{15360}-\frac{\Delta C_1^8}{122880}+\ldots\right)-\frac{\varepsilon C_1^2 C_4}{4}\Bigg]\cos\Omega t- \\ -\Bigg[\frac{\Delta C_1^3}{24}-\frac{\Delta C_1^5}{384}+\frac{\Delta C_1^7}{15360}-\frac{\Delta C_1^9}{1105920}+\ldots+C_2\left(\frac{\Delta C_1^2}{8}-\frac{5\Delta C_1^4}{384}+ \\ +\frac{7\Delta C_1^6}{15360}+-\frac{\Delta C_1^8}{122880}+\ldots\right)+C_3\left(\Omega^2-1-\Delta+\frac{\Delta C_1^2}{4}-\frac{\Delta C_1^4}{64}+\frac{7\Delta C_1^6}{15360}- \\ -\frac{39\Delta C_1^8}{5160960}+\ldots\right)+\varepsilon\left(1-\frac{C_1^2}{2}\right)C_4-\frac{\varepsilon C_1^2 C_5}{4}\Bigg]\cos 3\Omega t-\Bigg[-\frac{\Delta C_1^5}{1920}+ \\ +\frac{\Delta C_1^7}{46080}-\frac{\Delta C_1^9}{2580480}+\ldots+C_2\left(\frac{\Delta C_1^5}{384}+\frac{7\Delta C_1^7}{46080}-\frac{\Delta C_1^9}{386720}+\ldots\right)+ \\ +C_3\left(\frac{\Delta C_1^2}{8}-\frac{\Delta C_1^4}{96}+\frac{\Delta C_1^6}{3072}-\frac{57\Delta C_1^8}{10321920}+\ldots\right)-\frac{\varepsilon C_1^2 C_4}{2}+ \\ +\varepsilon\left(1-\frac{C_1^2}{2}\right)C_5\Bigg]\cos 5\Omega t-\Bigg[\frac{\Delta C_1^7}{322560}-\frac{\Delta C_1^9}{10321920}+C_2\left(\frac{7\Delta C_1^6}{46080}-\frac{\Delta C_1^8}{1146880}\right)+ \\ +C_3\left(-\frac{\Delta C_1^4}{384}+\frac{\Delta C_1^6}{7680}-\frac{\Delta C_1^8}{368640}+\ldots\right)\Bigg]\cos 7\Omega t-\Bigg[\frac{\Delta C_1^9}{92897280}-\frac{\Delta C_1^8}{10321920}C_2+ \\ +C_3\left(\frac{\Delta C_1^6}{46080}-\frac{\Delta C_1^8}{1290240}\right)\Bigg]\cos 9\Omega t+\frac{\Delta C_1^8}{10321920}C_3\cos 11\Omega t- \\ -\left[\frac{\varepsilon}{4}(C_1^2-4)\Omega C_1+\frac{\varepsilon}{2}C_1^2\Omega(C_2-C_3)\right]\sin\Omega t- \\ -\frac{\varepsilon}{4}C_1^2\Omega(C_1+2C_2)\sin 3\Omega t+\frac{\varepsilon}{2}C_1^2\Omega C_3\sin 5\Omega t=0 \tag{9.192}$$

The solution of Eq. 9.192 chosen so that do not contain secular terms, requires that the coefficients of $\cos\Omega t$ and $\sin\Omega t$ to disappear, i.e.:

$$C_3 = C_2 + \frac{C_1^2 - 4}{2C_1} \tag{9.193}$$

$$\begin{aligned}\Omega^2 = \frac{1}{C_1 + C_2}\Bigg[& C_1\left(1 + \Delta - \frac{\Delta C_1^2}{8} + \frac{\Delta C_1^4}{192} - \frac{\Delta C_1^6}{9216} - \frac{\Delta C_1^8}{737280}\right) + \\ & + C_2\left(1 + \Delta - \frac{3\Delta C_1^2}{8} + \frac{5\Delta C_1^4}{192} - \frac{7\Delta C_1^6}{9216} - \frac{\Delta C_1^8}{81920}\right) + \\ & + C_3\left(-\frac{\Delta C_1^2}{8} + \frac{5\Delta C_1^4}{284} - \frac{7\Delta C_1^6}{15360} - \frac{\Delta C_1^8}{122880}\right) + \frac{\varepsilon}{4}C_1^2 C_4\Bigg].\end{aligned} \tag{9.194}$$

Assuming the initial conditions:

$$u(0) = 2, \quad \dot{u}(0) = 0 \tag{9.195}$$

from Eq. 9.192 we obtain the first-order approximate solutions :

$$\begin{aligned}u_1(t) = {} & 2\cos\Omega t + \frac{\alpha_{13}}{8\Omega^2}(\cos\Omega t - \cos 3\Omega t) + \frac{\alpha_{15}}{24\Omega^2}(\cos\Omega t - \cos 5\Omega t) + \\ & + \frac{\alpha_{17}}{48\Omega^2}(\cos\Omega t - \cos 7\Omega t) + \frac{\alpha_{19}}{80\Omega^2}(\cos\Omega t - \cos 9\Omega t) + \\ & + \frac{\alpha_{11}}{120\Omega^2}(\cos\Omega t - \cos 11\Omega t) + \frac{\varepsilon C_1^2}{32\Omega}(3\sin\Omega t - \sin 3\Omega t) + \\ & + \frac{\varepsilon C_1^2 C_3}{48\Omega}(5\sin\Omega t - \sin 5\Omega t).\end{aligned} \tag{9.196}$$

where Ω is given by Eq. 9.194 and the parameters α_{1i} are :

$$\begin{aligned}\alpha_{13} = {} & \frac{\Delta C_1^3}{24} - \frac{\Delta C_1^5}{384} + \frac{\Delta C_1^7}{15360} - \frac{\Delta C_1^9}{1105920} + C_2\left(\frac{\Delta C_1^2}{8} - \frac{5\Delta C_1^4}{384} + \frac{7\Delta C_1^6}{15360} - \right. \\ & \left. - \frac{\Delta C_1^8}{122880}\right) + C_3\left(\Omega^2 - 1 - \Delta + \frac{\Delta C_1^2}{4} - \frac{\Delta C_1^4}{64} + \frac{7\Delta C_1^6}{15360} - \frac{39\Delta C_1^8}{5160960}\right) + \\ & + \varepsilon\left(1 - \frac{C_1^2}{2}\right)C_4 - \frac{\varepsilon C_1^2 C_5}{4},\end{aligned}$$

$$\begin{aligned}\alpha_{15} = {} & -\frac{\Delta C_1^5}{1920} + \frac{\Delta C_1^7}{46080} - \frac{\Delta C_1^9}{2580480} + C_2\left(-\frac{\Delta C_1^4}{384} + \frac{7\Delta C_1^6}{46080} - \frac{\Delta C_1^8}{286720}\right) + \\ & + C_3\left(\frac{\Delta C_1^2}{8} - \frac{\Delta C_1^4}{96} + \frac{\Delta C_1^6}{3072} - \frac{57\Delta C_1^8}{10321920} + \dots\right) - \frac{\varepsilon C_1^2 C_4}{2} + \varepsilon\left(1 - \frac{C_1^2}{2}\right)C_5,\end{aligned} \tag{9.197}$$

$$\alpha_{17} = \frac{\Delta C_1^7}{322560} - \frac{\Delta C_1^9}{10321920} + C_2\left(\frac{\Delta C_1^6}{46080} - \frac{\Delta C_1^8}{1146880}\right) +$$
$$+ C_3\left(-\frac{\Delta C_1^4}{384} + \frac{\Delta C_1^6}{7680} - \frac{\Delta C_1^8}{368640}\right) - \frac{\varepsilon}{4} C_1^2 C_5,$$

$$\alpha_{19} = \frac{\Delta C_1^9}{92897280} - \frac{\Delta C_1^8}{10321920} C_2 + C_3\left(\frac{\Delta C_1^6}{46080} - \frac{\Delta C_1^8}{1290240}\right)$$

$$\alpha_{1,11} = -\frac{\Delta C_1^8}{10321920} C_3.$$

The parameters $C_1, C_2, \ldots, C_5$ can be optimally determined, putting the condition that the residual functional J, given by :

$$J = \int_0^T \left[\ddot{u}_1 + u_1 + \varepsilon\left(u_1^2 - 1\right)\dot{u}_1 + \Delta \sin u_1\right]^2 dt, \quad T = \frac{2\pi}{\Omega} \tag{9.198}$$

be minimum:

$$\frac{\partial J}{\partial C_i} = 0, \quad i = 1, 2, 4, 5 \tag{9.199}$$

and also using the Eq. 9.193.

From Eqs. 9.193, 9.194 and 9.199 the parameters and the frequency Ω are determined. In this way the solution (9.196), in the first approximation, is well-determined.

The initial approximation and the functions $\alpha(t)$, $\beta(t)$ into Eq. 9.196 are not unique. We can consider other expressions for u_0, $\alpha(t)$and $\beta(t)$,

$$u_0 = C_1' \cos \Omega t + C_2' \cos 3\Omega t, \quad \alpha(t) = C_3' \cos \Omega t, \quad \beta(t) = C_4' \cos 3\Omega t$$

and so on. The new constants $C_1', C_2', \ldots, C_5'$ can be determined in the same way from Eqs. 9.199 and 9.193.

In what follows we present some examples with analytical solutions and frequencies to show the efficiency of the method described above for solving Eq. 9.180. We consider different values of ε and we compare our results with numerical integration results obtained using a fourth-order Runge-Kutta method.

Case (a) When $\varepsilon = 0.2$ and $\Delta = 2$, from Eq. 9.199 we obtained the following parameters:

$$C_1 = 2.66323168751, \quad C_2 = -1.68784445425, \quad C_3 = -1.10719594442,$$
$$C_4 = -193.447730074, \quad C_5 = 29.3463553689, \quad \alpha_{13} = 0.704744879839,$$
$$\alpha_{15} = -0.174581669433, \quad \alpha_{17} = 0.0660397909103,$$
$$\alpha_{19} = 0.508616351595, \quad \alpha 1, 11 = -0.00689002888449$$

$$T = 4.38111631932, \quad \Omega = 1.434204312$$

and the first-order approximate solution (9.195) under the form

$$\begin{aligned} u_1(t) = {} & 2.043022598 \cos \Omega t - 0.0212347494 \sin \Omega t - 0.042827221 \cos 3\Omega t - \\ & - 0.0309137 \sin 3\Omega t + 0.003536432 \cos 5\Omega t - 0.000668871 \cos 7\Omega t - \\ & - 0.003090852 \cos 9\Omega t + 0.000027913 \cos 11\Omega t \end{aligned} \tag{9.200}$$

Case (b) For $\varepsilon = 0.4$ and $\Delta = 2$, we obtained

$$C_1 = 2.73599246755, \quad C_2 = -1.33246558502, \quad C_3 = -0.69546551579$$
$$C_4 = -88.8010396268, \quad C_5 = 14.2051940821, \quad \alpha_{13} = 0.769760657104$$
$$\alpha_{15} = -0.010667132613, \quad \alpha_{17} = -0.122242192892$$
$$\alpha_{19} = 0.558572214112, \quad \alpha_{1,11} = 0.401862599992$$

$$T = 4.42073214804, \quad \Omega = 1.421302197$$

and the first-order approximate solution (9.196) under the form

$$\begin{aligned} u_1(t) = {} & 2.051264747 \cos \Omega t + 0.044884636 \sin \Omega t - 0.047631347 \cos 3\Omega t - \\ & - 0.0365834475 \sin 3\Omega t + 0.000220021 \cos 5\Omega t - 0.030523738 \sin 5\Omega t + \\ & + 0.001260686 \cos 7\Omega t - 0.00345634 \cos 9\Omega t - 0.001657766 \cos 11\Omega t \end{aligned} \tag{9.201}$$

Case (c) For $\varepsilon = 0.6$ and $\Delta = 2$, we obtained

$$C_1 = 2.80282194949, \quad C_2 = -1.10874387685, \quad C_3 = -0.420899456115$$
$$C_4 = -48.9597206575, \quad C_5 = 8.03360867586, \quad \alpha_{13} = 0.569412682511$$
$$\alpha_{15} = 0.406915100929, \quad \alpha_{17} = -0.0531640773192$$
$$\alpha_{19} = 0.54051554035, \quad \alpha_{1,11} = 0.708213166982$$

$$T = 4.43236399108, \quad \Omega = 1.417610032$$

and the first-order approximate solution (9.196) under the form

$$u_1(t) = 2.049602464 \cos \Omega t + 0.0165936117 \sin \Omega t - 0.03547963 \cos 3\Omega t - \\ - 0.103904776 \sin 3\Omega t - 0.008436824 \cos 5\Omega t + 0.029155642 \sin 5\Omega t + \\ + 0.000551141 \cos 7\Omega t - 0.003362053 \cos 9\Omega t - 0.002936764 \cos 11\Omega t \tag{9.202}$$

Case (d) For $\varepsilon = 1$ and $\Delta = 2$, we obtained

$$C_1 = 2.73692623845, \quad C_2 = -0.706276028914, \quad C_3 = -0.0685596765921$$
$$C_4 = -30.7921182188, \quad C_5 = 4.56819087016, \quad \alpha_{13} = -0.547828772763$$
$$\alpha_{15} = 1.35134428124, \quad \alpha_{17} = 0.484031973056$$
$$\alpha_{19} = 0.493905284112, \quad \alpha_{1,11} = 1.06513438702$$

$$T = 4.44502154653, \quad \Omega = 1.41400123$$

$$u_1(t) = 2.006482621 \cos \Omega t + 0.458814565 \sin \Omega t + 0.034249582 \cos 3\Omega t - \\ - 0.165548946 \sin 3\Omega t - 0.028161461 \cos 5\Omega t + 0.007566654 \sin 5\Omega t - \\ - 0.005043514 \cos 7\Omega t - 0.003087835 \cos 9\Omega t - 0.004439392 \cos 11\Omega t \tag{9.203}$$

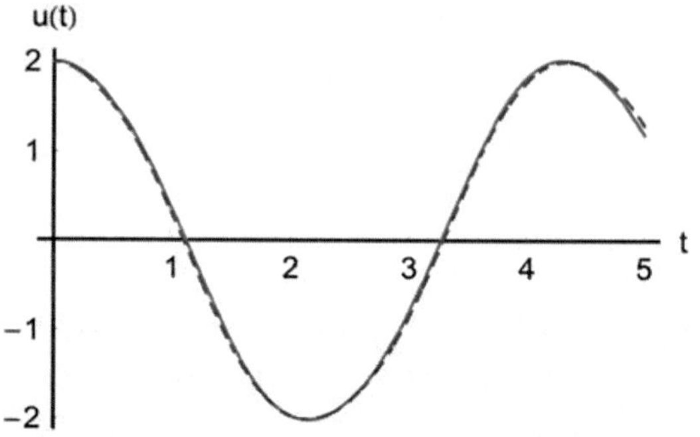

Fig. 9.28 Comparison between the results obtained for Eq. 9.180 in case (a) for $\varepsilon = 0.2$ and $\Delta = 2$, ________ numerical solution; _ _ _ _ _ _ approximate results obtained from (9.200)

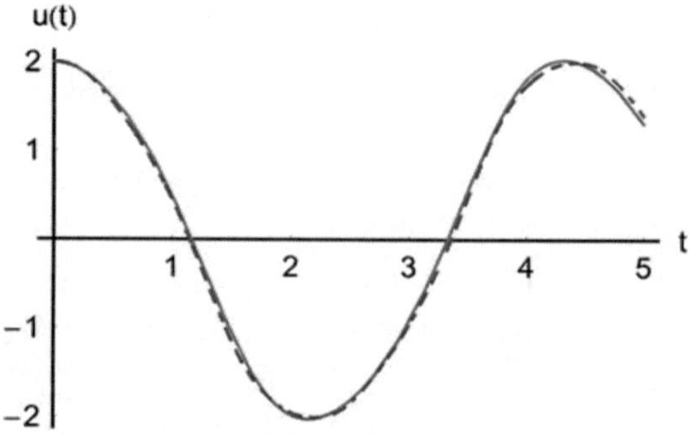

Fig. 9.29 Comparison between the results obtained for Eq. 9.180 in case (b) for $\varepsilon = 0.4$ and $\Delta = 2$, ________ numerical solution; _ _ _ _ _ _ approximate results obtained from (9.201)

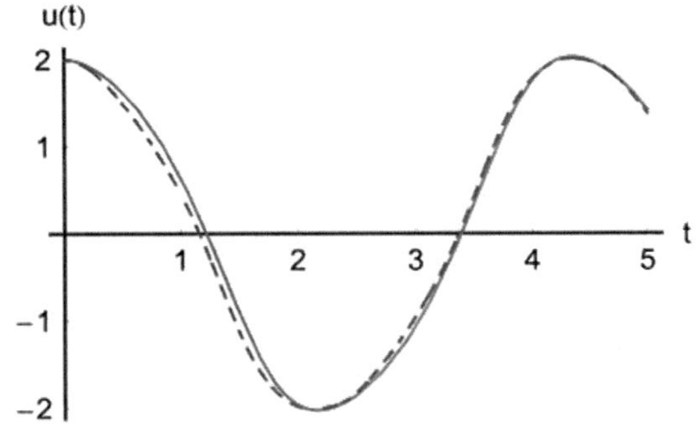

Fig. 9.30 Comparison between the results obtained for Eq. 9.180 in case (c) for $\varepsilon = 0.6$ and $\Delta = 2$, ________ numerical solution; _ _ _ _ _ _ approximate results obtained from (9.202)

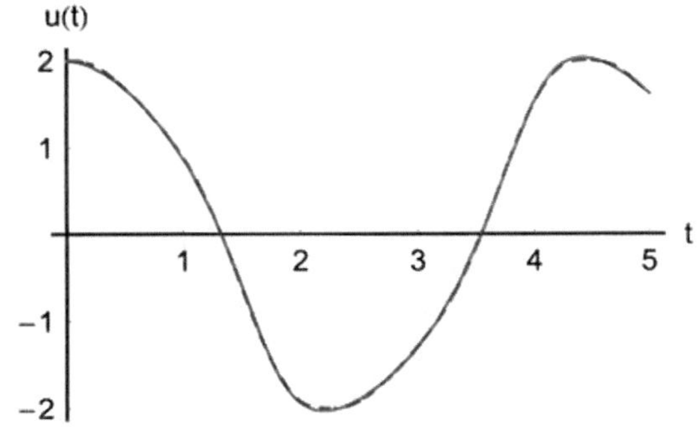

Fig. 9.31 Comparison between the results obtained for Eq. 9.180 in case (d) for $\varepsilon = 1$ and $\Delta = 2$, ________ numerical solution; _ _ _ _ _ _ approximate results obtained from (9.203)

It is easy to verify the accuracy of the obtained results if we graphically compare these analytical solutions with numerical ones. Figures 9.28–9.31 show the comparison between the present solutions and the numerical integration results obtained using a fourth-order Runge-Kutta method.

9.8 Volterra's Population Model

The study of Volterra integral equations originated with the work of Volterra in population dynamics [190]. Scudo indicates that Volterra proposed a model for a population $\hat{u}(t)$ of identical individuals which can be written in the form:

$$\frac{d\hat{u}}{d\hat{t}} = a\hat{u} - b\hat{u}^2 - c\hat{u}\int_0^{\hat{t}} \hat{u}(\tau)d\tau; \hat{u}(0) = \hat{u}_0 \tag{9.204}$$

where $\hat{u} = \hat{u}(t)$ and $a > 0$ is the birth rate coefficient, $b > 0$ is the crowding coefficient and $c > 0$ is the toxicity coefficient. The coefficient c indicates the essential behaviour of the population evolution before its level falls to zero in the long run. The last term contains the integral that indicates the "total metabolism" or total amount of toxins produced since time zero.

Several scales and population scales may be employed [191]. However, we shall scale time and population by introducing the non-dimensional variables

$$t = \frac{c\hat{t}}{b}, \quad u = \frac{b\hat{u}}{a}$$

producing the following differential equation:

$$\beta\dot{u} = u - u^2 - \int_0^t u(x)dx, \ \mathrm{u}(0) = \alpha \tag{9.205}$$

where $\beta = \frac{c}{ab}$, $\alpha = \frac{b\hat{u}_0}{a}$.

In [191] Small showed that if β is large, where the populations are strongly sensitive to toxins, the solution is proportional to $\mathrm{sech}^2(t)$. In this case the solution $u(t)$ has a smaller amplitude. Furthermore, for small β, where populations are weakly sensitive to toxins, the author showed that a rapid rise occurs along the logistic curve (curve given by Eq. 9.204 with $c = 0$) that will rich a peak and then followed by a slow exponential decay.

Noticing

$$N(u) = u^2 - u + \int_0^t u(x)dx \tag{9.206}$$

Equation 9.96 can be written for $n = 0$ in the form

$$\beta\dot{u}_1 + N(u_0) + \alpha(t, C_i)N_u(u_0) = 0 \tag{9.207}$$

where

$$\begin{aligned} u_0 &= \frac{C_1}{(\lambda t + 1)^2} + \frac{C_2}{(\lambda t + 1)^3}, \\ \alpha(t, C_i) &= \frac{C_3}{(\lambda t + 1)^2} + \frac{C_4}{(\lambda t + 1)^3} + \frac{C_5}{(\lambda t + 1)^4} \end{aligned} \tag{9.208}$$

with λ, C_1, C_2, ...C_5 unknown constants at this moment.

The initial condition (9.205_2) becomes

$$u_1(0) = \alpha \tag{9.209}$$

Equation 9.207 becomes

$$\beta \ddot{u}_1 + N\left(t, \frac{C_1}{(\lambda t+1)^2} + \frac{C_2}{(\lambda t+1)^3}\right) + \\ + \left[\frac{C_3}{(\lambda t+1)^2} + \frac{C_4}{(\lambda t+1)^3} + \frac{C_5}{(\lambda t+1)^4}\right] N_u\left(t, \frac{C_1}{(\lambda t+1)^2} + \frac{C_2}{(\lambda t+1)^3}\right) = 0 \tag{9.210}$$

where

$$N\left(t, \frac{C_1}{(\lambda t+1)^2} + \frac{C_2}{(\lambda t+1)^3}\right) = \left(C_1 - \frac{2C_1^2 + C_1C_2}{2\lambda}\right)\frac{1}{(\lambda t+1)^2} + \\ + \left(C_2 + \frac{2C_1^2 - 2C_1C_2 - C_2^2}{2\lambda}\right)\frac{1}{(\lambda t+1)^3} + \left(-C_1^2 + \frac{2C_1C_2}{\lambda}\right)\frac{1}{(\lambda t+1)^4} + \\ + \left(-2C_1C_2 + \frac{C_2^2}{\lambda}\right)\frac{1}{(\lambda t+1)^5} - C_2^2\frac{1}{(\lambda t+1)^6} \tag{9.211}$$

$$N_u\left(t, \frac{C_1}{(\lambda t+1)^2} + \frac{C_2}{(\lambda t+1)^3}\right) = \\ = 1 + \frac{2C_1 + C_2}{2\lambda} - \frac{C_1}{\lambda}\frac{1}{\lambda t+1} - \left(2C_1 + \frac{C_2}{2\lambda}\right)\frac{1}{(\lambda t+1)^2} - 2C_2\frac{1}{(\lambda t+1)^3} \tag{9.212}$$

By using Eqs. 9.211, 9.212, Eq. 9.210 can be expressed in the form

$$\beta \ddot{u}_1 - \left(C_1 + C_3 + \frac{2C_1C_3 + C_2C_3 - 2C_1^2 - C_1C_2}{2\lambda}\right)\frac{1}{(\lambda t+1)^2} - \\ - \left(C_2 + C_4 + \frac{2C_1^2 - 2C_1C_2 - C_2^2 + 2C_1C_4 + C_2C_4 - 2C_1C_3}{2\lambda}\right)\frac{1}{(\lambda t+1)^3} - \\ - \left(C_5 - 2C_1C_3 - C_1^2 + \frac{4C_1C_2 - 2C_1C_4 - C_2C_3 + 2C_1C_5 + C_2C_5}{2\lambda}\right)\frac{1}{(\lambda t+1)^4} \\ - \left(-2C_1C_2 - 2C_1C_4 - 2C_2C_3 + \frac{2C_2^2 - 2C_1C_5 - C_2C_4}{2\lambda}\right)\frac{1}{(\lambda t+1)^5} + \\ + \left(C_2^2 + 2C_1C_5 + 2C_2C_4 + \frac{C_2C_5}{2\lambda}\right)\frac{1}{(\lambda t+1)^6} - 2C_2C_5\frac{1}{(\lambda t+1)^7} = 0 \tag{9.213}$$

From the Eq. 9.213 and from the initial condition (9.209), the first order approximate solution becomes

$$u_1(t) = \alpha + A\left[\frac{1}{\lambda t+1} - 1\right] + B\left[\frac{1}{(\lambda t+1)^2} - 1\right] + C\left[\frac{1}{(\lambda t+1)^3} - 1\right] + \\ + D\left[\frac{1}{(\lambda t+1)^4} - 1\right] + E\left[\frac{1}{(\lambda t+1)^5} - 1\right] + F\left[\frac{1}{(\lambda t+1)^6} - 1\right] \quad (9.214)$$

where

$$A = \frac{1}{2\beta\lambda^2}\left(2C_1^2 + C_1C_2 - 2C_1C_3 - C_2C_3 - 2\lambda C_1 - 2\lambda C_3\right)$$

$$B = \frac{1}{4\beta\lambda^2}\left(2C_1C_2 + C_2^2 + 2C_1C_3 - 2C_1^2 - 2C_1C_4 - C_2C_4 - 2\lambda C_2 - 2\lambda C_4\right)$$

$$C = \frac{1}{6\beta\lambda^2}\left(2C_1C_4 + C_2C_3 + 2C_1C_5 + C_2C_5 - 4C_1C_2 + 2\lambda C_1^2 + 4\lambda C_1C_3 + 2\lambda C_5\right)$$

$$D = \frac{1}{8\beta\lambda^2}\left(C_2C_4 + 2C_1C_5 + 4\lambda C_1C_4 + 4\lambda C_1C_2 + 4\lambda C_2C_3 - \lambda C_2^2\right) \quad (9.215)$$

$$E = \frac{1}{5\beta\lambda^2}(C_2C_5 + 4\lambda C_1C_5 + 2\lambda C_2^2 + 4\lambda C_2C_4$$

$$F = \frac{1}{3\beta\lambda}(C_2C_5)$$

The parameters λ, C_1, C_2, C_3, C_4 and C_5 can be determined optimally, i.e. the residual functional J given by

$$J = \int_0^\infty \left[\beta\ddot{u}_1 - u_1 + u_1^2 - u_1\int_0^t u_1(x)dx\right]^2 dt \quad (9.216)$$

be minimized:

$$\frac{\partial J}{\partial \lambda} = \frac{\partial J}{\partial C_1} = \frac{\partial J}{\partial C_2} = \frac{\partial J}{\partial C_3} = \frac{\partial J}{\partial C_4} = \frac{\partial J}{\partial C_5} = 0 \quad (9.217)$$

From Eq. 9.217 the parameters λ and C_1, C_2, C_3, C_4, C_5 are determined.

We can also consider other expressions for u_0 and α, such as:

$$u_0(t) = C_1' \exp(-\lambda t), \quad \alpha(t, C_i) = C_2' \exp(-2\lambda t) + C_3' \exp(-3\lambda t) + \ldots \tag{9.218}$$

and so on. The new constants C_1', C_2',C_3',... can be determined in the same way, using Eq. 9.217.

In order to show the efficiency of the method described above, we present two examples of obtaining some analytical solutions. We consider two values for β and we compare our results with the numerical integration results obtained using a fourth-order Runge-Kutta method.

Case (a) When $\alpha = 0.1$, $\beta = 0.1$, we obtain from Eqs. 9.214, 9.215, 9.216 and 9.217 the following results:

$$\lambda = 1.1766462692; \quad C_1 = 2.7532136036; \quad C_2 = -1.6994761551;$$
$$C_3 = 0.6455659763; \quad C_4 = -1.9115831695; \quad C_5 = -0.8723926404 \tag{9.219}$$

From Eq. 9.214, the first-order approximate solution is given by

$$\begin{aligned} u_1(t) = {} & -0.00384197 + \frac{0.1078859759}{1.176646269285 \cdot t + 1} - \\ & - \frac{4.15358281}{(1.176646269285 \cdot t + 1)^2} - \frac{51.3816309336}{(1.176646269285 \cdot t + 1)^4} + \\ & + \frac{17.717637583152}{(1.176646269285 \cdot t + 1)^5} + \frac{4.200102526143}{(1.176646269285 \cdot t + 1)^6} \end{aligned} \tag{9.220}$$

Case (b) When $\alpha = 0.1$, $\beta = 1$, following the same procedure, we obtain

$$\lambda = 0.0704414; \quad C_1 = -0.0246497; \quad C_2 = 0.0101116;$$
$$C_3 = -1935.64; \quad C_4 = 1658.46; \quad C_5 = 2242.31$$

In this case, the first-order approximate solution becomes

$$\begin{aligned} u_1(t) = {} & -207.07 + \frac{1397.26}{(0.0704414 \cdot t + 1)} - \frac{3689.750012}{(0.0704414 \cdot t + 1)^2} + \\ & + \frac{4707.21}{(0.0704414 \cdot t + 1)^3} - \frac{2791.44}{(0.0704414 \cdot t + 1)^6} + \\ & + \frac{476.608}{(0.0704414 \cdot t + 1)^5} + \frac{107.292}{(0.0704414 \cdot t + 1)^6} \end{aligned} \tag{9.221}$$

It is easy to verify the accuracy of the obtained results if we graphically compare these analytically solutions with the numerical ones. Figures 9.32 and 9.33 show the

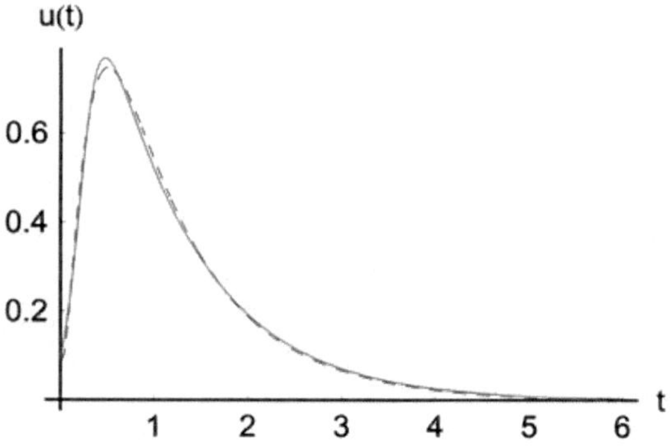

Fig. 9.32 Comparison between the results obtained for Eq. 9.205 in the case (a), $\alpha = \beta = 0.1$, _____ numerical simulation; _ _ _ _ present method (9.220)

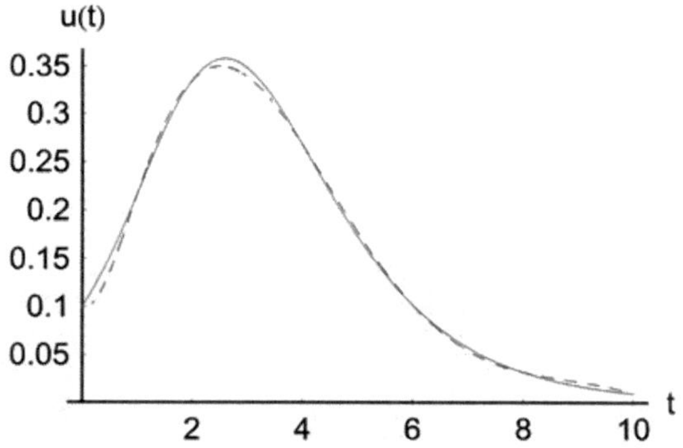

Fig. 9.33 Comparison between the results obtained for Eq. 9.205 in the case (b), $\alpha = 0.1$; $\beta = 1$, _____ numerical simulation; _ _ _ _ present method (9.221)

comparison between the present solutions and the numerical integration results obtained using a fourth-order Runge-Kutta method.

It can be seen from Figs. 9.32 and 9.33 that the solutions obtained by the proposed procedure are nearly identical with the numerical solutions.

9.9 Thomas-Fermi Equation

We consider the nonlinear differential equation called the Thomas-Fermi equation for the electron density around the nucleus of an atom which may be written as [192, 193]

$$y''(x) = \sqrt{\frac{y^3(x)}{x}}, \quad y(0) = 1, \quad y(\infty) = 0 \tag{9.222}$$

This equation is also frequently used for calculating form factors and for obtaining effective potentials which can be used as initial trial potentials in self-consistent field calculations.

Because of the singularity at $x = 0$, it is known that the solution of the Thomas-Fermi equation is very sensitive to the value of the first derivative at $x = 0$, which ensures a smooth and monotonic decay of the solution from 1 at $x = 0$ to 0 at $x{=}\infty$.

It is very important to determine a highly accurate value for the initial slope of the potential $y'(0)$. The importance of the initial slope is that it plays a major role in determining the energy of neutral atom in the Thomas-Fermi approximation

$$E = \frac{6}{7}\left(\frac{4}{3}\pi\right)^{\frac{2}{3}} Z^{7/3} y'(0) \tag{9.223}$$

where Z is the nuclear charge.

Equation 9.222 can be written in the form

$$x y''^2 - y^3 = 0 \tag{9.224}$$

which can be rewritten in the more general form

$$y'' - \lambda^2 y - N(x, y, y'') = 0 \tag{9.225}$$

where

$$N(x, y, y'') = x y''^2 + y'' - \lambda^2 y - y^3 \tag{9.226}$$

is a nonlinear function.

If we consider the second alternative of OPIM:

$$\begin{aligned} &y_0 = C_1 e^{-\lambda x}, \ \alpha(x) = C_2 e^{-\lambda x} + C_3 e^{-2\lambda x} \\ &\gamma(x) = C_4 e^{-\lambda t} + C_5 e^{-2\lambda x} + C_6 e^{-3\lambda x}, \ \beta(x) = 0 \end{aligned} \tag{9.227}$$

then for $n = 0$ into Eq. 9.96, we obtain

$$\begin{aligned} &y''_1 - \lambda^2 y_1 - N(x, C_1 e^{-\lambda x}, \lambda^2 C_1 e^{-\lambda x}) - \\ &\quad - (C_2 e^{-\lambda x} + C_3 e^{-2\lambda x}) N_y(x, C_1 e^{-\lambda x}, \lambda^2 C_1 e^{-\lambda x}) - \\ &\quad - (C_4 e^{-\lambda x} + C_5 e^{-2\lambda x} + C_6 e^{-3\lambda x}) N_{y''}(x, C_1 e^{-\lambda x}, \lambda^2 C_1 e^{-\lambda x}) = 0 \end{aligned} \tag{9.228}$$

$$y_1(0) = 1, \quad y_1(\infty) = 0 \tag{9.229}$$

From Eqs. 9.226 and 9.227_1 we obtain

$$N(x, C_1 e^{-\lambda x}, \lambda^2 C_1 e^{-\lambda x}) = \lambda^4 C_1^2 x e^{-2\lambda x} - C_1^3 e^{-3\lambda x} \tag{9.230}$$

$$N_y(x, C_1 e^{-\lambda x}, \lambda^2 C_1 e^{-\lambda x}) = -\lambda^2 - 3C_1^2 e^{-2\lambda x} \tag{9.231}$$

$$N_{y''}(x, C_1 e^{-\lambda x}, \lambda^2 C_1 e^{-\lambda x}) = 1 + 2\lambda^2 C_1 x e^{-\lambda x} \tag{9.232}$$

Substituting Eqs. 9.230, 9.231 and 9.232 into Eq. 9.228 we obtain

$$\begin{aligned}
& y''_1 - \lambda^2 y_1 - \left[C_4 - \lambda^2 C_2\right] e^{-\lambda x} - \left[C_5 - \lambda^2 C_3 + (\lambda^4 C_1^2 + 2\lambda^2 C_1 C_4) x\right] e^{-2\lambda x} - \\
& - \left[2\lambda^2 C_1 C_5 x - C_1^3 - 3C_1^2 C_2 + C_6\right] e^{-3\lambda x} - \left[2\lambda^2 C_1 C_6 x - 3C_1^2 C_3\right] e^{-4\lambda x} = 0, \\
& y_1(0) = 1, \quad y_1(\infty) = 0
\end{aligned} \tag{9.233}$$

From Eq. 9.233 we obtain the first-order approximate solution of Thomas-Fermi equation

$$\begin{aligned}
y_1(x) = (Ax + B) e^{-\lambda x} + (Cx + D) e^{-2\lambda x} + \\
+ (Ex + F) e^{-3\lambda x} + (Gx + H) e^{-4\lambda x}
\end{aligned} \tag{9.234}$$

where

$$\begin{aligned}
& A = \frac{\lambda}{2} C_2 - \frac{1}{2\lambda} C_4 \\
& B = 1 - \frac{4\lambda}{9} C_1^2 - \frac{8}{9\lambda} C_1 C_4 + \frac{1}{3} C_3 - \frac{1}{3\lambda^2} C_5 - \frac{3}{16\lambda} C_1 C_5 + \\
& \quad + \frac{3}{8\lambda^2} C_1^2 C_2 + \frac{1}{8\lambda^2} C_1^3 + \frac{1}{5\lambda^2} C_1^2 C_3 - \frac{16}{225\lambda} C_1 C_6 - \frac{1}{8\lambda^2} C_6 \\
& C = \frac{\lambda^2}{3} C_1^2 + \frac{2}{3} C_1 C_4, \quad D = \frac{4\lambda}{9} C_1^2 + \frac{8}{9\lambda} C_1 C_4 - \frac{1}{3} C_3 + \frac{1}{3\lambda^2} C_5 \\
& E = \frac{1}{4} C_1 C_5, \quad F = \frac{3}{16\lambda} C_1 C_5 - \frac{3}{8\lambda^2} C_1^2 C_2 - \frac{1}{8\lambda^2} C_1^3 + \frac{1}{8\lambda^2} C_6 \\
& G = \frac{2}{15} C_1 C_6, \quad H = \frac{16}{225\lambda} C_1 C_6 - \frac{1}{5\lambda^2} C_1^2 C_3
\end{aligned} \tag{9.235}$$

The parameters λ and $C_1, C_2, C_3, C_4, C_5, C_6$ can be determined optimally, which means the residual functional J given by

$$J = \int_0^\infty \left[x y_1''2 - y_1^3\right]^2 dx \tag{9.236}$$

can be minimized:

$$\frac{\partial J}{\partial \lambda} = \frac{\partial J}{\partial C_1} = \frac{\partial J}{\partial C_2} = \ldots = \frac{\partial J}{\partial C_6} = 0 \tag{9.237}$$

The initial approximation y_0 and the functions $\alpha(x)$ and $\beta(x)$ are not unique. We can consider other expressions for y_0, $\alpha(x)$ and $\gamma(x)$, such as:

$$\begin{aligned} y_0(x) &= C_1' e^{-\lambda x} + C_2' e^{-2\lambda x} \\ \alpha(x) &= C_3' e^{-\lambda x} + C_4' e^{-2\lambda x} + C_5' e^{-3\lambda x} \\ \beta(x) &= C_6' e^{-\lambda x} + C_7' e^{-2\lambda x} + C_8' e^{-3\lambda x} + C_9' e^{-4\lambda x} \end{aligned} \tag{9.238}$$

and so on, with the constants λ and $C_1', C_2', \ldots, C_9'$, which can be determined in the same way from Eq. 9.237.

Also, Eq. 9.224 can be written as

$$y'' - \frac{2}{(\lambda x + 1)^2} y - N^*(x, y, y'') = 0$$

where

$$N^*(x, y, y'') = xy''^2 + y'' - \frac{2}{(\lambda x + 1)^2} y - y^3$$

In this case, the initial approximation and the functions $\alpha(x)$ and $\beta(x)$ can be chosen in the form

$$y_0 = \frac{C_1}{\lambda x + 1}, \quad y_0'' = \frac{2\lambda^2 C_1}{(\lambda x + 1)^3}$$

$$\alpha(x) = \frac{C_2}{\lambda x + 1} + \frac{C_3}{(\lambda x + 1)^2}$$

$$\beta(x) = \frac{C_4}{\lambda x + 1} + \frac{C_5}{(\lambda x + 1)^2} + \frac{C_6}{(\lambda x + 1)^3}$$

We follow the procedure described above to find numerical results:

$$\begin{aligned} &\lambda = 0.8142096804743827; \; C_1 = -0.1463669113967; \\ &C_2 = -417.8208014694533; \; C_3 = -855.9241157162435; \\ &C_4 = -281.4804345074634; \; C_5 = -735.7393406374861; \\ &C_6 = 95.01941123786105 \end{aligned} \tag{9.239}$$

From Eqs. 9.239 and 9.235, the coefficients A, B, C, D, E, F, G and H are determined as follows:

$$\begin{aligned} &\mathrm{A} = 2.7581329751336514;\ \mathrm{B} = -11.451371899334289;\\ &\mathrm{C} = 27.47101530791185;\ \mathrm{D} = -39.64505180499373;\\ &\mathrm{E} = 26.92197372053835;\ \mathrm{F} = 47.77913460422296;\\ &\mathrm{G} = -1.8543596994158142;\ \mathrm{H} = 4.317289100105105 \end{aligned} \tag{9.240}$$

In this way, the first-order approximate solution (9.234) of Thomas-Fermi equation is well determined.

From Eq. 9.234 we obtain

$$y_0' = A + C + E + G - \lambda(B + 2D + 3F + 4H) \tag{9.241}$$

From Eqs. 9.240 and 9.241, the approximate initial slope $y_1'(0)$becomes

$$y_1'(0) = -1.5880659888022421 \tag{9.242}$$

and therefore the energy neutral atom given by Eq. 9.223 is well-determined.

Kobayasi [192] gave the numerical result for the same problem as $y'_1(0) = -1.588071$. A comparison between different approximations of the initial slop $y'_1(0)$is presented in Table 9.4. It is clear that the present result for the initial slope is better than those given by Liao's method [193].

Figure 9.34 presents a comparison between our first-order approximate solution (9.234) and Liao's results [193]. The first-order analytic approximate solution y(x) for $x \in [0,\infty)$ given by Eq. 9.234 is compared with the 60 th-order approximation obtained by Liao [193]. From Fig. 9.34 it can be seen that the solution obtained through the proposed procedure (after only one iteration) is nearly identical with the 60 th-order approximation of the series solution obtained by means of the homotopy analysis method for the same problem.

Table 9.4 Comparison between different approximations of the initial slope $y'_1(0)$.

Order of approximation	Liao [81]	Liao [193]	Present work (first-order approximation)	Kobayashi [192]
10	−1.28590	−1.50014		
20	−1.40932	−1.54093		
30	−1.46306	−1.55595		
40	−1.49236	−1.56373		
50	−1.51063	−1.56848	−1.5880659888	−1.588071
60	−1.52309	−1.57168		
70	−1.53211	−1.57399		
80	−1.53895	−1.57572		
90	−1.54430	−1.57708		
100	−1.54860	−1.57816		

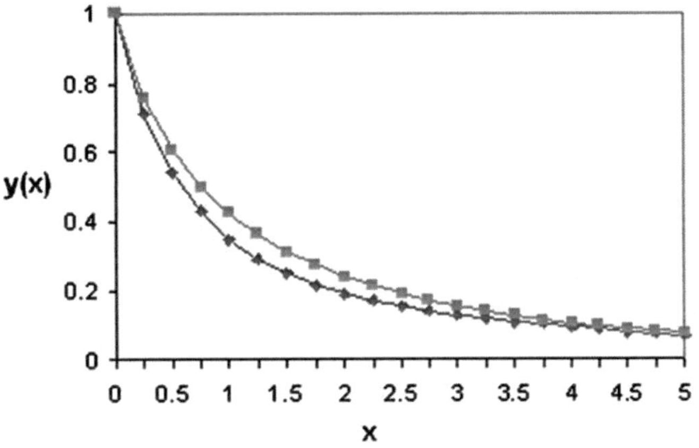

Fig. 9.34 Comparison between the first-order approximate solution (9.234) and Liao's 60th-order approximation [193]: ◆-◆-◆-◆- present results; ■_■_■_■_ Liao's results [193]

To the best of our knowledge, it is the first time then such an explicit analytic approximate solution to the Thomas-Fermi equation is given. Some series solutions were reported, without providing an explicit analytical solution in the traditional meaning of the word "analytical".

We remark that very good approximations are obtained in only a few terms. Our procedure provides a very accurate solution after only one iteration, unlike Liao's procedure [81, 193], which converges to an accurate solution upwards of the 60-th order of approximation.

The proposed solution and Kobayashi's solution [192] are both more accurate that Liao's [81, 193] in terms of the initial slope. Besides, the present solution is an explicit analytic approximate solution and to the best of out knowledge this is the first explicit analytic solution reported in the literature to the Thomas-Fermi equation.

9.10 Swirling Flow Downstream of a Turbine Runner

Theoretical analysis of swirling flows can employ tools ranging from simplified axisymmetric, inviscid steady or unsteady flow models to full 3D laminar or turbulent numerical simulation. Several simplifications must be admitted and the results must be interpreted accordingly. We consider a steady mean flow with axial and circumferential velocity profiles derived from experimental data. An inviscid incompressible fluid is considered, since our swirling flow representation does not account for the boundary layers near the wall.

If ψ is a perturbation of the base of a Francis turbine runner and k is the axial wave number of this perturbation, then the equation

$$\frac{\partial^2 \psi}{\partial y^2} - F_0(y)\psi = \frac{k^2}{2y}\psi \tag{9.243}$$

defines a generalised eigenvalue problem. The eigenvalue k^2 can be computed numerically once the problem is discretised [194]. The Eq. 9.243 is equivalent with the equation

$$\ddot{\psi}(y)y^2 - \left[F(y) + \frac{k^2}{2}y\right]\psi(y) = 0 \tag{9.244}$$

where the expression of $F(y)$ can be evaluated once an analytical swirl representation is available. The boundary conditions are

$$\psi(0) = 0, \quad \psi(a) = 0 \tag{9.245}$$

and the function F is given by the expression

$$F(y) = F_0(y)y^2 = \frac{\lambda_1^2 u_1 e^{-\lambda_1 y} + \lambda_2^2 u_2 e^{-\lambda_2 y}}{u_0 + u_1 e^{-\lambda_1 y} + u_2 e^{-\lambda_2 y}} y^2 - \frac{2\left[\Omega_0 y + \frac{\Omega_1}{\lambda_1}\left(1 - e^{-\lambda_1 y}\right) + \frac{\Omega_2}{\lambda_2}\left(1 - e^{-\lambda_2 y}\right)\right]\left[\Omega_0 + \Omega_1 e^{-\lambda_1 y} + \Omega_2 e^{-\lambda_2 y}\right]}{\left(u_0 + u_1 e^{-\lambda_1 y} + u_2 e^{-\lambda_2 y}\right)^2} \tag{9.246}$$

where a, Ω_0, Ω_1, Ω_2, u_0, u_1, u_2, λ_1 and λ_2 are known from experimental investigations.

Equation 9.244 is a linear differential equation with variable coefficients. In this case the optimal parametric iteration method is applied in a particular form, as follows.

Consider the homogenous linear differential equation whit constant coefficients:

$$a_0 u^{(n)}(y) + a_1 u^{(n-1)}(y) + \ldots + a_n u(y) = 0 \tag{9.247}$$

where all the coefficients a_i are constants. The so-called characteristic equation can be written in the form.

$$a_0 k^n + a_1 k^{n-1} + \ldots + a_{n-1} k + a_n = 0 \tag{9.248}$$

Thus, the general solution of Eq. 9.247 is of the form:

$$u(y) = \sum_{i=1}^{m}\left(C_{0i} + C_{1i}y + C_{2i}y^2 + \ldots + C_{\alpha_i - 1}y^{\alpha_i - 1}\right)\exp(k_i y) \tag{9.249}$$

where C_{si} are arbitrary constants, m is the number of distinct roots k_i of the characteristic Eq. 9.248 and α_i is the multiplicity of the root k_i.

The problem of integrating homogeneous linear differential equations (with variable coefficients), of the n th-order, reduces to choosing n or at least n-1 linearly independent particular solutions. However, particular solutions are readily selected only in exceptional cases. In more involved cases, particular solutions are sought in the form of a sum of certain series $\sum_{i=1}^{\infty} \beta_i \phi_i(x)$, often in the form of the sum of products of coefficients β_i and power series $\varphi_i(x)$. The conditions, under which there exist solutions of this form, are ordinarily established by methods of the theory of functions of a complex variable. In concrete problems the most used procedure is to choose such series that formally satisfies the differential equation. Having formally obtained a solution in the form of a series, the next step is to investigate the convergence and the possibility of n-fold termwise differentiation. In the region where the series converges and admits an n-fold termwise differentiation, it not only formally satisfies the equation, but its sum is indeed the desired solution.

In this context, we try to choose an approximate solution of Eq. 9.244 in the form of a truncated series:

$$\bar{\psi}(y) = \sum_{i=1}^{m} \left(C_{i1} + C_{i2}y + C_{i3}y^2 + ... + C_{in}y^n\right) \exp(-h_i y) \tag{9.250}$$

where m, n are two arbitrarily integer positive numbers, h_i and C_{ij} are unknown parameters. The approximate solution $\bar{\psi}$ given by Eq. 9.250 is not unique, because the integers m and n are not unique.

The parameters C_{ij} and h_i that appear in Eq. 9.250 can be determined optimally i.e. by considering the residual functional

$$J(h_i, C_{ij}) = \int_0^a \left[\ddot{\bar{\psi}}(y)y^2 - \left(F(y) + \frac{k^2}{2}y\right)\bar{\psi}(y)\right]^2 dy \tag{9.251}$$

and minimizing this functional:

$$\frac{\partial J}{\partial h_i} = 0, \frac{\partial J}{\partial C_{ij}} = 0, \quad i = 1, 2, ..., m, j = 1, 2, ..., n \tag{9.252}$$

The parameters h_i and C_{ij} can also be determined optimally by using other procedures. For example, if $y_i \in (0,a)$, i = 1,2,. . .,2 m + n, then by substituting y_i into the residual of the initial equation obtained for the approximate solution (9.250),

$$R(y) = \ddot{\bar{\psi}}(y)y^2 - \left(F(y) + \frac{k^2}{2}y\right)\bar{\psi}(y) \tag{9.253}$$

we have a system of nonlinear algebraic equations

$$\begin{aligned} &R(y_1, C_{ij}) = R(y_2, C_{ij}) = \ldots = R(y_{2m+n-2}, C_{ij}) = 0, \\ &i = 1, 2, \ldots, m, \; j = 1, 2, \ldots, n, \quad \psi(0) = 0, \quad \psi(a) = 0 \end{aligned} \tag{9.254}$$

which can be solved to obtain the optimal values of the parameters h_{i} and C_{ij}. Alternatively, the least squares method, the Galerkin method, the Ritz method, the Kantorovich method or the collocation method can be used.

The solution of Eq. 9.244 can immediately be determined by using the optimal values of the parameters C_{ij}, once these are known.

In short, the idea of the proposed procedure is to construct a solution (9.250) by means of some unknown constants C_{ij} and h_{i}, which can be determined optimally so that the convergence of the approximate solution can be easily controlled.

We illustrate the accuracy of our procedure on some particular examples. The analytical results given by Eq. 9.250 are compared whit the numerical integration results obtained using a fourth-order Runge-Kutta method.

Case (a) For $\Omega_0 = 0.26675$, $\Omega_1 = -0.79994$, $\Omega_2 = 3.3512$, $u_0 = 0.31501$, $u_1 = 0.07324$, $u_2 = -0.29672$, $\lambda_1 = 15.14557522$, $\lambda_2 = 231.0418175$, $k^2 = 125$, $a = 0.55125$ and the discharge coefficient 0.360, from Eq. 9.254 we obtain:

$$\begin{aligned} &h_1 = 105.43721461, \mathrm{h}_2 = 125.1513718, \mathrm{C}_{11} = 0.0001, \\ &\mathrm{C}_{12} = 0.021646321, \mathrm{C}_{21} = 0.000006613, \mathrm{C}_{22} = -0.009531734, \\ &\mathrm{C}_{23} = 3.272949511, \mathrm{C}_{24} = -361.9985675, \mathrm{C}_{25} = 7090.637654 \end{aligned} \tag{9.255}$$

From Eqs. 9.250 and 9.255, an approximate solution in this case becomes

$$\begin{aligned} \bar{\psi}(y) = {} & (0.021646321y + 0.0001)e^{-105.43721461 \cdot y} + (7090.637654y^4) - 361.9985675y^3 \\ & + 3.272949511y^2 - 0.009531734y + 0.000006613)e^{(-125.1513718y)} \end{aligned} \tag{9.256}$$

Figure 9.35 shows a comparison between the present solution and the numerical integration results obtained using a fourth-order Runge-Kutta method.

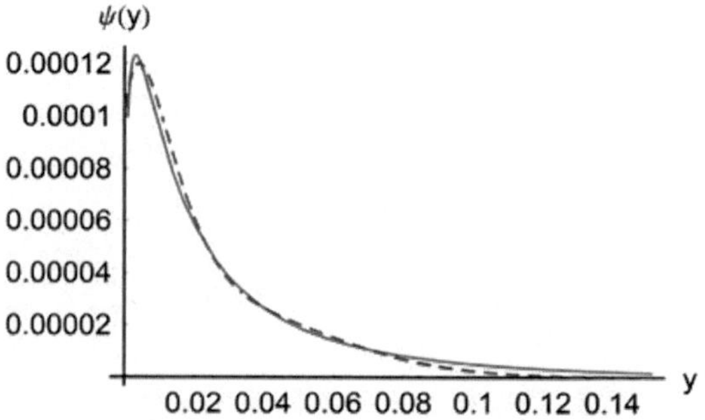

Fig. 9.35 Comparison between analytical and numerical results obtained for Eq. 9.244: _ _ _ _ _ _ _ analytical results given by Eq. 9.256; ________ numerical results

Case (b) In the second case $\Omega_0 = 0.27113$, $\Omega_1 = -0.80310$, $\Omega_2 = 3.4960$, $u_0 = 0.31991$, $u_1 = 0.08710$, $u_2 = -0.27350$, $\lambda_1 = 9.192995429$, $\lambda_2 = 117.420092$, $a = 0.55125$, $k^2 = -11$ we obtain:

$$\begin{aligned}
&h_1 = 4.8248639, \mathrm{h}_2 = 8.759791538, \mathrm{h}_3 = 14.2149713,\\
&\mathrm{C}_{11} = 0.000087706, \quad \mathrm{C}_{12} = 0.012776622, \quad C_{21} = 1.2 \cdot 10^{-7},\\
&C_{22} = -0.000066696, \quad C_{23} = 0.067607239, \quad C_{24} = -0.912298988,\\
&C_{25} = 4.087586517, \quad C_{26} = -6.039667426, \quad C_{31} = -0.00002459,\\
&C_{32} = 0.267102602, \quad C_{33} = -1.080928642, \quad C_{34} = 18.14814559,\\
&C_{35} = -157.9759977, \quad C_{36} = 766.5114418, \quad C_{37} = -2076.606273,\\
&C_{38} = 2915.08435, \quad C_{39} = -1639.370297
\end{aligned} \tag{9.257}$$

This yields an approximate solution

$$\begin{aligned}
\bar{\psi}(y) = {} & e^{-4.8248639 \cdot \mathrm{y}}(0.012776622\ \mathrm{y} + 0.000087706) -\\
& - 6.039667426\mathrm{e}^{-8.759791538\mathrm{y}}(\mathrm{y}^5 - 0.67679\ \mathrm{y}^4 + 0.151051196\ \mathrm{y}^3 -\\
& - 0.011193868\ \mathrm{y}^2 + 0.000011043\ \mathrm{y}) - 1639.370297\mathrm{e}^{-14.2148713 \cdot \mathrm{y}}(\mathrm{y}^8 -\\
& - 1.778173214\ \mathrm{y}^7 + 1.26670971\ \mathrm{y}^6 - 0.467564554\ \mathrm{y}^5 + 0.096363828\ \mathrm{y}^4 -\\
& - 0.011070193\ \mathrm{y}^3 + 0.000659356\ \mathrm{y}^2 - 0.000016293\ \mathrm{y} + 0.000000015)
\end{aligned} \tag{9.258}$$

Figure 9.36 presents the comparison between analytical and numerical results in this case.

Case (c) In the last case we consider $\Omega_0 = 0.31765$, $\Omega_1 = -0.62888$, $\Omega_2 = 2.2545$, $u_0 = 0.30697$, $u_1 = 0.01056$, $u_2 = -0.31889$, $\lambda_1 = 14.38208782$, $\lambda_2 = 289.9684333$, $a = 0.55125$, $k^2 = 10000$, we obtain:

$$h_1 = 1994.365369, \mathrm{C}_{11} = 3.07 \cdot 10^{-9}, \mathrm{C}_{12} = 0.734753857 \tag{9.259}$$

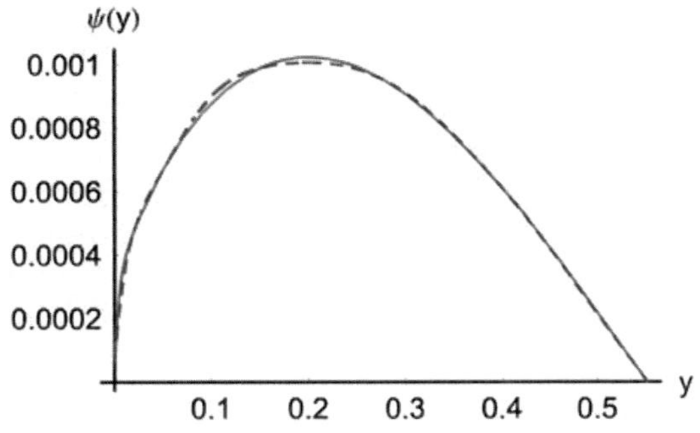

Fig. 9.36 Comparison between analytical and numerical results obtained for Eq. 9.244 in case (b): _ _ _ _ _ _ _ analytical results given by Eq. 9.258; ________ numerical results

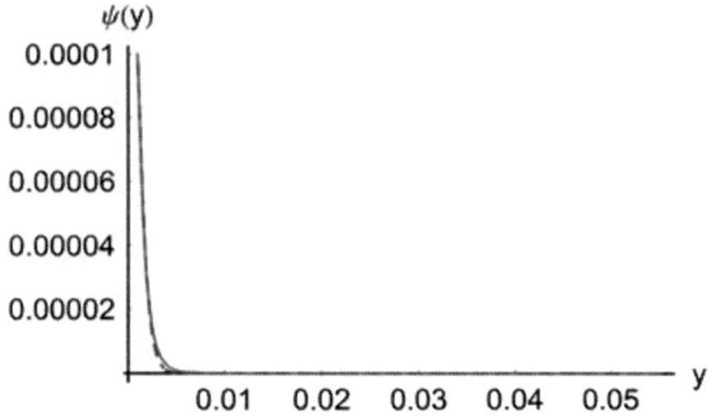

Fig. 9.37 Comparison between analytical and numerical results obtained for Eq. 9.244 in case (c): _ _ _ _ _ _ _ analytical results given by Eq. 9.260; ________ numerical results

In this case the approximate solution of Eq. 9.243 is given by:

$$\bar{\psi}(y) = 0.734753857ye^{-1994.365369 \cdot y} \tag{9.260}$$

Figure 9.37 shows the graphical comparison between analytical and numerical results in this case.

It can be seen from Figs. 9.35–9.37 that the solutions obtained through the proposed procedure are nearly identical with the numerical solutions obtained using a fourth-order Runge-Kutta method.

9.11 Lotka-Volterra Model with Three Species

Systems of nonlinear differential equations arise in many scientific models such as biological systems and are used in various fields as engineering, chemistry, ecology and so on. In 1925, A.J. Lotka [195] developed the motion of an evolutionary system based on two fundamental changes, those involving matter between components of a system and those involving exchanges of energy. In 1926, V. Volterra [196] developed the well-known mathematical models of multispecies interaction. These models, the predator, prey and competition models are known today as Lotka-Volterra models. Lotka-Volterra equations describe variations of population densities of few species that compete for the same resources. However, the ecological system is often affected by environmental changes and other human activities. In many practical situations, it is often the case that one of the species maybe suffers a significant loss or increase in density for some reason at some transitory time slots. These models, for instance can describe the competing fish species which are exploited by human activities, can also describe the dynamics of normal and tumour cells in a changing environment under the effects of the chemotherapy. These models are also applicable in case we are interested in the existence and stability of tumour-free solution and how treatment affects the interaction of tumour and normal cells. Although simplistic, these few models are still used as the foundation for mathematical models in biology. These models can also describe the time history of a biological system and are used in various fields as engineering, chemistry, biology or mathematics. In fact the Lotka-Volterra model is one of the most popular ones to demonstrate a simple nonlinear control system. The accurate solutions of the Lotka-Volterra equations may become a difficult task either if the equations are stiff or when the number of species is large.

In what follows we consider a nonlinear differential equation described by the following time dependent Lotka-Volterra model with three species:

$$\begin{aligned} \dot{x} &= x(1 - x - \alpha y - \beta z) \\ \dot{y} &= y(1 - \beta x - y - \alpha z) \\ \dot{z} &= z(1 - \alpha x - \beta y - z) \end{aligned} \tag{9.261}$$

were dot denotes derivative with respect to time t, α and β are known parameters not necessary small. Initial conditions are

$$x(0) = a, \quad y(0) = b, \quad z(0) = c \tag{9.262}$$

In accordance with OPIM procedure in the first alternative, we consider the following iteration formulas:

$$\begin{aligned} \dot{x}_{n+1} &= f(x_n, y_n, z_n) + m_1(t, C_i) f_x(x_n, y_n, z_n) + \\ &\quad + n_1(t, C_i) f_y(x_n, y_n, z_n) + p_1(t, C_i) f_z(x_n, y_n, z_n) \\ \dot{y}_{n+1} &= g(x_n, y_n, z_n) + m_2(t, C_j) g_x(x_n, y_n, z_n) + \\ &\quad + n_2(t, C_j) g_y(x_n, y_n, z_n) + p_2(t, C_j) g_z(x_n, y_n, z_n) \\ \dot{z}_{n+1} &= h(x_n, y_n, z_n) + m_3(t, C_k) h_x(x_n, y_n, z_n) + \\ &\quad + n_3(t, C_k) h_y(x_n, y_n, z_n) + p_3(t, C_k) h_z(x_n, y_n, z_n) \end{aligned} \tag{9.263}$$

where

$$\begin{aligned} f(x, y, z) &= x(1 - x - \alpha y - \beta z) \\ g(x, y, z) &= y(1 - \beta x - y - \alpha z) \\ h(x, y, z) &= z(1 - \alpha x - \beta y - z) \end{aligned} \tag{9.264}$$

and $f_x = \partial f / \partial x$, $n = 0,1,2\ldots$, and C_{i}, C_{j}, C_{k}, i, j, $k = 1,2,\ldots$ are unknown constants at this moment. There are many possibilities to choose the auxiliary functions $m_e(t, C_i)$, $n_e(t, C_j)$ and $p_e(t, C_k)$, e $= 1,2,3$. Basically, the shape of m_e, n_e and p_e must follow the terms appearing in Eq. 9.263. Therefore, we try to choose the auxiliary functions so that in Eq. 9.263 the products $m_e F_x$ be of the same shape with the function F_x, where $F \in \{f, g, h\}$.

We present an example to show the efficiency of the method described in the previous section for solving Eq. 9.26.

Taking into account the initial conditions (9.262) we choose the initial approximations as functions of the form:

$$x_0 = ae^{-t}, \quad y_0 = be^{-t}, \quad z_0 = ce^{-t} \tag{9.265}$$

For $n = 0$ into Eq. 9.263 we can construct the following iteration formulas:

$$\begin{aligned}
\dot{x}_1 &= x_0 - x_0^2 - \alpha x_0 y_0 - \beta x_0 z_0 + m_1(t, C_i)[1 - (2x_0 + \alpha y_0 + \beta z_0)] + \\
&\quad + n_1(t, C_i)(-\alpha x_0) + p_1(t, C_i)(-\beta x_0) \\
\dot{y}_1 &= y_0 - \beta x_0 y_0 - y_0^2 - \alpha y_0 z_0 + m_2(t, C_j)(-\beta y_0) + \\
&\quad + n_2(t, C_j)[1 - (\beta x_0 + 2y_0 + \alpha z_0)] + p_2(t, C_i)(-\alpha y_0) \\
\dot{z}_1 &= z_0 - \alpha x_0 z_0 - \beta y_0 z_0 - z_0^2 + m_3(t, C_k)(-\alpha z_0) + n_3(t, C_k)(-\beta z_0) + \\
&\quad + p_3(t, C_k)[1 - (\alpha x_0 + \beta y_0 + 2z_0)]
\end{aligned} \tag{9.266}$$

For the auxiliary functions m_i, n_i, p_i, $i = 1,2,3$ we choose the expressions:

$$\begin{aligned}
&m_1(t, C_i) = (C_1 t + C_2)e^{-t}, \quad n_1(t, C_i) = C_3 t + C_4, \\
&p_1(t, C_i) = (C_5 t + C_6)e^{-t}, \quad m_2(t, C_j) = C_7 t + C_8, \\
&n_2(t, C_j) = (C_9 t + C_{10})e^{-t}, \quad p_2(t, C_j) = (C_{11} t + C_{12})e^{-2t} \\
&m_3(t, C_k) = (C_{13} t + C_{14})e^{-t}, \quad n_3(t, C_k) = (C_{15} t + C_{16})e^{-2t}, \\
&p_3(t, C_k) = (C_{17} t + C_{18})e^{-t}
\end{aligned} \tag{9.267}$$

Alternatively, we can consider either other expressions for the initial approximations and the auxiliary function, such as:

$$\begin{aligned}
&x_0 = ae^{-2t}, \quad y_0 = b, \quad z_0 = ce^{-t}, \quad m_1(\mathrm{t}, C'_\mathrm{i}) = C'_1 e^{-2t}, \\
&n_1(\mathrm{t}, C'_\mathrm{i}) = (C'_2 t + C'_3)e^{-t}, \quad p_1(\mathrm{t}, C'_\mathrm{i}) = (C'_4 t + C_5)e^{-2t}, \\
&m_2(\mathrm{t}, C'_\mathrm{j}) = (C'_6 t + C'_7)e^{-t}, \quad n_2(\mathrm{t}, C'_\mathrm{j}) = C'_8 e^{-2t}, \\
&p_2(\mathrm{t}, C'_\mathrm{j}) = (C'_9 t + C'_{10})e^{-t}, \quad m_3(\mathrm{t}, C'_\mathrm{k}) = C'_{11} e^{-t}, \\
&n_3(\mathrm{t}, C'_\mathrm{k}) = C'_{12} t + C'_{13}, \quad p_3(\mathrm{t}, C'_\mathrm{k}) = (C'_{14} t + C'_{15})e^{-2t}
\end{aligned} \tag{9.268}$$

and so on.

Now, substituting Eqs. 9.265 and 9.267 into Eq. 9.266, we obtain the equations in $\dot{x}_1$, $\dot{y}_1$ and $\dot{z}_1$, respectively:

$$
\begin{aligned}
\dot{x}_1 &= (-\alpha a C_3 t + a - \alpha a C_4)e^{-t} + [(C_1 - \beta a C_5)t + C_2 - \beta a C_6 - \\
&\quad - (a^2 + \alpha ab + \beta ac)]e^{-2t} - (2a + \alpha b + \beta c)(C_1 t + C_2)e^{-3t} \\
\dot{y}_1 &= [(C_9 t - \beta b C_7)t + b - \beta b C_8 + C_{10}]e^{-t} - [(\beta a + 2b + \alpha c)C_9 t + \\
&\quad + \beta ab + b^2 + \alpha bc + (\beta a + 2b + \alpha c)C_{10}]e^{-2t} - (\alpha b C_{11} t + \alpha b C_{12})e^{-3t} \\
\dot{z}_1 &= (C_{17} t + C_{18} + c)e^{-t} - \{[(\alpha a + \beta c + 2c)C_{17} + \alpha c C_{13}]t + \alpha ac + \beta bc + \\
&\quad + c^2 + \alpha c C_{14} + (\alpha a + \beta c + 2c)C_{18}\}e^{-2t} - (\beta c C_{14} t + \beta c C_{16})e^{-3t}
\end{aligned}
\tag{9.269}
$$

The solutions of Eq. 9.269 with the initial conditions

$$x_1(0) = a, \quad y_1(0) = b, \quad z_1(0) = c \tag{9.270}$$

can be written in the form

$$
\begin{aligned}
x_1(t) &= a + \alpha a C_3 t e^{-t} + (a - \alpha a C_3 - \alpha a C_4)(1 - e^{-t}) - \\
&\quad - \frac{1}{2}(C_1 - \beta a C_5)te^{-2t} + \frac{1}{4}[(C_1 - \beta a C_5 + 2(C_2 - \beta a C_6) - \\
&\quad - 2(a^2 + \alpha ab + \beta ac)](1 - e^{-2t}) + \frac{1}{3}(2a + \alpha b + \beta c)C_1 t e^{-3t} - \\
&\quad - \frac{1}{9}(2a + \alpha b + \beta c)(C_1 + 3C_2)(1 - e^{-3t})
\end{aligned}
\tag{9.271}
$$

$$
\begin{aligned}
y_1(t) &= b + (b - \beta b C_8 + C_{10})(1 - e^{-t}) + (\beta b C_7 - C_9)te^{-t} - \\
&\quad - \frac{1}{4}[(\beta a + 2b + \alpha c)C_9 + 2(\beta ab + b^2 + \alpha bc) + \\
&\quad + 2(\beta a + 2b + \alpha c)C_{10}](1 - e^{-2t}) + \frac{1}{2}(\beta a + 2b + \alpha c)C_9 t e^{-2t} - \\
&\quad - \frac{1}{9}\alpha b(C_{11} + 3C_{12})(1 - e^{-3t}) + \frac{1}{3}\alpha b C_{11} t e^{-3t}
\end{aligned}
\tag{9.272}
$$

$$
\begin{aligned}
z_1(t) &= c + (C_{17} + C_{18} + c)(1 - e^{-t}) - C_{17} t e^{-t} - \frac{1}{4}[2(\alpha ac + \beta bc + c^2) + \\
&\quad + 2\alpha c C_{14} + 2(\alpha a + \beta c + 2c)C_{18} + \alpha c C_1 + (\alpha a + \beta c + 2c)C_{17}](1 - e^{-2t}) + \\
&\quad + \frac{1}{2}[\alpha c C_{13} + (\alpha a + \beta c + 2c)C_{17}]te^{-2t} + \frac{1}{3}\beta c C_{15} t e^{-3t} - \\
&\quad - \frac{1}{4}\beta c(C_{15} + 3C_{16})(1 - e^{-3t})
\end{aligned}
\tag{9.273}
$$

By substituting Eqs.9.271–9.273 into Eq. 9.261 it results the residuals:

$$R_1(t, C_i) = \dot{x}_1 - x_1(1 - x_1 - \alpha y_1 - \beta z_1) \tag{9.274}$$

$$R_2(t, C_i) = \dot{y}_1 - y_1(1 - \beta x_1 - y_1 - \alpha z_1) \tag{9.275}$$

$$R_3(t, C_i) = \dot{z}_1 - z_1(1 - \alpha x_1 - \beta y_1 - z_1) \tag{9.276}$$

Making collocation in the arbitrary points t_i, $i = 1,2,\ldots,6$

$$R_1(t_1, C_i) = R_1(t_2, C_i) = \ldots = R_1(t_6, C_i) = 0 \tag{9.277}$$

$$R_2(t_1, C_i) = R_2(t_2, C_i) = \ldots = R_2(t_6, C_i) = 0 \tag{9.278}$$

$$R_3(t_1, C_i) = R_3(t_2, C_i) = \ldots = R_3(t_6, C_i) = 0 \tag{9.279}$$

we obtain the optimal values of the constants C_1, $C_2,\ldots,C_{18}$ and therefore the solution (9.271–9.273) in the first approximation is well-determined.

In the case when $a = 0.2$, $b = 0.3$, $c = 0.5$, $\alpha = 0.1$, $\beta = 0.1$, from Eqs. 9.277–9.279 we obtain

$$\begin{aligned} &C_1 = 1.669841466, \quad C_2 = -0.452081708, \quad C_3 = 0.899181851, \\ &C_4 = 2.361528973, \quad C_5 = -1.940901217, \quad C_6 = 0.582517076, \\ &C_7 = 1033.008542, \quad C_8 = -540.7564632, \quad C_9 = 30.90993616, \\ &C_{10} = -14.59841284, \quad C_{11} = 29.8078082, \quad C_{12} = 94.01456893, \\ &C_{14} = 53.21346396, \quad C_{15} = -22.21139178, \quad C_{16} = -35.8538265 \\ &C_{17} = -0.345181562, \quad C_{18} = 0.993490199 \end{aligned} \tag{9.280}$$

By substituting Eq. 9.280 into Eqs. 9.271–9.273 we obtain after only one iteration, the following expression for the first-order approximate solution of Lotka-Volterra equations in three species (9.261):

$$\begin{aligned} x_1(t) = {} & 0.8281811 - (0.452081708t + 1.669841466)e^{-t} + \\ & + (2.361528973t - 0.899181851)e^{-2t} + (0.582517076t + 1.940902217)e^{-3t} \end{aligned} \tag{9.281}$$

$$\begin{aligned} y_1(t) = {} & 0.83333001 + (0.08032009t - 1.924281055)e^{-t} + \\ & + (1.715501457t + 0.342446025)e^{-2t} + (0.298078082t + 1.04850505)e^{-3t} \end{aligned} \tag{9.282}$$

$$z_1(t) = 0.836102311 + (0.345181562t - 1.148308637)e^{-t} + \\ + (0.001606241t + 1.532769033)e^{-2t} - (0.370189863t + 0.720960396)e^{-3t} \tag{9.283}$$

It is easy to verify the accuracy of the obtained solution if we graphically compare the analytical solution with the numerical one. Figures 9.38–9.40 show

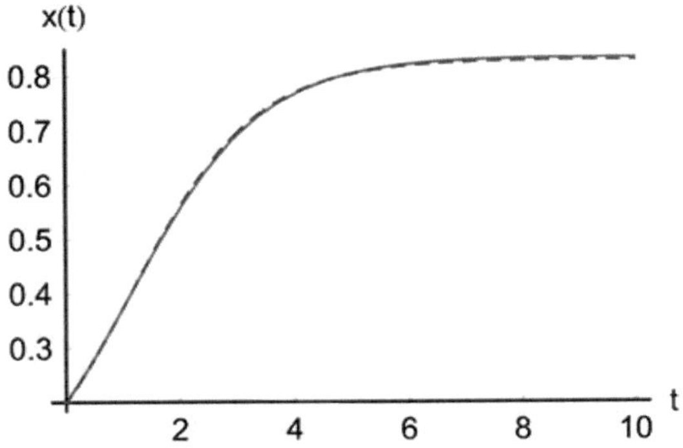

Fig. 9.38 Comparison between the approximate solution (9.281) and numerical results obtained for Eq. 9.261 in case $a = 0.2$, $b = 0.3$, $c = 0.5$, $\alpha = 0.1$, $\beta = 0.1$: ______ numerical solution; _ _ _ _ approximate solution

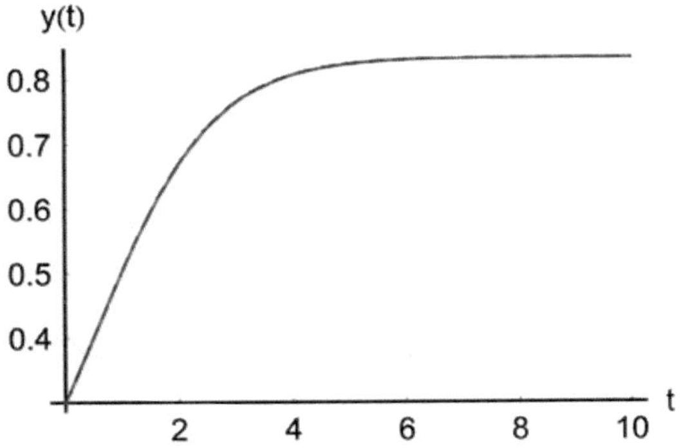

Fig. 9.39 Comparison between the approximate solution (9.282) and numerical results obtained for Eq. 9.261 in case $a = 0.2$, $b = 0.3$, $c = 0.5$, $\alpha = 0.1$, $\beta = 0.1$: ______ numerical solution; _ _ _ _ approximate solution

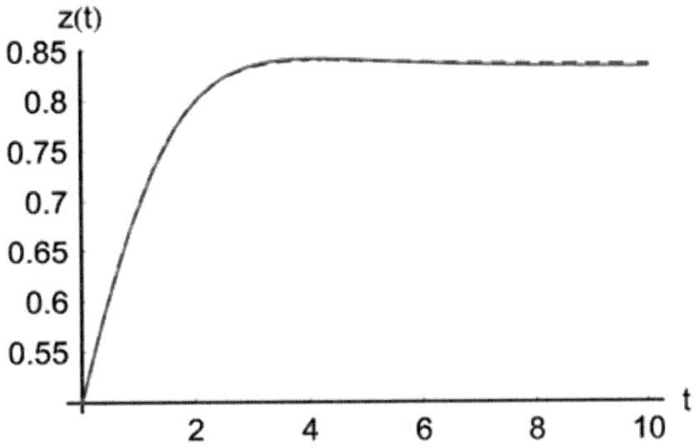

Fig. 9.40 Comparison between the approximate solution (9.283) and numerical results obtained for Eq. 9.261 in case $a = 0.2$, $b = 0.3$, $c = 0.5$, $\alpha = 0.1$, $\beta = 0.1$: ______ numerical solution; _ _ _ _ approximate solution

the comparison between the present solutions and the numerical integration results obtained by a fourth-order Runge-Kutta method.

It can be seen from Figs. 9.38–9.40 that the solution obtained by OPIM is in very good agreement with numerical integration results. Our method gives analytic solutions valid globally in time unlike other known methods, for instance Adomian decomposition method, which unfortunately does not guarantee analytic solutions valid globally in time as proved by Repaci [197].

References

1. H. Poincaré, Les nouvelles methods de la mécanique céleste, Tomes 1, 2 et 3 (1892, 1893, 1894) (Paris)
2. A.M. Lyapunov, *Stability of Motion* (Academic Press, New York, 1966) (translated from Russian)
3. G.D. Birkhoff, G.D. Birkhoff's collected mathematical papers, Dover, 1968 (three volumes)
4. J. Rayleigh, *The Theory of Sound* (Dover, New York, 1945)
5. B. Van der Pol, On relaxation oscillations. Philos. Mag. **7**, 978–992 (1926)
6. A. Liénard, Étude des oscillations entrenues. Rev. Génerale de l'Électricité. **23** 901–912 (1928)
7. N. Bogolyubov, Yu Mitropolski, A.M. Samoilenko, *Method of the Acceleration of Convergence in Nonlinear Mechanics* (Naukova Dumka, Kiev, 1969) (in Russian)
8. Yu Mitropolski, *The Average Method in Nonlinear Mechanics* (Naukova Dumka, Kiev, 1971) (in Russian)
9. A.A. Andronov, A.A. Witt, S.E. Khaïkin, *Theory of Oscillations* (Pergamon Press, Oxford, 1966)
10. I.G. Malkin, *Some Problems of the Theory of Nonlinear Oscillations* (Gostehizdat, Moscow, 1956) (in Russian)
11. C. Hayashi, *Nonlinear Oscillation in Physical Systems* (Mc-Graw-Hill, New York, 1964)
12. N. Minorsky, *Introduction in Nonlinear Mechanics* (Edwards, Ann Arbor/Michigan, 1947)
13. J.J. Stoker, *Nonlinear Vibration in Mechanical and Electrical Systems* (Wiley, New York, 1950)
14. N. Mc Lachlan, *Theory of Vibration* (Dover, New York, 1991)
15. N. Kauderer, *Nichtlineare Mechanik* (Springer, Berlin/Götingen/Heidelberg, 1958)
16. G. Sansone, R. Conti, *Nonlinear Differential Equation* (Pergamon, Oxford, 1964)
17. M. Roseau, *Vibrations nonlinéaires et théorié de la stabilite* (Springer, Berlin/Heidelberg/New York, 1966)
18. A. Einstein, *The Theory of Brownian Movement* (Dover, New York, 1956)
19. S.N. Crandall, W.D. Mark, *Random Vibration in Mechanical System* (Academic Press, New York/London, 1963)
20. I. Gumovski, Oscillating Evolution Processes, (Manchester University Press, Manchester 1989)
21. C. Mira, Some historical aspects of nonlinear dynamics possible trends for the future. J. Franklin Inst. **334**(5–6), 1075–1113 (1997)
22. A.N. Nayfeh, D. Mook, *Nonlinear Oscillations*. (Wiley, New York 1979)
23. S.N. Chen, Y.K. Cheung, A modified Lindstedt-Poincaré method for strongly non-linear two degree-of-freedom system. J. Sound Vib. **193**, 751–762 (1996)
24. V.T. Coppola, R.H. Rand, Averaging using elliptic functions – approximations of limit cycle. Acta Mech. **81**, 128–142 (1990)

V. Marinca and N. Herisanu, *Nonlinear Dynamical Systems in Engineering*,
DOI 10.1007/978-3-642-22735-6,

25. L. Cveticanin, Analytical methods for solving strongly nonlinear differential equations. J. Sound Vib. **214**, 325–328 (1998)
26. Y.Y. Wu, S.J. Liao, X.Y. Zhao, Some notes on the general boundary element method for highly nonlinear problems. Commun. Nonlin. Sci. Numer. Simul. **10**, 725–735 (2005)
27. V.P. Agrwal, N.N. Denman, Weighted linearization technique for period approximation in large amplitude nonlinear oscillations. J. Sound Vib. **99**, 463–473 (1985)
28. J.I. Ramos, An artificial parameter-decomposition method for nonlinear oscillators: Application to oscillators with odd nonlinearities. J. Sound Vib. **307**, 312–329 (2007)
29. S.J. Liao, *Beyond Perturbation-Introduction to the Homotopy Analysis Method* (Chapman and Hall, CRC Press, Boca Raton, 2004)
30. J.H. He, Homotopy perturbation technique. Comput. Methods Appl. Mech. Eng. **178**, 257–262 (1999)
31. J.H. He, Variational iteration method – a kind of non-linear analytical technique; some examples. Int. J. Non-linear Mech. **34**, 699–708 (1999)
32. R. Krivec, S. Mandelzweig, Numerical investigation of quasiliniarization method in quantum mechanics. Comput. Phys. Commun. **138**, 69–79 (2001)
33. P. Hagedorn, *Non-linear Oscillations* (Clarendon Press, Oxford, 1981)
34. A. Nayfeh, *Problems in Perturbation* (Wiley, New York, 1985)
35. F. Dincă, C. Teodosiu, *Nonlinear and Random Vibrations* (Editura Academiei R.S. România – Academic Press, New York, 1973)
36. P.G.D. Barkham, A.C. Soudack, Approximate solutions of non-linear, non-autonomous second order differential equations. Int. J. Control **11**, 101–114 (1970)
37. S.H. Chen, X.M. Yang, Y.K. Cheung, Periodic solutions of strongly quadratic non-linear oscillators by the elliptic perturbation method. J. Sound Vib. **212**, 771–780 (1998)
38. Y.K. Cheung, S.H. Chen, S.L. Lau, A modified Lindstedt-Poincaré method for certain strongly non-linear oscillators. Int. J. Non-linear Mech. **26**, 367–378 (1991)
39. S.Q. Dai, Generalization of the method of full approximation and its applications. Appl. Math. Mech. **12**(3), 255–264 (1991)
40. J.H. He, A new perturbation technique which is also valid for large parameters. J. Sound Vib. **229**, 1257–1263 (2000)
41. J.H. He, Modified Lindstedt-Poincaré methods for some strongly non-linear oscillations Part I: expansion of a constant. Int. J. Non-linear Mech. **37**(2), 309–314 (2002)
42. J.H. He, Modified Lindstedt-Poincaré methods for some strongly non-linear oscillations Part II: A new transformation. Int. J. Non-linear Mech. **37**, 315–320 (2002)
43. J.H. He, Modified Lindstedt-Poincaré methods for some strongly non-linear oscillations. Part III – double series expansion. Int. J. Non-linear Sci. Numer. Simul. **2**, 317–320 (2001)
44. H. Hu, A classical perturbation technique which is valid for large parameters. J. Sound Vib. **269**, 409–412 (2004)
45. F.F. Seeling, Unrestricted harmonic-balance- a several method to evaluate periodic structures in time and/or space of arbitrary stability for nonlinear chemical-reaction and reaction-diffusion systems. Theory and computer-program for time-dependent systems. Z. Naturforsch. A **5**, 1054 (1980)
46. R.E. Mickens, A generalization of the method of harmonic balance. J. Sound Vib. **111**(3), 515–518 (1986)
47. K. Huseyin, R. Lin, An intrinsic multiple-scale harmonic balance method for nonlinear vibration and bifurcation problems. Int. J. Non-linear Mech. **26**(5), 727–740 (1991)
48. S.B. Yuste, Comments on the method of harmonic-balance in which Jacobi elliptic functions are used. J. Sound Vib. **145**(3), 381–390 (1991)
49. J.L. Summers, M.D. Savage, Two timescale harmonic-balance. I. Application to autonomous one-dimensional nonlinear oscillators. Phil. Trans. R. Soc. Lond. Ser A **340**, 473–501 (1992)
50. B.S. Wu, W.P. Sun, C.W. Lim, Analytical approximations to the double-well Duffing oscillator in large amplitude oscillations. J. Sound Vib. **307**, 953–960 (2007)
51. B.S. Wu, C.W. Lim, Y.F. Ma, Analytical approximation to large – amplitude oscillation of a nonlinear conservative system. Int. J. Non-linear Mech. **38**, 1037–1043 (2003)

52. V. Marinca, N. Herişanu, *Rational Harmonic Balance Method for the Forced Duffing Oscillator with Hardening Non-linearity* (Bul. Şt. Univ. Tehn, Iassy, 2005)
53. X.M. Tian, L.J. Feng, C. Delin, Non-linear vibration by a new method. J. Sound Vib. **215**, 475–487 (1998)
54. J. Guckenheimer, P.J. Holmes, *Non-linear Oscillations, Dynamical Systems and Bifurcations of Vector Fields* (Springer, Berlin, 1983)
55. N.N. Bogolyubov, Yu.A. Mitropolsky, *Asymptotic Methods in the Theory of Nonlinear Oscillations* (Gordon and Breach/Hindustan Publ., New York/Delhi, 1961)
56. Yu.A. Mitropolsky, *Problems of the Asymptotical Theory of Nonstationary Oscillations* (Izdat Nauka, Moscow, 1964) (in Russian)
57. V. Marinca, An approximate solution for one-dimensional weakly nonlinear oscillations. Int. J. Nonlin. Sci. Num. Simul. **3**, 107–120 (2002)
58. M.P. Cartmell, S.W. Ziegler, R. Khanin, DIM Forehand – multiple scales analyses of the dynamics of weakly nonlinear mechanical systems. Appl. Mech. Rev. **56**(5), 455–492 (2003)
59. R. Khanin, M.P. Cartmell, Parallelization of perturbation analysis: Application to large-scale engineering problems. J. Symbol. Comput. **31**(4), 461–473 (2001)
60. C.L. Lee, C.T. Lee, A higher order method of multiple scales. J. Sound Vib. **202**(2), 284–287 (1997)
61. Z. Rahman, T.D. Burton, A higher – order methods of multiples scales in non-linear oscillations-periodic steady-state response. J. Sound Vib. **133**(3), 369–379 (1989)
62. A. Luongo, G. Rega, F. Vestroni, On nonlinear dynamics of planar shear indeformable beams. ASME. J. Appl. Mech. **53**(3), 619–624 (1986)
63. A. Hassan, Use of transformations with the higher order method of multiple scales to determine the steady state periodic response of harmonically excited nonlinear oscillators. Part.I Transformation of derivative. J. Sound Vibration. **178**, 21–40 (1994), Part. II: Transformation of detuning. J. Sound Vibration. **178**, 1–19 (1994)
64. B.A. Huberman, J.P. Crutchfield, Chaotic states on anharmonic systems in periodic fields. Phys. Rev. Lett. **43**, 1743–1747 (1979)
65. Z. Rahman, T.D. Burton, Large amplitude primary and superharmonic resonances in the Duffing oscillator. J. Sound Vib. **110**, 363–380 (1986)
66. A.H. Nayfeh, N.E. Sanchez, Bifurcations in a forced softening Duffing oscillator. Int. J. Non-Linear Mech. **24**(6), 483–497 (1989)
67. N.E. Sanchez, A.H. Nayfeh, Prediction on bifurcations in a parametrically excited Duffing oscillator. Int. J. Non-Linear Mech. **25**, 163–176 (1990)
68. L.D. Zavodney, A.H. Nayfeh, N.E. Sanchez, The response of a single-degree-of freedom system with quadratic and cubic non-linearities to a principal parametric resonance. J. Sound Vib. **129**(3), 417–442 (1989)
69. C.A. Holmes, R.H. Rand, Coupled oscillators as a model for nonlinear parametric excitation. Mech. Res. Commun. **8**, 263–268 (1981)
70. S. Yano, Parametric excitation in a self-excited vibration system (3 rd report, the influence of cubic non-lineary). Bull. Jpn. Soc. Mech. Eng. **27**, 1264–1271 (1984)
71. A.H. Nayfeh, Response of two-degree-of-freedom systems to multifrequency parametric excitations. J. Sound Vib. **88**, 1–10 (1983)
72. L.D. Zavodney, A.H. Nayfeh, The response of a single-degree-of-freedom system with quadratic and cubic nonlinearities to a fundamental parametric resonance. J. Sound Vib. **120**, 63–69 (1988)
73. A.H. Nayfeh, K. Asfar, Nonstationary parametric oscillations. J. Sound Vib. **124**, 529–537 (1988)
74. K. Asfar, K.K. Masoud, Damping of parametrically excited single-degree-of-freedom systems. Int. J. Non-Linear Mech. **29**(3), 421–428 (1994)
75. K. Asfar, A.H. Nayfeh, K. Barash, A nonlinear oscillator, Lanchester damper. J. Vibr. Acoust. Stress Reliability Des. **109**, 343–347 (1987)
76. T.M. Wu, A new formula of solving nonlinear equations by Adomian decomposition method and homotopy methods. Appl. Math. Comput. **172**, 903–907 (2006)

77. A.P. Morgan, Solving Polynomial Systems using Continuation for Engineering and Scientific Problems (Prentice Hall, Englewood, NY 1987)
78. C.B. Garcia, W.S. Zangwill, *Pathways to Solutions, Fixed Points and Equilibrum* (Prentice-Hall, Englewood Cliffs, 1981)
79. E.L. Allgower, K. Georg, *Numerical Continuation Methods, an Introduction* (Springer, New York, 1990)
80. S.J. Liao, The proposed homotopy analysis technique for the solution of nonlinear problems, Ph.D. thesis, Shanghai Jiao Tong University, 1992
81. S.J. Liao, *Beyond Perturbation: Introduction to the Homotopy Analysis Method* (Chapman and Hall/CRC Press, Boca Raton, 2003)
82. S.J. Liao, Notes on the homotopy analysis method: Some definitions and theorems. Commun. Nonlin. Sci. Numer. Simul. **14**, 983–997 (2009)
83. V. Marinca, N. Herişanu, I. Nemeş, Optional homotopy asymptotic method with application to thin film flow. Cent. Eur. J. Phys. **6**, 648–653 (2008)
84. V. Marinca, N. Herişanu, Application of optimal homotopy asymptotic method for solving nonlinear equations arrising in heat transfer. Int. Commun. Heat Mass Transf. **35**, 710–715 (2008)
85. N. Herişanu, V. Marinca, T. Dordea, G. Madescu, A new analytical approach to nonlinear vibration of an electrical machine. Proc. Rom. Acad. A **9**, 229–236 (2008)
86. V. Marinca, N. Herişanu, C. Bota, B. Marinca, An optimal homotopy asymptotic method applied to the steady flow of a fourth-grade fluid past a porous plate. Appl. Math. Lett. **22**, 245–251 (2009)
87. N. Herişanu, V. Marinca, Accurate analytical solutions to oscillators with discontinuities and fractional-power restoring force by means of the optional homotopy asymptotic method. Comput. Math. Appl. **60**, 1607–1615 (2010)
88. V. Marinca, N. Herişanu, Determination of periodic solutions for the motion of a particle on a rotating parabola by means of the optimal homotopy asymptotic method. J. Sound Vib. **329**, 1450–1459 (2010)
89. N. Herişanu, V. Marinca, Explicit analytical approximation to large-amplitude nonlinear oscillations of a uniform cantilever beam carrying an intermediate lumped mass and rotary inertia. Meccanica **45**, 847–855 (2010)
90. R.A. Shah, S. Islam, G. Zaman, T. Hussain, Solution of stagnation point flow with heat transfer analysis by optimal homotopy asymptotic method. Proc. Rom. Acad. A **11**(4), 312–321 (2010)
91. A.A. Joneidi, D.D. Ganji, M. Babaelahi, Micropolar flow in a porous channel with high mass transfer. Int. Commun. Heat Mass Transf. **36**, 1082–1088 (2009)
92. J. Ali, S. Islam, S. Islam, G. Zaman, The solution of multipoint boundary value problems by the optimal homotopy asymptotic method. Comput. Math. Appl. **59**, 2000–2006 (2010)
93. M. Esmaeilpour, D.D. Ganji, Solution of the Jeffery-Hamel flow problem by optimal homotopy asymptotic method. Comput. Math. Appl. **59**, 3405–3411 (2010)
94. S. Islam, R.A. Shah, I. Ali, Optimal homotopy asymptotic solutions of Couette and Poiseuille flows of a third grade fluid with heat transfer analysis. Int. J. Nonlin. Sci. Numer. Simul. **11**, 1123–1135 (2010)
95. M. Babaelahi, D.D. Ganji, A.A. Joneidi, Analytical treatment of mixed convection flow part vertical flat plate. Therm. Sci. **14**(2), 409–416 (2010)
96. M. Idrees, S. Haq, S. Islam, S. Islam, Application of optimal homotopy asymptotic method to special fourth-order boundary value problems. World Appl. Sci. J. **9**(2), 131–137 (2010)
97. M. Idrees, S. Haq, S. Islam, Application of optimal homotopy asymptotic method to special sixth order boundary value problems. World Appl. Sci. J. **9**(2), 138–143 (2010)
98. R.A. Shah, S. Islam, A.M. Siddiqui, Couette and Poiseuille flows for fourth grade fluids using optimal homotopy asymptotic method, World Appl. Sci. J. 9(11), 1228-1236 (2010)
99. S. Iqbal, M. Idrees, A.M. Siddiqui, A.R. Ansari, Some solutions of the linear and nonlinear Klein-Gordon equations using optimal homotopy asymptotic method. Appl. Math. Comput. **216**, 2898–2909 (2010)

100. M. Idrees, S. Islam, S. Haq, S. Islam, Application of the optimal homotopy asymptotic method to squeezing flow. Comput. Math. Appl. **59**, 3858–3866 (2010)
101. P. Amore, A. Aranda, Improved Lindstedt-Poincaré method for the solution of nonlinear problems. J. Sound Vib. **283**, 1115–1136 (2009)
102. T. Hayat, M. Sajid, On analytic solution for thin film flow of fourth grade fluid down a vertical cylinder. Phys. Lett. A **361**, 316–322 (2007)
103. M. Sajid, T. Hayat, S. Asghar, On the analytic solution of the steady flow of a fourth grade fluids. Phys. Lett. A **355**, 18–26 (2006)
104. R.R. Rajagopal, On the boundary conditions for fluids of the differential type, in *J. Navier-Stokes Equation and Related Non-linear Problems*, ed. by A. Sequira, (Plenum, New York, 1995), pp. 273–278
105. S. Abbasbandy, The application of homotopy analysis method to nonlinear equations arising in heat transfer. Phys. Lett. A **360**, 109–113 (2006)
106. S. Abbasbandy, Homotopy analysis method for heat radiation equations. Int. Commun. Heat Mass Transf. **34**(3), 380–387 (2007)
107. A. Rajabi, D.D. Ganji, N. Taherian, Application of homotopy perturbation method in nonlinear heat conduction and connection equations. Phys. Lett. A **360**, 570–573 (2007)
108. H. Schlichting, *Boundary Layer Theory* (McGraw Hill, New York, 1968)
109. H.K. Kuiken, A backward free-convective boundary layer. Q. J. Mech. Appl. Math. **34**, 397–413 (1981)
110. J.H. He, Approximate analytical solution of Blasius equation. Commun. Nonlin. Sci. Numer. Simul. **3**(4), 260–263 (1998)
111. A.M. Wazwaz, The variational iteration method for solving two forms of Blasius equation on a half-infinite domain. Appl. Math. Comput. 188, 485–491 (2007)
112. S. Abbasbandy, A numerical solution of Blasius equation by Adomian's decomposition method and comparison with homotopy perturbation method. Chaos Solitons Fractals **31**, 257–260 (2007)
113. L. Howarth, On the solution of the laminar boundary layer equations. Proc. R. Soc. Lond. A **164**, 547–579 (1938)
114. S.J. Liao, An explicit, totally analytic approximate solution for Blasius viscous flow problems. Int. J. Nonlin. Mech. **34**, 759–778 (1999)
115. J.H. He, A simple perturbation approach to Blasius equation. Appl. Math. Comput. **140**, 217–222 (2003)
116. L. Cveticanin, I. Kovacic, Parametrically excited vibrations of an oscillator with strong cubic negative nonlinearity. J. Sound Vib. **304**, 201–212 (2007)
117. M.N. Hamdan, N.N. Shabaneh, On the large amplitude free vibrations of a restrained uniform beam carrying an intermediate lumped mass. J. Sound Vib. **199**, 711–726 (1997)
118. M.N. Hamdan, M.H.F. Dado, Large amplitude free vibrations of a uniform cantilever beam carrying an intermediate lumped mass and rotary inertia. J. Sound Vib. **206**, 151–168 (1997)
119. M.I. Qaisi, N.S.A. Huniti, Large amplitude free vibration of a conservative system with inertia and static non-linearity. J. Sound Vib. **242**(1), 1–7 (2001)
120. P. Fleming, J.V. Poplawski, Unbalance response prediction for rotors on ball bearings using speed and load-dependent nonlinear bearing stiffness. Int. J. Rotating Machinery **11**, 53–59 (2005)
121. Y.P. Chang, S.C. Jen, S.H. Tu, S.S. Shyr, Y. Kang, Mode-locking quasi-period and chaos of rotors mounted on nonlinear bearings. Int. J. Rotating Machinery **6**, 191–200 (2000)
122. W.R. Finley, M.M. Hodowanec, W.G. Holter, An analytical approach to solving motor vibration problems. IEEE Trans. Ind. Appl. **36**, 1467–1480 (2000)
123. P. Sun, B.S. Wu, C.W. Lim, Approximate analytical solutions for oscillation of a mass attached to a stretched elastic wire. J. Sound Vib. **300**, 1042–1047 (2007)
124. L. Xu, Application of He's parameter-expansion method to an oscillation of a mass attached to a stretched elastic wire. Phys. Lett. A **368**, 259–262 (2007)
125. H.P. Gotlieb, Harmonic balance approach to limit cycles of nonlinear jerk equations. J. Sound Vib. **17**, 243–280 (2006)

126. B.S. Wu, C.W. Lim, W.P. Sun, Improved harmonic balance approach to periodic solutions of non-linear equations. Phys. Lett. A **354**, 95–100 (2006)
127. X. Ma, L. Wei, Z. Guo, He's homotopy perturbation method to periodic solutions of nonlinear jerk equations. J. Sound Vib. **314**, 217–227 (2008)
128. J.C. Sprott, Simplest dissipative elastic flow. Phys. Lett. A **228**, 231–244 (1987)
129. F.O. Zengin, M.O. Kaya, S.A. Demirbag, Application of parameter-expansion method to nonlinear oscillators with discontinuities. Int. J. Nonlin. Sci. Numer. Simul. **9**, 267–270 (2008)
130. L. Cveticanin, Oscillator with fraction order restoring force. J. Sound Vib. **320**, 1064–1077 (2008)
131. I. Kovacic, On the motion of nonlinear oscillators with a fractional-order restoring force and time variable parameters. Phys. Lett. A **373**, 1839–1843 (2009)
132. A. Belendez, C. Pascual, M. Ortuno, T. Belendez, S. Gallego, Application of a modified He's homotopy perturbation method to obtain higher-order approximations to a nonlinear oscillator with discontinuities. Nonlin. Anal. RWA **10**, 601–610 (2009)
133. J.I. Ramos, Piecewise-linearized methods for oscillators with fractional power noninearities. J. Sound Vib. **300**, 502–521 (2007)
134. S.P. Timoshenko, *Vibration Problems in Engineering* (McGraw-Hill, New York, 1954)
135. Y. Chen, On the vibration of beams or rods carrying a concentrated mass. J. Appl. Mech. **30**, 310–311 (1963)
136. E. Esmailzadeh, G. Nakhaie-Jazar, Periodic behavior of cantilever beam with end mass subject to harmonic base excitation. Int. J. Non-linear Mech. **33**, 567–577 (1998)
137. A.A. Joneidi, G. Domairry, M. Babaelahi, Three analytical methods applied to Jeffery-Hamel flow. Commun. Nonlin. Sci. Numer. Simul. **15**, 3423–3434 (2010)
138. J.H. He, A coupling method of a homotopy technique and a perturbation technique for nonlinear problems. Int. J. Non-linear Mech. **35**(1), 37–43 (2000)
139. J.H. He, Determination of limit cycles for strongly nonlinear oscillators. Phys. Rev. Lett. **90**(17) (2003). Art.no.174301
140. J. Biazar, M. Eslami, H. Ghazymi, Homotopy perturbation method for systems of partial differential equations. Int. J. Nonlin. Sci. **8**, 413–418 (2007)
141. D.D. Ganji, A. Sadighi, Application of He's homotopy perturbation method to nonlinear coupled systems of reaction-diffusion equations. Int. J. Nonlin. Sci. **7**, 411–418 (2006)
142. J.H. He, New interpretation of homotopy perturbation method. Int. J. Mod. Phys. B **20**, 2561–2568 (2006)
143. L. Xu, Determination of limit cycle by He's parameter-expansion method for strongly nonlinear oscillators. J. Sound Vib. **302**, 178–184 (2007)
144. J.I. Ramos, Series approach to the Lane-Emden equation and comparison with the homotopy perturbation method. Chaos Solitons Fract **38**(2), 400–408 (2008)
145. G. Domairry, M. Ahangari, M. Jamshidi, Exact and analytical solution for nonlinear dispersive K(m, p) equations using homotopy perturbation method. Phys. Lett. A **368**, 266–270 (2007)
146. V. Marinca, Application of modified homotopy perturbation method to nonlinear oscillations. Arch. Mech. **58**, 241–256 (2006)
147. J.H. He, Recent developments in asymptotic methods for nonlinear ordinary differential equations. Int. J. Comput. Num. Annal. Appl. **2**, 127–190 (2002)
148. G.M. Moremedi, D.P. Mason, V.M. Gorringe, On the limit cycle of a generalised Van der Pol equation. Int. J. Non-Linear Mech. **28**(2), 237–250 (1993)
149. I.S. Gradshteyn, I.M. Ryzhik, *Table of Integrals, Series and Products* (Academic Press, New York, 1980)
150. V. Marinca, N. Herişanu, Optimal homotopy perturbation method for nonlinear differential equations. Nonlin. Sci. Lett. A **1**(3), 273–280 (2010)
151. V. Marinca, N. Herisanu, Nonlinear dynamic analysis of an electrical machine rotor-bearing system by optimal homotopy perturbation method. Comput. Math. Appl. 61, 2019–2024 (2011)

152. C.W. Chang-Jian, C.K. Chen, Chaos and bifurcation of a flexible rotor supported by porous squeeze couple stress fluid film journal bearings with non-linear suspension. Chaos Solitons Fract **35**, 358–375 (2008)
153. J.H. He, Variational iteration method-some recent results and new interpretations. J. Comput. Appl. Math. **207**(1), 3–17 (2007)
154. J.H. He, X.H. Wu, Variational iteration method: new developments and applications. Comput. Math. Appl. **54**, 881–894 (2007)
155. J.H. He, G.-C. Wu, F. Austin, The variational iteration method which should be followed. Nonlin. Sci. Lett. A **1**(1), 1–30 (2010)
156. A.A. Soliman, Numerical simulation and explicit solutions of KdV-Burgers and Lax's seventh-order KdV equations. Chaos solitons Fract **29**(2), 294–302 (2006)
157. E.M. Abulwafa, M.A. Abdou, A.M. Mahmood, The solution of nonlinear coagulation problem with mass loss. Chaos Solitons Fract **29**(2), 313–330 (2006)
158. D. Slota, Direct and inverse one-plane Stefan problem solved by the variational iteration method. Comput. Math. Appl. **54**, 1139–1146 (2007)
159. S. Momani, S. Abuasad, Application of He's variational iteration method to Helmholtz equation. Chaos Solitons Fract **27**(5), 1119–1123 (2006)
160. N. Bildik, A. Konuralp, The use of variational iteration method, differential transform method and Adomian decomposition method for solving different types of nonlinear partial differential equations. Int. J. Nonlin. Sci. Numer. Simul. **7**(1), 65–70 (2006)
161. G.E. Drăgănescu, V. Căpălnăşan, Nonlinear relaxation phenomena in polycristaline solids. Int. J. Nonlin. Sci. Numer. Simul. **4**(3), 219–225 (2003)
162. V. Marinca, N. Herişanu, Periodic solutions for some strongly nonlinear oscillation by He's variational iteration method. Comput. Math. Appl. **54**, 1188–1196 (2007)
163. R.E. Mickens, *Oscillations in Planar Dynamic Systems* (World Scientific, Singapore, 1966)
164. N. Herişanu, V. Marinca, A modified variational iteration method for strongly nonlinear problems. Nonlin. Sci. Lett. A **1**(2), 73–82 (2010)
165. L. Elsgolts, *Differential Equations and the Calculus of Variations* (Mir Publishers, Moscow, 1980)
166. V. Marinca, N. Herişanu, D. Bălă, Some optimal approximate methods with application to thin film flow. WSEAS Trans. Syst. **7**(9), 744–753 (2010)
167. V. Marinca, N. Herişanu, Analytical approach to the dynamic analysis of a rotating electric machine. Comput. Math. Appl. **58**, 2320–2324 (2009)
168. H.P.W. Gottlieb, Frequencies of oscillators with fractional-power nonlinearities. J. Sound Vib. **261**, 557–566 (2003)
169. H.K. Kuiken, On boundary layers in fluid mechanics that decay algebraically along stretches of wall that are not vanishingly small. IMA J. Appl. Math. **27**, 387–405 (1991)
170. L. Xu, He's homotopy perturbation method for boundary layer equation in unbounded domain. Comput. Math. Appl. **54**, 1067–1070 (2007)
171. A.M. Wazwaz, The modified decomposition method and Padé approximants for a boundary layer equation in unbounded domain. Appl. Math. Comput. **177**, 737–744 (2006)
172. S.J. Liao, On the homotopy analysis method for nonlinear problems. Appl. Math. Comput. **147**, 499–513 (2004)
173. R.E. Mickens, Iteration procedure for determining approximate solutions to nonlinear oscillation equation. J. Sound Vib. **116**, 185–188 (1987)
174. R.E. Mickens, A generalized iteration procedure for calculating approximations to periodic solutions of "truly nonlinear oscillators". J. Sound Vib. **287**, 1045–1051 (2005)
175. C.W. Lim, B.S. Wu, A modified procedure for certain non-linear oscillators. J. Sound Vib. **257**, 202–206 (2002)
176. H. Hu, Solutions of a quadratic nonlinear oscillator: Iteration procedure. J. Sound Vib. **298**, 1159–1165 (2006)
177. Y.M. Chen, J.K. Liu, A modified Mickens iteration procedure for nonlinear oscillators. J. Sound Vib. **314**, 465–473 (2008)

178. H. Hu, A classical iteration procedure valid for certain strongly nonlinear oscillators. J. Sound Vib. **29**, 397–402 (2007)
179. J.H. He, Some asymptotic methods for strongly nonlinear equations. Int. J. Mod. Phys. B **20** (10), 1141–1189 (2006)
180. V. Marinca, N. Herişanu, A modified iteration perturbation method for some nonlinear oscillation problems. Acta Mech. **184**, 231–242 (2006)
181. J.H. He, Variational iteration method – a kind of non-linear analytical technique: Some examples. Int. J. Non-linear Mech. **34**, 699–708 (1999)
182. N. Herişanu, V. Marinca, An iteration procedure with application to Van der Pol oscillator. Int. J. Nonlin. Sci. Numer. Simul. **10**(3), 353–361 (2009)
183. R.E. Mickens, Perturbation procedure for the Van der Pol oscillator based on Hopf bifurcation theorem. J. Sound Vib. **127**, 187–193 (1988)
184. S.H. Chen, Y.K. Cheung, S.L. Lau, On perturbation procedure for limit cycle analysis. Int. J. Non-linear Mech. **26**, 237–250 (1993)
185. E.M.A. Elbashbeshy, M.F. Dimian, Effect of radiation on the flow and heat transfer over a wedge with variable viscosity. Appl. Math. Comput. **132**, 445–454 (2002)
186. M.E.M. Ouaf, Exact solution of thermal radiation on MHD over a stretching porous sheet. Appl. Math. Comput. **170**, 1117–1125 (2003)
187. V. Marinca, N. Herişanu, An optimal iteration method for a class of strongly nonlinear oscillators. Sci. Bul. Politehnica Univ. Timişoara Trans. Mech. **54**(68), 1–8 (2009)
188. V. Marinca, G. Drăgănescu, Construction of approximate periodic solutions to a modified Van der Pol oscillator. Nonlin. Anal. Real World Appl. **11**(5), 4355–4362 (2010)
189. M.N. D'Acunto, Determination of the limit cycle for a modified Van der Pol oscillator. Mech. Res. Comm. **33**, 93–98 (2000)
190. F.M. Scudo, Volterra and theoretical ecology. Theor. Popul. Biol. **2**, 1–23 (1971)
191. R.D. Small, Population Growth in a Closed Model, in *Mathematical Modelling. Classroom Notes in Applied Mathematics*, ed. by M.S. Klamkin (SIAM, Philadelphia, 1989)
192. S. Kobayashi, T. Matsukuma, S. Nagai, K. Umeda, Some coefficients of the TFD functions. J. Phys. Soc. Jpn. **10**, 759–765 (1955)
193. S.J. Liao, An explicit analytic solution to the Thomas-Fermi equation. Appl. Math. Comput. **144**, 495–506 (2003)
194. S.R. Resiga, G.D. Ciocan, I. Anton, F. Avelan, Analysis of the swirling flow downstream a Francis turbine runner. J. Fluid Eng. **128**, 177–189 (2006)
195. A.J. Lotka, *Elements of Physical Biology* (William and Wilkins, Baltimore, 1925)
196. V. Volterra, Variazioni e fluttuazioni del numero d'individui in specie animali conniventi. Mem. R. Acad. Naz. Lincei **2–3**, 30–111 (1926)
197. A. Repaci, Nonlinear dynamical systems: On the accuracy of Adomian decomposition method. Appl. Math. Lett. **3**, 35–39 (1990)

Index

V. Marinca and N. Herisanu, *Nonlinear Dynamical Systems in Engineering*, DOI 10.1007/978-3-642-22735-6, © Springer-Verlag Berlin Heidelberg 2011